# THE WORST OF EVILS

# THE WORST OF EVILS

## THE FIGHT AGAINST PAIN

THOMAS DORMANDY

YALE UNIVERSITY PRESS
NEW HAVEN AND LONDON

For information about this and other Yale University Press publications, please contact:
U.S. Office: sales.press@yale.edu www.yalebooks.com
Europe Office: sales@yaleup.co.uk www.yaleup.co.uk

Set in Minion by J&L Composition, Filey, North Yorkshire
Printed in Great Britain by St Edmundsbury Press Ltd, Bury St Edmunds

Library of Congress Cataloging-in-Publication Data
Dormandy, Thomas.
The worst of evils: man's fight against pain: a history/by Thomas Dormandy.
p.cm.
ISBN 0–300–11322–6
1. Pain—Treatment—Cross-cultural studies. 2. Pain—Treatment—History—Cross-cultural studies. I. Title.
[DNLM: 1. Pain—history. 2. Analgesia—history. 3. Anesthesia—history. 4. Pain—therapy. WL 11.1 D712w 2006]
RB127.D67 2006
616′.0472—dc22

2005034835

A catalogue record for this book is available from the British Library.

10 9 8 7 6 5 4 3 2 1

For Howard and Margaret Chapman with love

Pain is perfect misery, the worst
Of evils, and, excessive, overturns all patience

Milton, *Paradise Lost*, Bk vi, l. 462

# CONTENTS


List of Illustrations ix

Acknowledgements xi

INTRODUCTION 1

PART I: THE MISTS OF HISTORY 7

1. A gift of the gods 9
2. The grape and the poppy 16
3. Roots, barks, fruit and leaves 25
4. Pain denied 34
5. Pain ignored 41
6. The heresies 46
7. Healing and holiness 51
8. Islam 61
9. The age of the cathedrals 73
10. Pain exalted 83

PART II: SCIENTIFIC STIRRINGS 87

11. Rebirth, rediscovery and reform 89
12. Going to war 103
13. Foundations 111
14. Heavenly dreams 124
15. Animal magnetism 137
16. Pneumatic medicine 151
17. Laughing gas 162
18. The terror of the knife 168
19. Hospital disease 182

PART III: PAINLESS SURGERY 187

20. To the threshold 189
21. A gentleman from the South 202
22. 'This Yankee dodge' 208
23. In Gower Street 227
24. And beyond 231
25. Chloroform 243
26. The shape of dreams 249
27. Mr Anaesthetist 261
28. Conflicting views 276
29. The rights of pain 282
30. Who needs an anaesthetic? 288

PART IV: THE BEGINNING OF THE MODERN 299

31. The new physiology 301
32. The new pathology 314
33. The acute abdomen 331
34. Old drugs, new drugs 339
35. The bark of the willow 350
36. Cocaine 365
37. High Victorian pain 377
38. The power of pain control 390

PART V: YESTERYEARS 397

39. Seminal Years 399
40. The gift of St Barbara 412
41. Tic douloureux 422
42. Twilight Sleep 429
43. Dolorism 437
44. Renoir 444
45. Pills and poisons 453
46. The surgery of pain 466
47. The schism 473
48. Pain mechanisms 480
49. Pain clinics 488
50. Hospices 491

EPILOGUE 500

BIBLIOGRAPHY 503

INDEX 528

# ILLUSTRATIONS

*Girl with her Arm in a Sling* (Roman), Vatican Museum, Rome.

Jean Antoine Watteau, *L'Indifférent* (*c.* 1717), Musée du Louvre, Paris (©Photo RMN).

Francisco de Goya y Lucientes, *Self-portrait with Dr Arrieta* (1820), Minneapolis Institute of Art, Minneapolis.

The Laocoön, *Polydoros and Athendoros of Rhodes* (first century), Vatican Museum, Rome.

Matthias Grünewald, *Crucifixion* (*c.* 1515), Unter den Linden Museum, Colmar.

Ferdinand Hodler, *Valentine Godé-Darel im Krankenbett* (1914), Kunstmuseum, Solothurn.

Ejnar Nielsen, *The Sick Girl* (1896), Statens Museum for Kunst, Copenhagen.

Honoré Daumier, *Voyons . . . ouvrons la bouche!* (1864), from the series *Les moments difficiles de la vie.*

Richard Tennant Cooper, *Henry Hill Hickman Performing Experimental Surgery on an Animal Under 'Suspended Animation' Induced by Carbon Dioxide, c. 1826* (1912), Wellcome Library, London.

Gianlorenzo Bernini, *The Vision of St Theresa* (1645–52), Cornaro Chapel, Santa Maria della Vittoria, Rome.

Pierre André Brouillet. *Une leçon de Charcot à la Salpetrière* (1887), Faculty of Medicine, Lyon (Bridgeman Art Library).

Jean-Francois Millet, *L'Angélus* (1857–9) Musée du Louvre, Paris (©Photo RMN).

H.B. Hall, photograph of a reproduction of his stipple engraving of Horace Wells, Wellcome Library, London.

B.W. Richardson, autotype of photograph of John Snow M.D. (1856), Wellcome Library, London.

*Animal magnetism* (*c.* 1845), Wellcome Library, London.

Southwood and Hawes, *Operation under Ether Anaesthesia at the Massachusetts General Hospital* (1847), Wellcome Library, London.

Photograph by Bingham of Sir James Young Simpson, Wellcome Library, London.

Catherine Goodman, *Dame Cicely Mary Strode Saunders* (2005), National Portrait Gallery, London.

Sir Luke Fildes, *The Visit* (*c.* 1891), Tate Gallery, London (©Tate 2005).

# ACKNOWLEDGEMENTS

Without Elizabeth there would be no book and, for me, no life.

Michael Dormandy has been my guide along many of the byways of Antiquity and the early Church. He has made it a delightful exploration.

This book has been a niggle at the back of my mind for more decades than I care to remember: and many friends have discussed and have often illuminated it. They are too numerous to name individually but I must mention Miklos Ghyczy who has sent me some important papers which otherwise I would have missed and Ian Douglas-Wilson, a truly constructive listener and critic as well as an old friend. Dr. A.J. Tookman has provided valuable information on palliative care.

In a fast-changing world the courtesy and efficiency of the staff of the libraries of the Royal Society of Medicine and of the Wellcome Institute remain splendid constants. This book could not have been written without their help.

It has been a real pleasure to work with Heather McCallum, an inspiring editor, and the team at the London branch of Yale University Press. I would particularly like to thank Michael Tully and Candida Brazil. The critical editing by Beth Humphries has been outstanding: I am most grateful to her.

# INTRODUCTION

On the battlefield of Eylau Surgeon-General Dominique Larrey is trying to operate on the injured foot of a colonel of the Imperial Guard.[1] It seems, as so often in battlefield surgery, like the moment before disaster. The cannonade, the cacophony of the fighting and the ground trembling under the hoofs of the cavalry Larrey had long learnt to ignore. What seems to make disaster inevitable is that the injured man is racked by painful spasms. It is impossible to steady the limb. In desperation Larrey administers a blow to the colonel's chin. The colonel passes out in a faint. On recovering, he splutters indignantly: 'Monsieur, you have hit me, taking cowardly advantage of my temporary incapacity.' 'Colonel,' Larrey replies, 'I beg your pardon. I had to remove a bullet to save your life. I knew that the insult would temporarily distract you. The operation is over. Here is the bullet. Please shake my hand.' Or so the story goes. Its message is undoubtedly true. A knock-out blow was one of the earliest methods of anaesthesia. It was popular with the priests of the Old Kingdom of Egypt. It was used to render young mandarins unconscious at the emperor's court in twelfth-century China in preparation for the operation necessary for their entry into the ranks of imperial eunuchs. Travellers have described similar tales in twentieth-century Ethiopia. It is one of an astonishing variety of ways in which man has tried to conquer pain.

Pain itself, of course, long antedates man. Animals may not have a soul, as Descartes maintained, but they certainly react to pain. The sabre-toothed tiger must have licked its wounds. Injured caveman probably did the same. St Francis of Assisi went further. He never deliberately crushed a plant or broke a branch off a tree. Even plants, he said, could feel pain. When he was injured and a stick might have helped him to hobble, he preferred to crawl. Five

1. On 6 February 1807. The battle of Eylau was one of Napoleon's costliest victories, won over the combined armies of Russia and Prussia. Larrey, not yet raised to a barony, was already a legendary figure. He describes the scene in his *Mémoires* (see Bibliography).

hundred years later Goethe took up the idea in a charming (and slightly erotic) little ditty. Schubert transformed it into an immortal *Lied*.[2] Poetic fancy? Perhaps.

The world today is awash with drugs; but psychological and physical ways of fighting pain were tried and were partially at least successful long before chemicals. Initiation ceremonies in primitive societies all involved inflicting pain. Where they survive, sometimes in a symbolic form, they still do. It is an essential part of such rituals that they must be borne without flinching. Are they also borne without pain? Did the martyrs who begged their tormentors to inflict on them more and more severe tortures suffer pain? Did Christ?[3] It is one of the profound questions in the history of pain relief which touches on faith as well as on physiology.

Initiation ceremonies, both their infliction and their suffering, have always been male pursuits. The nearest female equivalent is the agony of childbirth. Unlike male rituals, in most societies it was not expected that labour pains would be endured without clamour. Yet under certain conditions even labour pains could be suppressed. When Hercules apologised to his mother, the nymph Alcmene, for the discomfort he had caused her in labour – prodigiously vigorous even at birth, he must have been a nightmare in the womb – the nymph assured him that the experience had been one of unalloyed joy. Whether Jupiter, the father, had a hand in this is uncertain; but happiness can undoubtedly mitigate pain.

Simple physical means, like selective pressure on sensory nerves, were a trick of the doctors' trade in ancient Egypt and have been rediscovered from time to time since. In 1902 Sir Victor Horsley, surgeon to University College Hospital, London, noted the frequency of young couples presenting in his outpatient clinic on Monday mornings with loss of sensation in one hand as well as a wrist drop. Horsley correctly diagnosed pressure on the brachial plexus, the criss-crossing of sensory and motor nerves from and to the upper limb, bunched together in the armpit. He also guessed, perhaps recalling his own youthful escapades, that the pressure on this usually well-protected site must have been caused by the upper horizontal bar of a park bench. Welcome as these structures are to elderly travellers of sedentary habits, they were not designed for loving couples meeting at nightfall and sitting on them side by side locked in a prolonged embrace. Happily the condition, called by Horsley 'Saturday-night palsy' but later portentously

2. In the third stanza of *Heidenröslein*: 'Und der wilde Knabe brach's / Röslein auf der Heiden, / Röslein wehrte sich und stach / Half ihn doch kein Weh und Ach, / Musst es eben leiden, / Röslein, röslein, röslein rot, / Röslein auf der Heiden.'

3. Not according to the Synoptic Gospel of St Peter. See Chapter 4.

renamed 'neuropraxia', is almost invariably transient. Happily too, by 1902 there were simpler ways of anaesthetising the hand.

The simpler ways included a battery of newly discovered chemical analgesics, among the most loudly trumpeted blessings of the new scientific age. But some of them were derived from the oldest of anodynes. The magical properties of the latex of the ripening seed pod of the poppy were known long before written records. So was the curious effect of fermented fruit juice. The analgesic power of the bark of the willow tree – or *salix* – was familiar to Dioscorides two thousand years before acetylsalicycilic acid, newly nicknamed aspirin, became the most commercially successful proprietary drug ever.[4]

The purpose of such century-hopping is not to warn off readers with an orderly cast of mind. It is a preamble to an apology. Dates and chronology are rarely riveting; but they usually help to make history comprehensible. One difficulty with any account of man's fight against pain is that the subject does not fit into a neat chronological framework. In political, military, economic and cultural history one event tends to lead to the next. But there was no continuity between the magic potions of the Homeric gods or even of Renaissance surgeons and the skills of the modern operating theatre. Nitrous oxide, ether and chloroform emerged in the mid-nineteenth century not as the culmination of a stepwise process traceable back to biblical times but as a barely believable new beginning. And even that is not entirely true. Anaesthetic gases were not unknown to the Oracle of Delphi. The *Arabian Nights* is replete with vapours, odours, scents and smokes which put inhalers to sleep or render them mad, bad, stupid, wise, amorous, cowardly or reckless. Reversible anaesthesia was old hat to Shakespeare's Friar Laurence.[5] Hypnosis and self-hypnosis too, in various more or less bizarre guises, have a habit of appearing and disappearing at unpredictable intervals.

In the summer of 1904 the Grand Duchess Xenia Alexandrovna visited the famous Grotto del Cane near Naples. The site was easier on her feet than Pompeii and less smelly than Capri and she dedicated an entry to it in her otherwise skeletal diary. The cave had been attracting visitors since the Renaissance but for centuries nobody knew its secret. Since carbon dioxide is heavier than air, it tends to accumulate at the bottom of suitably shaped holes in the ground. Wild beasts dropped into such holes rapidly lose consciousness and fall into a deep sleep. They can usually be aroused after being returned to a cage in the fresh air. Visitors who keep their heads above the carbon dioxide level can walk about with impunity, even nudge the sleeping beasts with their

4. Hailed as the first successful treatment of rheumatic fever, it later became emblematic of the jazz age. It acquired a new lease of life in the 1980s in the prevention of coronary thrombosis. See Chapter 45.
5. See *Romeo and Juliet* and Chapter 10.

toes. But they must not stoop to stroke them as Countess Tolstoy, one of the Grand Duchess' ladies-in-waiting, whose Italian did not encompass 'è pericoloso', discovered when she collapsed in a faint next to the dog. But the astonishing anaesthetising effects of certain gases had been known since biblical days and the elder Pliny, a sceptical Roman aristocrat, described how, to his own personal knowledge, the smell of the burning skin of a pregnant crocodile renders patients insensitive to the knife. The subject then lapsed but was explored again by Henry Hill Hickman in the early nineteenth century.[6] Like earlier attempts by starry-eyed philanthropists the exploration ended in failure.

When the breakthrough did come, it came from an unexpected quarter. Few in Britain and nobody on the Continent of Europe expected that the safe and simple abolition of pain in surgery, one of the golden moments of Western civilisation, would be achieved on another Continent, still inhabited by colonials, illiterate slaves and tribes of noble, perhaps, but often ill-tempered savages. Nor did anybody expect on either side of the Atlantic that the moment would be fathered by a disreputable crew of greedy adventurers. But anaesthesia seemed to have an odd effect even on its most civilised champions. In an only slightly more respectable vein, Sir James Young Simpson, discoverer of the sleep-inducing properties of chloroform, contrived to write the first entry on anaesthesia for the *Encyclopaedia Britannica* without mentioning the rival anaesthetic agent, ether. The action harks back to the no-holds-barred polemics of Paracelsus and his cronies.

The difficulty of such ups and downs is compounded by the fact that not only pain relief but also pain itself changes. At least the perception of pain does. Preanaesthetic surgery is inconceivable without assuming that ideas about what could and what could not be endured were different from what they are today. And, while such statements are usually made with the ghost of a self-satisfied smile, in two hundred years' time current attitudes to many 'silent epidemics' – the chronic 'non-lethal' backaches are but one example – which cripple more people today than they have ever done in the past, will probably seem just as incomprehensible. All this frustrates a strictly chronological narrative. Since this book sets out to be a history, and since historians have a vested interest in dates, an effort will be made to impose an order of sorts, but blips, bumps, backtrackings and digressions will be inevitable.

...

While there is probably little point in telling readers of an Introduction what the book they are holding is about – they may find that out all too soon – there

6. See Chapter 20.

is perhaps some merit in saying what it is not. In the present case one limitations deserves advance notice.

Pain and ways of overcoming it are deeply embedded in civilisation and culture. They are also inescapably part of everybody's personal experience. There have been European and American scholars who have spent a lifetime immersing themselves in Chinese, Indian, pre-Columbian and other non-European lores and could write about those with authority. The present writer is not one of them. Of course traditions from other cultures have often impinged on what can still be conveniently (if nonsensically) called the Western world: acupuncture, yoga, the use of cocaine and many features of early Islamic medicine are but a few examples. These will be discussed or at least mentioned. But, from necessity rather than from choice, this book is about the fight against physical pain as, for thousands of years, it has been perceived, feared, resisted, endured, suppressed and sometimes exalted by Western man and woman.

# PART I

# THE MISTS OF HISTORY

## Chapter 1

# A GIFT OF THE GODS

In times past sleep, joy, hope, happiness and relief from physical pain were part of the same package, a blessing that the gods of Mount Olympus (or wherever) occasionally bestowed on mortals. The idea that any part of this bounty could be separated from the rest would not have been understood by prehistoric man. It would have puzzled the more sophisticated healers of early civilisations. The Persian physician Abu Ali al-Hussayn ibn 'Abdallah ibn Sina revered in Christian Europe for centuries as Avicenna, laid down that potions must accomplish three things. They must alleviate pain. They must calm the mind. And they must induce restful sleep.[1] One without the others was meaningless, indeed inconceivable. He was summarising the wisdom of ages.[2] Similar objectives were enunciated by Hippocrates[3] and, a little later, by the infinitely wise and venerable Chinese doctor, Hua T'o.[4] But the package no longer exists.

Physical pain is of course still perceived as being related to sleeplessness, anxiety, grief, worry and other states of mind. A shelf-ful of weighty textbooks and busy university departments are now dedicated to exploring the relationships. But the basic facts have always been known. Fear can transform a niggle into torture. Discomfort bravely borne during the day becomes agony at night. The reverse is also true. Glad tidings can dispel even a serious bout of indigestion. But since the mid-nineteenth century there has been a tendency to separate physical pain from mental anguish. There is a reason for this – or

1. See Chapter 8.
2. Most of the great doctors of Antiquity whose names survive were probably great compilers of established wisdom – textbook writers, in modern parlance – rather than innovators.
3. Little is known about Hippocrates – the name means a driver of horses – except that he practised on the Greek island of Cos in the fifth century BC and that he was famous even in his lifetime. The Hippocratic writings, including the famous Oath, are ancient texts of varying provenance collected centuries later.
4. See below, p.14 and note 16.

at least an explanation. Suppression of the acute pain of surgical operations has been one of the triumphs of the past 150 years. Advances in the treatment of mental states like grief, fear and anguish have been significant but less spectacular. To some, therefore, the separation seems to make sense today. But it would have made no sense to Homer and other chroniclers of ancient civilisations.[5]

When Telemachus, son of Ulysses, visits King Menelaus in Sparta, the memory of all the sons, fathers, brothers and friends the assembled company have lost under the walls of Troy makes them deeply dejected. Many also still suffer from the after-effects of the wounds they had sustained in the war. Helen, lovely daughter of Zeus, is ready to help. She pours a drug, nepenthe, which had been given to her by Polydamma, wife of Thos of Egypt, into the wine they are drinking. At once all the pains and grief subside and all day long no tear is shed by the guests 'even if a brother or beloved son was killed before their own eyes'.

What was the drug which not only instantly banished pain but also calmed tempers and lulled those who drank it into slightly inane but happy conviviality? There have been many guesses but no certain answer.[6] Almost a thousand years after Homer, Pliny speculated that the active principle may have been Helen's delightful conversation or physical charms rather than her concoction.[7] Romans of his class were incorrigible sceptics. Yet 'helenium', a Roman remedy which was believed to have originated in Helen's tears and which in Pliny's lifetime was still grown on the Greek island of Helena, was highly prized for its healing power. It was, Pliny wrote, similar in its effect to the *nepenthes* praised by Homer as producing forgetfulness of all sorrows. Sadly, the exact nature of the plant and the composition of the precious tears remain obscure.[8]

5. Nor, at a sub-Homeric level, does it make sense to the present writer (see Chapters 48–50).
6. *Nepenthe* (νηπενδής) means no sorrow rather than no pain. Some classical scholars have argued in favour of cannabis; but Helen's nepenthe had a far more opium-like effect. Jean Cocteau, an expert user, was convinced that it was opium. An opium derivative by that name survived in pharmacopoeias and official formularies till the mid-twentieth century and was routinely prescribed as a preanaesthetic sedative till the 1950s.
7. Unless otherwise stated, Pliny in the present book refers to the elder of the two celebrated authors, uncle of the younger Pliny. He was an immensely erudite soldier and administrator who served under the emperors Vespasian and Titus in the first century and who wrote on a vast range of topics. Sadly, the only work of his to survive is the *Historia Naturalis*, which itself is encyclopaedic in scope, discoursing (tersely and sometimes a little obscurely) on art, science, political institutions and human inventions. It may have been his insatiable curiosity which kept him in Pompeii during the eruption of Vesuvius in AD 79 when he perished by suffocation, aged fifty-six.
8. It may have been related to the helenium or sneezeweed which still provides a splash of gorgeous autumn colour in English gardens but which is no longer grown for its

Similar uncertainty surrounds other balms whose effect was extolled by poets, philosophers and playwrights. What was the magical powder whose formula Chiron the Centaur bequeathed to his adopted son, Aesculapius, and which sent Pindar into iambic raptures?[9] What ointment did Patroclus apply to the wound of his friend Eurypylos, prepared from a root which 'slayeth pain, stays all the pangs of suffering and stops bleeding'?[10] What was the nepenthe which Aphrodite gave to her son Aeneas? And what remarkable constituent of the honey-sweet fruit of the lotus made those who drank it 'forgetful of home and worries but happy to remain and ignore all bad tidings'?[11]

Greek dramatists gave incomparable accounts of the different pains suffered by their heroes and heroines. Poor Deïaneira, distressed by the suspected fickleness of her husband Hercules, sends him a shirt she has soaked in a liquid that she hopes will restore his affections. But the shirt also carries traces of the poisoned blood of Nessus, one of Hercules' victims. Nessus now has his revenge. Donning the shirt the contaminated material literally burns off Hercules' skin. Extensive burns are among the most painful injuries known. Sophocles' description is searing, the greatest sustained scream in literature. Elsewhere the dramatist describes 'herbs of tried virtue wherewithal to mollify suffering'.[12] Unfortunately, neither he nor any of his fellow dramatists gives a hint of what the herbs of tried virtue are. This does not mean that they did not exist. Sophocles was not writing a manual of home doctoring. Perhaps what he was referring to was clear to his audience.

Virgil followed Homer in dropping hints without going into pedantic detail. Having survived the journey across the River Styx in Charon's flimsy boat, Aeneas and his guide, the Cumaean Sybil, approach the

medicinal properties. The episode of Helen's tears is related in Book IV of Homer's *Odyssey*: 'Then Helen, daughter of Zeus . . . cast a drug into the wine of which they drank, a drug to lull all pain and anger and bring forgetfulness of every sorrow. Whoever should drink a draft . . . would let no tear fall down his cheeks, even if his mother or father died, even if men slew his brother or dear son in front of him' (Homer, *Odyssey*, IV, 219).

9. Pythian Odes 3.1 refers to Aesculapius as the 'gentle artificer of drugs that dispel all pain'.
10. Homer, *Iliad*, XI, 847.
11. The chief and most characteristic alkaloid in both the white and the blue lotus (*Nymphaea caeruleum* and *N. ample*) is apomorphine, for long recognised as a powerful emetic, i.e. agent to induce vomiting. Its equally powerful psychogenic action has been recognised only in the last ten years. It has now been endorsed by the United States Food and Drug Administration as a good corrective of erectile dysfunction in male impotence. This effect could explain the appearance of the flower in Egyptian and Mayan as well as in Greek wall and vase paintings and in Homeric texts. (See E. Bertol *et al.*, *Journal of the Royal Society of Medicine*, 2004 (97) p. 58.)
12. Sophocles, *Philoctetes*, translated by F. Storr (London, 1924), line 649.

Underworld. They are suddenly faced with a frightening apparition. The savage three-headed dog Cerberus guards the gates of Hades. No problem.

> To Cerberus on whose necks bristle the snakes, the seer [Sybil] flung a bait mixed in honey and drugged meal. Opening his triple throat in ravenous hunger, Cerberus catches it when thrown, then relaxes and sinks to earth, succumbs to sleep and stretches his bulk all over the cave.[13]

There is no suggestion that the beast was irreversibly poisoned. It was anaesthetised as if for a routine hernia repair. What was in the drugged meal?

. . .

Graeco-Roman lore is not unique: the Greeks were in fact late developers. In 1874, rummaging around the royal tombs of Thebes, the eminent German Egyptologist Georg Moritz Ebers discovered a papyrus dating from about 1500 BC. (Or so he later claimed. In fact he probably bought the papyrus from Edwin Smith, an American dealer in looted antiquities.) A stunning 20 metres long, the scroll may be the oldest surviving medical document.[14] It deals with more than 150 ailments of the skin, the eye, the abdomen, the chest, the head and the limbs as well as a wide range of injuries, congenital malformations, and even such seemingly minor complaints as loss of hair. The treatments recommended involve chants, incantations, spells and supplications to Ra, the falcon-headed sun god, Thoth, the ibis-headed god of wisdom, Horus, the god of health and numerous lesser deities, but also more than 800 herbal, animal and mineral remedies. Some are of immense complexity and have a faintly exotic ring today. A drink prepared from the testicles of black asses is recommended to inflame sexual appetite. A mixture of hippopotamus, lion, crocodile, goose, snake and ibex fat is counselled for night blindness.[15] The drugs are variously prescribed as pills, ointments, poultices, fumigations, inhalations, gargles and suppositories and even to be blown into the urethra through a catheter. Many are credited to particular gods.

The alleviation of pain was a major concern in Egypt as elsewhere. Herodotus wrote in the fifth century that in Egypt 'one physician is assigned to the treatment of one disease'.[16] The Ebers papyrus suggests the existence of specialists whose expertise seems to have been in easing aches and pains of

13. Virgil, *Aeneid*, Book VI.
14. Ebers not only claimed to have discovered the scroll but also to have deciphered and translated it. When ill health forced him to retire from his chair in Jena he turned to writing exciting historical novels for young people set in ancient Egypt. He died in 1893, aged sixty-five.
15. This particular preparation, containing lashings of vitamin A, may have been effective.
16. Herodotus, 'father of history', remains a valuable source of medical lore not only on Greece but also on Egypt and Mesopotamia.

various kinds. The document also mentions several renowned healers by name. Iri, Keeper of the Royal Rectum, was presumably the Pharaoh's enema expert. Invented by the ibis-headed Thoth, colonic irrigations were widely used both in Egypt and in Mesopotamia to relieve abdominal pain. Another healer, the blessed Imhotep, 'he who comes in peace', chief minister of Pharaoh Zozes in the twenty-seventh century BC, had already become a figure akin to the Greek Aesculapius. One especially effective remedy for the succour of those wounded in battle was ascribed to him. It was a vapour generated by mixing together and burning the fruit of two different plants. Unfortunately, though the plants bearing large berries are illustrated, they are impossible to identify.

Similar accounts of pain-killing potions survive from other civilisations. (This is in contrast to love potions, which are largely a nineteenth-century operatic standby.) Abu'l Kasim Mansur or Hassan who composed his epic under the name of Firdausi describes the birth of Rustem, the Persian Hercules. Zal, the doctor who attends the Princess Rondabah, Rustem's mother, has remained on friendly terms with the Griffin which nurtured him. At their parting the Griffin had given him some of her feathers and had instructed him to burn one in a crisis. The size of Rustem in the womb now presented such an emergency. After burning a feather Zal is visited by a nurse who shows him how to prepare an intoxicating drink which will put the Princess to sleep and render her insensible to pain. Having done so, Zal delivers the outsize infant through an incision in the Princess's side. He then sutures and dresses the wound. The lady is soon restored to health and beauty. Caesarean sections may have been performed on dead and dying women for thousands of years; but there is no credible record for at least a thousand years after Firdausi's tale of a woman surviving the operation.[17]

17. By tradition, Julius Caesar was delivered in this way: indeed his name may be derived from the word *caesum* (from *caedo*, to cut), used by Pliny to denote delivery by cutting the mother's abdomen. Yet Caesar's mother, Aurelia, was alive during Caesar's youth; and such a happy outcome was not known for another 1,600 years. (The operation itself on dead women was allowed by the *Lex Regia*, the law promulgated by the second king of Rome, Numa, in about 500 BC: 'si mater pregnans mortua sit, fructus quam primum caute extrahatur'). Survival of the mother after Caesarean section was claimed by Jacob Niefer, a German sow-gelder, who said that he had carried out the operation on his own wife in 1500; but the case was not published till 1586, when it appeared as an appendix to an obstetric treatise by François Rousset of Montpellier. (Edward VI may have been born by Caesarean section but by then Jane Seymour was either dead or dying.) Rousset recorded a series of sixteen cases of maternal survival, all but two based on hearsay and some frankly incredible. The term 'Caesarian section' seems to have been first used in English by John Crooke in his 'Body of Man' in 1615 and in French by Théophile Reynaud of Paris in 1637. The first credibly documented case of survival of both mother and child was reported by James Knowles of Birmingham in 1836; but the operation did not become generally safe till after the introduction of anaesthesia and antisepsis in the middle of the nineteenth century.

...

In China myth merges into medical history during the Han dynasty (202 BC to AD 220). Hua T'o lived at the end of this period. Until then Confucian doctrine, which declared the human body to be sacrosanct, prevented the development of surgery. Hua T'o must have possessed a honeyed tongue as well as extraordinary skills. He overcame the prohibition and started to operate for a wide range of diseases. Eventually, it is said, no part of the human body was beyond the reach of his scalpel. What, apart from his technical mastery and knowledge of anatomy, enabled him to perform his operations was an effervescent powder. Dissolved in wine, it rendered the patient insensible to pain for several hours. Using this anaesthetic he opened abdominal cavities, removed diseased kidneys, amputated limbs, sutured wounds and treated open fractures. His followers built numerous shrines in his memory; but, despite much later guesswork, the secret of his powder died with him.[18]

The gods' gift did not always come in the form of powders, pills and potions. Acupuncture was also first codified during the Han dynasty, a practical offshoot of the wisdom of Lao-Tse.[19] In Taoist doctrine a life force or energy, *qi*, circulates through all organs of the body. The acupuncture points were (and are) located along fourteen invisible lines or meridians, each point controlling a specific organ or function. Since all diseases are the result of an imbalance in the flow of *qi*, they can be corrected by acupuncture needles inserted, twirled and vibrated in the correct way at the right insertion points. The oldest existing catalogue of these points is contained in the section entitled 'Ling shu' of the *Inner Canon*, one of the divinely inspired core texts of classical Chinese medicine. It dates from about 100 BC.[20] There were over 360

18. Apart from his skill as a surgeon, Hua T'o also earned fame as the great and early propagator of callisthenics and health-giving exercise:

> The body needs exercise, only it must not be to the point of exhaustion, for exercise expels the bad air from the system, promotes free movement of the blood, and prevents sickness. The used doorstep never rots, and it is the same with the body. That is why the ancients practised the bear's neck, the fowl's twist, swaying the body and moving the joints to prevent old age.

He described a system of exercises named after five animals – the tiger, the deer, the bear, the monkey and the bird – calling them 'frolics'.

> They remove disease, strengthen the spirit and ensure health. When out of sorts, one should practise the appropriate frolic which will give a feeling of lightness, alert the mind and increase appetite.

19. The 'Venerable Philosopher' – one meaning of the name – lived and taught compassion and humility in the sixth century BC.
20. See D. Hoizey, *A History of Chinese Medicine* (Edinburgh, 1993); M. Porkert and C. Ullmann, *Chinese Medicine: Its History, Philosophy and Practice* (New York, 1988; P.U. Unschuld, *Medicine in China: A History of Ideas* (Berkeley, 1985).

acupuncture points by the second century and the number continued to grow. Although acupuncture is a comprehensive system of treating all bodily ailments it was its power to abolish pain which, in the late eighteenth century, began to grip the imagination of Western Europe. It is this capacity which still commands a cult and a following.[21]

. . .

Ancient Indian medicine is a vast and labyrinthine field chronicled in many tongues; but it began to crystallise about the sixth century BC around the semi-divine figures of Sushruta and Charaka. The former is supposed to have learnt the art of healing from Dhanwantari, the Indian Aesculapius, and administered anaesthetic drugs by inhalations, generically referred to as *dhuma*. Since he is credited with effectively severing nerves for neuralgia and repairing torn intestines without inflicting pain (among other feats), the substances inhaled must have been powerful.[22] Inhalations had in fact been popular in Indian medicine before him: the physician Jiwaka is said to have administered an aperient to the great Buddha himself disguised in the smell of the lotus. Both Sushruta and Charaka warned against the blandishments and greed of unskilled practitioners, evidence of the emergence of organised medicine everywhere in all ages.

. . .

Leaving Asia, a tour tracing the mythological roots of pain relief could levitate to Peru, to Mexico, to Central Australia, to prehistoric Africa and to the misty world of the Celts.[23] Myth fatigue tends to set in because, with different casts and local variations, the legends follow a similar pattern. Among recurring motifs, a form of reversible anaesthesia was part of most early mythologies. Then as now, the reputation of tribal medicine men often depended on their skill in translating the divine gift into practice.

21. See Chapter 48.
22. N. Gallagher, 'Islamic and Indian Medicine', in *The Cambridge World History of Human Disease*, ed. K.F. Kiple (Cambridge, 1993); K.G. Zysk, *Religious Healing in the Vedas* (Philadelphia, 1993).
23. Learnedly discussed by E.S. Ellis in *Ancient Anodynes* (London, 1946).

## Chapter 2

# THE GRAPE AND THE POPPY

Many nepenthes of Antiquity will always remain a mystery or at least controversial. Others have been identified; and their active principles are still used in everyday medical practice. Two of the most ancient – the products of alcoholic fermentation, 'dispellers of cankering cares', and the inspissated juice of the poppy, 'bringer of happiness and sublime dreams' – have become fixtures of civilised life.

Which of the two came first is uncertain. Given time and sunlight, alcoholic fermentation can occur in any organic pot-pourri which contains sugar or honey. Much indirect evidence suggests that the product was discovered long before man progressed from food-gathering to settled farming. Even if the taste did not immediately appeal, the happy state engendered may have sent some cave dwellers back for another experimental sip – and then another. Yet the urge remains peculiar to man. As Pliny observed, only the gods and humans drink for reasons other than to quench their thirst.

The ingredients were widely available. The technical breakthrough was probably the discovery that the admixture of saliva can initiate fermentation even in the cold and the dark. Within the general enzymic pattern of any species, individual enzymic profiles are genetically determined. Some cave-dwelling kinships with a high yield of fermentation-promoting secretions may have been cherished by their tribe much as wine-growing regions especially blessed by Nature are cherished by wine-lovers today.

The cultivation of crops specifically for the manufacture of alcoholic drinks was everywhere the sign of a developing social order. By the third and fourth millennium BC fermented juices played a key part in the religious festivities and family jollifications of people in the Fertile Crescent. In Egypt the god Osiris was credited with instructing his acolytes how to brew beer from malted barley. By 2000 BC the drink, easier to manufacture than wine, was the most popular drug prescribed by Egyptian physicians. But order begets disorder. The first extant laws restricting the manufacture of alcoholic beverages were issued by Hammurabi, king of Babylon, around 1770 BC. These were concerned with the products of natural fermentation. Arrack, the first spirit to be distilled, was discovered in China around 800 BC. It increased the kick of the parent liquid five-fold.

But it was wine above all other alcoholic beverages whose praises were sung and treacheries deplored. The wild grape, *Vitis sylvestris*, was domesticated in Egypt around 4000 BC. During the next thousand years the new variety, *Vitis vinifera*, was welcomed along the shores of the Mediterranean. It is still the plant cultivated today. How Noah fits into this story is uncertain. Genesis relates that, after emerging from the Ark and giving thanks to the Lord, he took up viticulture, and, on one occasion, became insensibly drunk on the juice of his own growth. Lying shamefully naked in his tent, he was mocked by one of his sons, covered up by another and indifferently gazed at by the third.[1] These are still the prevailing attitudes to drunkenness. What is not clear is whether the Patriarch actually discovered the stupefying effect of fermented grape juice or took advantage of an effect already known.

The Greeks had a colourful mythology centred on Dionysus and his rumbustious followers, both male and female, and inebriation became a staple of their poetry and drama. ('Wine-coloured' was one of Homer's favoured adjective to describe the calm sea.) Whilst the Greeks seem to have drunk more for the effect than for the taste, the Roman Empire lasted long enough to breed a class of connoisseurs and even wine bores. A modern wine buff could teach Proconsul Lucius Lucullus little that is worth knowing about vineyards and vintages. By general consent, the most memorable growth had been harvested in 121 BC, the year in which Gaius Gracchus was killed and Lucius Opimius was consul. A few casks of Vinum Opinianum were still a prized possession of Pliny two hundred years later. Like their owner but perhaps mourned more widely, they perished in the eruption of Vesuvius which buried Pompeii.

. . .

Only God and love can compare with wine as inspirers of poetry throughout the ages: no important work of literature is without at least passing reference to the drink. But nobody has bettered Horace in summarising its varied pharmacological effects.

1. The mockery is implied rather than stated. The episode was a favourite with Renaissance painters. It was treated as a tragic subject by Michelangelo on the ceiling of the Sistine Chapel or as a comical one, as in a woodcut by Cranach. But the greatest 'Drunken Noah' (in the writer's opinion) is Bellini's little-known masterpiece, one of his last paintings and perhaps a self-portrait, in the Musée de l'Art in Besançon.

> What wonders does not wine perform? It discloses secrets, confirms our hopes; thrusts the coward forth to battle; relieves the anxious mind of its burden; instructs the vulgar in the arts. Whom has not a cheerful glass made eloquent! Whom not quite free and easy from pinching poverty?[2]

Significantly perhaps, the poet does not specifically mention relief from physical pain. Though some alcoholic brews soon became standard ingredients of pain-killing potions – Tacitus wrongly stated that *all* anodynes were vinous in character – it was rarely recommended by professional healers as a remedy on its own. In this respect it was surpassed by its rival, opium.

...

It is not always clear which variety of the poppy Greek, Roman and later Islamic writers refer to. In 1753 in his great *Genera plantarum* Linnaeus (or Linné) distinguished 10 genera and coined the name 'somniferum' or bringer of sleep for one. Today at least 28 genera and over 250 varieties are known.[3] Only two are of medicinal importance. The red wild poppy, *Papaver rhoeas*, is a weed, beautiful when painted by Monet or a contemporary artist such as Richard Robbins, and deeply evocative of the sacrificial fields of Flanders, but a comparatively poor source of opium. It is *Papaver somniferum*, the white garden poppy, which was and is deliberately grown for its juice. It requires a moderately rich soil and a temperate climate but no excessive irrigation or expensive fertilisers. It has comparatively few pests. This does not mean that its harvesting and the collection of opium is easy.

The delicate flower is short-lived; and, as the petals drop, they expose the pod, about the size of a pea. Their tapping has to be done by hand, using a special tool. It is a backbreaking task, requiring skill and experience. If the blade cuts too deeply, the opium will flow too quickly, drip into the soil and be lost. Too shallow an incision and the pod will weep internally and block the lacteals. The timing of the tappings – a pod can be tapped several times over a few days – is critical to the nearest hour. The fresh juice which oozes out contains no opium. It must be left to dry – that is to oxidise – for just the right length of time. The cloudy, white, mobile liquid then becomes a dark, brown, sticky and viscous gum. After it has been scraped off, it has to be sun-dried to achieve the right consistency, about that of beeswax. It then needs processing, every bit as demanding as the processing of vintage wines.

...

2. Horace, *Odes*, 3, 21.
3. How the opium-yielding *papaver somniferum* evolved remains controversial. It was probably a natural mutation but it is just possible that it was deliberately cultivated.

In his pioneering study of the remains of the settlements of prehistoric lake dwellers in today's Switzerland, Ferdinand Keller reported in 1868 that he had found seeds of the cultivated garden poppy among the fruit of the villages. Indeed, he identified the fossilised remains of cultivated poppy-seed cakes.[4] Others have cast doubt on this. Keller was an admirer of the prehistoric Swiss – rightly so, for they seem to have been a peaceable and industrious lot – but it is not clear how he distinguished between two kinds of petrified buns.

But there is no doubt that both varieties of poppy were grown in ancient Egypt and Mesopotamia; and in fourth-century Greece Xenophon put an endorsement into the mouth of his friend Socrates:

> Wine tempers the spirit while opium lulls the body to rest. It revives our waning joys and is oil to the dying flame of life.[5]

The word opium itself is Greek, from ὄπιον, meaning the fresh juice of the poppy. A little confusingly, the Greek name for the dried juice was meconium (from μήκων, the poppy), the same word as was used for the discharge of intestinal contents of the newborn.[6] Opium in the Hellenic world remained a largely upper-class indulgence. It was popular in Alexander's army but in the officers' messes rather than in the barracks or the taverns. It also provided an upper-class escape from the sometimes burdensome tribulations of wealth, fame and power. Forsaken and exiled by his countrymen and harassed by Rome, Hannibal took his own life in 183 BC with a dose of Egyptian poppy juice which he carried with him. Agrippina the younger, the Emperor Claudius' wife, may have used a similar preparation to poison her stepson Britannicus and ensure the succession of her son Nero in AD 55.

In the first century the Emperor Claudius' personal physician, Scribonius Largus, gave detailed instructions on how the pod of the poppy should be split and how the escaping juice should be first allowed to dry and then scraped off.[7] This is still the method used in countries where labour is cheap and the plant is cultivated as a means of livelihood. (The procedure is too labour-intensive to be profitable elsewhere.) Today few of the peasant cultivators use it or can indeed afford medicine of any kind; but this was not so in Imperial Rome. A bed of poppies graced the gardens of many modestly prosperous

4. F. Keller, *The Lake Dwellings of Switzerland* (London, 1866).
5. *The Banquet of Xenophon*, translated by James Welwood (London, 1750), Book 2: 24.
6. The term was used by Hippocrates for the bowel contents of the newborn perhaps because of the resemblance to crude poppy juice.
7. The knife he described is similar to the special implement still used in opium-growing countries.

citizens. Theophrastus, Aristotle's successor as director of Athens' famous School of Philosophy and later acclaimed as the 'Father of Botany', gave a faultless account of the actions of the drug, 'both magically beneficial and occasionally dangerous'.[8] His description of the hierarchy of effects is especially impressive.

> The higher moral faculties are affected first . . . then the processes of logical thought . . . Base animal urges and essential but uncontrolled bodily functions are preserved longest.[9]

Had Theophrastus' learnt his Latin as part of the classical education provided by a *fin-de-siècle* Vienna *Gymnasium* rather than as a toddler to express his basic bodily needs, he might have anticipated the Freudian id, ego and superego.[10]

Even many erroneous ideas have ancient origins. Pliny among others warned against the disturbing effects of the oil prepared from dried poppy seeds. Today the consensus is that the seeds contain no active ingredient; indeed, even the fresh juice is virtually inactive.[11] Yet the fear lingered. In the early years of the eighteenth century Paris was gripped by panic that oil prepared from seeds imported from the Protestant – i.e. godless – Netherlands was poisonous. The city's Medical Faculty was asked to investigate, using convicted felons as guinea pigs. They took three years to conclude that the seeds were harmless. Their conclusions were of course disbelieved and Louis XV issued a decree prohibiting the sale of poppy seeds and poppy-seed oil, whether of Protestant or of Catholic provenance. The law was not formally revoked till Aristide Briand, a noted gourmet as well as a crafty politician, persuaded the Chamber to do so in 1924. Despite the scientific evidence, poppy seeds are still added to milk in villages in Eastern Europe to silence hungry infants.

. . .

8. Theophrastus' *De historia plantarum* gives a splendid description of 550 different species of plants from Spain to India (the data for the latter gathered mainly from participants in Alexander's campaign). His *De causis plantarum* describes the common characteristics of plants; and it is in the first volume of this work that he gives an account of the preparation and properties of opium. He died in 287 BC at the age of eighty-three. He was comparatively neglected by Latin authors, but his rediscovery during the Renaissance started a craze for botany in Europe and led to the foundation of the first botanical gardens in Italy and France. Henri IV's queen, Marie de' Medici, was a fan.
9. Theophrastus, *Enquiry into Plants*, translated by Sir Arthur Hort, Bart (London, 1916), Book 9, Chapter 8.
10. Being modestly proficient in kitchen Latin, carried a social cachet in the Vienna of the 1880s: it distanced upwardly mobile Jewish families from more recently arrived 'Ostjuden'. Freud never missed a chance to display his familiarity with the Classics.
11. It is during drying that oxidising enzymes convert inactive precursors not only into morphine but also into codeine and other active alkaloids.

All poets of the Augustan Age praised opium. Even after becoming the darling of Roman society, Virgil remained a country boy at heart, happiest when singing the simple joys of rural life and labours. Or not so simple. In one of his sublime *Georgics* he extols '*Lethaeo perfusa papavero somno*', the 'poppy steeped in sweet dreams'.[12] Ovid in *Fasti* describes the goddess of sleep with 'her calm brow wreathed in poppies, bringer of delectable dreams'.[13] Elsewhere he praises the drug which 'induces deep slumber and steeps the vanquished suffering eyes in Lethean night'.[14] Nobody would ever surpass him in coining the memorable phrase. Catullus was a fan. A hundred years later Lucian in his satirical *True Story*, a splendid spoof of all travellers' tales (as *Don Quixote* would be of knightly romances), describes his arrival on the Isle of Dreams and the City of Sleep, which is surrounded by a wood 'in which the trees are tall poppies and mandragoras'.[15] He and his companions spend thirty wonderfully luxurious days and nights there – sleeping.

In contrast to the poets and their lady friends, doctors and administrators tended to disapprove. Pliny, a grand admiral among his other offices, deplored the growing use of poppy 'even in sober households: it is in any form highly dangerous'. His fears were endorsed by the two leading medical figures of the Empire.

Aurelius Cornelius Celsus was a landed gentleman who cultivated medicine as a hobby. His six-volume *De medicina*, completed in about AD 30, is the only part of a more comprehensive scientific encyclopaedia that survives.[16] It is unique among medical textbooks in being a literary masterpiece. No other

12. Virgil, *Georgics*, 1. 77–8.
13. Ovid, *Fasti*, 4. 661.

    Elsewhere – *Fasti*, 531–2 – Ovid describes the goddess Ceres putting her child Triptolemus to sleep with poppies before placing him over the fire to purge him of mortality. (Unfortunately she is interrupted by his mother, Metanira.) Before she enters the little house she picks a gentle and sleep-bearing poppy and, as she tastes it, it 'unwittingly dispels her hunger'.
14. In Graeco-Roman mythology Lethe was one of the rivers of the Underworld whose water, when drunk by the dead, caused them to forget all the evils and fears of the past. It became the personification of forgetfulness in Hesiod and appears in the *Frogs* by Aristophanes in the fifth century BC.
15. Lucian, *True Story*, in Volume I of *Works*, p. 337, with an English translation by A.M. Harmon (Cambridge, Massachusetts and London, 1961).
16. Though little read during the Middle Ages, in 1478 *De medicina* was one of the first medical books to be printed (under the title *De re medicina*) and went into at least fifty editions. It then remained widely used until the middle of the eighteenth century.

    Among his other accomplishments, it was Celsus who virtually created modern medical Latin. Some terms he simply adopted from the Greek and Latinised (e.g. *stomachus, bracchium*); but more often he translated into Latin the vivid Greek imagery. The Greek *kynodontes* (dog's teeth) became the Latin *canines* and the Greek *typhlon* (blind gut) became the Latin *caecum.* He also preserved in a Latinised form the Greek tradition of likening the shapes of anatomical structures to familiar objects like

writer could have summarised the fearsome complexities of acute inflammation in a sonorous sequence of four two-syllable words that still roll off the tongue and remain valid two thousand years later: *dolor* (pain), *rubor* (redness), *tumor* (swelling) and *calor* (heat).[17] His contemporaries recognised his genius. Valerius Tarsus, another medical amateur, referred to him as the 'Cicero of Medicine'. (This was clearly meant as a compliment, though Celsus' epigrammatic style seems far removed from the ponderous cadences of the great orator.) Almost repetitiously Celsus emphasises that it is the cause of pain rather than pain itself which doctors should treat, or try to. More specifically, in Volume III he states that 'pills and potions that relieve pain are numerous; but, unless there is an overwhelming necessity, it is improper to use them'. He does not specify his reasons, except to say that they are 'alien and harmful to the stomach'. Speaking of the juice of the poppy, he notes that 'it has been used to calm tempers and to induce pleasant dreams since the Trojan War and is still popular'; but 'doctors should use it with circumspection'. He warns that 'dreams can be sweet; but the sweeter they are, the rougher tends to be the awakening'.

In contrast to Celsus, Galen was the complete professional. Indeed, no doctor in the history of medicine has exercised such a profound influence on the practice of the profession for so long.[18] Fourteen hundred years after his death the statement that 'Galen says so' still settled academic disputes. Paracelsus in the sixteenth century inaugurated modern medicine (or so he claimed) by burning a few Galenic books in public. Today the secret of the famous Roman is not easy to fathom. The son of a wealthy architect, he was born in 131 in Pergamum, then a flourishing commercial and cultural centre in Asia Minor. He enjoyed, by his own account, a pampered childhood and an expensive education. After travelling widely, he settled in Rome. He was a consummate name-dropper: 'when I was treating the Emperor' and 'the Emperor's brother (or cousin or aunt or uncle) often assured me . . .' crop up in all his books at regular intervals. By contrast, he frequently refers to his professional brethren as 'my dim-witted colleagues . . .'. He fancied himself as a supreme medical scientist as well as a practitioner of bottomless wisdom,

musical instruments (e.g., *tuba*, trumpet or *tibia*, flute), armour (e.g. *thorax*, breastplate), everyday tools (e.g. *fibula*, needle), familiar plants (e.g. *uvea*, grape or *glans*, acorn) or animals (e.g. *helix*, snail or *musculus*, mouse).

17. Thomas Sydenham is thought to have added the fifth *functio laesa* (impaired function), breaking the two-syllable sequence.
18. Unlike the work of many of his contemporaries, a significant part of his writings survived the collapse of the Western Roman Empire and was available for translation by Islamic scholars in the ninth and tenth centuries. He is still being translated and his doctrines continue to command lively interest among medical historians. (See Bibliography.)

writing on anatomy, physiology, medical chemistry, surgical instruments and the soul with equal authority. He had in truth much going for him, at least as a compiler of other people's ideas. He was immensely industrious and never in the slightest doubt about anything, an indispensable trait in writers of textbooks. He never needed praise from others. He himself summed up his life's achievement:

> I have done as much for medicine as the glorious emperor Trajan did for the Empire when he conquered lands and built roads and bridges. Hippocrates paved the way . . . but it is I and I alone who have revealed the true path of medicine.[19]

Like many men extravagantly successful in their prime, he became embittered in old age. The world's follies weighed on him heavily. Though patronising towards Celsus ('a good stylist but lacking in decisiveness'), about pain-killing potions he expressed himself even more strongly. 'More than anything else, I abhor carotic drugs', the term applied at the time to all remedies which produced stupor or sleep.[20] Relenting slightly, he recommended pain relief in older patients, praising (with reservations) the juice of the poppy (when carefully prepared) as relatively safe. Indeed, he mentions 'several eminent senators' who 'have benefited from potions prepared for them with my own hands'. His potions rarely contained fewer than twenty ingredients, most of them exotic, expensive and useless; but in that respect he was not significantly different from other Roman prescribers. But neither he nor any other respectable Roman doctor would prescribe opium to young people; especially not to children.

The reasons are both obvious and puzzling. Like Celsus, Galen hints at digestive upsets and specifically warns against constipation.[21] This was much feared in Rome, a legacy of its Eastern conquests. After the annexation of Egypt the regularity of bowel movements became a prime concern of every Roman

19. Quoted in V. Nutton (ed.), *Galen: Problems and Prospects* (London, 1981), p. 56.
20. The term derives from the Greek καφοῦν, to stupefy and καφωτίδες carotid arteries, the arteries of the neck. Rufus of Ephesus says: 'The ancients called the arteries of the neck καφωτίδες carotid arteries because they believed that when they are pressed hard, the animal becomes drowsy.' The ancients were right and their observations apply to humans as well. There are two relevant mechanisms. First, cutting off the blood supply to the brain leads to oxygen deprivation and carbon dioxide accumulation. The combined effect is stupefying before killing. More dramatically, the carotid body, a small nodule in the neck near the bifurcation of the common carotid artery into an inner and an outer branch, is richly supplied with specific nerve endings, the stimulation of which can initiate powerful, so-called vasovagal reflexes which slow the heart. Pressure on the body can lead to a feeling of faintness, sudden fainting or death. The term *carotique*, meaning stupefying, is still used in modern French.
21. Still one of the complications of morphine therapy.

citizen. Galen also states that too big a dose can kill patients, though he does not specify respiratory depression as the likeliest cause of death. But neither he nor any other Greek or Roman writer refers to the danger of addiction, that is the need for increasing doses to obtain the same or even a diminishing effect. (That alcohol can be addictive in this way was well known.) Of course nothing is easier in medicine than to ignore the obvious; and Galen and his colleagues may have been unaware of much that is familiar to the point of tedium to every teenager today. But perhaps, for some reason, addiction was not a problem. The later Roman Empire was as over-regulated – or civilised, depending on one's point of view – as any member state of the European Union is today; but no legislation forbidding the growth, marketing or use of the drug survives.

Yet even expertly prepared, the crude poppy cake must have been dangerously unpredictable. This was known and widely deplored. Diocles of Carystos, one of Galen's contemporaries, practised in Asia Minor and regularly prescribed an opium-based paste for toothache. He praised it for its efficacy but castigated the plant, a fashionable literary conceit among scientific writers of the late Empire, for not 'giving up its secrets'.

> There is much that is still unknown about this wondrous flower. The potion prepared from its capsule will soothe some but cast others into melancholy. In some its effect will be immediate. In others it may be delayed for many hours. It some the medication will be well tolerated. In others it will have unpleasant side-effects. Yet even when it does not abolish pain, the pain no longer preys on the person's mind.[22]

...

Lamentations do not solve mysteries. An efficient way of tapping the plant had been developed. A product was available, if not cheap. The flower continued to be held in affection. It often appeared on coins, tiles and ornamental seals. (Later it would be carved on medieval church pews.) Nobody thought of venturing beyond that. The Greeks were ingenious thinkers and wonderful poets. Romans were accomplished administrators and builders of roads. Neither had a feel for analytical chemistry. But perhaps this is unfair. Opium was introduced to Persia, India, China, Japan and perhaps the East Indies. There were no significant advances in understanding its mode of action anywhere. Not until the development of analytical organic chemistry in Europe at the end of the eighteenth century – an explosion of creative practical science – would the picture change. It would then change beyond recognition.[23]

22. Diocles of Carystos, quoted by M. Madai, *Romai orvostudomany* [Roman Medicine] (Budapest, 1936).
23. See Chapter 13.

# Chapter 3

# ROOTS, BARKS, FRUIT AND LEAVES

The grape and the poppy were famous and would remain so; but they were not the only pain-killing plants. The leading pharmacopoeia of the late Roman Empire was compiled by Pedatius Dioscorides of Anazarba in Cilicia (today's Syria), a military surgeon who trained in Alexandria and for a time served in Nero's army. His style, like that of many military writers, was brisk. The five volumes of his *Materia medica*, written in Greek despite the Latin title, contains detailed descriptions of the appearance, harvesting, preparation, properties, uses and contraindications of 342 medicinal herbs, as well as chapters on metals (like copper and lead), minerals (like sea salt) and spices (like cassia and cinnamon). The compilation became the Roman physician's bible. Rediscovered in the twelfth century, it was to be revered by the teachers of Salerno, Padua and Montpellier.[1] The diseases for which it recommended specific remedies ranged from a runny nose and menstrual disorders to fevers, snake bites and internal inflammations. Almost a third of all prescriptions were intended to combat pain of some kind or another.

As light relief from practical prescribing, Dioscorides allowed himself a few historical digressions. Relying on Apollodorus' *Bibliotheka*, a slightly scurrilous compilation of Greek myths, he called the black hellebore, *Helleborus niger*, the oldest of all herbal medicines. According to Apollodorus, it was the shepherd Melampus who 'first devised treatments with plants'.[2] Arcadian shepherds tended to be well connected as well as handsome. When the three daughters of Proteus, king of Argos, lost their reason and began to dance naked around his palace, the guardian of the sheep was sent for. His treatment with an extract of hellebore re-established the princesses' sanity or at least their sense of decorum. After that the plant became famous as a calmer of

1. The *Urtext* is the beautifully illustrated *Codex Anicia*, presented to Juliana Anicia, daughter of Anicius Olibrius, Emperor of the West in 462.
2. Apollodorus (attributed to but probably not by him), *Bibliotheka*.

nerves. It also became the laxative most favoured by the followers of Hippocrates. Or was it a different plant bearing the same name? Mythological medicine is full of uncertainties, as of course will be modern formularies to historians of the fourth millennium.

Even before Dioscorides, henbane or *Hyoscyamus niger*, a member of the great family of *Solanaceae*, was recommended by Celsus. 'It is,' he said, 'both reliable and unlikely to cause permanent damage to the higher faculties.' Humans were not the only potential beneficiaries. The name 'hyoscyamus' suggests that the plant was invigorating to hogs, while the Anglo-Saxon 'henbane' warns against its poisonous effect on fowl. Despite its obnoxious smell, it continued to be used as a sedative and is still grown for that purpose in some countries. The active principle prepared from the dried seeds, hyoscyamine, is an optical isomer of atropine, still an essential aid to ophthalmology and, until comparatively recently, routinely used as a premedication before general anaesthesia.[3]

The milk of the stalk of the mulberry tree, especially of the black variety (*Morus niger)*, was a soporific recommended by Dioscorides for women in labour. Celsus considered it too unpredictable: it sometimes, he suggested, encouraged recklessness rather than tranquillity. But that was a risk with most anodynes. The garden lettuce gathered when young and tender had an established reputation as a mollifier of grief. But it could also encourage frenzy. In Ovid's *Metamorphoses*, after the cruel death of Adonis, Venus seeks solace by throwing herself on a bed of lettuces. (The plant is not specified in Shakespeare's reworking of the story.) Historically better documented is the fact that the great Galen himself found it a relatively harmless soother of 'unseemly and unsettling rages' in his own troubled old age.[4] Rage control was a major Roman preoccupation.

The narcotic and pain-killing properties of the twining hop plant, *Humulus lupulus*, were appreciated independently of the brewing of beer. According to Theophrastus, the discovery was made when the scented air of hothouses where hops were stored overcame those who inhaled it for more than a few minutes. They fell into a deep sleep marked by delightfully amorous dreams. Hop-pillows were still used many centuries later to combat the painful agitation of King George III during his spells of insanity.[5]

3. In addition to calming nerves, both atropine and hyoscyamine are useful for drying up secretions in the mouth and the respiratory tract, reducing the danger of vomiting and making inhalation anaesthesia easier.
4. He died about 216 in his late seventies.
5. Now widely attributed, on well presented but flimsy evidence, to the rare metabolic disorder of acute intermittent porphyria.

As still happens today, demand for individual remedies tended to fluctuate. During the first century dittany or *dictamnus*, a plant grown in the mountains of Dicte in Crete, enjoyed a vogue. August patronage may have had something to do with it. According to Suetonius, a potion prepared from the root was Caligula's favourite tipple before performing as a dancer for the delectation of his petrified guests. They felt it prudent to marvel at the potion's effect. By then extracts from the leaves had been used for centuries. Their pain-killing potency was discovered by wild goats who, according to Dioscorides, 'when wounded by arrows, go in search for it to ease their suffering'. Venus was among their disciples, at least according to Virgil.

> Smitten by her son's [Aeneas'] cruel pain, with a mother's care, she plucks from Cretan Ida a dittany stalk clothed with dawny leaves and purple flower . . . This Venus bore down, her face veiled in dim mists . . .[6]

After adding a few less specific ingredients (such as ambrosia), the aged Iapyx applied the mixture to Aeneas' wound; and lo,

> all the pain fled from [Aeneas'] body, all blood was staunched deep in the wound. And now, following his hand, without constraint, the arrow fell out, and newborn strength returned as of yore.[7]

. . .

The effects of *Cannabis indica* were extolled both in the Greek world and in the East. In a famous passage Herodotus, writing in the fifth century BC, describes how a healing bath was administered to wounded barbarians.

> The Scythians are given the seed of this hemp, creep under rugs and then throw the seeds on red-hot stones. The seeds begin to smoulder and send forth so much steam that no Greek vapour bath could surpass it. The wounded Scythians howl with relief and joy.[8]

In Indian lore the plant was given such epithets as 'leaf of delusion', 'exciter of desire', 'increaser of pleasure', 'cementer of friendship', 'mover of laughter', 'causer of the reeling gait' and 'soother of pain'.

. . .

Though Dioscorides took some credit for its discovery – at least in early illustrated manuscripts he is shown receiving the roots from the goddess of healing – mandrake or mandragora, the *Atropa mandargor* of Linné, is among

6. Virgil, *Aeneid*, Book 12.1. 411, 414, 416–17 and 421–4.
7. Ibid.
8. Quoted by V. Robinson, *Victory over Pain* (London, 1947), p. 17.

the most ancient of medicinal plants. Perhaps Dioscorides merely wished to claim that he was the first to exploit its full potential: it is an ingredient of eighty-five of the potions he recommends. But the name is Assyrian in origin, and wreaths made of its leaves were found in Tutankhamun's tomb in Egypt, dating from the fourteenth century BC. Even earlier perhaps, in Chapter 30 of Genesis Reuben, eldest son of Jacob, goes 'in the days of wheat harvest, and finds mandrakes in the field, and brings them unto his mother, Leah'. Their objective in that case was probably to 'open Leah's womb' to let her bear more children to Jacob; and it was probably the fruit which was consumed. It worked for Leah (much to Rachel's grief); and the use of the 'love apple' as a fertility agent and aphrodisiac continued. In the Latin literature of the Silver Age Venus is sometimes apostrophised as *Mandragontis*.

But the main use of the plant was as a painkiller, healer of intractable ailments and, above all, a general inducer of tranquillity. This was not of course always thought desirable. In Athens Demosthenes remonstrated with his fellow citizens for behaving like 'men drunk on mandragora' instead of preparing to defend their city to the death against Philip of Macedon. The elder Cato also thundered against young men 'befuddled by mandrake' when they should have been cultivating manly virtues. Both orators were admired for their elevated sentiments rather than for their common sense.

It is in the passage describing this famous herb that Dioscorides uses the term 'anaesthesia' in the modern medical sense for the first time.[9] 'One cyathus of the wine of mandragora is given to those who cannot sleep and such as are in grievous pain and those to be cut or cauterised, when it is wished to produce anaesthesia.'[10] A little reluctantly perhaps, even Galen mentions its 'calming effect well known to the ancients'. Pliny claimed that the mere smell of the leaves could put people to restful sleep.

No other medicinal plant became so encrusted with fear, awe and nonsense. Similar tales crop up in the literature of three continents separated by centuries. Apuleius in the *Golden Ass* tells the tale of a doctor who, using an extract of mandrake, treats a mortally sick boy. The boy falls into a deep sleep and is mistaken for dead. Happily (for the doctor as well as for the patient), the boy rises from the coffin just before being interred. He sings a happy tune.

9. Literally the term means without feeling, from αν privative ἀισδησις sensation. More loosely, it had been used even earlier. In *Timaeus* Plato's interlocutor suggests that severe physical pain usually originates in bones and describes 'anaesthesia' as the reason why certain bones but not joints have been covered by the gods with dense protective flesh.
10. Pedatius Dioscorides 'Manuscript Anicia', English edition 1655 by John Goodyer. Gunther edition (Oxford, 1934), Book 4, p. 456.

A *cyathus* was a twelfth of a *sectarius* which was a sixth of a *congius* or pint. A *cyathus* was therefore about 10 millilitres.

While apparently dead, all his ailments had vanished.[11] Benedek relates the same story recorded on a Nepalese scroll half a millennium later.[12] A passage in Frontinus recalls the Carthaginian officer and his army under Hannibal who pretend to be beaten and flee, leaving behind in their camp barrels of wine doctored with mandrake. When the enemy set on the camp and gorge themselves on the loot, the Carthaginians return and slaughter their insensible victims.[13] Almost the same tale about Visigoths and Vandals is retailed by Isidorus, Bishop of Seville in the seventh century. The bishop devoted a massive tome to this 'incomparable slayer of pain'.[14] He states in his Introduction that he himself could not have completed his labours without frequent recourse to it, a dubious recommendation perhaps for a therapeutic manual.

Not everybody everywhere was so enthusiastic. St Hildegard in the twelfth century suggested that the plant was formed from the soil from which Adam was created and that 'it will arouse man's desire, for good or evil, and lull him into vain and unwholesome dreams'.[15] A hundred years later Bartholomaeus Anglicus maintained that, while superb in 'abating all manner of soreness, so that man feeleth unneath though he can be cut', it must be warily used 'for it too readily kindles lascivious desires'.[16] In this vein the fruit was sometimes called the Devil's Testicles and Joan of Arc was accused of having carried one in her bosom. (She denied it.) Shakespeare refers to mandrake at least seven times (in contrast to the poppy, which he mentions only once), most notably when Cleopatra, temporarily abandoned, asks for a drink of mandragora. 'Why, Madam?' 'That I might sleep out this great gap of time.'[17]

The fear that surrounded the harvesting of the herb and the belief that the most potent variety grew under the scaffold, nourished by the sweat and blood of malefactors, was also ancient. Theophrastus quotes 'numerous authorities' who advise that those who dig it up must not have the wind blowing in their face and should, to be safe, trace three circles around it with a sword before embarking on the task. Also, while cutting the soil, 'say as many things as possible about the mysteries of love'. The vaguely anthropomorphic root, not unlike the friendly little figure which today points in the direction of the male lavatory, was mentioned by Dioscorides long before it aroused the

11. Apuleius, *The Golden Asse of Apuleius*, done into English by W. Adlington (London, 1908), p. 134.
12. I. Benedek, *Mandragora* (Budapest, 1979).
13. Sextus Julius Frontinus, *The Stratagems*, translated by Charles E. Bennett (Cornell, 1950), p. 230.
14. Quoted by E.S. Ellis, *Ancient Anodynes* (London, 1946), p. 71.
15. Quoted by C.J.C. Thompson, *The Mystic Mandrake* (London, 1934), p. 86.
16. Quoted in Ellis, *Ancient Anodynes*, p. 74.
17. William Shakespeare, *Antony and Cleopatra*, I. v. 4.

misgivings of St Hildegard. When and where the belief of its dread 'scream' originated is uncertain: it is recorded in Caelius Aurelianus' third-century herbal as an established fact. By the early Middle Ages it was widely believed that, when pulled from the ground, the root emits a bloodcurdling sound that can strike its hearer dead. To escape such a fate the plant was often tied to a dog and the animal was lured away with a bait. When kindly disposed, the dog's master sounded a horn to deafen the creature. More than one otherwise puzzling depiction of this ritual survives in Western Books of Hours, most famously perhaps in the sumptuously illustrated volume prepared for the duc de Berry.

. . .

The best ways of using medicinal plants were much debated. In this respect too mandrake led the field. The optimal positions of the moon for various stages of the processing were laid down in commanding detail by Dioscorides. The juice was most commonly drunk or boiled and the vapour inhaled; but William Turner, Dean of Wells during the reign of Henry VIII, advised that, for best effect, the whole plant should be ground into a paste and 'put into the fundament'. He promised that this would send even those in terrible pain into a deep slumber. Too deep, sometimes. 'If the pacient be too much slepi put stynkynge thynges unto hys nose to waken hym therewith,' he recommended.[18] Two hundred years later Peter the Great returned from his travels in Western Europe with a supply of the powdered root. He regularly dosed himself and his empress, the future Catherine I, in times of stress. These were not infrequent. The draught may have helped her to live with the severed head of her lover placed in a jar and positioned on her bedside table by her aggrieved husband during the last fraught weeks of his reign. In 1695 Augustus II, King of Poland and Elector of Saxony, who probably suffered from diabetic gangrene of the foot, was painlessly operated on after a dose administered by his favourite surgeon, Albertus Weiss, a pupil of Petit of Paris. (The condition is still common; and the painlessness may have been due to loss of sensation secondary to diabetic neuropathy rather than to the anaesthetic draught.) Armed with the knowledge and techniques of modern analytical chemistry, Sir Benjamin Ward Richardson, the eminent Victorian doctor, undertook to investigate extracts in his private laboratory. He found that both the root and the leaves were rich sources of atropine, scopolamine, stramonium and hyoscine, all of them powerful pharmacological agents affecting heart action, vascular tone, bowel movements, pupillary response, bladder function, sweating and other vital activities not under voluntary control. He also carried out experiments on animals and confirmed their overall anaes-

18. William Turner quoted in Ellis, *Ancient Anodynes*, p. 75.

thetic effect.[19] By then, with both ether and chloroform available, this was of academic interest only; but Benedek still devoted two large volumes to the use of the plant in India and Tibet as recently as 1972.

...

It might be thought that, with such a plethora of effective painkillers, nobody in the Graeco-Roman world ever had to suffer the agonies of a tooth extraction or an amputation. Though this is implied even in some modern histories of anaesthesia, there are no firm grounds for the belief. Patients in pain or about to undergo some painful procedure were no doubt liberally dosed with opium, alcohol and herbal cocktails (as they would continue to be for two thousand years); but there is no example in classical literature of herbal narcotics being employed to suppress the *acute* pain of planned surgical procedures. The reasons are not far to seek.

In Greek mythology and Hellenistic sculpture Hypnos and Thanatos are represented as two chubby twins. Both hold torches pointing earthward. The torch of Hypnos, the god of sleep, is flickering. The torch of Thanatos, the god of death, is extinguished. The twins were inseparable. All the plant extracts recommended as anodynes or narcotics could also be fatal poisons. It was never possible to be sure which of the twins would prevail. Some hypnotic agents were indeed used for relatively humane executions.

The prettily fronded green leaves of the hemlock (*Conium maculatum*) had been popular in Greece at least since the days of Homer. Even their smell when fully grown was languorous and heavy enough to induce a feeling of calm and well-being. But the plant's pain-relieving power varied with the weather, the soil, the rainfall and the technique of extraction. So did its propensity to kill. The former was said to be most dependable before, the latter after, the flowering. But nobody was certain. The plant extract demonstrated the kinship between sleep and death. The Athenians in their cultural heyday regarded a large dose as the most satisfactory means of judicial murder. It is in this capacity that it takes centre stage in one of the most celebrated death scenes in world literature.[20]

By 399 BC Socrates, a respected craftsman, probably a stonemason, of Athens, who had formerly served in the army, had become a popular debater and teacher of philosophy. He was clever and articulate, with an obsessional dislike for any kind of humbug. Unfortunately for him, Athens had become a democracy over the previous hundred years, perhaps the first in the world; and without a soupçon of humbug democracies cannot function. Socrates was incapable of hiding his dislike of spin and double-talk and soon became

19. Robinson, *Victory over Pain*, p. 134.
20. Plato, *Phaedo*, 118.

a thorn in the flesh of the political establishment. Eventually he was accused of inciting impressionable young men – nobody worried much about impressionable young women – to immoral behaviour. He had led astray even such formerly level-headed citizens as Plato. His impious references to the soul were particularly resented. (Despite his professed devotion to clarity, classical scholars still argue whether he was an atheist, an agnostic or a believer in a personal God.) He was duly tried, initially probably with no objective more virulent than to admonish him to curb his tongue. But by now he had become a showman with a death-wish. At his trial he so infuriated his judges with provocative questions about their credentials that he was sentenced to death. To the exasperation of most right-thinking citizens, he refused to take advantage of several opportunities to escape. Plato later described his last hours.

Friends and disciples chat in the death cell. The executioner enters carrying a cup of hemlock. He is affably received by the prisoner, who asks him for guidance how to use the poison most effectively. 'It is quite simple,' the executioner explains. 'You have only to drink this cupful and then walk about until your legs feel heavy. Then lie down and allow the poison to act slowly and painlessly.' Socrates thanks him and cheerfully drains the cup: he is confident that the end will be painless. For a time the light-hearted badinage continues; but, as predicted by the executioner, the teacher's legs soon begin to feel leaden and he lies down. Sensation in the feet then disappears. (Greeks and Romans were aware of the separation of motor and sensory functions but the two kinds of nerves were not clearly distinguished until the researches of Charles Bell in the early years of the nineteenth century.)[21] The assembled company continue to chatter while the executioner ascertains that the anaesthesia is duly spreading upwards. Socrates touches himself from time to time and confirms the executioner's findings. When the effect reaches the heart, he explains, he will be leaving his friends. But first he turns to his youngest disciple, Crito, and utters his famous instruction: 'Crito, we owe a cock to Aesculapius.' The meaning of the enjoinder has been argued about but Socrates was surely referring to the debt he owed to medical science for the painlessness of his parting. Crito promises to perform the sacrifice. He then asks the teacher if has any other wishes or instructions. There is no reply.

...

After the Athenian experience Pliny was reluctant to refer to hemlock as a therapeutic agent; but 'since in the proper manner it can be employed for many useful purposes, it cannot be entirely omitted'. He therefore mentioned

21. See Chapter 31.

its pain-relieving properties especially when applied as a poultice and in allaying discomfort and pain in the eyes.

But what was true of hemlock was to some extent true of most painkillers. Some, like the 'deadly nightshade' (*Atropa belladonna*), were familiar to Greeks and Romans but are barely mentioned by Latin authors. (Dioscorides refers to it once in passing.) The risks when handled carelessly were too great. Or perhaps the desire of Roman matrons to look irresistible with their pupils alluringly dilated was not as compelling as it would be to the *belle donne* of Venice a thousand years later.[22]

In contrast to any planned surgical anaesthesia, numerous descriptions survive of operations without painkillers. Though a scion of the ruling classes, Gaius Marius rose to dictatorial power as champion of the *plebs* in the second century BC. As part of his image he cultivated a reputation for toughness; and it was probably this rather than any medical need which induced him to order his army surgeon to repair some 'prominent and painful' vessels on his legs. (The condition sounds like varicose veins but it is difficult to be certain.) Plutarch relates that, though warned of the pain, he scorned inebriants and refused to be bound. Without a groan and 'with a steadfast countenance' he endured the operation on the right leg. But he then refused to have his left leg operated on, 'judging that the cure was worse than the illness.'[23] Patients miraculously rendered insensible to the knife were probably more common in fiction than in real life.

22. The blurred vision caused by the paralysis of the pupil induced by atropine (the active alkaloid in belladonna) was a small price to pay to the Venuses of Giorgione and Titian, all of whom seem to have their eyes permanently adapted to night vision.
23. Plutarch, *Lives*, translated by B. Perrin (London, 1914), p. 564.

# Chapter 4
# PAIN DENIED

All modern anaesthetics and most currently popular painkillers aim at *suppressing* pain. In this they are remarkably successful. Their success tends to overshadow another possible approach to pain. Pain can be *denied* as well as suppressed. Of course denial is less generally applicable. It is out of fashion. It is also often described as scientifically unsound. Perhaps it is. But it has existed for longer than any modern painkiller and it has never entirely disappeared. The fakir sleeping peacefully on his bed of nails remains both the tired cartoonist's standby and a fact. But it was in Antiquity and then again in the sixteenth century, sometimes referred to as the First and Second Ages of Martyrs, that pain denial achieved its most memorable miracles.

The term 'miracle' is here used in two senses. The denial of pain by the martyrs of the early Church has been admired as an act of superhuman heroism for almost two thousand years. By definition, such acts are not only unexplained but also inexplicable in purely medical terms. The faithful still pray to some of these martyrs. Such prayers are often answered. It in no way detracts from the loftiness of their sacrifice that their actions can also be approached at another level. The terms 'hypnosis' and 'autohypnosis' are comparatively recent but the states they represent are ancient.[1] Using modern techniques and in a sceptical frame of mind, neuroscientists have extensively investigated them over the past fifty years. That such states can eliminate the sensation of pain is no longer in doubt. The underlying neuronal mechanism remains almost totally obscure. 'Self-hypnosis', in other words, is a scientific tag which makes miracles acceptable to those who do not believe in miracles.

Even more than attempts at the suppression of pain, the denial of pain by the martyrs has to be seen against an historical background. For almost two hundred years after Nero's crazed rampage (in which both the Apostles Peter

1. See Chapter 20.

and Paul perished) Imperial Rome remained tolerant of foreign faiths.[2] A bizarre variety of cults, many of them originating in the East, were openly practised by the professions, in the army and throughout the civil service. Some became chic.[3] They did not usually conflict with the perfunctory veneration of ancestral guardian spirits. The toleration extended to the Christian communities. It was, by and large, reciprocated. During the Pax Romana, as majestically portrayed in the opening pages of Gibbon's *Decline and Fall of the Roman Empire*, most followers of Christ saw a strong imperial power as their guarantee of safety.[4] 'Render unto Caesar' was a command by which they could abide. In return, they were exemplary citizens. Justin, future saint and martyr, could describe his coreligionists as 'ideal subjects, embracing chastity instead of debauchery, charity rather than greed, paying their taxes and tributes more punctiliously than most, and praying for their enemies instead of plotting acts of unlawful revenge'.[5]

There had nevertheless been sporadic outbreaks of persecution before the last and fiercest. Domitian's in 89 was provoked by the mutiny of his army commander in Germany, allegedly a Christian, and led to the execution of several Christian members of the Emperor's own family. After three years he himself was stabbed to death on the orders of a niece whose principal lover he had had crucified. The persecution instituted by his successor, Decius, followed the millenary celebrations of the foundation of Rome in 248.[6] Christians disapproved of religious festivities devoted to ancestral deities; and

2. Nero used Christians as scapegoats for the conflagration which destroyed much of the old, timber-built Rome. He may have started the fire himself to inspire him in the composition of an ode describing the burning of Troy. The persecutions form the backdrop to Henryk Sienkiewitcz's famous novel, *Quo Vadis?*
3. During the first century the most popular foreign faith was the worship of Mithras, originally a subordinate spirit of the Persian Zoroaster but promoted in Rome to be the principal god, more or less identified with the sun. According to Plutarch, the creed was imported by the troops of Pompey, Caesar's rival, returning from the East. Mithras was widely portrayed as a handsome youth wearing a Phrygian cap and plunging his dagger into a prostrate bull.

   Neither the Greek nor the Roman pantheon was exclusive. Olympian deities were always ready to assume new and additional identities. Underlying this flexibility was the firm belief that humans could never fully understand the world (or worlds) of the gods. The Greek and Roman pantheons represented just one echelon above the human condition, not necessarily the ultimate truth. It was the certainty professed by the Jews and even more by the Christians which most offended pagan sensibilities.
4. See H. Chadwick in *The Oxford History of Christianity*, ed. John McManners (Oxford, 1990), p. 21.

   Trajan wrote to proconsul Pliny the Younger in Asia Minor to tell him that Christians should be encouraged to take their malicious and false accusers to court. Hadrian had several trusted Christian advisers.
5. Quoted by B. Moynahan, *The Faith. A History of Christianity* (London, 2003), p. 41.
6. Tradition held that Rome was founded by the twins Romulus and Remus in 753 BC, year 1 in the Roman calendar.

in times of peril their disapproval angered pagan traditionalists. In the mid-third century along the northern frontiers of the Empire barbarians were on the move. Eventually fear gripped the Emperor himself and he ordered that sacrifices be made to the old gods by all his subjects (except the Jews, whose traditional exemption he respected).[7] Many Christians refused. Fabian, Bishop of Rome, already called *Papa*, was among the first to be martyred. Hundreds, though probably not thousands, were beheaded or burnt at the stake. Decius was slaughtered the following year when his army was ambushed by the Goths in the swamps of the Danube delta. When the Emperor Valerian revived the Decian persecutions in 258 a few Christians recanted but once again several hundred, including another pope, Sixtus II, were martyred. A year later Valerian was taken prisoner by the Persians and starved to death in captivity.

The background to the last and bloodiest wave of terror remains controversial. Diocletian, an uneducated, perhaps illiterate, soldier from Illyricum, the modern Dalmatia, was declared emperor by a few of his drunken chums in the legion. Such unseemly goings-on, mocking the traditional power of the Senate, were getting deplorably common. Happily, they were usually quickly forgotten. The new emperor was expected to disappear without a trace, lucky to escape with his life. But some senators in Rome recognised him as a potentially useful tool for dealing with a restive populace and his election was approved. As often happens, the tool became the master. Diocletian was ruthless – he strangled an argumentative prefect of the Praetorian Guard with his own hands – cool, calculating and extremely able. He reorganised the administration of the provinces. He secured the frontiers of the Empire. The rich waxed richer. The poor were given their circuses.

As under previous strong emperors, the Christian communities prospered for the first twenty years of the new reign. They themselves sought neither power nor wealth: they were therefore model subjects. The Emperor's wife, Prisca, and her daughter, Valeria, were thought to be Christian sympathisers. But apparent security often engenders false confidence. In 295 Maximilianus, the twenty-five-year-old son of a veteran of one of the Augustan legions in North Africa, refused to bear arms. 'You may cut off my head,' he told his surprised commanding officer, 'but I will not be a soldier of this world because I am a soldier of God.' He was tortured but laughed in the face of his torturers. 'Suffering seemed to touch him not . . . The torments seemed to fill him with joy.' He embraced death happily. To his persecutors this was a new and bewildering experience.

7. Jews were exempt from sacrificing to Roman gods on the tacit understanding that they would refrain from proselytising among the Gentiles. This was consonant with the Jewish view of themselves as God's exclusively chosen people.

It also proved to be contagious. A few months later a centurion first class by the name of Marcellus threw away his sword during a parade in honour of the Emperor's birthday. 'I am a soldier of Jesus Christ,' he proclaimed. 'From now on I cease to serve your emperors and I despise the worship of your gods of wood and stone.' He too was tortured, an ordeal which he suffered with equanimity, even joy. The cases of defiance multiplied. The crisis was compounded by a disastrous harvest in Egypt, the granary of the Empire, and by the virtual destruction of the Mediterranean fleet in a storm. After seeking the advice of Apollo's Oracle at Dydima, Diocletian issued his first anti-Christian decree on 23 February 303. Christian services were banned. Places of Christian worship were demolished. A few months later all privileges were ended for professing Christians, including the right to be executed by the sword. Early in 304 all those holding public office and prominent in the professions were ordered to offer at least a token sacrifice to pagan gods.

The First Age of Martyrs has sometimes been compared to the mass murders of the twentieth century. Such comparisons are misleading. Apart from a few localised outbreaks of mob hysteria, anybody under Diocletian could save themselves by a formal written statement of submission. Many such submissions – known as *libelli* – survive.[8] Indeed, so many asked for a certificate that the issuing authorities complained of overwork. But the many were not all. Flight and going underground in the cities were alternatives. The bishop of Neocaesaria in today's Turkey fled with his entire flock into the desert rather than commit apostasy. It was not necessarily a safe option. As Eusebius, first historian of Christianity and future martyr, wrote: 'a large number wandered over deserts and mountains till hunger, thirst, cold, sickness or wild beasts destroyed them'.[9] But the truly zealous made no attempt to hide, recant or flee. On the contrary, the more savage the persecutions, the louder they proclaimed their faith. Among them were two doctors.

Cosmas and Damian were twins born in Arabia around 240. They studied medicine in Alexandria and then settled in the busy seaport of Aegina in Asia Minor. Soon they acquired a reputation as wonderfully skilled healers.[10] It was

8. One, quoted by B. Moynahan, which presumably saved the life of a Christian woman reads: 'It has ever been my practice to sacrifice to the gods. Now in your presence, in accordance with the Command of the Emperor, I have sacrificed by pouring a libation and tasting the offering. I beg you to certify my statement, I, Aurelia Demos, have presented this declaration. I, Aurelius Irenius, husband, wrote for her as she is illiterate. I, Aurelius Sabinus the Commissioner, saw you sacrificing' (Moynahan, *The Faith*, p. 138).
9. Eusebius, Bishop of Caesarea, *Ecclesiastical History of the Martyrs of Palestine*, translated by H.J. Lawler and J.E.L. Oulton (London, 1927), Vol. I, p. 453.
10. They cured the blind and the paralysed, brought back the dead, and successfully attended sick animals (notably a fallen camel); but their most famous deed, frequently shown in medieval and early Renaissance illustrations, was to transplant successfully the lower limb of a dead Negro on to a patient who had lost his in battle.

their fame as doctors which made their conversion to Christianity and subsequent martyrdom a *cause célèbre.* After Diocletian's first anti-Christian decrees, the doctors were arrested and imprisoned on the order of the prefect Lysias. Proceedings quickly became a trial of strength. 'Bind them hand and foot and torture them till they sacrifice,' the Prefect ordered. But, to his henchmen's consternation, the more the doctors were tortured, the happier they seemed. 'Lycias, we ask you to torment us further, for we suffer not,' the victims pleaded. Burning and stoning proved ineffective. Ultimately they were beheaded, possibly on 27 September, their Saints' Day, in 279.[11]

Austere reference books describe Cosmas and Damian as 'entirely legendary'. Austere reference books are often right. Some of the deeds attributed to the doctors are certainly hard to believe. Indeed, as individuals they may have been wholly mythical. But they were not alone. The persecution and martyrdom of many is well documented. Justin at his trial before the prefect Rustinus confirmed his belief not only in 'the God of the Christians, whom alone we believe to have been the maker of the entire world from the beginning', but also 'in the Lord Jesus Christ, the Child of God'.[12] He was tortured but all he repeated was 'Thanks be to God for letting me bear witness to His Glory.' Ignatius of Antioch was a highborn Roman. When news of his arrest reached Rome, his friends intervened on his behalf. In a letter of tragic grandeur he begged them to desist. The moment of his martyrdom, when he will be mauled and torn by wild beasts in the arena, will be his happiest, he writes. It will bring him closer to Christ. 'How do I deserve such grace, oh God,' Symphorosa prays under torture. Apollonia is offered an escape route but refuses.

The martyrs kissed, laughed and embraced on the scaffold or in the circuses. When Bishop Polycarp was tied to the stake and the fire was lit he exclaimed in strong tones: 'I bless Thee that Thou hast deemed me worthy of this happy day and hour, that I should share in the fate of Thy martyrs. I rejoice.'[13] When the flames did not appear to touch him, a centurion had to be sent in to kill him with a dagger. The executioner succumbed to his burns.

The most convincing testimonies come from the enemies of the Christians. Tacitus hated those who practised the new religion: 'a group of fanatics known for their evil practices'.[14] Yet even he, in his well-bred scorn, had no doubt that

11. They became the patron saints of numerous medical bodies and organisations, most famously perhaps of the Paris Collège de Saint-Côme, since 1533 the Collège des Chirurgiens. They had – and still have – numerous shrines in France. Some Greek Orthodox churches are dedicated to them.
12. Quoted by Moynahan in *The Faith*, p. 125.
13. Ibid., p. 127. Many of the most horrendous and miraculous martyrdoms are related in detail in Jacobus de Voragine's *Legenda Aurea* and quoted at length by D. Le Breton in *Anthropologie de la douleur* (Paris, 2000).
14. Tacitus, *Annals*, translated by C.D. Fisher (Oxford, 1906), 15: 44.

they were innocent of the crimes of which they had been accused. And he was amazed at the serenity, even cheerfulness, with which they suffered their fate, which 'truly passes belief'. Like heroes of the legendary past, they knew no fear. And many seemed to feel no pain.

. . .

In centuries to come the painlessness of the early martyrs would raise profound theological questions. Even allowing for the devout gloss put on the events by monkish chroniclers, there had been too many instances like the torture of Cosmas and Damian to leave room for doubt. Yet the claim of the hagiographers that some at least of the martyrs 'suffered not' could be interpreted in two ways. It could be seen as a miracle, a manifestation of God's infinite mercy. It was in fact the *miracle*, not the suffering, which would raise the martyrs to sainthood in the Catholic Church. It was and still is the only ground for canonisation. But it could also be argued that there was no particular heroism in suffering tortures which the sufferers did not feel. The dilemma was most acute in the case of Christ Himself. According to St Peter's Synoptic Gospel He was miraculously saved by God from suffering on the Cross. 'And they brought two malefactors, and they crucified the Lord between them; but He held His peace, as having no pain . . .' This might have been presented as the supreme Christian miracle, but it was always rejected by the Church. How could God's Lamb bear the sins of the world if He did not suffer? In the case of the martyrs, the dilemma was debated at several early synods at length and with passion. Some of the Fathers argued that, even if saved from suffering, the martyrs could have had no foreknowledge of God's intervention. Nor was the hope of such an intervention the mainspring of their heroism. Such explanations were rejected. At the basest level – that is in terms comprehensible to today's political spin doctors – there was no merit in claiming a miraculous state of anaesthesia when, to the simple-minded, this might diminish rather than enhance the sacrifice of the sufferer.

. . .

But Cosmas and Damian's claim that 'we suffer not' has echoed down the centuries. The Second Age of Martyrs, the time of the great religious conflict within the Christian Church itself, is separated from the first by more than a thousand years but must be mentioned in the same chapter. Being closer to the present, the historical background is more familiar. There were martyrs and astonishing denials of pain on both sides of the religious divide. Jeremy Collier describes the death of Archbishop Cranmer, by no means a saintly character in his earlier career as Henry VIII's compliant cleric, as 'he burnt to appearance without motion or pain'. With a happy smile 'he seemed to repel the suffering and overlook the torture.'[15] And John Foxe begins his tremendous record of the burning of Bishop Hooper:

> In the tyme oof which fire, euen as he first flame, he prayed, saying mildely and not very loud but as one with no paines: 'O Jesus the sonne of David haue mercy vpon me and receaue my soule . . .'[16]

Thousands of such executions apparently 'with no paines' were documented at the time across Europe, both Catholic and Protestant, inspired by an event in a distant province of a long-extinct empire fifteen hundred years earlier.

15. J. Collier, *The Ecclesiastical History of Great Britain* (first published 1708–14). The burning of Archbishop Cranmer was first described in the 1840 edition.
16. John Foxe, *The Acts and Monuments of John Foxe*, ed. J. Prett (London, 1977), p. 235.

# Chapter 5

# PAIN IGNORED

The denial of pain was never confined to Christianity or indeed to any religion; and in the West the roots of the secular variety also reach back to Antiquity. The Stoic school of philosophy, founded by Zeno in the third century BC, was an intellectually powerful but essentially non-religious creed. In his book (which survives only in the fragmentary quotations of his followers) Zeno wrote that passions like anger, fear, grief and pain are no more than 'irrational and unnatural movements of the soul'. (Homer's Achilles was particularly taken to task for his childish sulk in his tent.) To achieve 'euroia biou', a smooth and happy flow of life – what in Cicero's and Seneca's writings became the *tranquillitas animi* (the tranquillity of the soul) – was simple. It merely required the suppression of emotions and passions, including the emotions and passions aroused by physical injury. Some of Zeno's followers refined the doctrine by distinguishing destructive from approved emotions. The latter included joy, watchfulness and caution. Pain and its outward manifestations remained in the first category.

Spread by Alexander's conquests to Asia and Egypt, Stoicism later became one of the most popular philosophies of the Roman Empire, a guide to happiness in this world without any reference to the next.[1] Though Zeno and his immediate followers wrote in Greek, most surviving texts are in Latin. In modern books outlining past philosophical systems, the Stoic philosophy is sometimes compared to Christianity in cutting through social and class barriers. The statement is largely based on the fact that Epictetus, one of the most approachable of Stoic writers, started life as a slave and, even after being

1. Zeno of Citium (in Crete), founder of the Stoic school of philosophy, is not to be confused with another and earlier philosopher by the same name, Zeno of Elea, known for his paradoxes, most famously for that of Achilles and the tortoise. The name Stoic derives from the Greek 'Stoa Poikile' or painted porch where the later Zeno had his academy.

given his freedom, remained a poor man.[2] But Epictetus was an exception. In contrast to Christianity, Stoicism was and would always remain the philosophy of the comparatively well-to-do. (As Anatole France acidly commented centuries later, it is easier to bear stoically one's good fortune than the slings and arrows of the outrageous kind.)[3] Stoics exalted reason, despised the pursuit of pleasure for its own sake and, most alluringly perhaps, promised victory over physical pain. The last was eloquently discussed by one of the school's august exponents, the Emperor Marcus Aurelius. He wrote:

> The pain which is intolerable carries us off; but that which lasts a long time is always tolerable; since the mind can maintain its own tranquillity by retiring into itself . . . If you are pained by any external agent, it is not the agent that disturbs you but your own perception and judgement of it.[4]

This sounds impressive; and though Marcus Aurelius enjoyed an exceptionally privileged childhood and a comparatively straightforward rise to power, he had later suffered enough infirmities to speak with authority. Nor was he without illustrious followers. Immanuel Kant of Königsberg was an honest man as well as a notable thinker. He was also a victim of gout.[5] When pressed to explain how he managed to write some of his most profound analyses of the human mind while suffering from an acutely inflamed big toe, he explained that all he had to do was to stare fixedly at his umbilicus and focus his mind on what he remembered of the works of Cicero.[6]

Yet reason alone, however radiant, has rarely been a strong anaesthetic agent. Paul Eichendorf was an erudite and cynical Viennese physician and a friend of Sigmund Freud. He also wrote extensively if rather superficially about philosophical systems. After praising the moral strength and intellectual

2. He was born in Phrygia in Asia Minor about AD 40 but spent his youth as a slave in Rome, tutoring the sons of his master Epaphroditus, a friend of Nero. He was expelled with other philosophers by Domitian, as hostile to Stoics as he was to Christians, and spent the rest of his life in Nicopolis in Epirus.
3. A. France, *Penguins' Island* (London, 1907), p. 45.
4. Marcus Aurelius, *Meditations* , 2 volumes, translated by A.S.L. Farquaharson (Oxford, 1944), p. 65. The Emperor died in 180, aged fifty-nine.
5. For reasons totally obscure, gout, a genetically determined metabolic disorder of uric acid metabolism, appears to have been far more common in the eighteenth century than it is today. Its main manifestation was the extremely painful swelling and inflammation of the big toe or big toes. There was no known cure. (Sufferers resting their heavily bandaged foot on a footstool are often shown in eighteenth-century paintings and engravings, most famously perhaps by Hogarth in *The Rake's Progress.*)

   Immanuel Kant spent all his life in the east Prussian town of Königsberg, now Kaliningrad and part of Russia. He died in 1804, aged eighty-one.
6. Those who do not share Kant's enthusiasm for Cicero may regard this as an instance of successful counter-irritation.

rigour of the leading Stoics, he concludes on a personal note: 'But speaking as a practising doctor, I would regard the advice of a physician to a patient suffering from a dental abscess to follow Marcus Aurelius's admonition and retire into himself and reason himself out of his pain as grounds for homicide.'[7] This was *fin-de-siècle* Viennese spleen. There is much that researchers into fashionable theories of pain might recognise as strangely modern in the writings of the learned emperor, even if his physiology is wrapped in currently unfashionable moralising.

Though lacking the universality of the Christian message, some Stoic texts are strangely Christian in spirit. In his *Hymn to Zeus* Cleanthes, Zeno's successor as director of the Academy of the Painted Porch, addressed Zeus in 262 BC as

> Almighty and everlasting, sovereign of nature . . . No work upon earth is wrought without thee, lord, not through the divine ethereal sphere, nor upon the sea; save only whatsoever deeds wicked men do in their own foolishness . . . Deliver men from fell ignorance. Banish it, father, and grant to them to obtain wisdom whereon relying thou rulest things with justice.[8]

And Seneca wrote:

> The body is not a permanent dwelling but a sort of inn (with a brief sojourn at that) which is to be left behind when one perceives that one is a burden to the host.[9]

But these were written texts. From the point of view of pain denial the two faiths were fundamentally different. Stoics, unlike the Christian martyrs, had no heavenly vision on which to focus their minds: indeed, classical Stoic teachers firmly rejected the idea of heaven or of any world outside the one which could be physically perceived. Also, unlike Christians to whom human life was God's gift and remained His property, they regarded suicide as an eminently rational way of escaping from intolerable suffering. It was the exit which Seneca chose when tutoring Nero began to outstrip his educational prowess.[10]

7. P. Eichendorf, *Eine kleine praktische Philosophie* (Vienna, 1910), p. 56.
8. Quoted in *Encyclopaedia Britannica*, 14th edn (London, 1932), p. 430.
9. Seneca, *Epistolae ad Lucilium*, CXX, translated by M.T. Griffin (Oxford, 1976).
10. Seneca's suicide was in fact only semi-voluntary: a centurion of Nero's called on the aged philosopher and informed him that he was no longer wanted among the living. Yet he could easily have escaped. He chose to re-enact the death of Socrates but with only partial success. When surgeons opened his veins, the bleeding was insufficient to kill him; and a drink of hemlock had no effect. Eventually he suffocated himself in a steam bath.

...

Strictly speaking, the Stoics of Antiquity did not deny pain. They simply maintained that pain was not important. At least not important enough to disturb the equanimity of a rational human being. It was in many ways an admirable precept. Even its critics tend to describe it as 'absurd but noble' or 'misguided but sublime'. A bizarre educational development has fostered this respect. Long after the political demise of the Roman Empire and for reasons still not entirely clear, the best kind of education available to young people in Europe became linked to the study of Latin literature. But Latin literature of the imperial period is permeated with Stoic ideas. For centuries, therefore, well-educated Europeans at an impressionable age have absorbed the Stoic prescription for a noble and happy life without realising it. Though historically questionable, the story of Mucius Scaevola has been one of many to provide moral uplift.

In the sixth century BC the still small and unimportant city of Rome was besieged by Lars Porsena, king of Clusium, a neighbouring Etruscan state. Porsena was a rich and powerful despot and his war aim was to reinstate his fellow tyrant, Tarquinius Superbus, justly expelled by the Romans.[11] A young Roman, Gaius Mucius, later to be known as Scaevola (or left-handed), made his way into the Clusian camp to assassinate Porsena. He was caught, hauled in front of the king and sentenced to death. To show his indifference, he asked for a brazier with a fire burning in it. When his wish was granted, he thrust his right hand into the flames. *Unflinchingly* he told the king that there were any number of young Romans who were as indifferent to pain as he in defence of their country, all determined to kill the King if he persisted in threatening their liberty. So impressed was Porsena that he released Mucius and lifted the siege.[12]

This not particularly edifying story may not have been familiar to every schoolboy in Rome; but it was revived and copied by Renaissance scholars, made into an oratorio by Alessandro Scarlatti, turned into an opera by Rameau, commented on by Goethe and, until comparatively recently, remained obligatory reading in every English public school, French *lycée* and German *Gymnasium*. Even today, while the names of great Stoic thinkers

11. Sextus, one of the sons of Tarquinius Superbus, seventh and last king of Rome, raped the virtuous lady Lucretia, who committed suicide. (The event was brilliantly painted by the aged Titian and set to music by Benjamin Britten.) The outraged population rose under Brutus the Elder and expelled Tarquinius and his horrible brood. Rome became a republic which survived for over 500 years.
12. The story is told in the *Annals* of Livy (II, 12). Titus Livius died in AD 17, aged sixty-six.

may be recalled only dimly if at all, their teaching survives in the English stiff upper lip and even in the fictitious 'Red Indian' world of continental authors like Karl May.[13] His Appache heroes might suffer agonies at the hands of the evil Kiowas but never will a moan or scream escape their lips and their bravery will be admired and lauded by their enemies.

13. Karl May (who died in 1912, aged seventy) was an excellent and highly moral writer of long adventure stories mostly set in the American 'Wild West' (which he had never visited). May's reputation may have slightly declined since (through no fault of his own) he became Hitler's favourite writer.

# Chapter 6

# THE HERESIES

Disgusted with the world that failed to recognise his selfless efforts at improving it, Diocletian retired to his native Dalmatia 'to grow cabbages'. The year was 305 and the First Age of Martyrs was over.[1] For some years the imperial succession was disputed. In 313 Constantine, adventurer and political survivor, won a decisive victory over his chief rival, Maxentius. The night before the battle a cross appeared to him in his dreams. A heavenly voice advised him that it would be under this sign that he would conquer. Like many professed sceptics, he was deeply superstitious. As quid pro quo, he established Christianity as one of his empire's official religions. (It did not become the Empire's *only* official religion for another sixty years.) The act inaugurated a new chapter in the history of the Church. It also inaugurated a new chapter in the history of pain.

To Christians the establishment of their faith as a state religion was in most respects a happy event. Its consequences for world history would not be fully revealed until the collapse of the Western Empire 150 years later. In Europe that collapse would mark the beginning of the Age of Faith.[2] It would create a new approach to healing, and to the relief of pain in particular. But even before that new approach, pain created problems for the newly established Church. It animated the first upsurge of what later became known as the heresies.[3]

1. See Chapter 4.
2. See Chapter 7.
3. The concept of heresy of course antedated the official recognition of the Church – it clearly emerges from the letters of St Ignatius, Bishop of Antioch, written during the first years of the second century – but orthodoxy could not be enforced while the Church itself was being persecuted. Later the most famous and most persistent heresies centred on the nature of the Trinity. They were linked to the names of Arius (the propounder of the heretical view) and Athanasius (the voice of what became orthodoxy); but, while they generated impassioned and learned disputations among bishops and theologians and inspired some of Edward Gibbon's most scathing comments, they were understood by only a tiny minority of the faithful. This is probably still true. The question of suffering by the innocent, by contrast, was understood by everybody.

In its original sense 'heresy' (from the Greek αἵρεσις choice, meaning to choose) did not imply blasphemy: indeed, Eusebius described the Christian faith as a 'most sacred Heresy'.[4] But establishment and power soon gave birth to a body of doctrines labelled 'right-thinking' or 'orthodox'; and it is in the nature of orthodoxies to breed heterodoxies. As early as 386 in Trier in present-day Germany Christians set fire to a group of their own faith for the first time. The accused had been charged with 'the vile Manichaean heresy' and the accusers had imposed on them the same penalty that Nero had imposed on their forefathers three hundred years earlier. Many of the heretics, then and later, suffered their fate with the same surpassing fortitude, even cheerfulness, as had the early Christians.

Though the numerous, complex and warring doctrines which sprang up in the wake of the Church's secular ascendancy provide a fascinating (if also a hugely depressing) field of study, they are outside the scope of the present work. Physical pain and its spiritual significance became one of the new heresies' chief preoccupations. The question of how an omnipotent and infinitely loving God could permit the suffering not only of the sinful but also of the virtuous, the pure and the innocent exercised the minds of early Christians, as it exercises the minds of many still. Of course, then as now, suffering could take many forms, but physical pain led the field. This may be difficult to appreciate at a time when it is often and sometimes fatuously referred to as a 'useful warning signal'. This deserves a brief digression.

Pain as a stop command is of great evolutionary significance. It still warns an animal with a broken leg to stop running. Not only would continued running be futile, but it would aggravate the consequences of the fracture. With rest, by contrast, there is always a chance of healing. The explanation is simple. A living body's natural response to injury is acute inflammation. This is a complicated set of changes both at the site of the injury and in other parts of he body.[5] Locally inflammation leads to the formation of 'granulation tissue', still sometimes referred to as 'proud flesh'. The lay term is more descriptive. Granulation tissue is bright red and shiny, superficially resembling a clot. In fact, it is a highly structured mesh of capillary channels. These begin to sprout from the edge of the wound within a moment or two of the injury. Growing rapidly, they quickly replace the clot.

Despite its seeming frailty, granulation tissue is a strong barrier against infection, ideal as a protective dressing. It has a second function: the capillaries

4. Eusebius, *The History of the Church*, translated by G.A. Williamson (London, 1965), X, 2.
5. The local response is mainly vascular and accounts for the heat, the pain, the redness and the swelling. These are accompanied by systemic changes such as a rise in body temperature.

will also act as a scaffolding for the eventual formation of a permanent scar.[6] But there is a proviso. Though highly resistant to invading organisms, its frailty makes granulation tissue vulnerable to mechanical injury. To fulfil its function – and future healing depends on it – it must be protected against tension, torsion, pressure and any other kind of physical violence. In other words, it needs *rest.* This was brilliantly expounded by the Victorian surgeon John Hilton in *Rest and Pain*, a forgotten classic.[7] Hilton pointed out that rest is normally assured in both animals and man by 'a most remarkable alarm signal: pain'. A fracture before immobilisation hurts because it moves. Movement would also destroy the granulation tissue. Once a cast is applied, the pain stops. Animals can achieve rest by keeping the limb immobile.

Pain continues to be the single most useful guide to progress. Once granulation tissue is beginning to be replaced by a scar, it is important to start exercising. Too little rest for too short a time will prevent the injury from healing. Too much rest for too long a time will prevent it from being restored to normal. There is no rule of thumb to indicate when one regime must replace the other. No rule of thumb is needed. If it hurts, the exercise is premature or excessive. (Of course a little discomfort is usually unavoidable; but a little discomfort is not pain, at least not as defined by Hilton.) In cases of physical injury, therefore, pain does have a 'useful' function. But a false and pernicious mythology has been built on this fact.[8]

Even today the practical relevance of 'useful pain' is limited.[9] It was even more so in the past. A Roman senator with a perforated peptic ulcer must have suffered agonies to no useful effect. The same was true of a Babylonian child with acute appendicitis, a Sumerian woman with an ectopic pregnancy or a medieval knight with bone cancer. Rest or no rest, nothing curative could be done about any of these. With few exceptions, the pain which preceded death meant gratuitous suffering. And most strikingly to many, the suffering was –

6. This depends on the arrival of specialised cells called fibroblasts which migrate along the capillaries. Still nobody knows for certain where they come from.
7. Hilton was surgeon to Guy's Hospital and eventually president of the Royal College of Surgeons. He was particularly celebrated as an anatomist. He published *Rest and Pain* in 1863 and died in 1878, aged seventy-four. Though inevitably out of date, some of his observations about the diagnostic and prognostic value of pain make his book one of the few medical historical works still worth reading for other than purely historical interest.
8. The 'usefulness of pain' and the severe limits on this usefulness are discussed more fully in Chapters 47–9.
9. Even some modern textbooks parrot the 'old wisdom' that 'cancer is dangerous because in the early stages it is silent' – i.e. painless. The truth is that cancer is dangerous because it kills patients, not because it is silent. Nevertheless, the benefit of an early diagnosis has become one of the dogmas of medicine and sustains several lucrative screening industries. How valuable they are remains uncertain.

or could be – wholly undeserved. How to reconcile this with an all-powerful and all-merciful God?

An apparent answer was provided by several heresies, but one in particular. Its founder, Mani, was a Persian, always represented as a man wearing a blue cloak and green trousers with a staff in one hand and a book in the other. He proclaimed himself Ambassador of Light in about 240. He taught that a primeval conflict existed between Light, the spirit of love and happiness, and Darkness, the spirit of pain and evil. The former belonged to God but the latter was the realm of Satan. The two had originally existed in complete separation but then Satan emerged to rob man of his share of Light and replace it with Darkness. The two forces were independent and almost evenly matched. Worship existed to expel the particles of Darkness, of which pain was the cardinal manifestation. Abraham, Noah, Buddha, Jesus and Mani himself had been sent by God to help man achieve this.[10] Like Mohammed after him, Mani claimed to be the last of the prophets. To help restore Light, he demanded severe austerity of the white-robed priesthood of the Elect (*electi*). His ten commandments forbade them from killing any living creature as well as demanding a monastic life of extreme self-denial. Less but still a great deal was expected from the Hearers (*auditores*) who would pay for their more pleasant life in this world with a certain amount of pain in the next. The doctrine, subtle and complex in detail but simple in outline, accounted for God's apparent capriciousness in distributing His blessings. The suffering of the righteous was a manifestation not of God but of the Devil, almost as powerful as God. Carried to its logical conclusion, this implied that God was not quite as omnipotent as He was claimed to be by Christian orthodoxy. (No major heresy claimed the alternative: that He was omnipotent but not infinitely good.) Manichaeans were confident that the all-powerful and infinitely loving God would eventually prevail, but, for the time being, He needed all the help He could get from the Elect, just as the Elect needed His.

Mani fell out of favour at the Persian court in 276 and was flayed alive; but by then he had travelled widely, as far as India and western China in the East and into Christian territories in the West. His disciples and superbly illustrated books for the instruction of the illiterate further spread his teachings. They offered a basically simple explanation for the pain as well as for the happiness which exist in this world. They seemed to reconcile God's benevolence with the apparent misfortunes of the virtuous and they offered heavenly rewards for those who suffered for the faith. God could never entirely vanquish the Devil, or perhaps He did not wish to. Life on Earth would

10. Two short and brilliant studies of Manichaeism are F.C. Burkitt, *The Religion of the Manichees* (Cambridge, 1925) and *Church and Gnosis* (Cambridge, 1932). A more up-to-date introductory study is K. Rudolph, *Gnosis* (Edinburgh, 1983).

remain a perpetual testing ground. That Good would eventually prevail was not in doubt. But until then Evil would threaten and strike. There would be pain, both spiritual and physical. It was the duty of the faithful to help God in His struggle against it.

Though the followers of Mani firmly rejected the jealous and vengeful God of the Old Testament, Jesus and Paul were honoured; and Mani drew on several passages of the Gospels and the Sermon on the Mount in his own writings. In contrast, however, to orthodox Christian doctrine, which denied the independent power of the Devil and which maintained that basically nothing that God, the only Maker of the Universe, created was not good, Manichaeism appealed to 'common sense'. This seemed to suggest – as it still does to many – that evil is real and that nothing shows its power more clearly than physical pain. On the other hand, the heresy also offered a gratifying sense of separateness to the Elect which would render them immune to suffering. Unlike Christ, Mani experienced no pain when put to death, despite the horrific ordeal of being flayed. To many such a creed was worth embracing and suffering for. It spread rapidly. North Africa, a stronghold of Christianity in the third and fourth centuries, became predominantly Manichaean. The young St Augustine, already marked out for an episcopal career, became an *auditor* for eight years before his conversion to orthodoxy.[11] Despite condemnation by successive popes and emperors, in 447 Pope Leo the Great still railed against 'the fount of Manichaean wickedness'. The spectre of the martyred Persian would resurface again in the explosive heresies of the late Middle Ages. They would be put down with a ferocity unparalleled even in the history of organised religion. Yet, resisting massacres as well as the anathemas of learned synods, the conundrum of physical pain suffered by the innocent under the auspices of an infinitely loving God would never entirely go away.

11. See F.W. Farrar, *The Life of St Augustine* (London, 1993); and Augustine's own *Confessions* (translated by H. Chadwick, Oxford, 1992).

# Chapter 7

# HEALING AND HOLINESS

Healing and holiness may have little in common today; but the words have the same root; and during the Age of Faith – more commonly called the Dark Ages – healing was a manifestation of holiness and holiness alone was capable of healing. This was a root-and-branch change from Antiquity. Romans believed in *mens sana in corpore sano*, a healthy mind in a healthy body, implying a kind of equality, even complementariness, between the two. To the medieval mind this was preposterous. Had classical Greek been widely read, the same would have been true of the Greek ideal of *kalos kai agathos* – physical beauty linked to moral elevation. Moral elevation existed but it was the function of the soul. It had nothing to do with beauty, at least not beauty of the physical kind: arguably the reverse.

Starting with the establishment of Christianity as a state religion, but especially during the terminal decline of the Western Empire,[1] the needs of the soul permeated every aspect of civilised life. (To many later historians, the use of the term 'civilised' to describe any aspect of that turbulent age was to appear self-contradictory; but to others the darkness of the Dark Ages always had about it a touch of the spirituality and stillness of a starlit night.) It was a cornerstone of the new faith that there was a divine design for everything; and the divine design for man was startlingly simple: every human being had a soul and that soul had to be saved. The guidelines for achieving this were the Sacraments. Their observance governed every stage of existence, from womb to tomb and, most importantly, beyond. Such a plan had been totally lacking in the pagan universe. It now took precedence over everything else and determined attitudes to pain.

1. Disintegration and political decline were almost continuous for a hundred years before the last Western emperor, Romulus Augustulus, was formally deposed by Odoacer, a Germanic chieftain, in 476. Odoacer reigned in much of Italy until overthrown by Theodoric, King of the Goths.

The rise of holiness did not entirely extinguish interest in bodily afflictions. True, a clear separation now existed between the flesh and the spirit, the one to be attended by the doctor, the other by the priest. (The two were often but not invariably the same person.) True also, that when the two were in conflict, there was no question which should or would take precedence. Of the many remarkable miracles attributed to St Hilary none was more admired than the act of bringing back to life a dead child for long enough to receive the Last Sacrament. To have healed her of her physical ailment might have given her a little more time in this Vale of Tears. To have received the Last Sacrament assured her of everlasting bliss.

But if the soul was to be saved, the human body too was a gift of God, fashioned in His image; and wilfully to neglect it would have been blasphemous. Extreme sects – and sometimes popular and pious movements – loudly disparaged man's mortal coil as no more than a temporary prison. Some enjoyed the initial approval of the Church; but, sooner or later, most became extreme and were declared heretical.[2] Nevertheless, a compromise between the two was not always easy.

Many traditions mingled, sometimes clashing and sometimes providing mutual support, in early Christian attitudes to physical illnesses and pain. Until the gradual re-emergence of classical teaching in the eleventh century, Jewish attitudes as embodied in the Old Testament were the most influential. (Several early heresies drew a sharp dividing line between the Lord of the Old Testament and the Christian God, even condemning the former as a 'demiurge' or an incarnation of Satan; but by the third century continuity between the two had become the bedrock of orthodoxy.) Although the Old Testament, unlike Graeco-Roman lore, made no mention of professional physicians, charismatic figures like King Solomon and some of the Prophets were revered as healers. It was following in their footsteps that Christ and His disciples performed their medical miracles: the New Testament mentions no fewer than thirty-five. Conversely, the wrath of God often expressed itself in terms of bodily afflictions. As recorded in Deuteronomy:

> it shall come to pass if thou will not hearken unto the voice of the Lord thy God . . . the Lord shall make the pestilence cleave unto thee, until he have consumed thee from off the land . . . The Lord shall smite thee with a consumption, and with a fever, and with an inflammation, and with an extreme burning, and with the sword, and with blasting, and with mildew; and they shall pursue thee until thou perish.[3]

2. See Chapter 6.
3. Deuteronomy, 28: 15–22.

Certain diseases in particular were associated with divine punishment, most notably *Zara'eth*, usually translated as leprosy though the identification is speculative. The belief that such illnesses were curable only by the Lord led to a minority tradition which rejected all forms of human medicine as usurping the place of the Almighty. St Bernard of Clairvaux, tireless fomentor of crusades and the most influential political networker of his age, expressed the view that 'to consult physicians and take medicines befits no religion and is contrary to purity'. Adherents of this school took pleasure in recalling the fate of King Asa who, in the ninth century BC, 'sought not the Lord but his physicians and whose sore foot consequently got worse until he died'.

Pain could indeed be a godsend and a trial of faith which believers should welcome as a chance to testify to their trust in the Lord. In the words of Job, whom God had singled out for suffering with a purpose:

> Happy is the man whom God correcteth: Therefore despise not thou the chastening of the Almighty: For he maketh sore, and bindeth up: he woundeth, and his hands make whole. (Job 5: 17–18)

Such a response would have been alien, indeed repugnant, to a pagan Roman gentleman. Stoics had to bear pain with fortitude and claimed to do so; but that was to overcome its noxious effects on intellect and character, not because suffering was inherently empowering. Even in the Age of Faith Job's injunctions, though admired, never became a mainstream recommendation. Pain had to be borne if necessary and never justified rebellion against God; but nor was it wrong to try to assuage it. Yet as a strand in the complex relation between God and man, Job's teaching would survive centuries of canonical dispute and unease. 'O how happy am I to be in pain!' Thérèse Martin, better known as St Thérèse of Lisieux, was to write in her diary in 1897. 'My every spasm brings me closer to my beloved Saviour.'

. . .

The identification of sin with sickness, another Jewish tradition, also resonates in the Gospels. 'Rabbi, who sinned?' Christ's disciples ask on meeting a blind person. Jesus immediately rebukes them. It is not the blind man or his parents who sinned. The blind man is blind 'so that the works of God might be displayed in him', meaning that God might be glorified through the miracle of having the blind man's sight restored. In this lambent passage two traditions meet. When Jesus meets the paralytic, he first tells the man not that he is cured but that his sins are forgiven. That is the man's greatest need. But curing him follows a few verses later. Luke was a physician; but in Acts healing is usually shown as an act of faith, involving prayer and the laying on of hands. 'Is anyone sick among you?' asks the Apostle James. 'Let him call for the elders of the church and let them pray over him, anointing him with oil in the name of the Lord.' It is the prayer and the anointing which will be curative and it is the Lord in His mercy who will restore the sick person to health. This too became an element in the teaching of the medieval Church and another element that survives.

But the healing tradition espoused by the early Church was never a simple mixture of the old Jewish and the new Christian. There are other, more remote strands. The world in which the Apostles lived and taught was no longer the desert of the Old Testament. Tiny but surprising incidents point to the infiltration of Hellenic ideas. The washing of the injured man's wounds with wine by the Good Samaritan raises Homeric rather than biblical echoes.

. . .

Human aspirations change, but human nature does not. Even in an age when the soul's ascendancy over the body was never questioned, to ordinary man and woman bodily malfunctions mattered. The mainsprings of prayer will always remain deeply mysterious; and one has to distinguish communal devotions from private supplications. Communal and formalised Christian rites have always given precedence to prayers for the forgiveness of sins and for the salvation of the soul over those for physical well-being. At least since the sixth century the opening *Kyrie* of every mass has proclaimed and reiterated this message (as it still does). The Lord's Prayer asks for our daily bread but has nothing to say about our daily good health. Yet there can be no doubt that, even in the Age of Faith, individually composed and privately uttered prayers were often inspired by the hope that God through one of His saints might cure a bodily ailment or at least help the worshipper to bear it.

Such supplications were canonically tolerated and sometimes actually encouraged. St Luke and St Michael were called in for a wide range of illnesses but many saints had special fields of expertise. St Anthony cured erysipelas, St Artemis genital sores, St Sebastian pains in the head, St Christopher epilepsy and St Roche the buboes of the plague.[4] In the second rank but still often

4. Erysipelas is now known to be a streptococcal infection of the skin. For centuries it was one of the 'hospital diseases', red, painful and often fatal. Soliciting the help of St Anthony, a third-century monk and martyr in Egypt, was usually the patient's only hope of survival. The disease was often called St Anthony's fire. Little is known about St Artemis except that he dwelt in Asia Minor and was martyred by being speared through his anus. St Sebastian has always been a popular saint with artists as a model of male beauty in agony but little is known about his life. He was martyred under Diocletian by being shot with arrows. Numerous legends attach to St Christopher, a martyr under Decius, but again, little is known about him. His veneration as patron of travellers (and today of car drivers) dates only from the nineteenth century. St Roche was a thirteenth century saint born to wealthy parents in Montpellier in France. After giving away his fortune, he went on a pilgrimage and devoted the rest of his life to tending victims of the plague. He himself died of the disease and was later painted by many great artists (including Titian, Bassano and Tintoretto) pointing to a bubo (or abscess) in his groin. Hospitals were often named after him and functioned under his patronage.

effective, St Blaise helped patients with painful goitres, St Lawrence dispelled backache, St Vitus vanquished chorea, St Fiacre eased irritation round the anus and St Apollonia palliated toothache.[5] St Margaret of Antioch was the principal female saint to comfort women in labour.[6] There were many more. The shrines of these holy men and women flourished largely, though not of course entirely, to heal illnesses and assuage bodily pain rather than to ensure salvation; and sacred relics became the first international currency in European history firmly based on their attested curative prowess. (For the same reason they also became a channel for laundering illicit and criminal gains. Crusading knights were not supposed to return from the Holy Land laden with loot; but a precious relic or two was within the bounds of propriety.) Once again, there was specialisation. A drop of the Virgin's milk would cure blindness and a splinter of St Luke's skeleton would heal painful feet. The remains of St Thomas à Becket in Canterbury attracted pilgrims from all over Europe because they were reputed to cure not only blindness and deafness but also bladder stones. Others embarked on arduous journeys to Bury St Edmunds, Santiago de Compostela or Roquamadour to be relieved of tormenting headaches, fits and rashes. The worship of relics was not strictly part of religious doctrine: indeed the official Church frowned on pilgrimages undertaken solely or even mainly for the purpose of bodily miracles. Not till the twentieth century did travels to healing shrines like Lourdes receive the official blessing of the Church as well as the patronage of the tourist industry.

In this, as in so many other respects, to modern man the Age of Faith remains an age of paradoxes. Ailments of the body and in particular pain and physical suffering were less but also in some respects more important than they had been in Antiquity. Hospitality to strangers in distress is a universal ethical command and charity to the poor was part of both pagan and Jewish tradition; but neither the gods of Mount Olympus nor the Lord of the Old Testament imposed an unequivocal obligation on their followers to tend and cure the sick. Nor of course did the Roman state, even at its most enlightened. Bread and circuses, yes. Free medical care, no. Hospital-like institutions did

5. St Blaise was a bishop in Sivas, modern Turkey, martyred under Diocletian. It is not known why he was often addressed by patients with painful goitres. The lesion was endemic in the Alpine regions of Europe and in the Balkans till the beginning of the twentieth century. St Lawrence of Rome was a deacon martyred with Pope Sixtus II. St Vitus was a fifth-century martyr in Sicily saved from torture by an angel. It is not known why painful involuntary movements associated with streptococcal infections (especially rheumatic fever, now rare) were named St Vitus's dance some time in the fourteenth century. St Fiacre was an Irishman who became a monk in France near Meaux in the seventh century. St Apollonia was a third-century Alexandrian virgin and martyr, tortured by having her teeth extracted.

6. St Margaret of Antioch, third-century virgin and martyr, was swallowed by a dragon which disintegrated when she made the sign of the Cross inside its stomach. She was the most popular of several saints prayed to in pregnancy.

exist; but they were either establishments for the isolation of the obnoxiously sick and the dangerously mad or expensive sanatoriums for the seriously rich.

...

Strangely perhaps, hospitals to look after the poor and sick as a religious duty were the invention of a religion and an age which frankly deplored too intense a preoccupation with the welfare of the body. The movement began even before the final collapse of the Western Empire. In the East Leontius, Bishop of Antioch around 340, set up hostels not only to shelter the homeless but also to treat and if possible to cure those in pain. With that in mind he employed numerous 'clerks skilled in the art of healing'. About the same time Bishop Eusthathius of Sebasteia built a hospital dedicated to the disabled poor; and St Basil erected outside Caesarea almost a new city for sufferers from diseases of the body. Some of the establishments were essentially leprosariums, designed for the isolation of the unclean (much as two thousand years later municipal sanatoriums would became establishments to segregate rather than to treat the tuberculous) but many were hospitals in the modern sense, the first of their kind.

The change came a little later in the West but the impulse overcame the political and social turmoil of imperial decline. In about 360 a hospital was founded in Rome by St Fabiola, a rich Christian convert, who, after an unhappy marriage, decided to devote the rest of her life to the poor and sick of the city. 'She assembled sufferers from the streets and highways,' wrote her mentor, St Jerome,

> and tended the impoverished and unhappy victims of disease. I have often seen her washing wounds which others, even men, could hardly bear to look at ... She founded a hospital where she gathered the sick and gave them all the attention of a nurse. Need I describe the many woes which can befall a human being? The cut-off noses, lost eyes, mangled feet, swollen bellies, withered thighs, the ailing flesh that is filled up by hungry worms? How often did she carry those dirty and poor sick on her shoulders ... how she washed the pus from sores that others could not even behold.[7]

In all this, Jerome says, 'she merely fulfilled her duty to God'. Perhaps so; but that duty was a new and essentially Christian one.

By the fifth century hospitals in the East – *nosokomeia* in Greek, or places for the caring of the sick – had become large and permanent. Jerusalem, still part of the Eastern Empire, had one with 200 beds. At St Sampson's in Constantinople surgical operations were performed regularly and there was a wing dealing specifically with diseases of the eyes. St Pantocrator, another

7. St Jerome, *The Desert Fathers*, translated by H. Waddell (London, 1936), p. 38.

imperial foundation on the Bosphorus, had a full-time staff of physicians, surgeons and apothecaries and even places where students could be taught. Edessa had a famous women's hospital to which patients travelled from distant provinces. All these institutions were free, unless those admitted or their families wished to make voluntary donations; and all prided themselves on helping those in pain. 'We labour not only to cure but also to relieve suffering', was the proud motto of St Sampson's, pre-echoing Trousseau's immortal *soulager toujours.*

In the West, struggle for sheer survival as wave after wave of barbarian hordes seemed to sever every thread of civilisation delayed but did not permanently prevent such a development. In 528, holy men who had for years lived and meditated in solitude in the caves and forests of southern Italy decided to band together for mutual support and the reinforcement of their individual prayers. They elected the learned Benedict of Nursia to become their leader or abbot. He converted the people of Monte Cassino from idolatry and founded there the first of the original twelve Benedictine monasteries. To guide his disciples he composed a set of Rules which, though they did not receive pontifical blessing for another seventy years, became the blueprint of monastic life and labours. They still are. Many of the ordinances could have been composed by a pious pagan philosopher or a Talmudic scholar; but one rule struck a new and specifically Christian note:

> The care of the sick is to be placed above every other duty, as if indeed Christ were being directly served by waiting on them.[8]

Thereafter no monastery of any size was without an infirmary, or usually two, one for sick monks and the other for the public. Probably even St Bernard of Clairvaux aimed his strictures at *lay* physicians, not at priests performing medical acts as part of their holy vocation. This was an important distinction.

What the care of the sick in these infirmaries actually involved is not clear. Nursing and feeding certainly, the administration of drugs probably. Most of these must have been locally grown: travel was hazardous and international trade at a standstill. In the crumbling Empire and during the centuries of political and social chaos which followed its collapse medicine as an academic discipline disappeared. The libraries had been looted; the academies lay in ruins. Formal training in the art and craft of healing did not exist for half a millennium. Books were still being written but they dealt with matters more momentous than coughs, catarrh and constipation. What even in the midst of political turmoil exercised scholarly minds were questions like the true state of the Body of Christ, the interpretation of the Sacraments and the nature of

8. St Benedict's Rules, quoted by R. Porter, *The Greatest Benefit to Mankind* (London, 1997), p. 111.

the Trinity. The fierce disputations of church councils were recorded in meticulous detail and read. Massive theological exegeses were being written, copied and circulated. They seem far removed from the scientific treatises of Antiquity, yet some are not without medical relevance and interest. In the detailed recounting of miracles, often the main evidence for doctrinal assertions, there are references to mandragora, poppy juice and henbane. One of the most influential theological works of the age and for centuries to come, though named and inspired by the Trinity, contains a wealth of medical information.

...

St Hilaire or Hilary of Poitiers was the scion of a noble Roman family in the South of France whose first interest may have been in medicine and the natural sciences and who came to Christianity through the philosophical questioning of the meaning of life. After his baptism he was elected bishop of his native Clerus and became involved in the raging dispute over Arianism.[9] Temporarily out of favour with the Emperor Constantius, he was banished to the Eastern Empire; and it was while in exile in Asia Minor that he wrote his great theological work, *De trinitate*. Though most of this deals with the physical nature of Christ's body, long passages review the nature of sensation and the occasional loss of painful sensations in the human body in general.

> The nature of our bodies is such, that when imbued with life and in conjunction with a sentient soul, they become more than inert matter. They feel when touched, suffer when pricked, shiver when cold, feel comfort in moderate warmth, waste with hunger and grow fat with eating. Above all, pervaded by the soul, but not when dead, they respond with pain and pleasure to their surroundings. Yet this is not true of the body as a whole or under all circumstances. Most of the sentient body responds with pain to being cut or pierced, but the nails and hair can be cut without experiencing any discomfort. Also, in certain diseases, even before a limb becomes withered and discoloured, it loses all sensation, so that it can be cut and burnt with impunity. Yet it is still part of a living body. Also, when through some grave necessity part of the body has to be cut or removed, the soul can be

9. Arianism, the greatest and most successful heresy of the ancient world, was launched in Alexandria by Arius, a presbyter of one of the city's churches, in 321 when he was already in his sixties. The kernel of his doctrine was that the Son was a created Being, not an incarnation of God, though He far surpassed all others. Though rejected by several early synods in favour of the Athanasian doctrine of the Trinity, Arianism was the form of Christianity adopted by many of the barbarian tribes which overthrew the Empire (including the Vandals). It continued to attract adherents and to arouse controversy for centuries.

> lulled to temporary sleep either as a whole or in part. Such a body will later return to life and remember no suffering . . . Some plants can produce death-like forgetfulness. After eating or drinking extracts of such plants, limbs can be cut off without pain.[10]

Hilary gives no details, but he clearly speaks from experience; and there are several practical therapeutic hints in the work. In another passage he refers to 'putting the soul to sleep' in a limb 'by prolonged pressure'. It is, he says, a way of lessening the pain 'known to the ancient Egyptians'.[11] By his own admission, his interest in the 'functioning of the human body and mind' had helped him to understand and to explain the mystery in Christ's incarnation and therefore the nature of the Trinity.

The bishop eventually returned to France and distinguished himself by preaching, his theological disputations and his profusion of miracles. In a particularly fierce dispute with the heretical Pope Leo, just when Hilary was threatened with being outgunned in argument, Leo fell to the ground and died miserably from dysentery. In medieval representations the saint is often shown triumphing over his smitten adversary. But most of Hilary's miracles were entirely benign. He cured the sick, calmed the mentally ill and resurrected more than one deserving dead. Perhaps it was his probings into the natural world which imbued his theology with unusual compassion and charity. 'Christ teaches us to bear wrong, not to avenge it,' he wrote. This might seem an unremarkably pious platitude today (to be uttered and then to be forgotten); but in doctrinal matters Hilary's age was deeply unforgiving. His near-contemporary, St Augustine wrote that 'kindly harshness' should be used to make unwilling souls yield. 'We have to consider their welfare rather than their inclination: for when we are stripping a man of the lawlessness of sin, it is good for him to be vanquished, since nothing is more hopeless than the happiness of sinners.'[12] Hilary's approach was different. 'Satan bursts in with an axe and sword but the Saviour knocks and speaks gently to the soul' is a comment by which he might wish to be remembered.[13]

10. St Hilary of Poitiers, *De trinitate*, translated by J. Balogh (Budapest, 1934), p. 341.
    Though now recognised as a doctor of the Church, in the light of the crystallised Athanasian doctrine and Nicene Creed, Hilary's belief was far from orthodox. In particular, to him the Body of Jesus was not an earthly body because the Lord had made it without human aid inside the body of a Virgin. It was this which, in Hilary's view, explained the walking on the water and the Transfiguration.
11. See Introduction.
12. Quoted in Moynahan, *The Faith*, p. 246.
13. But he is in fact best remembered by having the Oxford University term which starts on the Sunday nearest his feast day, 13 January, named after him.

Hilary was not of course the only exegete to write on medical and physiological as well as on transcendental topics. Yet ultimately it was not the saints and learned bishops of the early Church who rescued the achievements of Antiquity from oblivion and eventually rekindled scientific progress in the Christian West. Ironically perhaps, that was largely the result of the rise of Islam.

# Chapter 8

# ISLAM

The speed of Islamic conquests remains without parallel. The Prophet Mohammed fled from Mecca to Medina in the year 622 of the Christian calendar, year 1 AH in Islamic chronology.[1] By the time of his death ten years later his followers were masters of the Arab Peninsula. Within a century Islam had conquered the greater part of the Byzantine Empire in Asia, all of Persia, Egypt, North Africa and Spain. Some fragmentation was inevitable (in addition to the great schism between Sunni and Shiite in the late seventh century). In the East, under the Abbasid Caliph, Harun al-Rashid, a contemporary of Charlemagne, with whom he corresponded, Baghdad became the most populous and civilised city in the world. It had also become the centre of a unique cultural enterprise.

For centuries after the fall of the Western Roman Empire, Teutonic, Celtic, Frankish, Slav and Mongol tribes seemed to be competing in the destruction of the literary remains of classical Antiquity.[2] Enter scholarly explorers from the Abbasid court with instructions to acquire what ancient manuscripts were still available. It was often a dangerous mission, but one pursued with extraordinary dedication. The finds were carried back to Baghdad. There they were rendered into elegant Arabic by a specially trained academy of translators.[3] The texts included not only practical works of science, medicine and engineering but also poetry and philosophy. A Nestorian Christian from southern Iraq, Hunayn ibn Ishaq, became a familiar figure in the capital, roaming the streets and declaiming the *Iliad* in Arabic.

1. H for *Hijrah* or *Hegeira* meaning 'Flight'.
2. To the conquerors the works were of no value as well as being in incomprehensible tongues. After the sixth and seventh centuries religious houses began to preserve ancient manuscripts but their main interest was in biblical and theological texts.
3. They were more honoured and better paid than translators have ever been before or since and many of their Arab renderings are said to be superb.

From this movement Galen emerged as a supreme medical authority. His works were copied, analysed, quoted and abstracted. But during the golden age of Islamic medicine his views never went unchallenged. All Arab texts of the tenth to twelfth centuries contain chapters disputing his statements. Even more determined was resistance to the Galenic spirit of doctoring. Almost symbolically, Roman disapproval of analgesics found no echo in the Muslim world. A stunning array of new drugs – much of it the loot of military conquests – was transforming materia medica. Its treasures became the pride of Islamic medicine. The late Roman formularies listed about a thousand potentially useful medicinal preparations. The Arab pharmacopoeia of Ibn al-Baytar who died in 1248 contained 3,000 items based on 1,800 botanical species, 145 mineral preparations and 330 medicines of animal origin. They included hundreds of plants of oriental origin never heard of in Rome, including camphor, nutmeg, mace, tamarind, manna and ginger. Many are no longer easy to identify. No less important, Islamic chemists – or alchemists, as they were soon to be called in the Christian West – excelled where Greeks and Romans were sluggish. Filtration, crystallisation, sublimation and distillation became techniques used not only in institutions of learning but also in the big city pharmacies. In bustling urban centres like Cairo and Damascus these were – as they would remain for centuries – shady retreats where men of taste, literary interests and learning could gather to exchange information and gossip. Most of the popular prescriptions would be difficult to test today. How effective the distilled and recrystallised urine of a pregnant rhinoceros may have been (and might still be) against heatstroke will never be known. At a more basic level, however, the list of Arab terms still in use in chemistry – alkali, alcohol, syrup and sugar is a tiny sample – reflects the practical achievement of Islamic scientists. Fittingly enough, 'drug' itself is an Arab word.

...

Pharmaceutical practices may be impossible to revive, but much of the literary output of early Islamic medicine survives. The first in a succession of famous Persian healers, Mohammed ibn Zakariya-al-Razi, known in the West as Rhazes, wrote some 200 works on a wide range of topics. Though he declared himself a disciple of Galen, he vigorously denounced many Galenic ideas and devoted a whole book to criticising Galenic anatomy and physiology. In almost every instance his criticisms were later vindicated. He was among the first to dismiss the Galenic image of 'used' blood passing through invisible pores in the cardiac septum from the right to the left side of the heart. Rhazes rightly stated that no such pores existed. His most remarkable achievement was his commonplace book, widely quoted in the West as Rhazes' *Comprehensive Book of Medicine.* It covers in note form every part and organ of the body, most illnesses being illustrated with telling case histories and often recollections of the author's own ailments. The number and variety of those, faintly reminiscent of the afflictions of the permanently ailing Victorian savants, make his prodigious literary output (also reminiscent of the Victorians) even more remarkable. The tone of the case reports is extraordinarily modern, giving the patient's name, sex, age, symptoms, signs, treatment and the course and outcome of the illness, as well as many such trendy refinements as occupation and social history. Like Galen, he thought that important personages with sensitive palates were entitled to have foul-tasting medicines made more palatable by the addition of honey, syrup or grape juice, the last perhaps code for some form of alcohol. On the other hand and unlike Galen, he devoted an entire book to advice and recipes for 'those in pain but who cannot afford a physician'. Given a choice, the simple prescriptions in the second category seem infinitely preferable to the exotica destined for wealthier clients. Also unlike Galen, his treatment aimed primarily at presenting symptoms, mostly trying to relieve pain and discomfort, rather than wrestling with spectacular but speculative and often imaginary pathologies.

To Greek and Roman doctors medicine had always been an academic discipline as much as a useful skill. This was to be one of their legacies to the institutions of learning which would emerge in Western Europe in the later Middle Ages. To Rhazes and his followers, by contrast, medicine was essentially a means for relieving symptoms and curing the sick.[4] In the book intended for the poor such simple but effective remedies as gargling with warm salt water for sore throats make their first appearance. Three days' complete starvation for any ailment associated with diarrhoea is still eminently sensible advice. Among analgesics, a variety of preparations based on mandragora, henbane, poppy juice and what was almost certainly cannabis are the ones most warmly recommended.

Born near Tehran in 825, Rhazes spent most of his career in the service of the great Persian ruler, Mansur ibn Ishaq; but in middle life he was summoned to preside over the Abbasids' showpiece, the Al-Mu-tadidi Hospital in Baghdad. He had earlier written a treatise 'On Examining Physicians and on Appointing Them', an administrator's guide which must have caught the Caliph's eye. His last years were partly taken up with trying to treat his own painful glaucoma. No remedy proved curative but this did not stop him from keeping a record of the illness and the treatments tried. He was aware of the familial aetiology and of the aggravating effect of anxiety. It was one of the

4. The Western tradition continues. Most modern medical textbooks in English are not even meant to be bedside aids. Their contents are to be absorbed in the solitude of the night or the hushed tranquillity of a library. Their authors would resent being praised for having written a handy, practical companion. To guard against such a danger successive editions tend to be increasingly immovable by readers of average physique.

conditions in which his strong all-purpose purgatives may have been effective. Then as now, blindness in the Islamic world was a gateway to spiritual enlightenment; and the doctor died full of honours, aged sixty-five.

. . .

Rhazes' immediate successors were overshadowed by Abu Ali al-Hussayn ibn 'Abdallah ibn Sina, known in the West as Avicenna. The son of a tax collector, he was born in 980 in a village near Bokhara, an oasis of date trees in the arid desert of what is today oil-rich Uzbekistan. It was a centre of Islamic art and learning in the tenth century. According to pious legend he astonished his contemporaries by reciting the whole of the Qur'ān by the age of ten and was practising successfully as a doctor by the age of sixteen. In his autobiography he records that much of his writing was done on horseback during military campaigns, in hiding, in prison and even while recovering from stupendous drinking bouts.[5] The last may sound odd from the pen of a Muslim scholar; but, unlike Rhazes who was admired for his wisdom and sobriety, Avicenna cultivated the image of being one of the lads, boasting of his love of wine, women and song. His death was later ascribed to over-indulgence in one, two or perhaps all three. Yet the self-portrait is not entirely convincing. He wrote two monumental encyclopaedias, one of science and one of medicine. The latter, known in the West as the *Great Canon*, became, in Ambroise Paré's words, 'the most famous medical book ever written' and would secure for its author an honourable place in Dante's *Inferno*.[6] In over a million words it summarises and extends the writings of Hippocrates, Dioscorides, Celsus,

5. He also, at one time, served as the Vizier or chief minister of Bukhara's ruling Emir. He died in 1037, aged fifty-seven, having, in his own words, lived his years 'in breadth, not in length'.

He was, by several accounts, exceptionally handsome and witty but not much liked. His unsparing expressions of contempt for both earlier sages (like the great Rhazes) and fashionable contemporaries probably earned him a great deal of dislike. Some of his non-scientific poetry voices deep despair, like his famous quatrain (written in his native Persian):

How I wish I could know who I am
What it is in this world that I seek

6. It is in the First Circle of the *Inferno*, described in Canto IV, that Dante espies worthy and great men who, as Virgil, Dante's guide explains, 'are innocent of sin; however, / lacking in Baptism, they could not claim / its saving grace, and thus are doomed forever.' In this sad place 'of interminable sighs but no tortured groans' (where Virgil himself must dwell) the poet recognises Dioscorides, 'the scientist of herbal essence', as well as Galen, Avicenna, Hippocrates and Averroës 'who wrote the commentary'. They share their Underworld abode with a large and distinguished company which includes Plato, Socrates, Diogenes, Hector, Aeneas, Seneca, Orpheus, the Elder Brutus, Homer, Horace, Ovid and Julius Caesar. As usual, Dante is not entirely consistent, since the possibility of baptism was theoretically open to the Islamic doctors who must have known many Christians. (*Inferno* translation by Ciaran Carson, London, 2002).

Galen, Rhazes and a host of lesser lights. Every branch of medicine and medical science is explored, with all diseases listed by organ, from scalp to little toe. There are sections on ulcers, abscesses, tumours, injuries and fractures and a great deal about poisons and their antidotes. Much of it is rather dense (perhaps more so in translation than in the original); but some of the comments could have been written yesterday. 'The unhappy as well as the greedy eat too much. They should be made content rather than deprived of sustenance'; 'obesity affects the mind as well as the body'; and 'pain in the stomach is a bad counsellor'. (Was the mighty Emir a victim of duodenal ulcer?)[7] As might be expected from an Islamic scholar, materia medica takes up almost a quarter of the book, with lists of hundreds of medicinal plants and advice on their cultivation, harvesting, preservation and processing.

Two massive encyclopaedias might seem enough for a lifetime's achievement but Avicenna wrote a total of 270 works, including a slim volume in execrable verse containing useful mnemonics for students and 'for the busy practitioner in the market-place who might be called upon to give advice on infirmities without time to ponder the matter at leisure'.[8] He devoted much thought to the nature and management of pain. Celsus, Galen and Rhazes treated it non-specifically, classifying it into mild, moderate and severe. Avicenna described fifteen distinct types – boring, compressing, corrosive, dull, fatiguing, heavy, incisive, irritant, itching, pricking, relaxing, stabbing, tearing, constricting and throbbing – and illustrated each with vivid case histories. In a few the pathology is recognisable. The 'throbbing sensation which is not immediate after injuries but which may start a few days later, accompanied by heat and redness', clearly refers to wound infection; and 'the stabbing pain on breathing and especially on coughing' is still characteristic of pleurisy.

The elaborate classifications cultivated by Islamic doctors was more than a manic urge to be all-inclusive (which was gripping scholastic literature in Europe at the time and is doing again today). Different pains required

Chaucer's Doctour of Physic (in a country Avicenna could never have heard of) was also proud to have read the works of the great Persian, almost four hundred years after Avicenna's death. No medical work written today could aim at such longevity.

7. Quoted by S.M. Soheil in *Avicenna, His Life and Works* (London, 1958), a biography that combines scholarship and readability. Other useful works on Avicenna (but heavier going) are the collection of essays edited by G.M. Wickens, *Avicenna: Scientist and Philosopher, A Millenary Symposium* (London, 1952), O. Cameron Grüner's *A Treatise on the Canon of Avicenna* (London, 1930), and W.H. Grohlman's *The Life of Ibn Sina.*
8. Avicenna, *Urjuza fi'l-tibb* (A Medical Poem). Like early Greek philosophers (e.g. Empedocles) who wrote in verse, Avicenna did not aim at high-flying poetry. Both in verse and in prose Arab purists tend to dislike his style as being 'full of cumbersome new coinages' but praise his lucidity and precision. Words and language probably had no magic for him, and they did not particularly interest Aristotle either.

different management: rest for some (like acute inflammation), mild or moderate exercise for others (like stiffness). Avicenna's poultices would alleviate some complaints (like the 'fatigue' of joints); ice-cold compresses might ease the pain of sprains. The conditional tense is important. Nothing is certain in medicine. The scientific and godlike infallibility of Galen and his disciples, some of them still active, is entirely lacking in Islamic scholars. Most impressive are Avicenna's general insights, like the need for 'dispelling anxieties and distracting the mind' in seemingly intolerable irritation. 'It is the body which generates pain but the brain which decides how much of it is tolerable.'

Avicenna also distinguished three general approaches to pain. First, it is best to remove the cause. Here he includes the splinting of fractures, since 'it is not the fracture but the movement of the broken parts which is painful'. Second, more widely applicable are measures which 'counteract the acrimony of the humours'. To modern minds these humours remain an enigma. Their attributes are described at length in all histories of medicine; but does anybody really understand what the terms meant? Happily Avicenna does not dwell on them. For counteracting their 'acrimony' any drug which 'dulls sensitive faculties' and induces restful sleep will do. *Cannabis indica* is easily recognisable. Third and last, general inebriants like poppy juice and mandragora preparations will often make even severe pain bearable. 'For best effect let the libations be accompanied by good cheer.'

> If it is desirable to get a person unconscious quickly, without him being harmed, add 2 grains of sweet-smelling moss and half a grain of aloes wood to wine. For severe burning pain add ten drops of darnel water and administer fumitory opium and hyoscyamus, half a dram each, nutmeg and crude aloes wood, 4 grains each. Boil black hyoscyamus in water, with mandragora bark, until the mixture becomes red hot and then add it to the wine . . .[9]

However labyrinthine they may sound, Avicenna's prescriptions were clearly designed to be put into practice rather than to impress the reader. He acknowledges that 'no effective pain killer is without risk'; but this can be minimised by taking into account the patient's age, sex, general strength and sensitivity. 'Every healer knows that no two patients are entirely alike.' In some ways most remarkable is his humility. 'No medicine, however carefully prepared and however expensive, works in all patients or even in the same patient every time.' (Galen may have thought it but he would never have made such an admission.) In one memorable passage Avicenna says:

9. Avicenna, *The Canon of Medicine*, translated by O. Cameron Grüner (London, 1932), p. 242.

> If even double the recommended dose does not work, it should be abandoned. The patient should be encouraged to bear his affliction bravely and know that his efforts will be admired and ultimately rewarded. Where man fails, Allah will soon lessen his suffering . . . Most illnesses carry with them the seeds of their own cure and Nature must be given a chance . . . In every case the patient must be comforted with kind words and the prospect of future happiness . . . In some illnesses the cure may not be known; but no illness is without hope.[10]

The last statement points to a basic difference between the attitude of Islamic medicine and medieval Christian teaching. Suffering was often accepted in Christian societies as a manifestation of God's inscrutable will. Islamic doctors were encouraged to believe that Allah created no ailment without creating a remedy for it, even if it was beyond the wit of man to discover what that remedy was. Outside urban centres this often led to somewhat simplistic practices. Islam had no saints and holy relics but the text of the Qur'ān could perform no less miraculous feats. For women in labour village doctors might recommend that certain verses should be written on a slate, then cleaned off with water which was given to the sufferer to drink. Even the drinking was not always necessary. The mere touch of a healing text was sometimes enough. Glazed tiles with texts appropriate for particular ailments are still sold in souks (next to the Kalashnikovs); and allegedly antique specimens fetch idiot prices from better shod but no less gullible collectors in London and New York.

. . .

The Jewish contribution to Islamic medicine was important. Contrary to what is sometimes stated, anti-Semitism was not unknown in early Muslim societies; but in some Muslim countries the Jewish communities flourished. Under Roman and Byzantine rule the Iberian Peninsula sheltered some of the most successful Jewish settlements of the Diaspora. This changed after the invasion by the newly Christianised Visigoths in the seventh century. Jews were forcibly baptised, their rites were forbidden, their property was confiscated. The Moorish invaders of 711 were welcomed. Cordoba became the capital of the Umayyad dynasty, among the most enlightened of Islamic rulers. Medicine was a field in which the skill of the Jews was recognised. The personal physician of the Umayyad Caliph Abd al Rahman III, Hisdai ibn Shaprut, enjoyed unique privileges at court; and the doctor himself become a patron of Jewish philosophers, poets and scientists. But conflict between sophisticated dynasties and fundamentalist tribes was deeply embedded in

10. Ibid., p. 358.

Islam and golden ages of tolerance rarely lasted for more than three or four generations. Cordoba fell to fanatical Berber Muslims in 1013 and the Umayyads disappeared from history. In Granada and elsewhere thousands of prosperous and educated Jews were killed or rendered destitute. Others fled to North Africa.

Among the refugees was the family of a gifted youth, Moshe ben Maimon, later to be known as Maimonides.[11] From North Africa new waves of persecution forced them on. Eventually they settled in Fustat on the outskirts of today's Cairo under the martial but tolerant rule of the formidable Salah-ed-din Yussuf ibn Ayub, better known in the West as Saladin.[12] Maimonides became a celebrated healer, his fame reaching the invading Crusading armies. In 1185, the year of his appointment as physician to the Sultan in Cairo, he was also invited to become the personal physician to 'the Frankish king', probably Richard the Lionheart. He wisely declined. His doctoring was of a special kind. As a contemporary Arab commentator recorded: 'Galen's medicine is only for the body; but that of Maimonides is for both the body and the soul.' One gets glimpses of his character from his correspondence.

> I seem to be in some demand with the high officers like the chief kadi, the emirs, the house of Al-Fadr and other nobles who are carried to my residence in Fustat in litters attended by slaves and servants; but ordinary people find it too far to come, especially when unwell. So I have to spend part of every day visiting and talking to them in Cairo. When I get home I am too tired to pursue for long my studies. I try to catch up at night but this is not always possible.[13]

A few years later to another friend:

> My new duties to the Sultan in Cairo are becoming heavy. I visit him every morning if he feels ill or if one of his children or harem are sick; and I may have to spend the greater part of the day at the palace. When I get back to Fustat, I am tired and hungry, yet I find the courtyard crowded with people . . . high and low, gentiles, theologians and judges . . . I dismount, wash my hands and beg them to wait while I eat in a hurry, my only meal of the day. Then I attend to the patients who are queuing until nightfall and

11. He was born in 1135 and died in 1204 aged sixty-nine.
12. Saladin (known in Islamic lands as 'The Magnanimous') suppressed the ineffective ruling dynasty of Fatmidis in Egypt and eventually became ruler of Egypt, Syria and most of the Holy Land. Though defeated more than once by the Third Crusade, he was left in possession of Jerusalem and the Crusaders were eventually forced to withdraw. He died in Damascus in 1193, aged fifty-six.
13. Maimonides, *Correspondence*, quoted in I. Twersky (ed.), *A Maimonides Reader* (New York, 1972), p. 357.

> sometimes till dawn. I talk to them sometimes lying on my back because I am too weak to get up . . . The Israelites can have a consultation with me on the Sabbath when they come to me after the service and I advise them what to do during the coming week.[13]

What was his secret? Despite his busy life, his literary output was vast. His fifteen-volume religious treatise, the *Mishneh Torah*, written in Hebrew, is said to be somewhat compressed and convoluted, but his medical writings, translated from the original Arabic, are transparent. 'Above all, a doctor must be understood' was one of his aphorisms; and he always was. His *Regimen of Health*, a short manual, is full of unpompous advice about everyday problems:

> How can a person be fit to discharge his duties and enjoy his pleasures if his intestines are allowed to become a stagnant and painful pool of rotting matter? To prevent this, if a young man, he should eat salty and spiced food cooked with olive oil and fish brine and eat a bowl of spinach or cabbage every morning. If old, he should drink honey mixed with hot water a few hours before taking his first meal . . . He will then be able to cope with onerous daily chores instead of becoming a burden to himself.[14]

He had no patience with silent suffering; and some of his views would have outraged Seneca and Marcus Aurelius.

> The Lord gave us tears to shed . . . Do not try to stem their flow. When potions and vapours fail to ease the pain, lamentations often relieve the suffering.[15]

And what about drowning one's pain in alcohol? As a Jew practising in a Muslim society he had to tread carefully. In his *Regimen of Health* he devotes several paragraphs to the beneficial and calming effects of wine and to its value as the basis of complex anodynes. He also describes how such potions can be prepared and how they should be administered. However, he concludes with the statement: 'But since the drink is forbidden by the Law and since the Law must be obeyed, no more needs to be said on the subject.'[16]

Probably more important than his writings was his personal style of doctoring. Fierce warnings and admonitions were not his language. Moral

13. Ibid., p. 370.
14. Maimonides, *Regimen of Health*. This was Maimonides' book most often quoted by later Western Christian authorities like Henri de Mondeville, Arnald of Villanova and Guy de Chauliac.
15. Quoted in R. Weiss and C. Butterworth (eds), *The Ethical Concepts of Maimonides* (New York, 1975), p. 193.
16. Quoted in I. Epstein (ed.), *Moses Maimonides. Anglo-Jewish Papers in Commemoration of his Eighth Centenary* (London, 1935), p. 357.

support was the key. Almost repetitiously he insists on the unhurried personal consultation. His approach to the doctor–patient relationship a thousand years before that pompous phrase was invented is summed up in his aphorism: 'May I never see in any of my patients anything but a fellow creature in pain.'[17]

The idea of 'fellowship' was in tune with Islamic medicine. It already differed from the medical tradition of the Christian West. There a proper professional distance between the doctor or *giver*, and the patient or *receiver* continued to be regarded as a precondition of successful practice. The difference still exists. Both traditions maintain that it is the whole patient rather than an isolated symptom or physical signs which should be addressed; but the principle is applied differently. When doctors in the West enquire about symptoms not directly related to the main complaint, the response expected is a litany of negatives.[18] Sensing the expectation, wise Western patients comply: there is no merit in aggravating the doctor. In the traditional Eastern consultation both the expectation and responses are the opposite. It is polite for a patient complaining of indigestion to respond to ritual questions like 'Have you also got a cough?', 'Do you also suffer from headaches?', 'Do you also get short of breath?' in the enthusiastic affirmative. The cough, the headaches and the shortness of breath may be at best incidental; but, far from being time-wasting distractions, the affirmative answers are a compliment to the doctor's perspicacity. Also, the more numerous and varied the symptoms on offer, the more easily will the doctor make a diagnosis and prescribe treatment.

. . .

Most but not all of the authoritative texts of Islamic medicine deal with what today would be classified as medical diseases. Among notable exceptions Abu'l-Qasim Khalaf ibn Abbas al-Zahravi, known in the West as Albucasis, practised in the Umayyad Caliphate and was the author of a comprehensive treatise on midwifery, child-rearing, dentistry and, above all, surgery.[19] Some 1,500 pages long, his textbook falls into thirty-five sections. Section 30 describes such common surgical conditions as haemorrhoids, anal fissures and tumours of the skin. He also deals with fractures and dislocations, insisting on prolonged and effective splinting. He gives good advice on efficient ways of opening abscesses and suturing wounds, taking into account the

17. Maimonides, *Aphorisms*, quoted in A.I. Herschel, *Maimonides*, p. 215. There are many other interesting works on Maimonides, including O. Leaman, *Moses Maimonides* (London, 1990) which reinterprets some of Maimonides' work in a wider than Jewish context; and F.G. Bratton, *Moses Maimonides, Medieval Modernist* (Boston, 1967).
18. Ideally it should be instantly translatable into a tick on a form or computer print-out.
19. He died in Cordoba in 1013, aged seventy-four.

direction of the natural skin creases and the disposition of tissue planes. He knew that some internal abscesses could track long distances before pointing at the surface.[20] There are step by step descriptions of operations for stones in the bladder, amputations and eye surgery. For virtually all these procedures the surgical instrument recommended is the cautery, or rather, a series of cauteries. These were not the heavy irons used to brand cattle and criminals. Finely honed scalpels, needles and trephines, many with handsomely carved ivory handles, were heated to incandescence and presumably wielded with skill. The sterilising effect of heat may have made them safer than cold steel, especially dirty cold steel. Their disadvantage was the pain of burning and the risk of excessive tissue destruction.[21] To lessen the first, Albucasis recommended a number of 'well-tried pain-killing potions', hinting at but not actually mentioning alcohol. He also suggested the inhalation of 'soothing vapours'. But, significantly perhaps, 'speed and skill are essential'.

His younger colleague, Abu-l-Walid Mohammed ibn Ahmad ibn Mohammed ibn Rushd, Latinised to Averroës, was a noted theologian as well as a doctor. In some countries his religious books were burnt in his lifetime: in others they are still venerated.[22] His chief medical compilation was a seven-volume encyclopaedia of anatomy, general health, pathology, dietetics, drugs and hygiene, intended to complement a shorter manual by Ibn Zuhr, known in the West as Avenzoar.[23] Both emphasise the need to render patients as insensible as possible before operations. Both mention soothing vapours. Neither gives details of what should be vaporised.

Though medical injunctions are comparatively few in the Qur'ān, one is the requirement to treat the insane with compassion. 'Pagan superstitions' about the mad being possessed by the Devil are rejected.[24] Specific remedies in Arab books for disturbed mental states are nevertheless few. Avenzoar is exceptional in giving details of 'calming diets' for the 'irrational and agitated', based on milk, goat's cheese and fresh fruit, especially dates. Twenty-two varieties of the last are listed in order of effectiveness. If those fail, inhalations of 'powerful soporific mixtures' often help. He commends several 'well-tried brews' based on herbs and wild fruit, none of them easy to identify today but all liberally

20. A tuberculous abscess of the spine could track along the psoas muscle from the back and point under the groin. Albucasis was aware of such a possibility.
21. Electrical cauteries are still widely used in surgery to stop bleeding points (as an alternative to ligatures) and excessive tissue destruction is still a risk.
22. He died in 1198, aged about seventy.
23. He died in 1162, aged seventy-one.
24. It was in Muslim countries that the first separate hospitals for the insane were set up (known as *maristans*) and European travellers marvelled at the humanity shown to inmates. The tradition had some impact in the Christian West: the first mental hospitals in Europe were set up in the former Muslim towns of Spain, starting with Granada in 1386.

laced with 'bhang', almost certainly *Cannabis indica.* 'The effect will be soothing and sleep will usually be instantaneous.'

. . .

Long before the invention of printing, these tomes were copied and widely distributed.[25] The famous hospitals in Cairo, Baghdad and Damascus prided themselves on their libraries.[26] The ready availability of hallowed texts, coupled with the drill of learning long passages by heart as part of the training of young doctors, has been blamed for the decline of Islamic medicine after the fourteenth century. Rightly, perhaps. Intensely creative periods in art and literature (whether Renaissance painting in Italy or Georgian architecture in England) often have a paralysing effect on succeeding generations. There is no reason why this should not be true of medical science.

. . .

Cordoba fell to the Christians in 1236 and Baghdad was sacked by the Mongols in 1258. Arab civilisation and with it Arab medicine began to decline after 1300. The Ottoman Turks who were the dominant Muslim power for centuries created their own art, crafts and architecture but showed little interest in science. Nor, more surprisingly, has there been much interest in the modern Arab world in exploring the medical past.

25. Thousands survive in museums and libraries in Turkey and other Muslim countries (both in Arabic and in translations into Hebrew, Persian or Latin), many more than manuscripts of medical texts in the Christian West.
26. For centuries the best known was that of the hospital in Cairo built in 1283 by al-Mansur Qualawun who dedicated it to 'all who need care – rich and poor, old and young, male and female of all faiths'. In addition to its library, the hospital had special wards for physical and mental diseases, lecture rooms and a chapel for Christians as well as a mosque. It was still in use when Napoleon invaded Egypt in 1802 and was much admired by French army doctors.

# Chapter 9

# THE AGE OF THE CATHEDRALS

In the West the end of the first millennium marks the end of the Dark Ages. Nothing dramatic happened in the year 1000, any more than anything dramatic happened in the year 2000; but the two non-events had the opposite effect. Celebrations in the year 2000 were inanely upbeat; yet the first major event of the new millennium was a criminal war. Before the year 1000 there was a widespread expectation of the Second Coming. Coronations were postponed less they might be regarded as vainglorious provocations.[1] When the heavenly trumpets failed to sound, the effect was invigorating. Human nature did not change; but the eleventh century was a century of hopeful beginnings. The age which witnessed the rise of the great cathedrals and the creation of the most beautiful books of Western civilisation can no longer be described as 'dark'. But, disappointingly, the new beginnings did not include dramatic advances in pain relief or even in the wider fields of medicine and surgery.

Medical stagnation during the High Middle Ages is sometimes attributed to the hostile attitude of the Church. This is at best only half true. Of course religion continued to permeate every walk of life; but not until the seventeenth century was there any angry confrontation between faith and medical science. Some medieval church ordinances are sometimes described as anti-progressive but this too is a misinterpretation. Frequent prohibition on clerics to shed blood was not an attack on surgery but an attempt to prevent priests from developing a lucrative sideline in cupping and venesection. Contrary to legend, the dissection of human bodies was never prohibited. A letter from Sixtus IV explicitly informed the University of Tübingen that there was no objection to

1. In 997 Stephen, King of Hungary (the future St Stephen), asked Pope Sylvester II for a royal crown. The Pope obliged and his emissaries with the headgear arrived at Stephen's Court in 999. The coronation, however, had to be postponed because there was a widespread belief that the year 1000 would mark the end of the world and that a ceremony so soon before the end might be interpreted as a blasphemous act of defiance. Not till 1001 was Stephen crowned first Apostolic king.

'anatomising' the dead provided they were ultimately given a Christian burial. Pain could ennoble and purify the Christian soul and hair shirts became part of the habit of the new mendicant orders; but the relief of pain remained a commendable Christian act.

But if the substance of medicine and surgery changed little, their form and social status did. Medicine once again became a respectable and profitable profession. South of the Alps the tradition of lay doctoring continued. Elsewhere those who aspired to become physicians usually took holy orders. In France and Germany monks and clerics were the only people learned enough to be called doctors. Professions traditionally spawn books. Textbooks and health manuals were once again being written. They included some beautifully illustrated medical works, the treasures of princely and episcopal libraries.[2] But unadorned plain texts were also produced and widely circulated. Petrus Hispanus' *Thesaurus pauperum* (Poor People's Treasury) was among the first to acquire a popular readership. It was, as the title implied, specifically written for those who could not afford expensive medical care. This was still a praiseworthy objective. On pain relief it contained some sound if not particularly novel advice, like the rubbing of wild boars' lard on painful joints. The fact that it also recommended a course of pigs' faeces for abdominal colic must not be held against it. Animal excreta remained highly advertised (and cheap) remedies for a variety of pains till the eighteenth century. The author's reputation as a doctor helped to gain him election as pope as John XXI in 1276, the only medically qualified pontiff in history. Unwisely perhaps for a cleric, he dabbled in architecture as well as in medicine. The roof of the audience chamber which he had designed for himself collapsed on him a year after his (and its) elevation, putting an end to a promising pontificate.[3]

. . .

The fight against pain was never a hospital or an institutional monopoly. New drugs and therapeutic practices had always been introduced in a variety of ways; and they reflected and would continue to reflect social and cultural changes. The re-emergence of hospitals in the West during the High Middle Ages was nevertheless a crucial event. The less that could be done in the way of a cure, the more central became the task of dealing with symptoms. Among those, pain would remain by far the most important. Most pain relief would still be carried out in people's homes and in the infirmaries run by monasteries; but as hospitals gradually became more medicalised, they increasingly set the pattern. The majority of the new institutions were church foundations,

2. Guy de Chauliac's superb *Chirurgia* in the Bibliothèque Nationale is one of several examples.
3. He was in his early sixties.

either functioning under the patronage of individual saints or, as in France, designated houses of God. In England St Bartholomew's of London was founded in 1134 and St Thomas's around 1215. By 1287 St Leonard's of York looked after 224 patients. All began as shelters for the infirm and resting places for exhausted pilgrims; but the specialised care of the sick soon emerged as their key objective. Specialised care required specialised skills, and lay doctors began to play a part. The speed of transformation varied in different countries. The *hôtels-Dieu* in France remained hostels for the poor for much longer than their English counterparts. As in many other fields, Italy led the way. By the end of the fourteenth century Sta Maria Nuova in Florence, calling itself 'the best hospital in Christendom', had a medical staff of ten doctors as well as professional nurses, apothecaries and, among part-time assistants, the first female surgeon on record. Florentines were never known for obsessively hiding their light; and they had much to be immodest about. The establishment also had eight private beds 'for the higher classes'.

In other countries the founding of hospitals was often the beneficial side-effect of the otherwise deplorable enterprises known as the Crusades. The military orders, the Knights of St John of Jerusalem (later the Knights of Malta), the Knights Templar and the Teutonic Knights all built hospices for the sick. By the fourteenth century non-military brotherhoods (like the Order of the Holy Spirit) were maintaining infirmaries in six European countries, while the Order of St John of God built the first insane asylums in Spain. (They would later put up over 200 hospitals in the New World.) Coloman the Bookish, King of Hungary, placed an obligation on all diocesan bishops to build hospitals for the sick and needy within half a day's travelling distance of their episcopal palaces and to visit them regularly. He himself set an example.

Crusading zeal had another unintended benefit. Opium had been virtually absent from Europe for five centuries. Now the Crusaders learnt its use from their Arab enemies. Soon it acquired mythical fame. Of course it was often confused with other drugs like hashish; but spellbinding stories of raptures, superhuman valour and, above all, the abolition of all pain and sorrow, spread from castle to castle and from monastery to monastery. Returning warriors recounted the tale of the Old Man of the Mountains who spurred on his zealous troops with potions of poppy juice and of the superhuman stamina of Tartar horsemen (and their horses) who banished exhaustion with the drug. Turkish soldiers steeled themselves with it before battle and the wounded suffered their trials without a murmur. Then the drug itself was arriving, mostly from Persia through Venice. By the mid-fourteenth century it was once again an essential though still expensive constituent of pain-killing and soothing concoctions. When, a century or two later, Columbus, Vasco da Gama, Magellan and others set sail across the oceans, finding new sources was almost as high a priority as bringing back gold and spices. But most opium continued to be transported by land: travellers to Central Asia recounted meeting caravans of fifty or more camels laden with nothing but opium destined for the West progressing in stately file across the desert.

...

The same confident spirit (coupled with new wealth) which founded hospitals and encouraged trade with distant lands also stimulated the establishment of the first Western universities. Paris was founded in 1100, Bologna in 1158, Oxford in 1167, Montpellier in 1181, Cambridge in 1209, Padua in 1222 and Naples in 1224. In all these institutions but especially in France, medicine became an academic discipline. This was – as it would remain – a mixed blessing. Later generations would mock at the time and effort wasted by doctors in philosophical disputations instead of in tending the sick. One dispute in the University of Paris on the topic, 'Can sleep be harmful to health?' rumbled on eruditely for twenty years before the judges decided to call it a day, ruling that it all depended on the sleeper and God's grace. But it was too easy to mock. Teaching by rote was inevitable in an age when books were still scarce and when it was familiarity with the laws of the universe rather than the efficient management of diarrhoea and vomiting which raised physicians above the common herd of charlatans, quacks, bonesetters and wise women. Surgeons were more practical; but they too liked to maintain standards. 'A firm grasp of universal principles is essential to the understanding of individual cases,' Guy de Chauliac proclaimed on the first page of his great surgical treatise. Unfortunately, in any age universal principles are more difficult to verify and disprove than humdrum practicalities and are therefore more likely to prove universal fallacies. Yet it was not all hot air. In Bologna attendance at the dissection of animal and, when available, human bodies was obligatory from about 1350. The Emperor Sigismond urged his doctors to try diligently to discover new remedies though he insisted that they be tried on horses before being administered to his family and himself. The acquisition of a medical degree was certainly not easy. In Montpellier the curriculum for the title Bachelor of Medicine took seven years, and a doctorate was rarely awarded in under ten.

As part of the academic trend not only Aristotelian philosophy but also medical mathematics and astrology were highly prized. Learned physicians in their flowing robes had to be familiar with 'critical days' and even 'crucial hours' when an illness might reach a crisis point or a pain might be amenable to treatment. Zodiacs and heavenly constellations helped to pinpoint the right time for blood-letting, cupping and other dramatic interventions. The importance of the stars in determining treatment and prognosis, a timeless belief, was rarely so dominant as in the fourteenth century. So was belief in supernatural forces, including the ubiquitous and malign power of witches.[4] The efficacy of painkillers in particular – or their frequent lack of efficacy – was generally attributed to the moon not only in folklore but also in learned texts. Chaucer's physician was proud of being well 'grounded in astronomye'.[5] Was this a manifestation of a faltering faith in religious miracles, in the wisdom of the universal Church and even in the omnipresence of a personal God? The soaring steeples of the great cathedrals were challenges as well as affirmations.

What academic curricula did not as a rule include was the teaching of practical procedures like the application of leeches, the pulling of teeth, the draining of abscesses, the stanching of haemorrhages and the curetting of fistulas. These were tasks for surgeons and barber surgeons, a breed a long way below physicians with university degrees. Towards the end of the High Middle Ages some surgeons nevertheless did acquire considerable renown; and it is largely their literary output which deals with pain relief.

Lanfranc was an Italian from Milan who settled in Paris and there wrote his *Chirurgia magna* and a slightly abridged (but still substantial) version, his *Chirurgia parva.* Translated into French, Italian, Spanish, German, English, Dutch and Hebrew, the smaller version became the surgical bestseller of the age. It still gives the impression of being a useful do-it-yourself guide. Although it devotes a hundred introductory pages to the basic sciences like anatomy, embryology and 'general principles' (almost incomprehensible today), its main attraction was to provide step by step instructions of how to stop bleeding, reduce fractures and dislocations, and how to wield the cautery for ulcers. It also had a long section on skin diseases including a chapter on cures for baldness. One might have thought that in a brutish world beset by unspeakable pestilences the lack of flowing locks might be among the less pressing concerns of ordinary folk; but this merely shows how insensitive modern historians can be to the priorities of past generations. Looks always mattered; and Lanfranc enumerates no fewer than eighteen different ointments to be rubbed into the scalp, depending on age, sex, hair colour (or, one presumes, the desired hair colour) in both complete and patchy alopecia. The management of haemorrhoids takes up less than half the number of pages. Lanfranc also includes a lengthy section on healing and pain-killing herbs but does not claim to add anything new to the repertory.[6]

Lanfranc's pupil, Henri de Mondeville, was a more innovative and grander personage. He adorned the courts of Philip III and Philip IV of France and travelled widely, lecturing at some of the prestigious universities of the age.

4. Some religious orders were dedicated to hunting them down.
5. 'With us ther was a Doctour of Phisyk,/ In all this world ne was ther noon him lik / To speke of physik and of surgerye; / For he was grounded in astronomye.'
6. He died in 1305 in his fifties.

His big and never-completed text gives considerable space to what appears to have been one of the few anaesthetic innovations of the High Middle Ages, the sleeping sponge or *spongia somnifera.* Primacy is always difficult to assign in such matters; but the first detailed description of the device seems to have been contained in a thirteenth-century work, the *Chirurgia* by Theodoric, Bishop of Cervia, who in turn attributes the discovery to his father (or mentor), Hugh de Lucca.

> Take of the juice of the unripe mulberry, of hyoscyamus, of the juice of hemlock, of the juice of the leaves of mandragora, of the juice of the wood ivy, of the juice of the forest mulberry, of the seeds of lettuce, of the seeds of the dock which has large round apples, and of the water hemlock – each an ounce; mix all these in a brazen vessel, and then place it in a fresh sea-sponge. Let the whole boil as long as the sun lasts on the dog-days so that the sponge consumes all the fluid.
>
> As often as there shall be need of it, place this sponge in hot water for an hour and let it be applied to the nostrils of him who is to be operated on, until he has fallen asleep, and then let the surgery be performed. When the surgery is finished, in order to awaken him, apply another sponge, dipped in vinegar to the nose, or throw the juice of the root of fenugreek into the nostrils. Shortly he will awaken.[7]

Or not. The danger of too deep a sleep is apparent from frequent injunctions to 'keep tweaking the patient's nose' and make him inhale vinegar whenever 'sleep appears too deep'. The actual ingredients of the concoction seemed to include every soporific known at the time but few if any which had not been tried for centuries. If too deep sleep was one ever-present danger, the surgeon is also urged to secure the patient with sturdy ropes to prevent any abrupt or uncontrolled movement. This might not only interfere with the operation but put the surgeon at risk. De Mondeville devotes a whole chapter to how such restraints should be applied for different procedures, including operations for hernia and bladder stones.

More innovative from the surgical point of view was De Mondeville's advocacy of the immediate closure of some wounds. This went against the wisdom of centuries. Ever since Hippocratic times and fully endorsed by the school of Salerno, wounds after operation were left open and indeed kept open with plasters and powders to promote the formation and free flow of 'laudable pus'. This was believed to remove poisoned matter and other noxious substances. Healing from the depth of the wound outward *per intentionem secundam* ('by second intention') would follow. There was a great deal of sense

7. Theodoric, *Chirurgia*, Vol. IV, quoted in Ellis, *Ancient Anodynes*, p. 138.

in this, however well disguised. If the wound was infected, as many, perhaps most wounds were, closing superficial tissues and the skin would merely bury the suppuration. The abscess might burst; but, even more dangerously, the infection might burrow its way deeper and deeper. Sooner or later, it would erode a vein. The result would be fatal pyaemia, that is actual globules of pus in the blood, giving rise to multiple abscesses in distant organs. Such an outcome could be virtually guaranteed if there was a foreign body like a piece of metal or dead bone left behind in a closed wound. It was not of course the pus itself which was laudable, but the free drainage of pus. If infection could be avoided wound healing *per intentionem primam* ('by first intention') without suppuration would be both quicker and less painful. Though the pathology of wound infection would remain mysterious for another five hundred years, De Mondeville and his colleagues had more than an inkling of the conditions likely to lead to it. They advocated the bathing of wounds with clean fresh spring water at the end of operations, they emphasised the need to destroy as little tissue as possible during the surgery, and they insisted that no dead matter should be left behind. 'With such precautions and with a gentle touch,' De Mondeville prophetically wrote, 'it is not always necessary for pus to be formed and if it is not formed the patient will be saved much pain.'[8] He also recognised that pain was the surgeon's best guide. If a wound was closed and no pain developed, all was well. 'But if there is pain, the wound should be reopened and the flow of pus encouraged.'[9]

De Mondeville's work was carried on by the man who, for a century, was acknowledged as the supreme master of his craft in the West. Guy de Chauliac was born in Gascony into a noble though apparently impoverished family and studied medicine at Montpellier and Bologna.[10] Little else is known about him but he himself proclaims in the Introduction to his monumental *Chirurgia magna* that his express purpose was to raise the status of his vocation to that of a learned profession. In this he probably partly at least succeeded; and in France the rise of surgery can be dated to his career. His book is oppressively comprehensive, containing no fewer than 3,299 references to other named works (including 890 quotations from Galen alone), foreshadowing the computer-generated monstrosities of today. It also emphasises in chapter after chapter the need for a surgeon to be learned as well as skilful – and learned not only in matters of immediate surgical relevance (like anatomy) but also in astrology, in Aristotelian logic and in mathematics. Not perhaps unwisely, in the matter of wound healing he sat on the fence, maintaining that it was sometimes but not always wise to encourage pus formation and the free

8. H. de Mondeville, quoted in E.J. Gurlt, *Geschichte der Chirurgie*, Vol. I, p. 150.
9. De Mondeville died in 1320, aged about sixty.
10. De Chauliac died in 1368, aged seventy.

discharge of noxious emanations, 'depending on the nature and causation of the wound'.[11] Perhaps such cryptic provisos were better understood by his contemporaries than by modern readers. Like De Mondeville, for operations he continued to advocate the application of the soporific sponge, partly no doubt to save the patient pain but also because for successful surgery 'it is of extreme importance to keep the patient immobile and preferably [though not perhaps essentially] insensible'. Like most of his contemporaries, he enlivened his instructions with case histories, some reflecting on his own astounding operative feats and others revealing the abominable practices of his colleagues. His unsparing terminology would keep an innful of libel lawyers in funds today. Perhaps in the same vein, the wonderful coloured illustrations in the original book (which was widely copied in less ornate versions) show patients almost enjoying having their perianal abscesses incised, their dislocated elbows reduced or their gangrenous foot amputated by the great surgeon. Yet, even if his self-advertising strains one's credulity, his first-hand account of the plague remains the most hauntingly vivid of all contemporary reports of this nearly unimaginable disaster.

Most medieval authorities do not dwell at length on pain as such or on pain relief; but Guy de Chauliac devotes a whole chapter to it. Broadly speaking, he envisages two mechanisms which can cause pain. First, there is 'discontinuity' of normally continuous structures, including cuts, wounds, fractures and other injuries. Second, harmful humours and substances may be generated and may accumulate in parts of the body. The recommended treatment for the first was obvious: every effort should be made to re-establish continuity by splinting, suturing and the application of healing ointments. Treatment of the second group provided the rationale for a variety of measures, all aimed at eliminating noxious matter. Bleeding and purging were the most generally useful; but there was also cupping, blistering, the encouragement of sweating, heroic drinking sessions to promote the flow of urine and, in some wounds, applications to stimulate the copious flow of pus. The general principle of trying to free patients of disease-causing substances by some route or other was to remain the central plank of medical and surgical treatment for centuries.

Guy de Chauliac's acclaimed English contemporary, John of Arderne, served under John of Gaunt in the Hundred Years War and then settled in Newark-on-Trent.[12] There, despite the distance from metropolitan centres, he seems to have been visited by the great and the good of the land with a variety of ailments. Indeed, he endeavoured by his own admission to confine his

11. E. Nicaise, translator and editor, *La grande Chirurgie de Guy de Chauliac, composée en l'an 1363* (Paris, 1890), p. 436.
12. John of Arderne died in 1370, aged sixty-three.

practice to the nobility, to wealthy landowners and to the higher clergy who could afford his fees. They were allegedly exorbitant but he was worth every penny. Arderne made an exception of 'genuine' mendicant brothers to whom he gave advice (but not medicines) free. He invented – or perhaps only modified, since the operation is ancient – a procedure for the treatment of *fistula in ano*; and described himself as a *chirurgus inter medicos* (surgeon among physicians). Unlike his French colleagues, he did not rate the anaesthetic sponge highly. Instead he recommended several 'pain-easing ointments', some 'infallible' and others only 'highly gratifying as a rule'. In the former category he noted that

> An ointment with which, if any man be anointed he shall suffer cutting in any part of his body without feeling or aching can be prepared by mixing equal quantities of the juices of henbane, mandragora, hemlock, lettuce, black and white poppy, and the seeds of all these herbs; and, if available, Theban poppies and poppy meconium, one or two drachms with sufficient lard. Braise them all together thoroughly in a mortar and afterwards boil them well and let them cool. If the ointment be not thick enough, add a little white wax and then preserve it for use. When you wish to use it, anoint the forehead, the pulses, the temples, the armpits, the palms and the soles of the feet. Immediately the patient will sleep so soundly that he will be insensible to any cutting.[13]

He also recommended opium and henbane in wine but warned that the brew might make the patient sleep for too long or too profoundly. To avoid this (and worse): 'Know that it is well to pull the nose, to pinch the cheeks or to pluck the beard of such a sleeper to quicken his spirit less he sleep too deeply.'[14]

Arderne was advanced in devoting much thought to the post-operative comfort of patients, not, it seems, high on the list of priorities of most of his colleagues. *Castorerum*, the pungent secretion of the perineal gland of the beaver given by mouth dissolved in wine, was a favourite stimulant which

> most comforteth the chilled sinews and relieveth the paralysis after operations. Also give the patient things to comfort the brain such as castor, musk, nutmegs, oil of roses, nufar, myrtle and sumace.[15]

...

13. Joannes Arderne, *De arte physicali et chirurgia*, 1412, translated by Sir D'Arcy Power (London, 1922), p. 56.
14. Ibid., p. 236.
15. Ibid., p. 326.

No reader can fail to be impressed by the confidence of these grand high medieval texts as well as by the extraordinary mix of treatments recommended. On one page a learned surgeon like Guy de Chauliac might recommend the inhalation of an opiate mixture for 'strangulation' in the chest which may well have been effective in an anginal attack. On the next he gives a detailed prescription for epileptics to write in their own blood on a piece of 'thrice-blessed parchment' the names of the Three Wise Men and to recite twenty-three Ave Marias daily for three months.[16] But perhaps that too helped.

16. Of all ailments, epilepsy or the 'falling sickness' perhaps attracted the most bizarre and elaborate remedies. John of Gaddesdon (died in 1349, aged seventy), physician to Edward II, recommended in his *Rosa anglica medicinae* (*The English Rose of Medicine*), reciting the Gospel over an epileptic patient while bedecking him with peony and chrysanthemum amulets or the hair of a white dog. Mistletoe was widely regarded as curative, harking back to the young King David seeing a woman collapse in a fit and then being instructed by an angel that 'whoever wears the mistletoe in a finger ring on the right hand, so that the mistletoe touches the hand, will never again be bothered by the falling sickness'. In Central Europe the stalk of mistletoe was hung around children's necks to prevent seizures and in Scandinavia the handle of a knife was cut from mistletoe.

A founder of modern science and founding member of the Royal Society, Robert Boyle, in the seventeenth century still endorsed pulverised mistletoe for epilepsy, 'as much as can be held on a sixpence coin, early in the morning, in black cherry juice, during several days around the full moon'.

# Chapter 10
# PAIN EXALTED

It was towards the end of the High Middle Ages that pain and heresy intersected for the last time, and in a most spectacular manner. Torturing the body to attain spiritual enlightenment had been practised by devout individuals since the beginning of Christianity; but public flagellation in groups did not become an expression of communal penance till the mid-eleventh century. Even then, the outbreaks were at first sporadic. In 1235, in response to the ravages of civil war a concourse of men and youths marched through Perugia in central Italy, with banners and burning candles, chanting while flogging themselves: 'Holy Virgin, take pity on us! Beg Jesus Christ to spare us!' From the start the ritual was a response to disasters and communal suffering. Some of these were plainly the result of human wickedness and greed. Others seemed to be somewhat capricious visitations from God. From central Italy the movement spread south and north. Its leaders in Germany claimed that God had written a letter on a marble tablet ablaze with light. In this He threatened to destroy sinful humanity. They also claimed that the Virgin Mary had interceded on man's behalf to save those who joined the flagellant procession. The duration of the processions was usually thirty-three days and a third of a day, corresponding to the number of years Christ spent on Earth. Outbreaks were eventually recorded in most countries; but for over two hundred years they remained numerically insignificant. What catapulted them into a dramatic historical movement was the Black Death.

A wide variety of ill omens – spontaneous conflagrations, drops of blood in freshly baked bread, church bells tolling without human agency – heralded the greatest medical disaster in the history of Europe.[1] The probable beginning was the landing of three Genoese galleys in the Sicilian port of Messina in

1. Outside Europe, in proportional terms, there have been greater ones: the indigenous populations of many islands (like the Canaries) were totally wiped out by diseases imported by the first European explorers.

January 1348.[2] Disease 'clung to the very bones' of the crew. Panic-stricken, the Messinese fled to the countryside and neighbouring cities, carrying the plague with them. In Catania the pestilence was so violent that after the first few days 'no man could succour another'. This was to become the common experience. People died in their homes, in the churches, in the streets or on carts trying to flee. Two weeks after the outbreak hit Naples two-thirds of the population, including the archbishop and most of his clergy, were dead. The plague spread at great speed along the trade routes of the Mediterranean.[3] During the winter of 1348–49 it crossed the Alps. Within a year virtually the whole of Western Europe was affected. About a third of the population died; and it was a painful and horrible death. 'The wailing and screaming could be heard day and night and the stench of bodies hovered over the countryside.' The previous centuries in Europe had not been times of ease and plenty. But this was a catastrophe beyond medieval understanding. Most people saw it as a divine punishment for unspeakable sins. Many sought escape by inflicting pain on their own bodies.

In its new formalised intensity the movement started in Hungary and spread through the towns of Germany, the Low Countries and eventually France. A band of flagellants might have 50 members or 500 hundred. At the height of the movement a new band arrived in Strasburg every week over a six-month period. More than 5,000 visited Tournai. They wore uniforms of white robes with a red cross front and back and hoods. They were forbidden to shave, bathe, change their clothes or have dealings with women. The worship of the *Mater dolorosa*, the Virgin of Sorrows, was at the centre of their cult. As they marched, they sang the words of the 'Stabat Mater':

Wound me with his wounds
Let me drink his cross
For the love of thy son.

The twice-daily and once-nightly ceremony in front of churches began with the brothers throwing themselves on the ground with arms outstretched like crucifixes. They then rose and flogged themselves with metal-tipped leather

2. Probably migrating from China, the Black Death broke out among the Tatars who were fighting Italian merchants in the Crimea. According to contemporary chroniclers, the Genoese took refuge in the citadel of Kaffa (Feodosia), where they were besieged. The plague forced the Tatars to raise the siege, but, before withdrawing, they invented biological warfare by catapulting the corpses of plague victims over the citadel wall. The disease duly broke out within the citadel and travelled with the fleeing Genoese to the Mediterranean.
3. Though there has been some controversy, the epidemic was almost certainly caused by a virulent strain of the bacillus *Yersinia pestis* (bubonic plague), transmitted from rats to humans via fleas but also capable of spreading from humans to humans by droplet infection and entering through the respiratory passages.

scourges, singing, praying and sobbing, until they collapsed or the master gave the signal to stop A few never rose, though records suggest that the number of fatalities was surprisingly small.[4] The ones who died were mostly children. (Women but not children were barred from most flagellant groups.) Their bodies were sometimes hacked to pieces and became private relics. Self-torture could escalate into sadistic frenzy, especially in Germany. Jews had long been the traditional 'Scripture-inspired' scapegoats for communal sufferings. In Frankfurt, Mainz and some cities in Eastern Europe the flagellants incited the most violent pogroms before modern times. Thousands perished.

The movement was never officially encouraged by the Church or the higher clergy; but to try and suppress it was dangerous. The flagellants called themselves 'saints' and 'brothers of Christ' and in many towns were fiercely protected by the citizens. Attempts at forcible restraint also tended to end in futile aggravation. Unsurprisingly, but as those in power would discover to their surprise in every age, it is difficult to punish those who rejoice in punishment. The conflict between established Church and flagellants came to a head in Avignon. Pope Clement VI, a Frenchman of advanced views and expensive tastes, was determined to enjoy his pontificate. The plague visited the city in March 1348. Clement felt that a modest procession of a few hundred flagellants could do no harm and might do some good. But about two thousand marched, lashing themselves, including on that occasion women and children. Three died from blood loss and exhaustion. Next day the ritual was repeated with twice as many participants. The number of casualties also increased. As elsewhere, the craze was beginning to get out of hand. The demand was heard that the Pope himself and some of his cardinals should lead the procession. That, Clement felt, was ludicrous. It was time to stop the madness. In any case, the plague was moving on. The papal palace, clean and obsessively ventilated, had been miraculously spared.[5] Clement issued his first pontifical bull condemning the rites: 'To suffer pain in the service of God was laudable. To inflict pain on oneself for the purpose of pleasing God was vain and sinful.' It was the first explicit pontifical ruling on the subject. It was followed by others. A specific encyclical condemned the persecution of Jews. 'By a mysterious decree of God', none had been spared the plague: Jews had been as much victims as Christians. The charge that Jews had spread the plague was 'without plausibility'. Though Philip V of France a little glumly endorsed the Pope's

4. The Austrian Chevalier Leopold Sacher-Masoch, whose name is enshrined in 'masochism,' pointed out that serious injury inflicted by masochists on themselves is extremely rare.

5. Clement was either extraordinarily prescient or lucky. By a variety of drastic but effective measures he rid the papal palace of rats and insisted on cleanliness and continuous ventilation. He also forbade townsmen to enter the papal buildings during the epidemic. All those quarantined inside the palace escaped the disease.

pronouncements, the effect was not immediate. In June 1348 a boatload of flagellants from the Continent still visited London and performed their bloody act in front of St Paul's Cathedral. They were warmly applauded by the populace 'as if they were some visiting fair' but made no local converts. Then, after a wave of persecutions in Westphalia, the movement fizzled out. As a contemporary chronicler recorded, 'the flagellants vanished as suddenly and as mysteriously as night phantoms or mocking ghosts'.[6] This was a pious exaggeration. It took the Inquisition more than a century to stamp out the last remnants. Hundreds of flagellants were burnt at the stake in Sangerhausen as late as 1414; and 'the last covert flagellant', almost certainly innocent, was beheaded in Mainz in 1480.

But if, officially, the Clementine bulls ended self-inflicted pain as a short-cut to Paradise, indeed declared the attempt an 'abominable heresy', the burning of the last flagellants was not the end. Salvation through suffering remained a powerful motif in the Catholic Church and in some other branches of Christianity; and the dividing line between self-inflicted pain (forbidden) and pain inflicted by the Almighty and borne with joy (allowed) remained fuzzy.[7]

6. Quoted in Moynahan, *The Faith*, p. 293.
7. The doctrine of salvation through pain was revived in the ardour of the Counter-Reformation. See Chapter 13.

# PART II

# SCIENTIFIC STIRRINGS

# Chapter 11

# REBIRTH, REDISCOVERY AND REFORM

The rebirth of Antiquity – or what the reborn liked to think of as its rebirth – began in Italy and spread slowly through Europe.[1] It was marked by a firework display of creativity in the arts, in architecture, in religious thought, in music and in literature. Its immediate effect on medicine was dire. Inspired by the Renaissance ideal of returning to the immortals of Greece and Rome, the current texts of Galen, Celsus and the Hippocratic texts were meticulously purged of medieval and Islamic accretions. Some of the new editions were published in splendour by the pioneering Aldine Press of Venice. Since even in the Renaissance few doctors spoke Greek, the originals were retranslated both into Latin and into European vernaculars. The translators were the leading humanists of the age, men of prodigious learning and industry.[2] Their enthusiasm was rewarded by the discovery of texts that had been lost or forgotten for over a thousand years.[3] 'Medicine has been raised from the dead,' wrote Johann Guinther von Andernach, a leading light of the Paris Faculty of Medicine, commenting on the replacement of 'spurious and faulty travesties' which had been in circulation throughout the Dark Ages. His colleague, Jacobus Sylvius, another medical humanist, approached the restored originals just as Luther and his followers reread the Bible, with religious awe.[4] Where Paris led, other academies followed.[5]

1. Dates are difficult. By the end of the fourteenth century the Renaissance was in full flower in Italy but it took another 100–150 years to reach England.
2. Thomas Linacre, physician to Henry VII, translated several Galenic texts into English. During the sixteenth century an astonishing 590 editions of Galenic books were published in Paris, Lyons, Venice and Basle.
3. His newly discovered *On Anatomical Procedures* was translated by Johannes Guinther von Andernach (who died in 1576, aged eighty-seven).
4. Sylvius' Galen worship culminated in his *Order and Way of Reading Hippocrates and Galen*, published in 1539. After his initial patronage of Vesalius he turned against the younger man. He died in 1555, aged seventy-six.
5. The most influential and perhaps the greatest of the Paris faculty, Jean Fernel, introduced the terms 'physiology' and 'pathology' in two books in which he was surprisingly critical of Galen. He died in 1558, aged sixty-one.

. . .

But here a book whose subject is pain must briefly digress. What was true of ancient texts also applied to antique artefacts. Never before or since has there been such a scramble for works of art that had been buried under centuries of rubble. New discoveries of the treasures of Greece and Rome were hailed as miracles. Princes and popes bartered away castles to finance the acquisition of major finds, or to have the rightful owners discreetly disposed of and dispossessed. Many of the resurfacing sculptures were figures of great beauty. There were Venuses, athletes, gods (including the Apollo Belvedere), heroes. Credible artistic representations of physical pain – as distinct from mental states like sorrow, grief, dismay or resignation – have been comparatively rare in any age. Greeks and Romans especially disliked depicting ugliness. Art existed to please. Their beautifully decorated vases and urns mostly show youth in full bloom. Even their sepulchral monuments usually commemorate scenes of domestic bliss.[6] Nobody expected the emergence of a sculpture that would become the grandest representation of pain in art.

But on Wednesday, 14 January 1506, perhaps using imported Turkish gunpowder, a market gardener in his vineyard on the Esquiline Hill in Rome opened a niche in the rocky ground. The cavity had been so effectively sealed that for more than a thousand years it had escaped both invading looters and home-bred tomb-robbers. In the space behind the opening was a large marble sculpture. As it was too heavy to move or to sell privately, the discoverer opted for a pontifical reward. He sent his son to the Lateran to report the find. Julius II was intrigued. An insatiable collector, he dispatched his antiquary, Giuliano da Sangallo, to investigate. Giuliano in turn invited his house-guest, Michelangelo Buonarroti from Florence to join the party. Giuliano's nine-year-old son later recorded that 'even in the half-light, as soon as my father saw the marble, he exclaimed: "God be praised! This is the Laocoön of Pliny!" Michelangelo was at first incredulous but then embraced my father.'

Nearly fifteen hundred years earlier Pliny had described a 'marble group carved out of one block by the most consummate of artists, perhaps Agesandros, Polydoros or Athenodoros of Rhodes. The workmanship is so magnificent that it is to be preferred to any other sculpture.' The figures depict

6. On an early Attic plate, now in Berlin, Achilles is shown bandaging the leg of his wounded friend, Patroclus. The procedure must have been at least uncomfortable but no unease is registered on the smiling face of either man. On a fresco painting, now in Naples, Iapyx is depicted as removing an arrowhead from Aeneas' thigh. The hero's face shows no emotion whatever. The carved reliefs on Trajan's Column include representations of first-aid stations behind the battle lines. In real life such places must have resounded with screams and groans; but none of this is apparent from the serene expressions of the wounded.

a scene from the Trojan wars. After laying siege to the city for ten years, the Greeks have departed. They have left behind a huge wooden horse, seemingly a gift. But the Trojan high priest, Laocoön, accompanied by his two sons, warns the citizens not to trust such presents. He is nearly successful. To silence him, Aphrodite, rooting for a Greek victory, commands two sea serpents to rise from the waves. They strangle Laocoön and his sons. The struggle is both physical and spiritual – Laocoön senses the wrath of the goddess as well as the deadly grip of the monsters – and it is depicted with a realism that is, in Pliny's words, 'without parallel'. One can almost hear the screams and sense the agony.[7] The group is now in the Vatican Museum.

By a strange coincidence (but Renaissance art abounds with them) the rediscovery of the sculpted Laocoön coincided within a year or two with the completion of arguably the greatest representation of physical pain in painting. Little is known about the artist. He was Matthias or Mathis Grünewald and even the dates of his birth and death are uncertain. He moved around German cities and courts and probably never ventured south of the Alps. His idiosyncratic art was rooted in the Middle Ages; but his vision is timeless. It was probably in his capacity as court painter to the Elector of Mainz that he decorated the ten panels of an altarpiece for the church of Isenheim. (It is now in the Unter den Linden Museum in Colmar.) Tens of thousands of Crucifixions have been painted since the Cross became the symbol of the Faith in the second century. None are remotely like Grünewald's central panel. Nothing has been spared of the horror of what must have been one of the most agonising deaths invented, inflicted and suffered by man. The end has come at last; but the signs of pain are still evident. The image is horrifying, uplifting and unforgettable.

. . .

Art does not age. Science does. The impact of Laocoön and of Grünewald's Crucifixion is as powerful today at it was at the time of their creation or rediscovery. This was not true of the ancient texts. For the antiquarian there is much to be said for reprinting old works unexpurgated or in faithful translations. To practising doctors the antique works proved no more reliable than they had been in their own day. Indeed, the effect of most Islamic and medieval 'accretions' had been, on balance, significant improvements. To many, this went against the grain. For several decades Galen continued to enjoy a prestige unequalled in his own time. Even when some of his

7. The Laocoön became a model and a challenge to Michelangelo. It influenced the four dying and writhing slaves which he carved twelve years later (two now in the Louvre and two in the Uffizi), his own contributions to the portrayal of pain. There are echoes in Picasso's *Guernica*.

anatomy was shown to be indisputably wrong, this was attributed by certain besotted fans to the degradation of the human body over the intervening thousand years.

Galen represented more than facts. His was a universal system, an overall philosophy into which individual facts fitted – or had to be fitted. The approach of medieval authorities to the complexities and ailments of the body was primitive perhaps but not without logic. After a bow to general principles, their work tended to begin with diseases of the scalp and end with painful toes. This would have been regarded by Galen – and was again regarded by his early Renaissance admirers – as hopelessly benighted. It took no account of the humours and the temperaments, of the universal systems and essences governing life. Galen himself referred to these as his *practica*: they represented the certainties, the foundations on which all successful treatment had to be based. The thought of error never approached Galen's mind. If the appropriate therapy had been applied and the patient was still in pain – by far the commonest complaint in any age – that could only be the patient's fault.

But the Renaissance was also a time of questioning. Galenic omniscience could not remain immune any more than could the moral authority of a flagrantly corrupt Church. The first attack came from a character who, after five hundred years, still fascinates and repels. Theophrastus Philippus Bombastus von Hohenheim, known to the history of science and medicine as Paracelsus, was born in Einsiedeln in Switzerland in 1493, the son of a physician (or perhaps an apothecary). The name, meaning above or greater than Celsus, was his own characteristic coinage as was the Bombastus. The latter was intended to be complimentary. He learnt some botany, chemistry and medicine from his father; but, apart from a few brief sorties to Italian universities, he had no formal medical or scientific education. How he convinced the world at large and in particular the rich and sober burghers of the city of Basle that he was not a loud-mouthed charlatan but God's gift to suffering mankind is the first of many mysteries that surround him. His contempt for academic medicine and his constant proclamation of his own genius might have landed him in a lunatic asylum or on the stake in an age and place other than Europe during the first decades of the sixteenth century. Instead, in 1526, he was appointed town physician and professor of medicine at a university that lay at the crossroads of Europe and had been one of the cradles of humanism. Here he lectured not in Latin but in vernacular German, not attired in flowing academic robes but in an alchemist's leather apron, to a vast crowd of doctors and laymen, both old and young, who came to hear him from every country in Europe. He famously began his course with the ceremonial burning of the works of Galen and Avicenna and he introduced his published works with the statement: 'When I saw that nothing resulted from current medical practice but pain, killing and laming, I decided to abandon this miserable art and seek the truth elsewhere.' Where exactly the elsewhere was is the second mystery about him. While he was undoubtedly right in much of his condemnation of medicine as practised in his time (and indeed since), his occasional representation by successive generations of fans as a founder of experimental scientific medicine is wholly without foundation. The details of his complicated universal system of 'three elemental principles' (which changed at regular intervals) are not worth rehearsing. He was a believer in mysterious and esoteric 'natural' doctrines as well as in gnomes, fairies, nymphs and cosmic emanations. Many of his remedies he had learnt from 'ordinary folk' but some had been revealed only to him.[8] His claim to figure in a book about the fight against pain is his discovery of a miraculous substance which he called *laudanum*. 'Ich habe ein Arkanum, heiss ich Laudanum; ist über alles gegen alle Schmerzen und zum Tod reichen will.' (I have a magic elixir which I call laudanum which surpasses all others in relieving pain and will last till death.) His pupil, Joseph Duchesne, known as Quercetanus, who later became physician to Henri IV of France, supported this claim and derived the word – almost certainly mistakenly – from the Latin *laudo*, I praise.[9] What exactly Paracelsus's elixir was is another mystery. Although he undoubtedly used opium in his various remedies, his 'Arkanum' was a solid rather than an alcoholic tincture. He was the archetypal dogmatic anti-dogmatist, the extraordinary ordinary chap, the book-burning author of unreadable tomes and the compulsive scribbler whose recurrent theme was that written texts are the root of all evils. The last claim may partly explain his enduring appeal to an astonishing assortment of fringe groups.[10]

Yet he was, for all his absurdities, the embodiment of the patient-centred rather than the doctrine-centred doctor. In *Man and his Body* he wrote:

8. He was and remains the patron saint of all the more bizarre forms of alternative medicine and of Nature cures of all kinds. In addressing the British Medical Association in 1982, the Prince of Wales turned his comprehensive expertise from architecture, horticulture, husbandry (both animal and human), education, religion, ethics and handful of other fields to medicine and helpfully advised the assembled doctors that 'we could do worse than to look at the principles Paracelsus so desperately believed in, for they have a message for our time; a time when science has tended to become estranged from nature – and that is the moment when we should remember Paracelsus'. Sadly, he did not disclose what the principles in which Paracelsus desperately believed actually were, other than his own infallibility. This perhaps struck a sympathetic chord.
9. It may have come from the Greek λάδανον *ladanon*, the name of an oriental shrub known to Pliny and perhaps regarded at one time as the source of opium. Michael Scott in 1200 used the term 'laudanum', referring to a heavy and soporific scent in the air, perhaps that of the poppy. The alcoholic tincture of opium was first prepared by Thomas Sydenham in 1669 (see Chapter 13).
10. He was popular not only with the Prince of Wales but also in Nazi Germany where his Germanic 'folk wisdom' was praised by leading quacks.

> Every physician must be rich in knowledge, not only of that which is written in books. His patients should be his book for they will never mislead him . . . and by them he will never be deceived. But he who is content with mere letters is like a dead man.[11]

The suggestion that patients never mislead or deliberately deceive doctors may be questionable; but the sentiment is one to which it is impossible to remain wholly unresponsive. Also on the credit side, he was probably convivial company if one's taste ran to the loud, boastful and boisterous. One admiring pupil wrote:

> In curing intestinal ulcers he did miracles where others had given up. He never forbade his patients to eat as much food and drink as much as they liked. On the contrary, he frequently stayed all night in their company, feasting with them. He said he cured them most quickly when their stomachs were full. He had pills which he called laudanum which looked like mouse shit but which he used only in extreme cases. He claimed that with these pills he could wake up the dead.[12]

His ascendancy in Basle did not last. After a year or two the traditionalist majority – lambasted professors and cowed town councillors – recovered their nerve. He was accused of a long list of abominations, deprived of his chair and finally expelled. He expected his flock of students to rise. None stirred. His last years were spent wandering between archiepiscopal, princely and royal courts in southern Germany, Austria and Switzerland, seeking pupils and patronage. Little is known about these peregrinations. On 21 September 1541 he arrived at the White Horse Inn in Salzburg, a sick man. He suffered a stroke. Next day he died. He was forty-eight. His death was the signal for new and ever more luxuriant legends to sprout and grow. They still do.

. . .

Paracelsus was not the only or even the most important rebel against ancient authority. For nearly two hundred years – never to be repeated – anatomy became the beacon of the medical sciences. This was not the dead and dread descriptive discipline of medical student mythology but the explosively exciting anatomy of the living, the key to life. Indeed, though it took centuries for the discoveries of the fifteenth and sixteenth centuries to bear fruit in practice, they alone made the practice possible.

The initiators of the anatomical Renaissance were not doctors but artists. In his influential *De statua*, Leon Battista Alberti, oracular theoretician of art,

11. Quoted by W. Pagel in *Paracelsus: An Introduction to Philosophical Medicine in the Era of the Renaissance* (London, 1982), p. 210.
12. Quoted ibid., p. 245.

stated that an intimate knowledge of the structure of the body and the relation of that structure to function was vital to any painter or sculptor.[13] A few years later Lorenzo Ghiberti, creator of the heavenly gates to the Baptistery of the Duomo of Florence, maintained that his triumphs in bronze owed everything to the time and care he had devoted to the study of human anatomy. Verrocchio, Mantegna, Botticelli, Luca Signorelli, and Dürer and Holbein north of the Alps, echoed his sentiments. All attended public autopsies and many performed dissections of human bodies themselves. Whether Leonardo da Vinci did is uncertain. He owes his seemingly unshakeable status as the sacred cow of Western civilisation partly to his 600-odd anatomical drawings, some of them exquisite minor works of art but, like virtually all his excursions into science, incorrect in every important detail. In so far as he claimed to show *human* parts he was probably deluding himself: like Galen, he showed structures and anatomical configurations which could have been observed only in pigs, sheep or dogs. Fortunately for the advancement of anatomy, his drawings remained unpublished for two hundred years.

Professional anatomists soon took over. In one respect they were better placed than their clinical colleagues. While they too had to pay lip-service to Galen, the great Roman himself could be quoted as enjoining readers always to 'look for themselves'. (The term 'autopsy,' meaning seeing for oneself, was probably coined by him.) Andreas Vesalius, who did most of his work in Padua in Italy, is generally regarded as the greatest of the Renaissance anatomists; but he was only one star in a constellation.[14] Part of the success of his great book, *De humani corporis fabrica*, one of the few printed works to transform Western civilisation, was due to the illustrations by the Flemish artist, Ian Stephan van Calckar,[15] and the craftsmanship of his printers, Joannes Oporinus of Basle. In six successive volumes every part of Galenic anatomy came under scrutiny and most parts were subjected to corrections. He brought his work disarmingly to a close, saying:

13. Alberti's influence as a writer was immense though his surviving architectural works are few. Born in Venice, he was for some time pontifical secretary in Rome. He died in 1472, aged seventy-one.
14. Vesalius was born Andreas van Wesele in Brussels where his father was pharmacist to the Emperor Charles V: his mother was English. He studied in Paris where he showed his anatomical zeal by robbing wayside gibbets, smuggling the bones back to his quarters and reconstructing the skeletons. In 1537 he moved to Padua where his major anatomical work and writing was done. He finished his ground-breaking five-volume *De humani corporis fabrica* in 1542 and died in 1564, aged fifty.

    Most of Vesalius' students and successors – Fallopio, Eustachius, Thomas Willis and many others – are commemorated in the names of anatomical structures. (This was the golden age of eponyms. At no other time would an entire continent have been called after the Christian name of an explorer of the second rank.)
15. Curiously, no other works of his are known.

> How much has been attributed to Galen by those physicians and anatomists who have followed him, often against reason and their own better judgement! I myself cannot but marvel at my own stupidity when I recall my excessive trust in the writings of Galen and other anatomists.[16]

It is impossible to recapture the excitement generated by these discoveries. Anatomy became the metaphor of the age: it was Robert Burton's obvious choice in *Anatomy of Melancholy* and John Donne's in 'Anatomy of the World'. Anatomy lessons – Rembrandt's is only the most famous of many – became emblematic of science probing the secrets of Nature, peeling away layer upon layer in an irresistible search for the truth. Every advance became news. The recent discovery of a passage between the middle ear and the throat, known today after its discoverer as the Eustachian tube, must have inspired Shakespeare in backward England to make Claudius administer his poison to his brother by the novel route of aural instillation.[17] It was the anatomical discoveries of the Renaissance which paved the way for Harvey's inspired insight into the circulation of the blood, the understanding of the mechanism of respiration and the first attempts at intravenous anaesthesia.[18]

. . .

Second only to anatomy in its future impact on pain relief was the revived study of botany and materia medica. There were new and accurate translations of classical authors, some superbly illustrated in the new realistic manner. Symphorien Champier's massive *Apothecaries' Mirror* was subtitled: *The Mirror of the Apothecaries and Druggists in Which is Demonstrated How the Apothecaries Commonly Make Mistakes in Several Medicines Contrary to the Intention of the Greeks . . . and on the Basis of the Wicked and Faulty Teachings of the Arabs*. The new herbal by Otto Brunfels, town physician of Berne, *Herbarum vivae eicones*, transformed the genre by its naturalistic descriptions and illustrations. It spawned others, including William Turner's compilation in England with its detailed prescription for how to prepare a soporific and pain-killing potion from mandragora. His tone is both cautious and confident.

16. Vesalius, *De humani corporis fabrica*, Vol. I, quoted in Porter, *The Greatest Benefit to Mankind*, p. 181.
17. The Eustachian tube which ensures that air pressure in the otherwise closed middle ear does not deviate too far from ambient atmospheric pressure (which might otherwise burst the tympanic membrane) was in fact first described by Giovanni Ingrassia (died in 1580, aged seventy) who had discovered it in 1546; but it was popularised by the widely read textbook of Bartolomeo Eustachio (died in 1574 aged seventy-four) published in 1564.
18. See Chapter 12.

> If they drink this drink they shall feel no pain, but they shall fall into a forgetful and sleepish drowsiness . . . But they that smell too much of the apples may become dumb.[19]

Botanical gardens opened under princely or municipal patronage in Pisa, Padua, Bologna, Leyden, Leipzig, Basle and Montpellier. Pier Andrea Mattioli, physician to the province of Gorizia in northern Italy, produced a lavishly illustrated edition of Dioscorides' *Materia medica.* Even the first printing ran to 1,500 pages; but, like Linné 200 years later, Mattioli had a gift of cajoling travellers to collect and present him with specimens of new and previously unknown species. These were included in successive later editions. For anyone in search of pain-relieving herbs and plants here was an embarrassment of riches.[20] Did any of them work? It is possible.

. . .

The age of rebirth, rediscovery and reform was also the age of explorations, the most heroic in history. In terms of health they were also the most catastrophic. Whole populations and flourishing civilisations were wiped out by diseases like chickenpox introduced into virgin territories by the conquistadores.[21] In return, Columbus and his crew or, more probably, Antonio de Torres whose two expeditions to the New World returned to Spain between Columbus's first and second voyage, imported syphilis to the Old World.[22]

19. William Turner, quoted in Ellis, *Ancient Anodynes*, p. 56.
20. Mattioli died in 1577, aged seventy-seven.
21. The indigenous people of the New World were not living in a golden age but they had never suffered from many Eurasian illnesses and were therefore entirely without natural resistance. In the mid-fifteenth century the approximately 100,000 Guanche inhabitants of the Canary Islands were wiped out by an epidemic imported by the Portuguese. The inhabitants of Hispaniola were almost extinguished by a disease, perhaps swine influenza, introduced by Columbus' ships. In 1518 chickenpox killed at least a third of the inhabitants of Puerto Rico and Cuba. Either smallpox or chickenpox accompanied Hernan Cortés and his 300 Spaniards to Montezuma's Mexico where two-thirds of the Aztec inhabitants of the capital, Tenochtitlan, including Montezuma and his heirs, succumbed before the city fell. 'When the conquerors entered, a man could not put his foot down without stepping on the corpse of an Indian.' Similar events marked Pizarro's conquest of the Inca Empire in Peru. The epidemics not only killed vast numbers: they also extinguished all will to resist. They were only the first in a long series of genocidal germ onslaughts – measles, various kinds of influenza, typhus and later tuberculosis – unleashed by Europeans in the New World. Guns helped but it was mainly micro-organisms that achieved the conquerors' 'incredible feats of bravery' which became enshrined in European history books.
22. Columbus on his first voyage brought back only six 'Indians'. Antonio de Torres' first convoy in 1494 returned with 26 of both sexes, his second in 1495 with around 300. (Another 200 died on the journeys.) Many of the women of the second journey ended up as prostitutes in Naples. But the main transporters of the disease were the sailors and troops, greedy for more than the Natives' gold.

On its arrival in Europe the new disease was not the comparatively sluggish though ultimately fatal illness of later centuries. Nor was it divisible into well-defined stages separated by symptomless intervals of months or years. It was a terrifyingly explosive and painful affliction, the like of which had not been seen since the Black Death.[23] It made its first public appearance among the Flemish, Gascon, Swiss, Italian and Spanish mercenaries of Charles VIII of France who, in the last months of 1494, had invaded Italy. Militarily the campaign was a walk-over. Meeting hardly any resistance, the army entered Rome on the last day of the year and, on 22 February 1495, Naples. Even by the standards of the times, it was an ill-disciplined rabble bent on looting and raping. In its train swarmed a rag-bag troop of hucksters, beggars and prostitutes. Within a month of the French king's entry to Naples, riding a chariot drawn by four white horses and dressed as a Byzantine emperor, his position had become precarious. A Spanish army had landed in Sicily. Above all, his own troops were demoralised by the new disease which had sprung up among them. It received its first description after the battle of Fornovo by Cumano, a military surgeon attached to the Venetian troops who soon pursued the retreating French army.

> I saw several men at arms who, owing to a ferment of the humours such as nobody had seen before, were covered in horrible pustules on their faces and all their bodies but especially on their genitals. Some days after the eruption the sufferers would be driven to distraction by pains experienced in their arms, legs and feet. Their entire body became so repulsive and their suffering so great that many begged to be killed . . . Some of the victims later lost their eyes, hands, noses or feet.[24]

Though the view that America was the source of the 'pox' was endorsed by the great Fallopio, it was also soon questioned. The first objections were based on readings of the Bible: Guy Patin of the Sorbonne maintained that Solomon and Job had both been victims of the 'pox'. The dispute still rumbles on and is now centred on archaeological finds around the Mediterranean. In the present writer's inexpert opinion the American interpretation has most evidence in its favour. The disease may have been mild among Native Americans, a mirror-image scenario of chickenpox and other comparatively mild European fevers which devastated the population of the Americas.

23. One famous sufferer, Joseph Grünpeck, wrote: 'I have seen scourges, horrible sicknesses and many infirmities affect mankind . . . among them has crept in from the western shores of France a disease which is so painful, so distressing, so appalling that until now nothing so horrifying, nothing more terrible or disgusting has ever been known on this earth.'

    Ulrich von Hutten described 'the loathsome stench . . . the densely packed ulcers and sores as large as acorns, producing a repugnant secretion, the colour between green and black . . . the victims feeling as if surrounded by flames.' Quoted in C. Quétel, *The History of Syphilis*, translated by J. Braddock and B. Pike (Baltimore, 1990), pp. 35, 38.

24. Quoted in ibid., p. 45.

From Italy the 'French illness' spread to Switzerland, Germany, France, England and Scotland (there known as 'grandgor').[25] By 1496 it surfaced in Denmark, Sweden, Poland, Hungary and Russia. Everywhere it elicited an outpouring of hysterical edicts, emergency legislation, proclamations, attempts at segregation, expulsions, pamphlets and cartfuls of votive pictures. It also inspired an unprecedented number of learned monographs, some in verse, some in prose, some satirical and some wallowing in the horrors.[26] The importance of sexual contact was quickly recognised but the disease could also spread by kissing, from wet-nurses to sucklings or the other way round, and perhaps by clothing.[27] Astrological medicine did not end with the star-gazing High Middle Ages. The most widely believed explanation for the disaster was the malign constellation of the planets. The influence of Uranus was considered particularly baleful.

Almost as bad as the illness were the remedies which were recommended. Mercury had been used for centuries in Islamic medicine externally for diseases of the skin. Now, in insane doses, it was being administered by mouth, as enemas and even by instillations into the bladder. It was combined with heroic blood-lettings, starvation and debilitating 'sweatings' in specially constructed 'diaphoretic chambers'. The German poet, humanist and soldier,

25. In *L'Homme aux quarente écus* Voltaire gives one reason: 'In every town in France, Germany and Italy fine public establishments had been set up by royal command called b. . .s to preserve the honour of the ladies . . . The Spaniards carried the venom into these privileged establishments whence princes and bishops withdrew the girls they required. In Constance there were seven hundred and eighteen girls to serve the needs of the prince archbishop, his court and the town council which not long before had devoutly burnt John Hus and Jerome of Prague at the stake for the abominable sin of heresy.'

    The affliction went by many names, one nation usually attributing it to another. It was 'Christian disease' in Turkey and 'Portuguese disease' in India. In 1770 Captain Cook, exploring Tahiti, was mortified to learn that the natives called it *Apa no Britannia* or British disease. As Voltaire summed it up: 'the pox, like the fine arts, owes its origin to no particular race'. Quoted in ibid., p. 95.

26. The most famous was the Venetian Girolamo Fracastoro's pastoral poem, *Syphilis sive Morbus Gallicus* (Syphilis or the French Disease), published in 1530. In the poem a swineherd who acquires the disease is called Syphilis (from *sue*, hog, and *philein*, to love). Apart from the title the name crops up only once in the text; but Fracastoro gave a more detailed and astonishingly accurate description of the disease in his great work, *De contagione* published fifteen years later

    Fracastoro trained in Padua, where he was a friend and contemporary of Copernicus. His reference to 'tiny living creatures' transmitting diseases like syphilis was prophetic. He died in 1553, aged sixty-nine.

27. The renowned physician, Diaz de Isla of Barcelona, called the 'new pox' 'the serpentine illness'. 'Like the hideous serpent, it ulcerates and corrupts the flesh, breaks and destroys the bones, cuts and shrinks the tendons, destroys speech and sight and causes the most horrible pains.'

Ulrich von Hutten, endured the agonies of ten courses over eleven years.[28] He then tried the newly imported wonder drug, guiacum, an extract of Caribbean beech wood. After repeated infusions he believed himself to be cured. His book describing the happy outcome created a sensation. It became less widely known that the cure did not save him from dying a few years after publication at the age of thirty-four. The demand for the magic wood, the hardest known at the time, made the fortune of the Fugger family of Augsburg who imported and controlled its distribution. Their success kick-started Protestant capitalism.

...

There were other and more benign importations. In 1530 Nicolas Monardes of Seville devoted the first herbal and embryonic pharmacopoeia to the medicinal properties of plants from New World. Cocoa was introduced to Europe by Hernan Cortés in 1529, becoming a popular drink as well as a remedy for 'general aches' and 'wasting diseases' and, in the form of cocoa butter suppositories, an infallible dispeller of gloom. Also from the Americas came jalap, sarsaparilla and sassafras, all useful for 'cleansing the blood' as well as for relieving pain. They were to be eclipsed by the plant whose leaves Columbus and his men observed being harvested and dried within a few days of landing on the island they called Ferdinanda in October 1497. A few years later the Portuguese explorer Pedro Alvarez Cabral reported from Brazil that these leaves were used by the natives for treating painful abscesses, ulcers and many other ailments. The Spanish missionary priest Bernadino de Sahagún wrote that in Mexico the inhaling of the smoke of the burning leaves relieved even the most intractable headaches. Such reports aroused high expectations. They were not entirely disappointed. The plant, soon to be known as tobacco, was eventually introduced to Spain by the physician Francisco Hernandes de Toledo and quickly acquired the reputation as a universal panacea. Widely referred to as the 'holy herb' or 'God's remedy', it cured ulcers, abscesses, fistulas, sores and polyps and, above all, every manner of aches and pains. So wondrous were the results both of chewing the dried leaves and of inhaling their smoke that Monardes devoted a whole book to it, entitled *Joyful News out of the New-Found World.*

In 1560 the French ambassador to Portugal, Jean Nicot, was presented with a cutting and planted it in his garden. It 'grewe and multipluyed marvellously'.

28. Among other side-effects the treatment rotted the teeth and attacked the kidneys. The loss of body hair may have been caused by contaminant elements like thallium. Mercury never cured syphilis but was almost certainly the immediate cause of death of many famous sufferers (e.g. Franz Schubert). Applied as a thick ointment it gave symptomatic relief in syphilitic ulcers and continued to be used until the early 1900s.

One of Nicot's pages had recently developed a painful ulcer on his face which was beginning to 'to take root in the gristles of his nose'. Nicot applied leaves to the lesion and prescribed a course of inhalations of the smoke. He was a man of scientific rigour and requested a physician from the king's household to monitor the youth's progress. Both Nicot and the royal doctor were astounded when the pain ceased at once and the ulcer began to heal. A few weeks later Nicot heard that two ladies-in-waiting to the Queen Mother of France as well as several gentlemen at court were suffering from similar ulcers. At once he dispatched a large consignment of the miraculous plant. So generous was he and so gratifying were the effects that the plant became known as Nicotaine, later corrupted to nicotine. Thereafter tobacco in various forms was used to cure headaches, toothaches, bad breath, chilblains, worms, rheumatism, poisoned wounds, kidney stones, carbuncles and consumption. Despite wet blankets like King James I, it remained a popular medication till the late seventeenth century.[29]

. . .

Compared to tobacco and other newly imported and extravagantly promoted herbs, the discovery of ether, one day to become the first dependable general anaesthetic, passed almost unnoticed. Valerius Corbus, born in 1515 in Erfurt in Germany, was the son of an apothecary-doctor, Euricius Corbus. The father taught him the rudiments of his trade and then sent him to enrol at the University of Wittenberg, home of the Reformation. There Valerius forged a friendship with Lucas Cranach, another young apothecary who would later take up painting. Unusually for a socially unimportant individual, several references to young Corbus survive, including a letter from Cranach describing him as modest and kindly as well as an 'ingenious alchemist'. He wrote two short books on chemical synthesis, neither of them published in his lifetime, and compiled, still as a student, the first German pharmacopoeia. At the age of twenty-eight he set out for a study tour of Italy and is known to have arrived in Rome. There he died in obscure circumstances. His tomb in the church of St Aloysius disappeared when the church was demolished early in the seventeenth century.

It was in the first of his two books that Corbus described the distillate of a mixture of alcohol and sulphuric acid under the name *dulci vitrioli oleum*, or sweet oil of vitriol. He noted that it was volatile and floated on water. It was an excellent solvent of organic matter and had a characteristic odour; but he also found it 'oily'. The last unether-like property has raised doubts about the nature of the material; but his synthesis has been repeated and it is virtually

29. See the excellent article by A. Charlton, 'Medical Uses of Tobacco in History', *Journal of the Royal Society of Medicine*, 2004 (97), 292.

certain that he had discovered ether contaminated with diethyl sulphate.[30] Although his findings were not published till 1561 and his discovery never aroused the excitement of many other new concoctions, it never disappointed either. Diluted with wine, it was found helpful by both Robert Boyle and Isaac Newton against coughs and chest pain. It was renamed 'spiritus aetherus' by Frobenius in 1737; and a few decades later the German chemist Friedrich Hoffmann popularised it as Hoffmann's Anodyne. In this form it was recommended to the Empress Maria Theresa by her medical factotum, Gerhard van Swieten, as an 'excellent soother of the mind'. At least it helped the Empress to come to terms with the serial infidelities of her husband, Francis of Lorraine, who, in turn hailed it as a 'capital remedy for the infernal ache in my groin'. Nobody, it seems, as yet suggested that it might be useful as an anaesthetic in surgical operations.

30. See T. Robinson, 'On the Nature of Sweet Oil of Vitriol', *Journal of the History of Medicine*, 14 (1959), 1231.

# Chapter 12

# GOING TO WAR

None of the spectacular advances in anatomy and medicinal chemistry had an immediate effect on the practice of medicine and surgery. The life of people – at least of people of leisure – during the Renaissance and Reformation must have been more fun than life during the Middle Ages but it was no healthier, no longer and no less painful. This did not stop physicians from improving their status by founding colleges, elaborating complicated curricula, and inventing sonorous qualifications.[1] Surgery advanced not as a result of any startling new discovery but in response to more effective ways of maiming people. Gunshot and cannonballs caused terrible wounds and made frequent amputations on the battlefield or in the dark dank holds of storm-tossed vessels a necessity. 'Those who wish to become surgeons should go to war,' Hippocrates had advised his disciples; and all the important surgical texts of the sixteenth and seventeenth centuries deal mainly with the treatment of wounds sustained in battle. In England William Clowes, surgeon to St Bartholomew's Hospital, and Thomas Vicary, first master of the Barber Surgeons' Company, learnt their craft during military operations in the Low Countries. John Woodall's *The Surgeon's Mate* became the field surgeon's bible. Richard Wiseman, honoured in his lifetime as 'the father of English surgery', wrote a rival and morally more uplifting *Treatise of Wounds*, jocularly known in the Navy as 'Wiseman's Book of Martyrs'.[2] Both authors

1. John Caius (died in 1573, aged sixty-three) was an illustrious example of the breed. A protégé of Thomas Linacre, a founder of the College of Physicians and of medical lectureships in Oxford and Cambridge, he himself became president of the college for three terms and refounded his old college in Cambridge, Gonville Hall, as Gonville and Caius College. He became its first master. He did much to reconcile the Galenic tradition with new humanist learning, introducing formal courses of anatomy and formulating statutes and regulations.
2. It was especially addressed to naval doctors who 'seldom burden their cabin with many books'. Both Woodall (died in 1643, aged eighty-seven) and Wiseman (died in 1767, aged forty-seven) learnt their craft in wars, the former in the Low Countries, the latter in the Civil War.

recommended 'pain-killing vinous potions' before treating wounds; but Wiseman also advised that at least four able-bodied assistants should 'grasp the patient' during amputations. On the Continent Hans von Gersdorff's *Feldbuch der Wundartzney* became popular for its vivid (and gruesome) illustrations of amputations and trephining. He also showed drawings of the most useful instruments in field surgery, including two ingenious self-supporting devices for keeping an 'anaesthetic sponge' affixed to the patient's face. Hyeronymus Brunschwig in his *Buch der Wound Artzney* laid down the law that all gunshot wounds were poisonous and needed extensive cauterising.[3] He may also have introduced animal bladders as a useful dressing for amputation stumps.

Surpassing them all in fame even in his lifetime was a son of poor Breton farmers who worked his way up from humble barber-surgeon's apprentice to surgeon to four kings of France.[4] Apart from attending royalty, Ambroise Paré divided his long life between tending the sick at Paris' *hôtel-Dieu* and following the French armies in a long series of futile but bloody wars. In surgery his most important innovation was his rejection of the practice of treating gunshot wounds with boiling oil and his use of cautery before embarking on any restorative surgery. Like many such advances, it was forced on him by necessity while serving 'still as a fumbling tyro' in Francis I's ill-fated Italian campaign of 1537. After a particularly bloody skirmish he ran out of oil and had to apply an improvised salve of egg white, rose oil and turpentine to the wounded. The next night he 'slept bad, haunted by the thought that I would find all those I had treated thus dead'. He continues:

> So I got up early to visit them. To my great surprise, those treated with the salve felt little pain, showed no inflammation and swelling and had passed the night rather calmly, while those treated with seething oil were in high fever with pain, swelling and inflammation . . . At this I resolved never again cruelly to burn people who had suffered gunshot wounds.[5]

3. Gunpowder caused much more extensive wounds than swords and arrows and the wounds often became infected. Gersdorf died in 1529, aged seventy-three; Brunschwig died in 1533 aged eighty-three. The first English work specifically devoted to gunshot wounds, published in 1573, was *An Excellent Treatise of Wounds made with Gunneshotte* by Thomas Gale (died in 1587, aged seventy-three).
4. By his mid-thirties Paré (died in 1590, aged eighty) was famous enough to be called in to try to save the life of Henri II when the king was wounded in his head in a friendly joust. Even Paré could not save him, though he 'rehearsed' the operation on two convicted felons who were given a royal pardon in exchange for acting as guinea pigs. (It is not known whether they survived.) Paré wrote several books in a notably clear and readable style, including a treatise on gunshot wounds and one on podalic version. Famously he concluded his successful case histories with the sentence: 'I dressed him, God healed him.' Apart from Henri II, he was surgeon to the short-lived kings François II, Henri III and Charles IX.
5. A. Paré, *Oeuvres complètes*, Malgaigne edn (Geneva, 1970), Vol. II, p. 35.

The new treatment eased the pain of wound dressings; and both the cautery and hot oil gradually lost popularity. (They had never been popular with English surgeons.) This was not Paré's only innovation. The tourniquet to stanch bleeding had been invented before him – the term, derived from *tourner*, was originally applied to the screw or stick used to tighten a bandage – but in the second of his *Cinq livres de chirurgie* he advised that the band around the thigh or arm should be left in place for four or five minutes before the amputation of the leg or forearm. This was bold counsel at a time when speed was all-important; but the pressure may have been effective in numbing the sensory nerves. The idea had ancient roots and would be taken up again two centuries later in England.[6] Paré, an alert observer and vivid reporter, was also the first to describe a phenomenon for which 300 years later US Army surgeon Weir Mitchell would coin the phrase 'phantom limb'.[7] In his *Treatment of Gunshot Wounds* (1539) Paré wrote:

> Verily it is a thing wondrous strange and prodigious, which will scarcely be credited, unless by such as have seen with their eyes and heard with their ears the patients who have many months after the cutting away of a leg grievously complained that they felt exceeding great pain in that leg.[8]

A hundred years after Weir Mitchell the difficulty of explaining this pain in classical Cartesian terms was to become one of the starting-points of modern pain research.[9]

. . .

About dependable anaesthesia poets and dramatists continued to dream and fantasise, now no longer in learned Latin but in their own language. Chaucer paid tribute to opium in 'The Knight's Tale':

> Of a clarre, made of a certain wine,
> With Narcotickes and Opie of Thebes fine,
> That all the night through that man would him shake
> The gailer slept, he mighte not awake.

6. See Chapter 48.
7. Though Weir Mitchell gained valuable experience and distinguished himself in the American Civil War, he was more than an army surgeon. He wrote juvenile and historical fiction, a good deal of poetry and pioneering books on health care for a lay readership. His *Rest in the Treatment of Disease*, published in 1871, was particularly popular, its message not unlike that of John Hilton's *Rest and Pain* (1863). His memorable description of phantom-limb pain is in *Lessons on Nervous Diseases* (1895). He died in 1914, aged eighty-four.
8. Paré, in *Oeuvres complètes*, Vol. III, p. 127.
9. See Chapter 48.

Boccaccio in the *Decameron* mentions a soporific draught prepared by Mazzeo della Montagna of Salerno:

> It happened that there came to the attention of the physician a patient with gangrenous leg, and when the master had made an examination he told the relatives that unless a decayed bone in the leg were removed either the entire leg would have to be amputated or the patient would die. If the bone were removed, the patient might recover . . . but the doctor was of the opinion that without opiate the man could not endure the pain. The affair was set for the evening . . . Before then he distilled a kind of drink after his own composition which had the faculty of bringing to the person who drank it sleep for as long a time as was deemed necessary to complete the operation.[10]

Shakespeare has Friar Laurence tell Juliet about the drink which will lull her into a sleep indistinguishable from death but from which she will awake after forty hours. So did Shakespeare's forerunner, Arthur Brooke, in his less familiar *Tragicall Historye of Romeus and Iuliett.* Apart from the language, the earlier verse has the more modern anaesthetic resonance.

> It does in halfe an howre / astonne the taker so,
> And mastreth all his senses that / he feeleth weale nor woe:
> And so it burieth up / the sprite and living breath,
> That even the skilful leeche would say / that he is slayne by death.
> One vertue more it hath / as meruellous as this;
> The taker by receiving it, / at all but greeved is;
> But painlesse as a man / that thinketh nought at all,
> Into a swete and quiet slepe / immediately doth fall;
> From which (according to / the quantitie he taketh);
> Longer or shorter is the time / before the sleeper waketh;
> And thence (the effect once wrought) / agayne it does restore
> Him that receaved unto the state / wherein he was before.

Sadly, such potions rarely worked off the stage. In practice a bewildering range of concoctions was recommended, shamelessly touted by some and fiercely condemned by others. Though still used in a variety of forms, mandragora was beginning to lose popularity. Paré wrote:

10. The climax of the tale was, in true Boccaccio fashion, not the operation but the consequences of an amorous gentleman downing the soporific drink by mistake and disgracing himself as a consequence when finally bedding his lady love. It is one of the *Decameron*'s better tales.

> Though it lulls the senses, taken in excessive quantity it is poisonous, both root and fruit . . . Doctors used it much formerly when they wanted to cut or burn a member but I regard it as dangerous.[11]

To replace it, he devised a cloth soaked in oil of roses, water-lilies, poppies, rose vinegar and opium. He also gave patients flowers of hyoscyamus to smell. The opium probably had the desired effect. But every surgical author had his own recommended infallible mixture. Except for their variable alcohol and opium content, their number and diversity suggest that none was effective. It has been claimed that the first English reference to an anaesthetic agent used in a common surgical procedure was in William Bullein's *Bulwarke of Defence against all Sickness*, published in 1562.

> The juice of this herbe pressed forthe, and kepte in a close earthen vessel, accordying to arte; this bryngeth slepe, it casteth men into a trauns on a deep terrible dream, until he be cutte on the stone . . . he will feel no paine.[12]

But what was the 'herbe'? Probably it was a variant of mandragora.

Surgical operations remained a terrible ordeal. Hyeronimus Fabricius ab Aquapendente, Vesalius' successor in Padua, left a graphic description of an amputation:

> I was about to cut the thigh of a man of forty years of age, and ready to use the saw and cauteries. For the sick man no sooner began to roare out, but all ranne away, except only my eldest Sonne, who was then but little, and to whom I had committed the holding of the thigh, for forme only; and but that my wife then great with child, came running out of the next chamber, and kept hold of the Patient's Chest, both he and myself had been in extreme danger.[13]

Paré was a humane as well as a skilful surgeon and he used opium- and alcohol-based potions lavishly, but his prescription for the successful surgeon speaks for itself:

> A Chirurgeon must have a strong, stable and intrepid hand, and a minde resolute and mercilesse, so that to heale him he taketh in hand, he be not moved to make more haste than the thing requires; or to cut lesse than is needful; but which doth all things as if he were nothing affected with their cries; not giving heed to the judgement of the common people, who speake ill of chirurgions because of their ignorance.[14]

11. Paré, *Oeuvres complètes*, Vol. IV, p. 423.
12. Quoted in Robinson, *Victory over Pain*, p. 65.
13. Quoted in Porter, *The Greatest Benefit to Mankind*, p. 283.
14. Paré, *Oeuvres complètes*, Vol. II, p. 342.

Yet he was perhaps the first surgical writer to emphasise the need to gain the cooperation of his patients.

> The indications of the patient's state of mind, determination and strength must take precedence over everything else. If he is weak or in terror, it is necessary to forsake all other things in order to be helpful to him. If the patient lacks the necessary strength of mind, operations should be postponed – if possible. Nothing can be gained from surgery if the patient is unwilling to face his ordeal.[15]

Unsurprisingly perhaps, most people put up with a great deal of suffering rather than submit to the surgeon's knife. Michel de Montaigne, supreme explorer of the human mind, was among them. A victim of the common ailment of bladder stones, he wrote:

> I am in the grip of one of the worst diseases – painful, dreadful and incurable except by surgery . . . This I will not have . . . For at least I have one advantage over the stone. It will gradually reconcile me to what I have always been loath to accept – the inevitable end. The more the disease presses and importunes me, the less I will fear to die.[16]

Yet cutting for stone was, by the end of the sixteenth century, one of the most frequently performed operations; and contemporary accounts suggest that experienced surgeons performed it with extraordinary skill. Montaigne was a rich aristocrat who could afford the best; but he chose to endure the agony instead.[17]

. . .

15. Paré, *Oeuvres complètes*, Vol. IV, p. 34.
16. M. de Montaigne, *Essais*, ed. B.Villey (Paris, 1926), p. 245.
17. Bladder stones were among the diseases vividly described by Paré: 'One will know the stones in the bladder by these signs: the patient feels a heavy weight at the seat and perineum, with repeated shooting pain which can extend as far as the tip of the penis, so much so that the patient keeps pulling on it and rubbing it; and though he does his best to empty the bladder, for the pain causes a great need to urinate, he cannot do so freely and sometimes manages only a slow and painful dribble . . . In urinating he often feels extreme pain, crossing his legs and crouching to the ground, crying, moaning and gripping anything to hand' (Paré, *Oeuvres complètes*, Vol. V, p. 76).

    Montaigne's fear was not universally shared or always justified. Shortly before starting his diary, Samuel Pepys was cut for stone by Thomas Hollyer, surgeon to St Thomas's Hospital. The stone emerged intact, 'as big as a tennis ball'. The proud and fortunate survivor had it set in a handsome glass-fronted case which cost him 25 shillings. 'This day,' he recorded on 26 March 1659, 'it is two years since it pleased God that I was cut for the stone at Mrs Turner's in Salisbury Court. And did resolve while I live to keep it a festival, as I did the last year at my house.' He never suffered from the stone after that.

Most people, even kings, had no choice. Philip II of Spain was Montaigne's almost exact contemporary; but in character the two men were opposites. The Frenchman was a sceptical and humane spectator. Philip was a bigoted monarch who, after the annexation of Portugal, personally administered – or tried to – the most extensive empire the world had ever seen. But both were victims of a painful chronic disease The exact nature of the king's illness remains a mystery.[18] Historians abhor a vacuum and even his scholarly biographers call it 'gout'. This is one of the few pathologies which can be confidently excluded. But he almost certainly had diabetes, some form of neuropathy and arthropathy and what today might be described as an 'autoimmune' disease. After middle age, movements in all his joints became increasingly painful. His skin became exquisitely tender and sensitive. Eventually he had to spend most of his days strapped to a padded movable frame designed for him by a French engineer, which was capable of being lowered or raised between the upright and the horizontal. His doctors treated him with an assortment of painkillers, including poppy juice, none of which had much effect. Ointments and special diets, one based on goat's milk and another on exotic fruit, were tried but to no avail. The Pope sent his personal physician – a Portuguese Jew whose identity and faith had to be kept a secret – but the doctor declared himself baffled. He nevertheless recommended 'diaphoretics', which drenched the king in sweat. The treatment only compounded the suffering.

Despite his physical limitations, Philip continued to deal with his mountainous files, attending to every detail of his paper domains. His relaxations were few. He would take time off to gaze at the latest Titian dispatched by the aged artist from Venice. He lovingly tended his roses. Most restfully perhaps, he presided over the *autos-da-fé* arranged by the Holy Inquisition. Over a hundred heretics might be solemnly burnt on those occasions.

In June 1598 he set out for his last journey from Madrid to the Escorial, the palace/crypt/church/monastery/reliquary which he had built for himself and his dynasty. The travelling which in the past had taken three hours on horseback now became a painful progression in a specially designed litter and lasted six days. A bout of high fever on arrival signalled the beginning of the end. But the passing lasted almost another three months.

> Sores and boils on his legs and feet had to be lanced. A huge festering abscess also developed below his right knee and had to be opened. Pus poured from it, filling two basins every day. Bedsores erupted all along his

18. Michel de Montaigne's dates were 1532–92, Philip II's 1527–98. Philip's illness is one of the best documented in terms of symptoms and signs; but characteristically of most ailments of the past, a firm retrospective diagnosis is almost impossible.

back: he could no longer move. The worst torment was his diarrhoea which developed after three weeks. Because of the pain caused by being touched or moved, it seemed best not to clean the ordure that he produced and not even to change the bed linen. So many times the bed remained fouled, creating an awful stench . . . Eventually a hole was cut into the mattress . . . but it was only a partial remedy. Despite his debased condition, the king continued to conduct official business, spoke repeatedly to his son and heir [the future Philip III] and witnessed the consecration of the new archbishop of Toledo. He never complained but prayed almost continuously.[19]

When the king was advised to receive the last rites all pain-killing potions and fumigations had to be stopped. An unclouded mind was required, however severe the suffering. He took three days and three nights to complete his confession, 'dwelling scrupulously on every detail of his life'. A complex ritual was then enacted, including the blessing of his coffin and the positioning of an assortment of holy relics around his bed. Bishops, archbishops and the friars of St Lawrence kept continuous vigil. Six doctors never left his side, 'attending to his ailments'. It is not clear what they actually did. However severe the king's suffering, they were instructed by the king's confessor to treat him as best they could, hoping and praying for a recovery. The end must under no circumstances be hastened. At last, on 13 September the king asked for a crucifix and a candle. They were placed in his hands. He died a few minutes later.

19. The king's illness and death are described in detail by Carlos M.N. Eire in *From Madrid to Purgatory: The Art and Craft of Dying in Sixteenth-Century Spain*, an interesting and scholarly work which also deals with the illness and death of St Teresa of Avila (Cambridge, 1995).

# Chapter 13

# FOUNDATIONS

Surgical anaesthesia was the achievement of the nineteenth century; but its foundations were laid by the scientific revolution two hundred years earlier. Two physiological systems in particular held the key to future advances. First, the movements of the blood had to be understood. Arteries and veins were demonstrably different in their structure. What was the significance of the difference? How did blood move from the right to the left side of the heart? Or did it? What happened to substances injected into a peripheral vein? Could they be carried to other organs, especially the brain? And second, respiration was vital in every sense of the word. It was also mysterious. Inspired air seemed to undergo a critical change in the body. What was the mechanism of that change? What was its chemical nature? Why was it essential? Could pain-killing agents be inhaled?

The scientific revolution did not come unheralded. When, in 1600, young William Harvey enrolled as a medical student in the University of Padua, key elements of his discovery were already talked about.[1] Forty years earlier Fabricius had succeeded Gabriello Fallopio as professor of anatomy. He had set out to answer a question that had stubbornly eluded his predecessors.[2] How did blood remain more or less evenly distributed in the living instead of settling by gravity in the most dependent parts? (Blood did settle in the most dependent parts, but only after death.) Since Fabricius could not rid himself of the Galenic view that blood was made in and distributed by the liver, he

1. Harvey was born in Folkestone, Kent, into a prosperous yeoman family and studied medicine at Caius College, Cambridge, before setting out on a continental tour. After travelling in the Low Countries and Germany, he proceeded to Padua, the most famous medical school in Europe at the time. There he attended Fabricius' lectures and demonstrations, performing dissections himself. He returned to London in 1602.
2. Fabricius (died in 1618, aged eighty-two) was a celebrated surgeon as well as an anatomist. He had succeeded Gabriello Fallopio (died in 1563, aged forty), commemorated in the Fallopian or uterine tubes. Fallopio had succeeded Vesalius, whose favourite pupil he had been.

was doomed to misinterpret his own findings; but, dissecting the veins, he stumbled on the all-important presence of the valves. In a short book, *De venarum ostiolis* (About the venous valves), he correctly argued that, situated at regular intervals but especially at the confluence of two veins, their role had to be to prevent blood flowing 'backwards'. But what was 'backwards'? And why were valves absent from the arteries? To him his findings merely compounded the conundrum.[3]

Another prophetic discovery Harvey must have encountered concerned the pulmonary circulation. To explain how dark venous blood on the right side of the heart became bright arterial blood on the left side Galen had postulated the presence of invisible 'pores' in the cardiac septum, the muscular membrane separating the two chambers on the right from the two chambers on the left. (It may be recalled that both sides of the heart contain two communicating chambers, the auricles which receive incoming blood and the ventricles which expel blood.) Nobody had ever seen these pores and attempts in the isolated heart to squeeze liquid from one side into the other had almost always failed; but the doctrine, like the blood-forming role of the liver, became an entrenched part of Galenic wisdom.[4] The man who un-entrenched it was a strange genius, half theologian and half doctor, whose capacity to antagonise both sides of the religious divide and irritate every academic establishment with whom he came into contact still amazes.

Michael Servetus, to call him by the best known of his many names, was born in Tudela in Spain in 1511. A devout Christian, he was, like Luther, repelled by the spiritual degradations of the Renaissance Church; and his main objective always remained to reform the Faith rather than to rewrite anatomy. In reinterpreting the Gospels, he went further than either Luther or Calvin, developing subtle but clearly heretical views about the Trinity. Trying perhaps to deflect him from entering dangerous waters, the kindly medical humanist, Symphorien Champier of Lyons advised him to immerse himself in medicine. Medicine was not yet the safe, respectable and slightly dowdy profession it became later but a pursuit bubbling with subversive ideas. Servetus duly enrolled as a medical student in Paris. Here he provoked both admiration and outbursts of fury. Unlike Paracelsus who tended to direct his venom against the safely dead, Servetus preferred to confront the living. There was always much in medical teaching to invite confrontation – or to deter the cautious. Servetus was not cautious. Outrage eventually forced him to leave

3. They were incompatible with the Galenic vision of ebb and flow movement.
4. The reason why the squeezing through did not *always* fail was probably the potential opening in the cardiac septum which is normally closed but which can be forced open. Abnormal communication between the two sides through this opening is one cause of 'blue babies'.

France. For a time he settled in Vienna. There, in 1553 he anonymously published his chief work, the 700-page *Christianismi restitutio*. In the book he argued against the Trinity on the ground that there was no evidence for it in the Bible. It was in the framework of this essentially theological treatise that he formulated his prophetic physiological doctrine.

The Bible taught (or could be interpreted as teaching) that the blood was the seat of the soul and that the soul was breathed into man by God. It followed that there had to be an interface between air, the vehicle of God's breath, and blood. This, Servetus rightly guessed, must be the lungs. There were several observations which lent support to his idea. Most persuasively, the pulmonary vessels connecting the heart and the lungs were too big to be concerned solely with the blood supply of two relatively small organs. They were in fact designed, Servetus suggested, to ensure that the entire blood volume should regularly pass through the lungs where air and blood – that is, God's breath and the human soul – could mingle. Though mistaken in detail, this was a brilliant insight and the future cornerstone of both respiratory physiology and anaesthesia by inhalation.

It was not seen as such by Servetus' contemporaries. While passing through Lyons and after delivering his usual incendiary sermon about the Trinity, he was arrested by the Inquisition, summarily tried and sentenced to death. He escaped from prison – or was allowed to escape, for he had many admirers – and made his way to Geneva. He and Calvin had formed a pleasantly disputatious friendship in their student days and Servetus firmly believed that the great reformer was longing for another verbal joust. Like many powerful intellects, in practical matters Servetus was a fool. Or perhaps his timing was unfortunate. Though firmly established as ruler of a modern theocracy, Calvin was in the throes of a medical crisis, almost certainly the result of strangulated haemorrhoids. He was in no mood for theological hair-splitting. Servetus was put in chains, probably tortured and then burnt at the stake.[5] But that was not the end. As he himself proclaimed, it was only his body which perished. Though all available copies of his book had been consigned to the flames both in Catholic and in Protestant countries, thousands had been printed. There can be little doubt that its message about the pulmonary circulation was well known in Padua during Harvey's sojourn there.

. . .

5. The crime remained a blot on Calvin's record (as recognised by the modest and well-hidden stone erected in Servetus' memory by the municipality of Geneva three hundred years later); but it should be added that Servetus was the only heretic burnt under Calvin's rule.

After his continental tour Harvey returned to London and established himself as a successful practitioner. His marriage to the daughter of the royal physician, Lancelot Browne, was a sound career move. In time he himself became medical adviser first to James I and then to Charles I.[6] He was also appointed physician to St Bartholomew's Hospital and Lumleyan Lecturer of the Royal College of Physicians, charged with lecturing on anatomy and with public dissections. He was a short, swarthy man who habitually carried a dagger and had a tongue to match; but he was rightly trusted by his friends and royal masters. His position at Court gave him unlimited access to a wide range of experimental animals, from hares and snakes to deer and ostriches, both dead and alive. He made good use of the facilities.

Harvey's iconic status in English medicine makes it difficult to assess his achievement. His deductions were based on slender evidence and his measurements were widely wrong. His experimental work would be regarded as woefully inadequate by any respectable medical journal today.[7] Yet his reasoning was faultless and his flashes of insight were glorious. In all essentials his scheme of the circulation of the blood with the heart acting as an extraordinarily efficient pump as set out in his book, *De motu cordis et sanguinis in animalibus* (The movement of the heart and blood in animals) was correct. He rightly argued that there had to be two circulations, not just one. With each contraction the right ventricle pumped dark deoxygenated blood into the lungs and the left ventricle pumped bright red oxygenated blood into the rest of the body. With each relaxation deoxygenated blood from the periphery flowed into the right auricle and oxygenated blood from the lungs flowed into the left auricle. The force of the heartbeat ensured that blood in the arteries would flow away from the heart, while the venous valves ensured that blood in the veins would flow towards the heart. Most momentous discoveries in medicine take time to be accepted. The beauty of Harvey's concept was recognised almost at once.[8]

6. His marriage was not a love match but a happy union. He remained loyal to his royal master throughout the ups and downs of the Civil War and was fined £2,000 and banished from London for his Royalist sympathies by the Commonwealth after Charles' defeat and beheading. He then lived in retirement until his death in 1657, aged seventy-nine.
7. He greatly overestimated the total volume of blood circulating. He was in fact in many respects a traditional Aristotelian, a believer in vital forces, and not in the body as a machine.
8. The adulation started shortly after his death with a ghastly celebratory ode, commemorative orations and medals and has continued unabated since. Even his famously illegible handwriting has become the affectation of young doctors possessing a perfectly well-coordinated script and only the dimmest idea of who Harvey was. Of course, like all great minds, he was wrong about many unimportant details and stuck to his guns over his mistakes as well as over his insights; but there can be no doubt that he was (with Laënnec, Semmelweis, Pasteur, Lister and Koch) one of the great creative thinkers of the medical sciences.

But there was one missing link. How did vessels carrying deoxygenated blood communicate with vessels carrying oxygenated blood either in the lungs or in the tissues? However numerous and careful the dissections and vivisection – like all other contemporary scientists Harvey regarded experiments on live animals as essential – no direct communicating channels could be demonstrated between the arterial and the venous systems. It is indeed remarkable that the absence of such channels did not prevent Harvey from coming to the right conclusions and from sticking to his guns when the gap in his scheme was pointed out. Two developments a few decades after Harvey's death provided the explanation.

In the spring of 1673 Antoni van Leeuwenhoek, a draper, minor city official and amateur lens grinder in the prosperous Dutch city of Delft addressed the first of many long letters to the Royal Society of London.[9] Delft had been known then for its 'Chinese' tin-glazed tiles rather than as a hub of science; but that would soon change. The letters were beautifully illustrated but in Dutch, the only language their author spoke. Fortunately, Henry Oldenburg, secretary to the Society, had been born in Germany and was a linguist; and he acted as a translator. In his letters Leeuwenhoek described what was in effect the first serviceable microscope, a single-lens instrument of superb craftsmanship.[10] As he peered through its lens a hitherto unknown and

He was also an accomplished stylist (as were many great doctors of the past), an attribute for which he is not always given credit. To quote just one memorable passage (whose pleasing discursiveness would bar him from any learned journal today):

> So in all likelihood it comes to pass that all the parts of the body are nourished, cherished and quickened with blood, which is warm, perfect, vaporous, full of spirit, and, that I may say so, alimentative: in the parts the blood is cooled, darkened and as it were made barren, from thence it returns to the heart, as to the fountain or dwelling house of the body, to recover its perfection, and there again by natural heat, powerful and vehement, it is melted and dispensed again through the body from thence, being fraught with spirits as with balsam, and that all the things do depend upon the motional pulsation of the heart. So the heart is the beginning of life, the Sun, the Microcosm, as proportionable the Sun deserves to be call'd the heart of the world . . . [from the 1653 translation]

9. The Royal Society of London for Improving Natural Knowledge, to give the body its full title, grew from an informal gathering of scientific enthusiasts in the 1640s to 1660 when the first minutes were recorded and 1662 when King Charles II, who had previously become a member, granted the Society its first charter. Christopher Wren and Robert Boyle were among the founding members. Newton was elected fellow in 1671 and was president from 1703 till his death in 1727.
10. All the 247 microscopes Leeuwenhoek made during his long life were of the single-lens variety, the secret of their magnification being their small size (often less than 2 centimetres) and their craftsmanship. The single lens minimised the problems of chromatic and spherical aberration, inescapable with the two-lens instrument. The strongest of his extant microscopes (lovingly preserved in the Utrecht University Museum) had an amazing linear magnification of 266 and a resolving power of 1.35x, allowing two features separated by only 0.00135 millimetres to be distinguished.

extraordinary world was revealed to him, a world teeming with a seemingly infinite variety of previously invisible 'animalcules'. The sight made him an insatiable voyeur. (Anybody who can recall the thrill of looking down a microscope for the first time at an apparently lifeless drop of pond water may understand his obsession.) During his long life the Dutchman studied the texture of plants, the structure of animal bodies, blood cells flowing in vessels, the crystals of uric acid from painful gouty deposits, muscle fibres, scrapings from gums, teeth and hair, 67 species of insects, 10 species of crustacea, eight different kinds of spider and uncounted spirogyra, protozoa, bacteria, infusoria and human spermatozoa.[11] He estimated that a single drop of 'pure' pond water contained more than a million different kinds of living creature. And most immediately relevant to Harvey's idea of the circulation, he noted minute channels, barely big enough to accommodate a single blood cell, widely distributed in both animal and human tissues.[12]

The last discovery was taken up with zest by Marcello Malpighi, professor of medicine in the University of Pisa.[13] In contrast to Leeuwenhoek who was essentially an observer, Malpighi was an interpreter. He had already made important discoveries when examining the microscopic structure of the skin, the kidneys, the brain, the tongue and the testis. Now he turned his attention to the lungs. In a single paragraph of a short but splendid book, *De pulmonibus* (About the lungs), he provided the clinching link to Harvey's hypothesis:

> I saw the blood flowing in minute streams through the arteries . . . and I might have believed that the blood itself escaped into an empty space and was collected up again by a gaping vein; but an objection to the view was afforded by the movement of the blood in a tortuous path and in different directions, and by it being collected again in a definite path. My doubt was changed to certainty by the dried lung of a frog which, to a marked degree, had preserved the redness of the blood in minute tracts which were afterwards found to be tiny channels . . . Thus it was clear that blood flowed along small sinuous vessels and not empty into open spaces . . .[14]

11. The list is far from complete.
12. Leeuwenhoek was elected foreign corresponding member of the Royal Society but never travelled to London in person. He died in 1723, aged ninety-one, active almost to the last. He firmly believed that the study of nature would eventually demonstrate the existence of an All-wise Creator and explode atheistic ideas of spontaneous generation.
13. Malpighi was a native of Bologna and eventually returned there. He was a prolific investigator and microscopist and a kindly and generous person. Numerous microscopic structures are named after him, including the basal layer of cells of the skin and the microscopic filtering units in the kidney. He was elected a fellow of the Royal Society of London, which published his collected works. He died in 1694, aged seventy-two.
14. M. Malpighi, *De pulmone* (Pisa, 1654), p. 87.

What Malpighi was describing were the capillaries, invisible to the naked eye but revealed under the two-lens microscope. His observations were soon confirmed by an international coterie dedicated to the wondrous new instrument.[15]

...

The only immediate practical result of Harvey's discovery was to make the embalming of dead bodies easier for undertakers; but, within a generation, the fact that blood moved in a circle (or two circles, to be pedantic) suggested all kinds of exciting possibilities. At least the possibilities excited a group of scientists in what was scientifically as well as politically the most excitable age in English history. Among them was young Christopher Wren, recently expelled from Cambridge for his Royalist utterances, and setting out on a new career in Oxford; the Honourable Robert Boyle, rich Irish aristocrat who at the time maintained a private laboratory in the city; Robert Hooke, a difficult character but an experimental wizard; and Robert Lower, a young Cornish doctor who had as a youth worked as an assistant to the pioneer neuro-anatomist, Thomas Willis.[16] One day in October 1667 they and friends gathered in Boyle's laboratory over a chemist's shop on the High Street, overlooking Brasenose College. The gathering was to witness an experiment Wren was to conduct on a large dog. The occasion was later described by Boyle.

> His [Wren's] Way (which is much better learn'd by Sight than Relation) was briefly this: First, to make a small and opportune Incision over that part of the hind Leg, where the larger Vessels that carry the Blood are most easy to be taken hold of: then to make a Ligature upon those Vessels, and to apply a certain small Plate of Brass . . . almost of the Shape and Bigness of the Nail of a Man's Thumb . . . [with] four litle Holes in the Sides, that by a Thread pass'd through them, it might well be fasten'd to the Vessel . . . This Plate being fastened on, he made a Slit along the Vein from the Ligature towards the Heart, great enough to put in the slender Pipe of a Syringe; by which I

15. By the end of the century capillaries had been observed and described in most organs.
16. At the time of the intravenous opium experiment Wren (died in 1723, aged eighty-nine) was Savillian professor of astronomy in Oxford: his architectural career was still ahead of him. Boyle (died in 1791, aged sixty-four), seventh son of the first Earl of Cork, had been a founding fellow of the Royal Society. Robert Hooke (died in 1703, aged sixty-eight) already had several ingenious mechanical inventions to his credit and had become secretary of the Royal Society. In contrast to Boyle, numerous portraits of whom exist, no likeness of Hooke survives. Richard Lower (died in 1691, aged sixty) had already written a book on the circulation and was to play the principal role in the first blood transfusion experiment. Thomas Willis (died in 1675, aged fifty-five) had also been a founding fellow of the Royal Society and had written a ground-breaking work on the anatomy of the brain. The circle of arteries on the underside of the brain is named after him (See also Chapter 31, note 2, and C. Zimmer, *Soul Made Flesh* (London, 2005).

> had proposed to have injected a warm Solution of *Opium* in Sack, that the Effect of our Experiment might be the more quick and manifest.[17]

After the 'dexterous Experimenter' had injected the opium in wine, the drug was quickly carried

> by the circular Motion of the Blood to the Brain, so that we had scarce untied the Dog (whose four Feet it had been requisite to fasten very strongly to the four corners of a Table) before the *Opium* began to disclose its *Narcotic* Quality, and almost as soon as he was on his Feet, he began to nod with his Head and faulter and reel in his Pace, and presently after appeared to be stupefied, that there were Wagers offered his Life could not be saved. But I, that was willing to reserve him to be whipped up and down a neighbouring Garden, whereby being kept awake, and in Motion, after some Time he began to come to himself again; and being led home, and carefully tended, he not only recovered, but began to grow fat so manifestly that 'twas admired.[18]

It was the first intravenous anaesthesia in the history of pain relief.

...

Other similar experiments were less successful. Some were horribly botched.[19] Samuel Pepys accompanied the Duke of York (later James II) to the Royal Society to see a demonstration of putting a dog to sleep by pumping opium into a vein in its hind leg, using a quill. He recorded that

17. Quoted in L. Jardine, *Ingenious Pursuits* (London, 1999), p. 119.
18. *Ibid.*
19. Of the 'similar experiments' the most notable and most disastrous were the first attempts at blood transfusion from lamb (or sheep) to man, first in France, then in Oxford. In France the experimenter was Jean Baptiste Denis and his first try may have been 'successful' in that the patient, a mad and delinquent young man, did not immediately die. However, it is unlikely that any lamb's blood actually entered the recipient's circulation. His second patient died after the second transfusion and the doctor had to stand trial for murder (he was acquitted). Not to be outdone by the French, the Royal Society of London dispatched three fellows to Bedlam, the London asylum, to find a patient whose mental health might be restored by a transfusion. Much to his credit (but to the dismay of the Royal Society delegation), the physician in charge, a Dr John Selwyn, refused to cooperate. Eventually Dr Richard Lower, leader of the Royal Society delegations and his contemporaries found a deranged young man, Arthur Coga (a graduate in divinity from Oxford, presumably prior to his derangement); and, on 23 November and then again on 14 December 1667, attempted to transfuse him with lamb's blood. One can only hope in retrospect that the transfusion was a failure: unlike the French victim, Antoine Mauroy, Coga survived. Blood transfusion into humans was eventually banned by law both in France and in England. The events are vividly described by P. Moore in *Blood and Justice* (London, 2003).

> Mr Pierce the surgeon and Dr Clerke did fail mightily at hitting the vein, and in effect did not do the business after many trials but with the little they got in, the dog did presently fall asleep and so lay quietly till we cut him up.[20]

Unsurprisingly perhaps, as a clinical method intravenous injections did not catch on for another two hundred years;[21] but nor were the early trials entirely forgotten.

...

With the circulation more or less mapped out, Boyle could turn his attention to respiration. More precisely, he could cause the attention of Robert Hooke to turn to it. Even among the varied talents which made up the young Royal Society, the two provided a striking and providential contrast. Boyle was a keen formulator of good questions, hypotheses and ideas but lacked the manual skills to build and handle instruments. He also hated the gore of animal experiments. Perhaps he actually *liked* the dogs, calves and lambs which had to be sacrificed in the sacred or at any rate fascinating cause of science. Hooke, by contrast, was not a sprig of the privileged classes and, unlike the emollient Boyle, was abrasive when he felt (with good reason, as a rule) that he was being taken for granted. On the other hand, he had an uncanny gift for designing new and ingenious apparatus and for carrying out complicated experiments dreamed up by less skilful operators.[22] At Boyle's instigation he now developed his air pump or 'pneumatic engine' which allowed air to be evacuated or introduced into a container or, through a tube, straight into an animal's windpipe. John Evelyn, the diarist, who attended one of the semi-public sessions at the Royal Society described a display put on by Boyle and Hooke:

> We first put in a snake but could not kill it by exhausting the air, only made it extremely sick; but the chick died of convulsions in a short space of time after the air was pumped out.[23]

20. S. Pepys, *Diaries*, transcribed and edited by R. Latham and W. Matthews (1970–83), in 11 volumes. Vol. V, p. 342.
21. Intravenous opium experiments were taken up on the Continent and the method was studied a few years after the experiment in Oxford by Johannes Sigmund Elsholtz, physician-in-ordinary to the Elector of Brandenburg; but there too it fell into disuse by the end of the century.
22. Beside the double-barrelled air pump, he invented the spirit level, the aerometer, the marine barometer, the metronome, the balance spring of watches, the anchor-escapement of clocks and a sea gauge. He was an accomplished microscopist and may have anticipated Newton's laws of gravity and celestial motions. The two men famously quarrelled over priority (among other issues).
23. Quoted by Jardine in *Ingenious Pursuits*, p. 113.

Having established that breathing air in and out was essential for survival, the next question was, why? More immediately, what happened to the air inside the body? After an impressive number of assorted animals had been sacrificed in the air pump (with fellows of the Royal Society solemnly timing the duration of their suffocation), the time had come to dissect live and breathing creatures.

As usual, it was Hooke who did the cutting while Boyle observed and recorded. Dogs were in fact the 'brutes' nearest to the experimentalist's heart and Hooke vowed not to repeat the experiments without 'stupefying the animal, because it was cruel'. However, having opened a dog's chest, cut all the ribs and opened its belly,

> I was able to fill his lungs with a pair of bellows and suffered them to empty again. I was able to preserve the dog alive as long as I could desire by intermittently pumping air into him; and the heart kept on beating for a very long time; but keeping the lungs empty with the air pump caused the dog to expire.[24]

Despite Hooke's resolve not to repeat the experiment, when the next two were mishandled by clumsy operators, causing even more canine suffering, he relented and conducted a few more public demonstrations. By the end of these the role the lungs played in 'revivifying' the blood in some way by exposing it to inspired air was established.

. . .

Ideas and basic concepts in pain relief have always been as important as experimental observations; and the nature of pain was being reinterpreted about the same time as the first experiments establishing the mechanisms of circulation and respiration. The chief reinterpreter, René Descartes, was born in 1596 in La Haye, a small town in Touraine, into a modestly prosperous family. Recognised as a bright boy, he received a good education at the famous Jesuit College of La Flèche, including a grounding in mathematics and physics.[25] His career choice then rested between the Church and the army; and, as an agnostic, he chose to become a gentleman soldier. He served under William of Orange in the Netherlands and then in the Catholic forces of the Elector of Bavaria. In an age of mercenary armies the fact that the two rulers were on opposing sides was no bar to the change of employment. Happily he saw no action but visited many lands and met interesting people, both ordinary and

24. Quoted by Jardine in *Ingenious Pursuits*, p. 231.
25. La Flèche was founded by Henri IV and became one of the most prestigious schools in France. The Jesuits were particularly advanced in teaching science, including the new and heretical discoveries of Galileo.

extraordinary. While in winter quarters in the pretty garrison town of Neuberg on the Danube he had time for reflection. A quasi-mystical experience induced him to embark on his first and greatest book, the *Discours de la méthode.* After its warm reception by established academics he decided to devote himself full time to thinking and writing. He settled in Amsterdam, an island of comparative sanity in a Europe divided by religious frenzy.[26] It was during his sojourn there that Frans Hals painted the portrait by which he is known today.[27]

Installed in modest lodgings near the butchers' quarter, Descartes had many opportunities to examine the inside of animal carcasses and to carry out dissections himself. Whether he dissected human bodies is uncertain. He never married but fathered a much-loved daughter, Françine, whose death at the age of five he mourned for the rest of his life. He wrote five books, not in Latin but in luminous French. His gift for expressing profound thoughts in short, clear sentences has never been surpassed. He corresponded with most scientific trend-setters of the day, including Queen Christina of Sweden.[28] She eventually persuaded him to move to Stockholm. Sadly, she was a nocturnal creature and arranged for him to give her and a few favoured courtiers a course of instructions, starting at the unfriendly hour of five in the morning and rarely lasting less than four hours. A few months after his arrival the French guest caught pneumonia and died. He was fifty-four.

. . .

Descartes revolutionised mathematics, geometry and philosophy at several levels and his thoughts on nerve function were only a tiny part of his output. They nevertheless provided the first reasonable mechanism of what had been referred to for thousands of years as *pain.* At the core of his belief was the duality of body (or matter) and mind (or soul or *esprit*). The body was extended, corpuscular and quantifiable. The mind was insubstantial, immortal and the source of consciousness. The two were separate – or almost

26. This was the golden age of the Netherlands, a haven of tolerance for refugees during one of the most bigoted and hag-ridden periods in European history.
27. During his long life – he died in 1666, aged eighty-six – Hals painted everybody who was anybody in the rich merchant city of Haarlem (and many nobodies as well); but Descartes was his only international celebrity sitter. In fact, though the two men almost certainly met, there is no record of Descartes actually 'sitting' for the painter. But the portrait was acclaimed in its day as a good likeness and remains a pictorial celebration of Gallic wit.
28. She was the daughter and only child of King Gustavus Adolphus, clever but even by royal standards slightly mad. She eventually abdicated in favour of her cousin, was received into the Roman Catholic Church and spent her last forty years in Rome, patronising Alessandro Scarlatti, Bernini and other composers and artists. She died in 1689, aged sixty-seven.

so. In *Traité de l'homme* he proposed a mechanical model of the human animal, drawing analogies with clocks and automata. Such an artificial man would have physiological functions identical to those of real humans, explicable in terms of 'matter in motion'. But man (unlike animals) also had a soul or mind or *esprit*, and Descartes never quite worked out, at least not to the satisfaction of his critics, how the two interacted. His guess, that the site of interaction was an insignificant anatomical structure in the mid-brain, the pineal gland, merely confused the issue.[29] Mind remained a ghost in the machine; and his notion that the passions in some way mediated between mind and body was perhaps more than his basic dualism could accommodate. His ideas were nevertheless a momentous advance on the old Aristotelian cosmos governed by humours and elements, let alone on the mindless verbosities peddled by Paracelsus. On pain his concept was clear and his model has eminent advocates to this day. Accompanied by a much-reproduced, though clumsily drawn, figure of a youth kneeling near a fire, his 1640 treatise said:

> If, for example, fire comes near the foot, minute particles of this fire move at great speed and have the power to set in motion spots on the skin of the foot which they touch. By this means they pull on the delicate threads which are attached to these spots and open up at the same time the pores against which the delicate threads end, just as by pulling on one end of a rope one can, in the same instant, strike a bell which hangs at the other end.[30]

Descartes was aware that he had to provide a transition zone where the sound of 'the bell' would be 'heard': where, in more philosophical terms, 'external sensations were able to impress themselves on the mind'. Today this transition zone is sometimes referred to as 'the liaison brain' and is now located in the cerebral cortex. There, 'the self-conscious mind has the function of integrating its selections from an immense patterned input . . . in order to build its experiences from moment to moment'.[31]

Over the past twenty years Descartes' dualism – one limb the thinking, talking, feeling, suffering, mental 'self', the other 'the body' which sends alarm messages to the 'self' – has come under critical scrutiny.[32] Is there really a system of 'pain-sensitive' structures sending signals to a 'pain centre', as generations of doctors have been taught and have been teaching? Perhaps not. But the idea has served as a working model for centuries.

29. What the function of this curious little excrescence is remains a mystery: perhaps it *is* the seat of the soul.
30. R. Descartes, *De l'homme* (Paris, 1953), p. 86.
31. Sir K. Popper, quoted in P.D. Wall and M. Jones, *Defeating Pain* (New York, 1991), p. 32.
32. See Chapter 48.

. . .

Paradoxically at a superficial level, the age of Descartes, Harvey and of the budding scientific revolution was also the age of the great missionary saints. It was the last time when millions of ordinary people in Europe believed that Christianity – and indeed their particular brand of Christianity – would become the universal religion of the world not in the dim distant future but within their lifetime. It was an inspiring vision. Religious ardour saw a revival of the medieval doctrine of salvation through pain.

St Marguerite Marie Alacoque was born in 1647 in Vérosèvres in Burgundy. After her father's death when she was eight, she was sent by her widowed mother to the Salesian Convent of Charolles. There she suffered many ailments and experienced mystical visions. Following an attack of polio or some other kind of paralysis, she had a succession of private revelations. At the age of twenty-eight, on 16 June 1675, she had another powerful vision, this time of the bleeding Sacred Heart of Jesus. She was instructed by a heavenly voice to introduce the date as a feast day in its honour. In the convent she had to endure little but humiliation; but two kindly Jesuit fathers took up her cause and helped her to spread her devotion. 'Nothing but pain makes my life supportable,' she later wrote. The religious order inspired by her went as far as possible in actually seeking the experience in remembrance of Christ's suffering. St Marguerite Marie died in 1690. Despite a certain amount of resistance within the Church, she was beatified in 1864 and canonised in 1920. It was in her memory, as well as commemorating the suffering of St Thérèse of Lisieux, that Pope John Paul II proclaimed in 1990:

> What we express with the word pain is essential to the nature of Man. Sharing in the sufferings of Christ is also suffering for the Kingdom of God. In the just eyes of God . . . those who share in Christ's sufferings become worthy of this Kingdom. Through their pain they pay back the boundless price of our redemption. Pain is an appeal to Man's moral greatness and spiritual maturity.[33]

. . .

Saint Marguerite's doctrine has echoed down the centuries, permeating even the harsh world of politics. In 1920 the Lord Mayor of Cork starved himself to death in protest against English rule, declaring that 'they will win who can suffer pain the most'. In many lands the idea resonates still. Its modern metamorphoses will be taken up in a later chapter.[34]

33. Quoted in P. Wall, *Pain: The Science of Suffering* (London, 1999), p. 45.
34. See Chapter 43.

# Chapter 14

# HEAVENLY DREAMS

New insights into the heart, lungs, circulation and nervous system gave satisfaction to those engaged in their mildly subversive pursuit. They led to lively discussions in the Royal Society. They provoked impassioned orations under the *coupole* of the Académie royale. Their immediate effect on the welfare of suffering mankind was negligible. Charles II did as much as any non-scientist to promote scientific enquiry; and, for all his moral frailty, deserved a painless departure. In fact, after he had suffered a stroke and a fit,

> sixteen ounces of blood were removed from a vein in his right arm with immediate good effect. The King was allowed to remain in the chair in which the convulsion had seized him, his mouth being held forcibly open to prevent him biting his tongue in case the convulsions recurred. He dozed off or might have done; but the approved regimen was to keep him from sleeping at all cost. Messages were dispatched to the King's physicians, who quickly came flocking to his assistance. They ordered cupping glasses to be applied to the shoulders forthwith and deep scarification to be carried out on the back and buttocks, by which they succeeded in removing another eight ounces of blood. A strong antimonial emetic was then administered, but, as the King could be got to swallow only a small portion, it was decided to render assurance doubly sure by a double dose of Sulphate of Zinc. Strong purgatives were also given and supplemented by a succession of clysters. The hair was shorn close to the skull and pungent blistering agents were applied all over the head. Then a red-hot cautery was requisitioned . . .[1]

1. Reconstruction by Sir Raymond Crawfurd, 1932, quoted in Porter, *The Greatest Benefit to Mankind*, p. 234.

As Dr Charles Scarborough, one of the royal torturers, commented: 'Nothing was left untried and the King graciously apologised for being an unconscionable time a-dying'.[2]

The Enlightenment or the Age of Reason that intervened between the Glorious Revolution in England and the Glorious-in-Parts Revolution in France was rich in theories to explain how and why pain was experienced and what should be done about it. It was at a loss how to translate these theories into practice. It was, in fact, as ages of reason tend to be, a golden age also of quackery, the ignorance and pretensions of the charlatans, many of them sporting medical degrees, being surpassed only by their greed. It needed the pen of Molière to do them justice. In one exchange between the patient, Geronte, and the pretend physician, Sganarelle, the former is firmly put in his place.

> *Geronte*: It was very clearly explained, thank you, but there was just one thing which surprised me. Should not the heart beat on the left and the liver be on the right?
>
> *Sganarelle*: Yes, it used to be so, but we have changed all that. Everything is quite different in medicine nowadays.[3]

But optimism reigned. Reason will eventually prevail, the *infâme* will be *écrasé* and physical suffering will disappear. The marquis de Condorçet declared that future medical advances, supported by the civilising process,

2. Quoted ibid., p. 235.
3. In *Le Médecin malgré lui*. In *Le Malade imaginaire* the candidate for a doctor's degree is subjected to stiff questioning. Why does opium work? The candidate replies:

    Mihi a docto doctore
    Demandatur causam et rationem quare
    Opium fecit dormire.

    A quioi respondeo,
    Quia est in eo
    Virtus dormativa
    Cujus est natura
    Sensus assoupire.

    [I have been asked by a learned doctor what is the cause and reason why opium induces sleep. To which I reply, it is because there is in opium a dormitive virtue the nature of which is to put asleep the senses.]

    The dazzling answer provokes loud approbation:

    Dignus, dignus est intrare
    In nostro docto corpore.
    Bene, bene respondere.

    [Worthy, oh worthy is he to enter into our learned corporation: very well answered!]

would extend the human life-span 'perhaps to the point of immortality'.[4] The qualifying 'perhaps' was a token of modesty rather than of doubt. Advances in science and in particular in mathematics would be partly responsible. Hermann Boerhaave, professor of medicine in the University of Leyden, believed that health and sickness would soon become quantifiable and capable of being accurately measured and expressed as forces, distances, weights and pressures.[5] With this in mind, he introduced the clinical thermometer, an instrument almost two feet long and weighing nearly a pound. As a step in the same direction Stephen Hales, an Anglican clergyman with a scientific bent, measured the force of blood – or blood pressure, as it was to become – by inserting first into the jugular vein and then into the carotid artery of a horse a goose's windpipe attached to a glass tube eleven feet long. He noted that arterial pressure was much higher than pressure in the veins. He also explored Descartes' idea of nerve action in decapitated frogs, observing nervous reflexes still in action in the truncated creatures. This was too much even for Dr Samuel Johnson, not normally an animal rights activist.

> These doctors are extending the arts of torture . . . I know not that by living dissection any discovery has been made by which a single malady is more easily cured.[6]

He was listened to but ignored. The great Harvey himself had declared that the dissection of living creatures was the only way forward in medical research.

. . .

Universal theories, some fantastical, flourished in Leyden, London, Vienna, Edinburgh, Philadelphia and a clutch of German universities. Albrecht von Haller in Göttingen, William Cullen and his circle in Edinburgh and Théophile de Bordeu and his vitalist friends in Montpellier developed systems of great complexity in which health depended on the coordination of the separate lives of each organ. Georg Ernst Stahl of Halle deplored this mechanical philosophy as well as the 'materialism' of Boerhaave and invented a God-given *anima* (a soul but not exactly a soul) which was the agent of consciousness and the physiological regulator of the body. *Anima* had its own language, the experience of many different kinds of pain the key part of its vocabulary. The concept, expounded in a book of 800 pages, was warmly approved by the

4. A brave if misguided man, he committed suicide in prison during the Terror in 1795 to escape the guillotine, aged forty-nine.
5. Herman Boerhaave was the most famous clinical teacher of his age and professor in Leyden, Holland, for thirty years. He described the sweat glands. His most famous publication, the *Institutiones medicae*, was published in 1704. He died in 1738, aged seventy.
6. Quoted in Porter, *The Greatest Benefit to Mankind*, p. 235.

pietist Lutheran establishment. Stahl's colleague, Friedrich Hoffmann, also widely read, inclined towards a coordinated mechanism, 'medicine . . . being the art of detecting and properly utilising physico-chemical principles in order to preserve the health of man and restore it when disturbed'.[7] Somewhere in between, Boissier de Sauvage, professor of medicine in Montpellier, accepted that the body was a machine, comprehensible in mathematical terms, while disease was an effort by nature or the soul to expel unwanted matter and obstacles to normal function. Physiology was, or should become, the science of that struggle. Few thinkers went so far as to reduce man entirely to a mechanism; but the *philosophe* Julien Offray de la Mettrie postulated in his suppressed but much-discussed bestseller, *L'Homme machine*, that properly assembled matter can think and that the body as a potentially perfect machine could wind its own springs.[8] Pain was a mechanical fault which would be abolished as soon as the machine mastered its own working.

Sadly, none of these ingenious schemes, often expounded with verve and literary grace, brought any noticeable improvement to everyday life. In many ways general health seemed to be deteriorating rather than improving. Growing cities like Paris, London and Vienna became hotbeds of epidemics. Some struck at regular intervals and carried off their victims in terrible agony. Syphilis showed no sign of abatement. Even the growth of hospitals was a mixed blessing. Childbed fever, that most horrible slaughter of healthy young mothers, soared in the most prestigious obstetric units of Europe and the New World.[9] It had been virtually unknown in the bad old days of casual deliveries by unschooled midwives. To say that the Age of Reason saw no practical medical advances would be an exaggeration, but not a gross one. The cinchona plant, imported by Jesuit missionaries and widely known as Jesuits' bark or Peruvian powder, was arguably the first specific remedy for a particular disease.[10] (Because of its popish provenance Oliver Cromwell and his closest friends refused to make use of it.) William Withering, a Shropshire general practitioner, stumbled on and ingeniously adapted the extract of foxglove to the treatment of heart failure. His digitalis would never pass safety regulations today but it proved a blessing.[11] Most welcome perhaps, Edward

7. Translated from Hoffmann's *Fundamenta medicinae* (Nürnberg, 1695), p. 156.
8. Albrecht von Haller died in 1771, aged seventy; William Cullen died in 1790, aged eighty; Théophile de Bordeu died in 1776, aged fifty-five; Georg Ernst Stahl died in 1734, aged seventy-two; Friedrich Hoffmann died in 1742, aged eighty; Boissier de Sauvage died in 1767, aged sixty; and Julien Offray de la Mettrie died in 1746, aged fifty.
9. See T. Dormandy, *Moments of Truth* (London, 2004), Part II on Ignác Semmelweis, discoverer of the causation of childbed fever.
10. Quinine remained the only medicine to control malaria for three hundred years.
11. Withering published his landmark *Account of the Foxglove* in 1785. He died in 1799, aged fifty-eight. His discovery prolonged millions of lives but often had toxic side-effects.

Jenner discovered vaccination against the curse of smallpox.[12] But in the field of pain relief there was only one momentous development. For the first time since Antiquity, opium became available in a safe and palatable form.

. . .

'Poppy juice' had of course been used in Western Europe since the Crusades.[13] John of Arderne added it to his salves for his highborn patients to make them 'slepe so that [they] shal fele no kuttyng'. It was probably the mainstay of Paracelsus' concoctions. Shakespeare inevitably referred to it.[14] Sylvius de la Boe, a Frenchman who became a famous professor in Leyden, maintained that he could not have practised medicine without it.[15] But most preparations were unpredictable, some were dangerous and all were unpalatable. In uncertain times their price soared so that only the rich could afford them.

Much depended on provenance. The drug came from several exotic regions of the world and its marketing was uncontrolled. Indian opium, grown mainly in the Ganges valley between Patna and Benares, was the most abundant. Shaped into balls, it was shipped round the Cape and sold in 40-ball chests. It was cheap but unreliable. Children and babies in particular sometimes died after a single small dose. Turkish opium was safer but more expensive. It was dispatched from Smyrna (today's Izmir), its strength and quality indicated in carats like gold on a 0–24 unit scale. It was dark brown and waxy, the surface sprinkled with the seeds of sorrel to prevent it from sticking. The consignments travelled in pretty calico bags neatly packed into hermetically sealed zinc-lined wooden caskets. (They are now collectors' items.) It was generally regarded as reliable but had several disadvantages. Its foul taste was almost impossible to disguise and it was said to be especially constipating. Opium from the eastern region of Anatolia and Mesopotamia, today's Iraq, but shipped from Constantinople was a redder brown and sold in small lens-shaped cakes. It came attractively wrapped in poppy leaves and had a distinctive taste. It was expensive but Coleridge and the Prince Regent valued it above all other brands. Persian trade was centred on Yezd and Isfahan. The material was grey and came in the form of sticks wrapped in grease-proof paper tied

12. Edward Jenner published *An Inquiry into the Causes and Effects of Cowpox* in 1798. He died in 1823, aged seventy-two.
13. See Chapters 8 and 9.
14. 'Not poppy, nor mandragora, / Nor all the drowsy syrups in the world, / Shall ever medicine thee to that sweet sleep / Which thou ow'dst yesterday.' Iago's soliloquy in *Othello*, Act III, scene iii.
15. Sylvius (de la Boë) is not to be confused with Sylvius (Jacques Dubois). The former was a famous clinician who described the Sylvian fissure of the brain and died in 1572, aged sixty-two; the latter was an anatomist who described the Sylvian aqueduct in the brain and died in 1555, aged seventy-seven.

with cotton twine or in cones weighing a quarter to half a kilo. Its price fluctuated. It was often the cheapest brand available but it deteriorated dangerously on storage. Egyptian opium, a minority taste but passionately advocated by connoisseurs like Thomas de Quincey, was formed into round, flattened buns like ice-hockey pucks. Bright red in colour, it was sun-dried at source to a harder consistency than other varieties. One of the oldest varieties known in Europe, the product of the poppy fields around Thebes had been cherished since the days of Herodotus.[16]

Dealers, merchants and aficionados were expert at judging the quality and strength of every cargo, their knowledge esteemed like the art of the coffee and tea blender. But in the raw state even the best opium was nauseating. Its absorption in different individuals remained unpredictable. Some seemed totally resistant to it. In others it paralysed bowel function for weeks. Paracelsus had already bandied the name 'laudanum' about. What exactly he was referring to remains uncertain. The preparation which later acquired that name and which remained for centuries one of mankind's great blessings was largely the invention of an English doctor rarely commemorated today.

. . .

The Sydenhams were old Somerset gentry with a younger branch in Dorset. In the Civil War the Dorset branch became staunch Parliamentarians; and at least one of Thomas' brothers was killed in battle. Another, William, was to fill influential positions under Cromwell. The boys' mother too may have been killed by Royalist troops. Thomas himself, born in 1624, was old enough to rise to captain of horse in the Puritan army. After the surrender of Oxford by the Royalists on Midsummer's Day, 1649, he registered at Magdalen Hall to study medicine. He found the teaching useless and the university comatose so he travelled on the Continent, visiting Leyden, Paris and Montpellier. Eventually he settled in Westminster in a house in Pall Mall, but continued to spend some time in the family seat in Wynford Eagle. He married and raised a family. Within a few years he became a sought-after doctor, his professional reputation saving him from unpleasantness after the Restoration. He took little or no part in public affairs but he was on intimate terms with many of the leading scientists of the age. John Locke, a doctor though not a practising one, was a particularly close friend. Sydenham died in 1689, aged sixty-five, almost certainly from renal failure secondary to gout. The comparative paucity of biographical data led his admiring early biographer, R.G. Latham, to draw a parallel between 'our greatest doctor and our greatest dramatist, the

16. Thebaine, one of the alkaloids of opium, was given its name to commemorate the poppy fields around Thebes.

names of both beginning with S'. Not many doctors today could instantly identify the first of the two.[17]

Greatest or not, Sydenham remains an engaging figure. He wrote well and a great deal. His theoretical and philosophical musings are interesting; but his strength lay in his bedside observations, his empirical approach to disease and the importance he attached to treatment.[18] It was the last which led him to formulate a comparatively simple preparation of opium, capable of being standardised and given even to children without fear of an accidental overdose.

> The laudanum tincture which I have mentioned as being given in daily draughts is quite simply prepared . . . one pint of sherry wine, two ounces of good quality Turkish or Egyptian opium, one ounce of saffron, a cinnamon stick and a clove, both powdered. Mix and simmer over a vapour bath for two or three days until the tincture has the proper slightly viscid but still easily poured liquid consistence, easy and pleasant to administer.[19]

Apart from wanting to have opium in a state which could be accurately measured, he also hoped to undermine the sales pitch of fashionable quacks who peddled complicated and expensive remedies all of which depended for their effect, if any, on their opium content. He used the tincture for a wide range of diseases like dysentery, childhood diarrhoea and other 'fevers' but, above all, he used it to relieve pain.

Like many great doctors, Sydenham was himself a victim of a painful disease of which he gave a classic description.

> The pain starts moderately at the junction of the big toe and foot and gradually becomes more intense. As it does so, the chills and shivers begin. After half an hour or an hour the pain reaches its peak. It spreads like hot irons buried under the skin to the metatarsus and the tarsus [foot and heel], attacking bones and ligaments as well as the flesh. It varies constantly. Now it is gnawing, now a pressure, now a burning.[20]

Although he recognised that gout was an affliction of the rich, he did not think that it was caused by red meat or port. He himself consumed little of

17. The first edition of the *Opera omnia* was published in 1683. The Sydenham Society published a translation by Dr Greenhill in two volumes, with a long and charming introductory biography by R.G. Latham.
18. Sadly, the Sydenham Society, founded in 1828, which published the *Collected Works*, no longer exists. The name used to be known to medical students from 'Sydenham's chorea', a complication of rheumatic fever. Both have virtually vanished. An excellent short biography with excerpts from the works was published in 1966 under the editorship of K. Dewhurst: *Dr Thomas Sydenham (1624–1689)* (Berkeley, 1966).
19. T. Sydenham, *Collected Works* translated by Dr Greenhill, Vol. II, p. 34.
20. Ibid., Vol. I, p. 346.

either. But he was not given to speculation: 'All will be revealed when the body's chemistry is better known.' (Not everything has yet been revealed: much of the natural history of gout in particular remains a mystery. Why, in an affluent world, has it become so rare?) Having known the pain he also knew that laudanum, properly prepared, was the only remedy which brought relief.

> And here I cannot but break out in praise of the great God, the giver of all good things who hath granted to the human race, as a comfort in their affliction, a medicine of the value of opium; . . . either in regard to the number of the diseases that it can control, or its efficiency in extirpating them . . . So necessary an instrument is opium in the hand of a skilful practitioner that medicine would be crippled without it; and whoever understands it well, will do more with it alone than he could well hope to do with any other single medicine.[21]

On the old duality of body and soul, Sydenham imposed a new one, more practical and in some ways more useful to doctors than more elaborate doctrines. There was the outer man, responsive and reacting to external stimuli, and the 'inner man' governed by the movements of the 'animal spirit', responsive to and causing changes inside the body. Both the accumulation and depletion of animal spirit in any organ could cause disease; and if the change exceeded a critical threshold, it was brought to the individual's attention by *pain.* This in turn could trigger 'reflex actions' like 'movements of alarm and retraction', an overflow of emotions and physical flight. The concept of this dual nature of man was taken up and elaborated by some of the leading thinkers and doctors of the age – Locke in England, Buffon, Cabanis and Bichat in France, Boerhaave in the Netherlands. All rejected the existence of 'innate ideas' and made sensation and experience the starting-point of all knowledge.

At a practical level, Sydenham perceived that universal theories, however eloquently expounded by men 'more clever than a mere doctor like myself', offered little to patients. This was true even of his own ideas. But there was one important general notion which was not widely enough appreciated. Different diseases had different aetiologies and medicine should strive to eliminate the specific causes, not the symptoms. This is now accepted as so basic that its novelty in Sydenham's time of universal panaceas is easily overlooked. The shining example was the effect of the cinchona bark on the course of malaria (or 'the aigue'). 'We need more such remedies,' Sydenham wrote. Eventually, they would come; but it would be a long wait.

21. Ibid., Vol. II, p. 68.

. . .

Opium in the new form of laudanum quickly made headway in England and was widely acclaimed as a blessing. In France and in other Catholic countries it continued to be viewed with suspicion. (The war of words between French and English doctors on the topic has been described as the first Opium War.) The decisive influence to make it acceptable was Boerhaave's medical best-seller, the *Institutiones medicae.* Published in 1708 and translated within a few years into at least six languages, it sold thousands of copies. Boerhaave was, like Sydenham, a practical clinician and advocated laudanum in moderate doses in any painful condition. There was no merit in suffering unnecessarily. Boerhaave's pupil, Gerhard van Swieten introduced the tincture to Vienna where opium had been officially forbidden. Of course prohibition was, as always, an incitement to its consumption. According to the French ambassador to the Court of Maria Theresa, Schönbrunn 'reeked of the Turkish weed'. At Van Sweiten's urging, laudanum prepared according to the rules now became available. Returning from his first triumphant visit to London, 'Papa' Haydn approved. He had been introduced to the drug in London and it was the only remedy which helped his occasionally painful micturition.

. . .

Sydenham's original tincture was followed by other opium-based preparations. Young Thomas Dover was living as an apprentice in Sydenham's house in Dorset when he contracted smallpox. Sydenham nursed him through the illness, liked him and taught him practical medicine. A kind of extended gap year was popular among adventurous youths at the time and Dover took time off for a spot of privateering. He soon commanded his own vessel, the *Duke*, and profitably raided the Spanish coast of South America. On 2 February 1709 he rescued the shipwrecked Alexander Selkirk from the Juan Fernández Islands. Selkirk later achieved fame as Defoe's *Robinson Crusoe.* Dover then settled in London to 'practice physic', although all the training he had had was what he had received from Sydenham during his convalescence. His specialty, 'Dover's Powder' nevertheless became famous, first as a 'diaphoretic', a substance designed to induce sweating, and later as a curative potion for all forms of abdominal pain and diarrhoea. The ingredients were 1 ounce each of opium, liquorice and ipecacuanha with 4 ounces each of saltpetre and vitriolated tartar. 'Drink from 40 to 60 grains in a glass of wine before going to bed and all pain will vanish by the next morning.' Occasionally – but not often – the patient's life would vanish with the pain. 'Some apothecaries,' Dover disarmingly wrote in his book, *The Ancient Physician's Legacy to his Country*, 'desired their patients to make their will before they ventured upon this remedy in the dose recommended. But nothing ventured, nothing gained.' The powder remained in the *London Pharmacopoeia* till 1898 and was a favourite with the famous surgeon and skin physician, Sir Jonathan Hutchinson.

A peripatetic French Capuchin monk by the name of Rousseau cooked up another mixture of an oily consistency known in England as Lancaster Black Drops and in France as *La Brune*. It contained crushed pearls (allegedly), nutmeg, saffron, crab apple and yeast as well as opium. Rousseau eventually settled in Paris and was patronised by the crotchety duc de Saint-Simon. Saint-Simon complained of the cost of the concoction but recommended it to all and sundry. From Paris Thomas Jefferson took samples back to the United States – 'a capital remedy for all aches and pains' – and the preparation was much used in the Civil War. It remained in the United States *Pharmacopoeia* (with the crushed pearls as an optional extra) till 1879.

But it was Sydenham's laudanum which led the field, sherry later being replaced by Canary wine. Benjamin Franklin, the young Duke of Wellington, George IV, Thomas Wedgwood and Dr Samuel Johnson among other celebrities used it regularly, though none became addicted. Robert Clive of India did: having taken it for some years for a bowel complaint, he killed himself with an overdose.[22]

. . .

The first medical monograph devoted to the drug was Dr John Jones' *Mysteries of Opium Reveal'd*, published in 1700. Jones, almost certainly an addict himself, was also among the first to utter clear warnings against habituation and addiction. He vividly described the rigours of abrupt withdrawal.

> Anguish and pain, followed by great, even intolerable Distresses, Anxieties and Depressions of Spirit, which, in a few days commonly end in a most miserable Death, attended with strange Agonies.[23]

Yet Jones was an enthusiast: nothing could compare with the drug as a means of dulling pain. It also lifted the spirits and induced not only serenity but also 'Promptitude, Alacrity, Expediteness in Dispatching Business, Assurance, Ovation of the Spirit, Courage, Contempt of Danger, Magnanimity, Euphory, Contentation and Equanimity'. What more could man desire? If, after taking opium,

22. Robert (later Lord) Clive was forty-nine when he took his own life in 1774, after much suffering from what must have been a form of ulcerative colitis.
23. J. Jones, *Mysteries of Opium Reveal'd* (London, 1700), p. 49.

> the person keeps himself in action and discourse of business, it seems like a most delicious and extraordinary refreshment of the spirits upon hearing good news or any other great cause of joy . . . It has been compared, not without good cause, to the gentle and yet intense degree of that pleasure which modesty forbids me to name.[24]

Jones the bashful was not alone: nor were his views confined to a few eccentrics. By the mid-century laudanum and a wide range of opium-based preparations had long entries in family herbals and books on home doctoring in most European languages. Even in France, where in Sydenham's day Philippe Hecquet, an eminent Paris doctor, was censured by the Académie royale for praising his English colleague, laudanum became widely recommended and consumed. Jean-Baptiste Pomet, chief apothecary to Louis XVI, praised it as a 'narcotic and hypnotic beyond compare'. It was particularly useful for 'composing the Hurry of the Spirits as well bodily diseases of the breasts and lungs . . . preventing coughs, spitting of the blood, colds, vomiting, looseness of the bowels' and a long list of other trying and dangerous ailments.[25] The complication most to be feared was not habituation but constipation.

. . .

As the eighteenth century entered its last quarter, the role and repertory of laudanum expanded. The drug continued to be used as an anodyne for every kind of pain; but it was becoming more. It emerged as a formative influence on the arts, on literature, even on ordinary lifestyle. It was the dawn of the Romantic Age. Laudanum more than anything else served its emerging spirit. Those who celebrated the new cult of freedom, fancy and fantasy found it an indispensable aid to overcoming inhibitions, banishing mundane worries. It inspired bold resolutions (rarely kept) and brave designs. Among its acolytes was the young Goethe, Schiller, Schlegel and Hölderlin in Germany, Chateaubriand and his circle in France, Pushkin in Russia, and Coleridge, Wordsworth, Walter Scott, Keats, Shelley, Byron and Thomas de Quincey in Britain. All indulged in the 'mighty brew', at least occasionally. It helped them to respond to the secret whisperings of nature, to kindle a new spontaneity of feeling, and to encourage the expression of passions, pathos and (in the case of the young Werther) *Weltschmerz.* Thomas de Quincey, champion of the cult in England, was a student at Worcester College, Oxford, when he made his first purchase to cure a toothache. He never forgot the experience:

24. Jones, *Mysteries of Opium Reveal'd*, p. 257.
25. Quoted in M. Booth, *Opium: A History* (New York, 1996), p. 46.

> In an hour, O heavens! What a revulsion! What a resurrection, from the lowest depth, of the inner spirit! What an apocalypse of the world within me. That my pain had vanished was now a trifle in my eyes; this negative effect was swallowed up in the abyss of divine enjoyment thus suddenly revealed. Here was a panacea . . . for all human woes; here was also the secret of happiness.[26]

He got the meaning of 'revulsion' wrong; but that often happened under the influence and it added extra piquancy to opium-inspired effusions.

But it was not only poets, artists and writers who embraced the new lifestyle. In 1807 in *A View of the Nervous Temperament* Thomas Trotter, a gifted medical practitioner and observer of the social scene, portrayed his entire generation as being 'afflicted by the nervous disposition, formerly the privilege of poets and aristocrats'. And if not actually afflicted, acting as if they were afflicted. The new sensitivity was percolating down the social scale 'to embrace the formerly sober middle classes and women as well as men'. Roy Porter called it the 'pressure-cooker society' in which ordinary citizens, especially the young, began to take stimulants as a daily necessity, first coffee and tea and then drugs and narcotics.[27] Habit and addiction led to insomnia, hypochondria and hysteria which in turn called for more urgent medications.

But to become an addict was becoming fashionable. Men and women of hypersensibility, too good for a wicked world, were intensely admired and copied. Even the ploddingly normal envied the exquisite folk blessed with a 'highly charged' nervous system. ('Highly charged' was a term borrowed from the fascinating new science of electricity.) France was in the throes of a revolution and much of the rest of Europe was smouldering; but everywhere the elegantly morbid was 'in' and the languorous and emaciated much admired. Both the languor and the emaciation were stoked by laudanum. Even the martial heroes of the hour – Napoleon and Wellington – were known to indulge; and the habit was one of the bonds between Horatio and Emma.

But there was more to the drug than Romantic chic. Chronic pain was one of the triggers and often at the root of the habit. Suddenly, since it could be abolished or at least mitigated, physical suffering was no longer as acceptable as it had been in the past. The availability of cheap laudanum coincided with an upsurge in tuberculosis, an ancient disease which was becoming devastatingly common.[28] The blessed tincture was the only defence against the lancinating cough that tormented young Keats, Shelley, Tom Girtin, Schiller, Chamisso, Hölty, Leopardi, Lermontov, Csokonay and hundreds of others. All

26. T. de Quincey, *The Confessions of an English Opium Eater* (1822; Oxford, 1930), p. 43.
27. Porter, *The Greatest Benefit to Mankind*, p. 342.
28. See Chapter 16.

were soaring talents struck down in youth. Time for them was short, life precious. The illness and the medicine became intertwined. They sometimes inspired poetry of sublime beauty. More often they inspired poetry of morbid mediocrity. But in German-speaking lands even the morbidly mediocre could sometimes be turned into a hauntingly beautiful *Lied* by a poor, young, myopic Viennese song-smith called Schubert.

'Vorüber, ach vorüber!
Geh, wilder Knochenmann!
Ich bin noch jung, geh, Lieber!
Und rühre mich nicht an.'[29]

Laudanum was no defence against the 'wild man of bones' but for a precious few hours it stilled the suffering – and oh, what heavenly dreams of spring and roses did it inspire!

29. 'Go, go, oh go on and pass me by/ You cruel scythe man/ I am stll young, oh go dear man/ I beg you do not touch me.' The words of *Death and the Maiden* are by Schubert's friend, Matthias Claudius, whose two daughters, Emilia and Franziska, had died of tuberculosis in their teens. Several of Schubert's circle in Vienna, including Schober, his closest friend, and Mayrhofer with whom he shared a room for some years, were regular users of laudanum; and Mayrhofer (who eventually committed suicide) was almost certainly addicted. For background see T. Dormandy, *The White Death* (London, 2000).

# Chapter 15

# ANIMAL MAGNETISM

The Age of Reason was also the Age of Quacks and Franz Anton Mesmer has sometimes been portrayed as their archetype. But he was not a quack. He was well qualified. He firmly believed in what he preached. He was generous with his time and money. He cured many of his patients. He became the only doctor in history whose name has passed into the common language. 'Mesmerised' is used by people who have never heard of him as a person.[1] Yet none of this would entitle him to a chapter in this book. What endows his half-baked doctrine of animal magnetism and his colourful career with interest is that they revolve around two problems at the very heart of understanding pain. Or not understanding it. They aroused fierce passions in his day and still provoke tetchiness among doctors. Tetchiness, but also fascination.

The first mystery is the extraordinary power of mind over matter or the spirit over the body. Of course the mystery was not new in Mesmer's day. There have been references to it in this book in connection with religious faith and philosophical wisdom. But it was Mesmer who transposed the relationship from the realms of religion, literature and myth into humdrum clinical practice. After him, welcome or not, hypnosis and self-hypnosis under a variety of names could never be quite banished from the medical scene.

1. The nearest is perhaps 'Freudian' as in 'Freudian slip'; but the provenance of Freudian is usually clearer to the user than when the words 'mesmerised' and 'mesmeric' are used.
   Mesmerism also passed into literature. Chauncy Hare Townshend's *Facts in Mesmerism* (1840) is the link between Mesmer and Edgar Allan Poe. Poe wrote three stories in which mesmerism is a theme, 'A Tale of the Ragged Montain', 'Mesmeric Revelation', and 'The Facts in the Case of M. Valdemar'. Edward Bulwer-Lytton wrote about mesmerism in 'A Strange Story'. Robert Browning and Elizabeth Barrett frequently referred to mesmerism in their correspondence. There are references to a 'mesmeric' state in Balzac, Victor Hugo, Ludwig Tieck, E.T.A. Hoffmann, Dostoevsky and even in Schopenhauer. There may be others. George du Maurier's bestseller, *Trilby* (1894) is the great Edwardian period-piece presentation.

Even more intriguing and more aggravating to many is the puzzle of the placebo effect. The term is ancient. Chaucer mockingly referred to it, quoting Psalm 116, verse 9, 'Placebo domino in regione vivorum'.[2] By the seventeenth century doctors used it to describe inactive medicines which nevertheless profoundly impressed patients.[3] Thomas Jefferson recalled that 'one of the most successful physicians I have ever known has assured me that he used more bread pills, drops of coloured water and powders of hickory ash than all the other medicines put together'.[4] Here Jefferson distinguished between 'pious fraud' (of which he disapproved) and the placebo effect of a therapy in which the therapists themselves believed. It was always the latter which provoked most incredulity and which was (and still is) the more insidious, the more powerful and the more difficult to quantify. It remains *the* nightmare of modern medical scientists and especially of drug researchers into analgesics. The placebo effect cannot disperse a swelling or mend a fracture (or not easily) but it can assuage even the most severe forms of pain. Or the opposite. Any treatment accompanied by high expectations and faith in its success is likely to prove successful and vice versa. Even the effect of a well-tried drug can be significantly enhanced or diminished by the extent of the patient's belief in its effectiveness. This too had of course been known – and sometimes exploited – for centuries; but it was Mesmer who put it on the therapeutic map.[5] Indeed, research into pain relief almost naturally divides into before-Mesmer and after-Mesmer.

...

2. 'I will please the Lord in the land of the living' (Psalm 116, verse 9). It was the first word in prayers for the dead which were said by priests and friars who pestered the bereaved for money to recite them. In the Middle Ages the term placebo came to denote a flatterer or sycophant. (Hocus-pocus also has a liturgical origin from 'hoc est corpus' – here lies the body.)
3. In 1621 Robert Burton wrote in his *Anatomy of Melancholy*: 'There is no virtue in some remedies but a strong conceit and opinion.'
4. Quoted in P.D. Wall, *Pain* (London, 1999), p. 152. The book contains an excellent chapter on the placebo effect.
5. Yet the placebo response was not studied with scientific detachment until comparatively recently: indeed, for long the very mention of a 'placebo trial' was regarded as 'hostile questioning' by both academics and enthusiasts of complementary medicine. Some doctors still think that a placebo is the same as no treatment. They are of course wrong – as Mesmer demonstrated two hundred years ago.

   In a prestigious London dental hospital ultrasound treatment for post-extraction pain and stiffness had been used with great success for some years. A carefully controlled trial was then designed in which treatments with the ultrasound apparatus turned off and turned on were compared. There was no difference in the beneficial effect; but both were superior to management without the machine. Numerous such experiments have now been reported, not least after surgical operations in which incisions were closed because on inspection the lesions were considered to be inoperable.

He was born on 23 May 1734 in Ignanz, a small village near Radolfzell on the German side of Lake Constance. His father was a forester in the service of the Prince Archbishop of Constance, a benign potentate who supported Anton's studies at the Jesuit College of Dillingen and then provided him with a stipend to attend the famous Jesuit University of Ingolstadt. His interests there were canon law and theology – he always remained a professing Catholic – and it is not known why, at the relatively advanced age of twenty-eight, he decided to study medicine. Once decided, Vienna was his obvious destination: it was the most famous medical school in German-speaking Europe.[6] Neither in Ingolstadt nor in Vienna did he shine. In 1765 he qualified with a thesis about the influence of the moon and the planets on health and sickness. This was not particularly prescient. The work rehashed Newtonian physics and briefly surveyed the seasonal incidence of certain illnesses.[7] He also married the not inappropriately named Frau Maria Anna von Posch, a rich and vaguely aristocratic widow, twenty years his senior, who had two daughters.[8] Her fortune and position enabled him to open a consulting room at the fashionable end of the Landstrasse. It also encouraged him to indulge his love of music and patronise Gluck, the Haydn brothers and the recently arrived Mozarts.[9] When court intrigue prevented the staging of the twelve-year-old Wolfgangerl's first opera, *La finta semplice,* Dr and Frau Mesmer commissioned him to write a *Singspiel* for their private theatre. There *Bastien and Bastienne* had its first performance and was enthusiastically received.[10] The friendship between the two families endured: twenty years later Mozart included an affectionate reference to Mesmer 'and his magnets' in *Die Zauberflöte.*[11]

6. The University of Vienna is, after the Charles University in Prague, the oldest university in Central Europe; but it did not acquire a medical faculty till 1760. Under the guidance of Gerhard van Swieten, the Empress Maria Theresa's medical adviser and confidant, it attracted a brilliant team of professors (collectively sometimes referred to as the First Vienna Medical School to distinguish it from the generation of Skoda, Rokitansky, Schuh and Feuchtersleben in the nineteenth century, who constituted the second).
7. This is still much of a mystery, as is the regular periodicity of certain ailments.
8. When Mesmer left Vienna Frau Mesmer stayed behind; she died in 1761.
9. The Mesmers met the Mozarts during the Mozarts' second visit to Vienna, which was not nearly so successful as the first. Leopold Mozart was hoping to promote the career of Wolfgangerl (12) and Nannerl (14) by staging concerts during the lavish festivities planned for the marriage of the Archduchess Maria Josepha. Sadly, the archduchess died from smallpox a few days before the planned wedding and the Mozarts themselves fled to Olomouc in Moravia to escape from the epidemic. Both children caught the disease and nearly died.
10. The book of this charming piece was a German translation of *Les Amours de Bastien et Bastienne,* itself a popular parody of Jean-Jacques Rousseau's lachrymose comedy, *Le Devin du village.*
11. 'Hier der Magnetstein/ Soll's Euch beweisen,/ Ihn brauchte Mesmer einst,/ Der seinen Ursprung nahm/ Aus Deutschland's Gauen/ Und so berühmt ward/ in Francia.' (Here the magnet shall be your proof: once used by Mesmer who originated the practice in German lands and famously in France.)

Magnets continued to exercise Mesmer's imagination even during these tranquil years. They were easy to produce since, in 1750, John Canton had discovered how to make artificial ones simply by rubbing iron or steel. They were entertaining: perhaps more. Their effect over a distance without anything visible or palpable was remarkable. Mesmer began to believe that 'all things in nature possess a power which manifests itself by acting on other bodies without chemical union or physical contact'. Gradually the idea of the *fluidum* took shape in his mind, not a real liquid but something akin to Newton's ether, a substance which impregnated all bodies – animals, plants, water, even stones – and which could move over considerable distances. For a time he toyed with the idea that it might have something to do with electricity. The new science was causing a stir. A young professor in Bologna was making dead frogs jump by touching their nerves with wires carrying an electrical current.[12] Or so Signor Galvani claimed. In the not entirely imaginary Viennese world of the *Rosenkavalier* such crude goings-on were frowned upon; but one did not need decapitated frogs to be fascinated by the wonders of electricity. In 1774 Mesmer heard that Father Maximilan Hell, one of the Empress Maria Theresa's court astrologers and a professor in the University of Vienna, was performing remarkable cures by placing steel magnets of varying sizes and shapes on diseased parts of sick people. He attended one of Hell's public healing sessions and was hooked. He remained hooked even when the medical establishment, led by the magisterial Gerhard van Swieten, rounded on Hell. The proper province of a priest, they insisted, was souls and, stretching the point, the stars, not human bodies. Hell valued his *Kuchen mit Schlag* and, cravenly in Mesmer's opinion, abandoned his magnets. But Mesmer was a doctor, not a priest; and nothing could stop him continuing Hell's researches. Above all, he addressed himself to trying to understand and if possible to cure the most mysterious but also the most terrible scourge of Creation: pain.

In his earliest experiments he observed that the application of magnets initially often increased the suffering; but that deterioration was followed by a period of 'restful slackness and then by gradual and often complete recovery'. He also realised that it was essential to enter into some kind of a 'relationship' with the patient. This did not necessarily involve physical contact: contact through a magnet, even eye contact, was sometimes enough.[13] But touching helped in some cases. It was also essential that the patients themselves should wish to be cured.[14] This was a penetrating insight. The whole field still needed painstaking experimentation; but it was infinitely promising.

12. Galvani was another scientist (but not a doctor) whose name passed into the language. He died in 1781, aged sixty-one.
13. In the Freudian *oeuvre* this would become the crucial stage of 'transference'.
14. This was to be one of Freud's early 'discoveries' too. But Freud resented being compared in any way to Mesmer.

His first notable success was with Franzl Osterlin, the daughter of a friend. Franzl suffered from attacks of 'hysterical fever' which doctors had tried for years to treat with 'conventional medication', mainly, it seems, heroic purges and sweatings.

> With the hysterical fever [Franzl] combined violent convulsions, attacks of vomiting, inflammation of the intestines, inability to pass urine, violent toothache, earache, depression, hallucinations, cataleptic trance, swooning, temporary blindness, breathlessness, attacks of paralysis and other terrible symptoms which usually lasted a few days.[15]

In 1774, after all other treatment had failed, Mesmer yielded to the imploring of the family and began to apply Hell's magnets. After some preliminary trials

> I decided to produce an artificial ebb and flow [of the *fluidum*] in the patient's body. When in July she had another attack I tied two magnets to her feet and hung another, heart-shaped one round her neck so that it just touched her breasts. Suddenly a hot piercing pain rose along her legs from her feet and ended with an intense spasm around the iliac bone . . . Here the pain was united with an equally agonising one which flowed from both sides of the breasts . . . At certain parts of the body the magnetic stream seemed to be interrupted, only to become even more effective a few minutes later . . . The entire paralysed side of the body perspired freely but then the pain gradually ceased.[16]

After several progressively milder attacks treated in the same way, there was complete recovery. Five years later Mozart wrote home to his father in Salzburg:

> Can you guess where I am writing this? In the garden of Dr Mesmer's house in the Landstrasse. But you will not guess who is sitting next to me. Fräulein Franzl, now Frau von Bosch, but, on my honour, you would barely recognise her. She is fat, healthy and never in pain, and, would you believe it, she has two children and is expecting the third![17]

By then Mesmer had had a string of similar successes; and he had slowly realised that the magnetic *fluidum* could flow directly from his own 'magnetised' body into patients even without the application of steel. Yet even after he concluded that the passage of magnetic force by direct contact was the crucial part of the treatment, he continued to believe that a 'magnetic atmosphere' and indirect contact between the patient and a 'magnetised object',

15. A. Mesmer, *Mémoire sur la découverte du Magnétisme Animal* (Geneva, 1799), p. 245.
16. Ibid., p. 247.
17. *The Letters of Mozart and his Family*, translated by E. Anderson (London, 1938), p. 68.

usually a bowl of water containing glass shards and iron shavings, and other 'magnetised objects', was equally important.

. . .

Mesmer's fame spread and he was invited to Munich by the scientifically smitten Maximilian III, Elector of Bavaria. There a priest, Father Gessner, had been using similar methods to Mesmer's with some success but attributed his results to 'exorcism'. Mesmer indignantly rejected the priest's interpretation: 'exorcism was medieval superstition' whereas animal magnetism was 'physical reality'.[18] Far from being a credulous dupe, Mesmer was a child of the Enlightenment, besotted with science: his weakness was his inability to accept that some phenomena were not instantly explicable in terms of what was known about the physical universe in his day.[19] He clinched his argument with Gessner by curing court councillor Peter van Osterwald who had for years suffered 'catarrh of the stomach, hernia, haemorrhoids, both causing attacks of agonising pain, paralysis of the legs and almost complete blindness'. Osterwald later declared that 'if anyone says that my restored eyesight is mere imagination, I do not believe that any other doctor in the world could have restored it and could have made me healthy and without pain'.[20]

Honours were heaped on Mesmer. The Academy of Sciences in Augsburg issued a report stating that

18. Crabtree, *From Mesmer to Freud*. Healing by the laying on of hands and exorcism have probably always existed and some exorcist healers became famous in the centuries before Mesmer. One of the most interesting, Nicolas Malebranche, a French metaphysician who was born in 1638, wondered in his extraordinarily modern book, *Recherche de la vérité*, whether people might not be made ill or recover under the domination of a personality more forceful than their own. See M. Goldsmith, *Franz Anton Mesmer. The History of an Idea* (London, 1934).

    In Britain the most celebrated case was that of Valentine Greatrakes of County Waterford in Ireland, a simple and transparently honest country gentleman who became famous first in his own country and then in London as 'the Stroker' or 'Touch Doctor'. In 1663, following an 'inner impulse', he began to heal people by touching them. Moving to London three years later, 'he stroked away successfully a varied list of ailments' when people crowded round him in Lincoln's Inn Fields. He was a person of independent means and never charged patients for his cures. He also declared that many sufferers were beyond his help. A devout Protestant, he attributed his gifts entirely to God.

    Though only a few years older than Mesmer, Father Johann Gessner was the last of the great exorcists. All he demanded from his patients, who came from far and wide, was a firm belief in God and a cheerful disposition. He always spoke to them in Latin: they reacted to his orders whether they understood him or not. He claimed to cure 'unnatural diseases' which were the evil machinations of the Devil, but not 'natural diseases'.
19. This is still the weakness of some of the brightest/silliest 'absolute materialists' who pontificate about the universe today.
20. Quoted in Goldsmith, *Franz Anton Mesmer*, p. 89.

> what Dr Mesmer has achieved in the way of curing the most diverse maladies leads us to believe that he has discovered a mysterious force in nature not previously recognised.[21]

The Electoral Academy in Munich elected him a member. Its citation read:

> There is no doubt that the activities of so outstanding a personality who has won fame by incontrovertible experiments and whose erudition and discoveries are as unexpected as they are useful must add lustre to our Institution.[22]

Of course, neither Augsburg nor Munich had to live with Mesmer. The medical establishment in Vienna who did were growing restive. Their main complaint was that he would not explain how his treatment worked. This was not strictly true but not entirely untrue either. He did explain, but his explanations never quite made sense. Nor did he deny that he was uniquely – at any rate exceptionally – adept at applying his own doctrines. A certain amount of mystification was acceptable. Far more lamentably, he had cured several important patients, including a Princess Lobkowitz, from painful spasms of the womb, a fashionable and lucrative complaint, after their own eminent doctors had declared her condition to be incurable. That showed a grave lack of respect. The final straw was the case of the musical prodigy, Maria Theresa von Paradies.

Fräulein von Paradies was the daughter of one of the Empress' recently ennobled private secretaries. At the age of three she suffered some kind of a fit and went suddenly blind. Bravely, she continued her musical education; and the Empress instructed Professor von Störck, considered to be the greatest oculist in Europe, to treat her eyes. He did so for ten years. Her vision did not improve. Blindness did not prevent her from becoming a virtuoso pianist. As such, she became the Empress's protégée. Child prodigies were the rage. Blind child prodigies were gifts from Heaven. Fräulein von Paradies travelled with her father and was fêted in London, Paris and Munich. In Vienna Mozart dedicated his delicious B flat major Piano Concerto, K 456, to her. Nobody, not even the composer, played the slow movement with so much *Gefühl.* Acutely he added that her blindness gave her playing an unique other-worldly quality. Now her parents consulted Mesmer. He explained to them that if their daughter's optic nerve was damaged, he could do nothing. But if the nerve was intact, he would cure her. The proviso was astonishing at a time when the function of the cranial

21. Quoted ibid., p. 124.
22. Quoted ibid., p. 136.

nerves was still almost totally obscure. She moved into the private hospital which had been built next to the Mesmer house.[23]

After a few sessions Fräulein von Paradies experienced a curious state of night vision. She began to see in the twilight but not in daylight. The nervous twitchings of her eyes ceased. She then developed sharp pains in her head. When the pains subsided she became acutely sensitive to light. But then gradually her vision improved. 'Dr Mesmer was the first human form she saw,' the delighted father wrote to a friend.[24] Her mother professed to be overjoyed. The daughter even recovered her lost sense of smell which in true Viennese fashion she valued more than her vision. Indeed, with her regained vision she herself was less than happy. She found human forms revolting, preferring the shape of horses. Noses were repugnant. Ears were comical. She could differentiate between colours but could not attach names to them. Black made her depressed. She fainted when relatives she had known but never seen came to visit her. She burst into a fit of laughter on seeing her two favourite uncles in uniform. She was often besieged by a curious crowd. She hated that. She made strange, sometimes shocking remarks when she first saw familiar objects. Her first glimpse of the starlit night sky was a 'deeply moving experience' (according to her father) but the flight of birds made her swoon. Most worryingly, she found the sight of a crucifix unbearably painful, 'bursting into uncontrollable and inexplicable tears' when one was held up to her. Was she a saint or a fool?

The gravest cause of her parents' anxiety was the effect of her cure on her playing: 'She now requires far greater concentration.' Formerly she could play difficult concertos without hesitation. Now she watched her fingers moving over the keyboard and missed many notes. She became frightened: the visible world was suddenly filled with menace. Her parents saw their livelihood threatened. She was growing up. Good-sighted clavier players were two a penny in Vienna, and the Empress could not be expected to subsidise them all. In a bold volte-face the prodigy's father now threatened to kill Mesmer (or, more conveniently, have him assassinated). A high-powered medical commission, headed by Professors Barth and von Störck, was charged to investigate the case. It was the moment the medical establishment had been waiting for. The Commission could not deny that Fräulein von Paradies who had been blind could now see, but 'since she could not name any colours, she must still be considered blind'. Dr Ost, one of the court physicians, handed the verdict to Mesmer. Animal magnetism was humbug. He must at once cease his fraudulent practices. Instead, in January 1778 he quietly departed.

23. No private nursing homes existed in Vienna or anywhere in Europe but fashionable physicians often had a residential wing for patients added to their homes or consulting rooms.
24. Quoted in Goldsmith, *Franz Anton Mesmer*, p. 146.

...

Paris, Mesmer's home for the next ten years, was the heart of the European Enlightenment. Apart from abundant local talent, it was visited by scientists, artists thinkers and men of letters from every part of the world, including France's expensive protégé, the young United States of America.[25] The *salons* of Mme du Deffand and Mme Geoffrin, where the likes of Diderot, d'Alembert and Grimm (with Voltaire an occasional guest) held forth, saw themselves as arbiters of taste, intellectual powerhouses of a dawning civilisation. Among foreigners Austrians were particularly numerous. Since the success of his new-style opera, *Iphigénie*, four years earlier, Gluck had been lionised and was a special favourite of the Queen, herself an Austrian princess.[26] Remembering past kindnesses, he welcomed Mesmer. Austria's ambassador, Count Mercy, was also glad to introduce at court a countryman who had no political axe to grind or infallible financial schemes to peddle. Through Gluck Mesmer also met Charles Leroy, president of the Académie des sciences. Leroy assured him of his benevolent interest, but the medical profession was divided. Some leading surgeons like Pierre-Joseph Desault were at least intrigued. Desault had an incisive mind and the possibility of operating on mesmerised patients passed through it. But the armour of the Académie de médecine was hard to pierce. The Académie could not be blamed. For decades Paris had been infested with adventurers, impostors and confidence tricksters. Memories of 'Count' Cagliostro's 'elixir of immortal youth' and the 'abbé' Monilant's 'pentagon to abolish original sin' were still fresh. The more outrageous the patter, the more easily brilliant *philosophes*, devout clerics and intellectual ladies seemed to be duped. Yet 'it is impossible to convey the passions aroused by Mesmer', wrote Baron Dupotet. 'No theological controversy of earlier ages could have been conducted with a greater furore.'[27]

Nobody could doubt Mesmer's success with patients. Talleyrand, still bishop of Autun, was to remark in old age that nobody who had not lived during the *ancien régime* knew what *douceur de vivre* meant.[28] He could

25. Financing the American War of Independence was probably the greatest drain on French finances in the years leading up to the Revolution.
26. Marie Antoinette was the youngest daughter of the Empress Maria Theresa and the sister of the Emperors Joseph II and Leopold II.
27. Quoted by V. Buranelli, *The Wizard from Vienna* (New York, 1975), p. 67. It was Dupotet who later introduced mesmerism to the United States.
28. He was one of history's great survivors and an expert on *douceurs* of every kind. Nobody ever amassed such a fortune accepting bribes from so many conflicting interests or sired such a gifted brood of illegitimate children. He ended his career as France's ambassador to the Court of St James when he successfully defused a crisis that threatened to plunge the two countries into war. His solution was the creation of Belgium. For this he must be forgiven. He was a great European as well as a great Frenchman. He died in 1836, aged eighty-four.

equally have said *douleur de vivre.* An interesting pain was *de rigueur* in fashionable society. A talking point, it helped to banish ennui, the terrible companion of too much *douceur.* But ennui did not stop sufferers from wanting to be cured: an interesting cure was as good as an interesting affliction. 'You do not have to be healthy, but change your illnesses from time to time,' Mme du Deffand admonished one of her admirers.

Mesmer opened a consulting room. Within weeks he found himself besieged. Such indeed was the onrush of patients that he had to devise ways of treating them in groups. This was the origin of the famous *baquet,* a wooden tub about five feet across and one foot deep, filled with water and in the water were iron objects and rods partially projecting from it. Patients clung to these rods and pressed them to their ailing organs. To facilitate the flow of magnetism, they also held hands. Later Mesmer magnetised a tree so that patients could be healed by holding on to ropes hanging from branches as well as embracing each other. The last, according to Mesmer's ill-wishers, could easily get out of hand. But it was not only the rich and bored who came. The poor travelled from every corner of France, especially when word got round that the doctor treated those unable to pay free of charge. They provided him with subjects for experimentation.

The most immediate and noticeable effect of 'animal magnetism' was the *crise.* This often took the form of violent convulsions but was sometimes manifested in painful spasms, tingling or even a difficult to describe sensation of 'something happening inside my body'. The objective then was to continue treatment until the patient was completely 'demagnetised'. This was usually signalled by an abrupt disappearance of the main symptom, almost invariably pain.

Mesmer's first accommodation soon proved inadequate and he was now able to afford a substantial house in Créteuil near Paris, deliberately inaccessible, where up to twenty patients could be accommodated. He rarely had empty beds. The rich continued to seek him out in search of a new thrill, the poor to find a cure. The former included the *crème de la crème* of French society. The queen sent for him and Mesmer was thrilled. The patient turned out to be Marie Antoinette's elderly poodle, Marionette. Mesmer did not mind: he liked to experiment on animals. Marionette was duly demagnetised and, according to the queen's friend, Mme d'Estrée, was rejuvenated. Then Mme d'Estrée herself became a patient. Having suffering for years from abdominal cramps, she spent two weeks in Mesmer's healing emporium and was cured.

But what Mesmer wanted above all was recognition from his professional brethren. The greater his success, the more unattainable this became. Complaints from eminent doctors, that he was 'seducing their patients', that he was a 'dangerous charlatan' and 'a secret revolutionary' flooded into the

Académie de médecine. A submission was eventually composed and sent to Versailles. It still exists. It could have been composed today. What caused the doctors' sleepless nights was not their loss of income. Perish the thought. They were deeply concerned that the quack was putting their patients' lives at risk. Louis XVI was not an impulsive man. With the economy in turmoil, the treasury empty and some of the provincial *parlements* in rebellion, animal magnetism was not perhaps the poor man's most pressing concern. But on 12 March 1784 he appointed a commission to investigate the allegations.

Starry-eyed as always, Mesmer was delighted. The Commission was to include some of the country's most respected figures. Lavoisier, the chemist, already recognised as an ornament of French science, was to be a member.[29] Jean Sylvan Bailly was a famous astronomer, soon to be Mayor of Paris. Both were to end their life on the guillotine. Dr Guillotine himself, a popular practitioner already known for his advocacy of humane executions, was also invited.[30] Among other appointees were Jean-Pierre de Jussieu, one of a famous family of botanists, Leroy of the Académie des sciences, Sallin, Majoult and d'Arcel, well-known members of the Académie de médecine, and Dr Benjamin Franklin, then seventy-two and ambassador of the United States.[31] The commissioners, Bailly reminded them, must limit themselves to establishing physical proof or the lack of it. As always, the outcome depended on formulating the questions to serve the desired result. Did Dr Mesmer's 'universal fluid' exist or not? Could one smell or taste it? What were its properties? This was the most vulnerable aspect of Mesmer's doctrine. He had never claimed for it physical attributes. But *fluidum* had to exist. He had discovered the extraordinary power of medical hypnosis; but he could not leave it at that. It smacked too much of religious hocus-pocus. There must be a physical explanation why his patients got better. The Commission was deliberately prevented from investigating whether Mesmer's treatment actually did any good. The 'illiterate and credulous' crowd who represented the majority of patients were obviously 'unreliable'. The minority, who included some aristocratic figures, 'could not be interrogated as if they were criminals'. Indeed, the testimony even of ardently enthusiastic clients – the Commission avoided calling them patients – was excluded.

The result was predictable. The report was damning. Bailly in particular had discovered 'by means of an electrometer and a needle' that Mesmer's

29. See Chapter 16.
30. Joseph-Ignace Guillotine did not, contrary to legend, die by the instrument he had invented. He lived till 1814 when he died in his bed, aged seventy-six.
31. It was suggested that Franklin should be chairman but he declined on account of his 'weakness'. He lived in Passy at the time, attended only one meeting, sat on the fence as usual, and expressed little interest in Mesmer and his ideas.

*baquet* contained neither electricity nor magnetism. Animal magnetism did not exist. Mesmer was a confidence trickster.[32] What was not predicted was the outcry in Mesmer's support. Outraged patients rallied round. Their saviour had become the victim of bigotry, obscurantism and greed. Molière was dead but in Beaumarchais' France dissatisfaction with the medical profession was still seething. The aggrieved had now found a martyr. One of the consequences was an upsurge of new associations of animal magnetism, known as *sociétés d'harmonie.* They would carry on Mesmer's work in spite of the doctors. Within a few months of the official report disciples published a *Receuille des pièces les plus intéressants sur le Magnetism animal*, filled with testimonies to the extraordinary accomplishments of Mesmerism. The marquis Dissert de Rover recorded the healing he himself had performed 'using the methods elaborated by M. Mesmer' on a fourteen-year-old kitchen boy who 'enjoyed his first painless day in years'. Father Gérard, an army chaplain, had performed a similar miracle in a military hospital, 'credit for which must revert to Dr Mesmer'. Charles Moline, a Protestant clergyman wrote that he himself 'owed his renewed vigour, faith and health to animal magnetism'.[33] As would happen to Freud in Vienna a hundred years later, Mesmer's lay following felt liberated by the hostile reaction of the medical profession. They became his most fanatical supporters.

But they also branched out in new directions. Mesmer may have disapproved of the branching; but rebels cannot reproach other rebels. Perhaps he dimly perceived the new developments as progress. Two of his most devoted followers, the marquis de Puységur and his brother, Count Maximilien Puységur, started a chain reaction. They recognised that what mattered in mesmerism was not the magic fluid or the magnetic emanations but the induction of a 'somnambulistic state'. (Somnambulism or sleepwalking had been known long before Shakespeare sent Lady Macbeth on her celebrated perambulation.) The Puységurs rationalised this by calling it a state of 'second consciousness'. They also recognised that such a state could be induced and that occasionally the 'second consciousness' could trigger off a healing process. Although devoted to Mesmer as a person, the brothers emancipated themselves from the heavy curtains, the soft music, the scent, and the hushed atmosphere. Most of their patients were simple country folk; and the Puységurs treated them out of doors under an old elm tree. 'Indoor consultation is contrary to a natural cure,' Count Maximilien wrote. The magnets and the *baquet* were quietly jettisoned. The patients tended to go quietly to sleep

32. De Jussieu submitted an intelligent and honest minority report. The surgeon Charles Deslon wrote his own dissenting report. They made little impact.
33. See P.F. Deleuze, *Histoire critique du magnétisme animal* (Paris, 1813), p. 234.

rather than develop convulsions. After numerous sessions, following the command of the healer, they simply recovered. Mesmerism had come of age.

. . .

The Revolution put a stop to scientific ventures in France but some of the émigrés introduced mesmerism to German courts, to Stockholm and to St Petersburg. Like most enlightened intellectuals, Mesmer at first welcomed the political upheaval. He was not averse to having ancient privileges clipped; he held no brief for academies and universities. Like the unorthodox Dr Philippe Pinel, he did not believe that mad or even hysterical people were possessed by the Devil. But the honeymoon did not last. He was known for his aristocratic clientele. He was Austrian, like the unpopular queen. Even the fact that the first Mme Danton had been one of his patients did not count in his favour. Forewarned of his impending arrest, in 1792 he left Paris, abandoning his house, books, notes, manuscripts and faithful hound.

In Germany the Margrave of Baden gave him a grant 'to pursue his humanitarian studies'. Encouraged by such patronage, in 1793 he returned to Vienna. The move was a mistake. He expressed politically incorrect views about ordinary people having been exploited by the *ancien régime*. Austria, humiliated by the French on the battlefield, was passing through a phase of malignant spy fever. Mesmer was arrested and spent a few days in jail. On his release he moved to a village in Canton Thorgau in Switzerland. There he spent the next few years rewriting his notes. In 1798, after the fall of Robespierre, he returned to Paris, hoping that the Directory and later the Consulate would pay him compensation. Eventually Bonaparte granted him a yearly pension of 3,000 francs, a modest but not a paltry sum.[34] Perhaps the First Consul was swayed by his wife, Joséphine, who was an admirer.[35] Mesmer then moved to Frauenfeld, a small village on Lake Constance, not far from where he was born. A visitor a few years later wrote:

> An elderly female relative keeps house for him: they live simply but the food is always good and he does not despise good wine. He likes discussing his 'system' . . . and is rewriting his notes; but he is out of touch with recent developments.[36]

In 1809 there was talk in Berlin about the Prussian Academy of Sciences inviting him for a series of lectures. The Napoleonic Wars intervened.[37]

34. A captain's pay in the army.
35. Her first husband, General Beauharnais, had been cured of his rheumatism by Mesmer before he became one of the last innocent victims of the Terror.
36. K. Bittel, *Der Berühmte Hr. Doct. Mesmer, 1734–1815* (Überlingen, 1938), p. 346.
37. Between 1805 and 1812 Napoleon was at the peak of his power. He crushed the Prussian army at Jena in 1807 and defeated the Austrians again at Wagram.

Interest in mesmerism continued, but more along the lines advocated by the De Puységurs than as the original 'animal magnetism'. Yet even those who departed from the original doctrine looked on him as an ancestral figure. A Dr Ernst Wolfart, a leading mesmerist in Germany and the author of the massive *Mesmerism, or the System of Reciprocal Influences: Theory and Practice*, went to see him in 1812.

> I found the doctor busy in his small sphere. The clarity of his insight and his ceaseless energy are as remarkable as the profound and elegant manner in which he expresses himself. Families come from far and wide to seek his advice as a doctor . . . One Schwabian miller brought his sick horse. It too was cured. He never charges them anything.[38]

But Mesmer had no wish to take part in the disputes that began to crackle around topics like 'somnambulism', 'autosuggestion' and 'hypnosis'. In 1814 he moved to a cottage even closer to his birthplace. A few months later he suffered a stroke. He died in his eighty-first year. 'Animal magnetism' died with him but not the idea that the power of the mind could overcome pain, even the terrible sufferings of surgical patients under the knife.

38. Bittel, *Der Berühmte Hr. Doct. Mesmer*, p. 348.

## Chapter 16

# PNEUMATIC MEDICINE

Pneumatic medicine carried the search for painless surgery to the threshold of surgical anaesthesia – and then stopped. The exact reason for this remains obscure.[1] By 1805 both nitrous oxide and ether, though not chloroform, had been discovered; and their anaesthetic properties had been glimpsed. Fifty years would pass before this was put to practical use. The actual discoveries nevertheless remain landmarks; and they helped to create a climate of opinion which eventually made anaesthesia acceptable.

Pneumatic medicine itself had a forgotten forerunner. Even in his lifetime John Mayow was overshadowed by the flamboyant cast of the scientific revolution. As a medical student in Oxford he hovered on the periphery of Boyle's circle and assisted him with his experiments but never became an intimate. After leaving university he set up in practice in Bath, not yet the fashionable spa of a century later, and seems to have pursued his experiments as a hobby. He died in 1679, at the age of thirty-eight. He published one short treatise on respiration. It seemed to have been read by few and understood by fewer. When, with considerable difficulty, Thomas Beddoes obtained a copy of *De respiratione* more than a century later, he was stunned. He sent his translation to his mentor and hero, Joseph Black in Edinburgh. Black, one of the great chemists of the day, had held the chairs of both medicine and chemistry first in Glasgow and then in Edinburgh. In 1750 he had discovered carbon dioxide. Called 'fixed air', it was the first gas to be isolated.[2] Though much admired and

1. See Chapter 18.
2. His most important discovery was the latent heat of compounds and chemical reactions in 1760. He died in 1799, aged seventy-three.

   'Gas' is one of the few words which have been invented rather than derived from some earlier root; but the inventor, the Flemish chemist Jean Baptiste van Helmont (died in 1644, aged sixty-seven), admitted the influence of the Greek χάος (chaos). Joseph Priestley suggested that van Helmont was more inspired by the German *Geist* (spirit) or *Gaascht* (an eruption of wind).

even loved by his pupils, he was an austere character, not given to theatrical gestures. But, after perusing Mayow's book, he, by his own account, 'lifted his hands to heaven in shock and amazement.'[3] Here were in essence most of the revolutionary discoveries of respiratory gas exchanges which had convulsed the scientific world in his own time and which would later inspire Beddoes to hail his age as the dawn of the chemistry of life. Characteristically generous, Beddoes not only translated Mayow's book but devoted much time and effort to having the author recognised as a pioneer.

The purpose of respiration, Mayow had argued, was to pass something in the air to the blood during the pulmonary circuit. That something – Mayow called it 'nitrous particles' – was essential to life. It was carried by the blood to all parts of the body where it sustained 'fermentation' and was responsible for keeping the body warm. It was responsible in particular for muscular contraction, including the pumping action of the heart. Lack of it would deprive blood of its power to maintain body heat and the heart of its capacity to contract. The result would be death. No breath, no life. Mayow also collected evidence to show that all combustion – whether the burning of a candle or the explosion of gunpowder – was a kind of 'fermentation', similar to the process which kept animals and humans alive. All depended not on air as such but on a portion of air – in some of his works Mayow calls it 'inflammable air' – which these processes both needed and consumed. Not all air was 'inflammable air' because when a candle was burnt in a closed vessel, only part of the air disappeared. The remainder would not support either flame or life. Mayow believed in accurate measurements. When he burnt camphor resting on a wooden platform floating on water in an enclosed bell jar, the water level rose by the time the flame went out. The flame had consumed about a fifth of the total air space. He then enclosed a mouse in the same manner. By the time the mouse died, the 'air space' had diminished by about the same volume. When a candle was burnt in the same bell jar as the mouse, the mouse died more quickly. Respiration, he concluded, abstracted 'inflammable air' from atmospheric air. As a dog was being deprived of 'inflammable air', it would start to pant and gasp. When it was transfused with blood from another dog, it quietened down for some minutes. Though Mayow did not actually isolate any gases, he also discovered that nitrous gas (NO) would absorb oxygen from air to form nitric oxide ($NO_2$). This was to become the basis of the pneumatic chemists' standard method of measuring gases.

. . .

Even in the face of mounting evidence, simple, attractive and plausible ideas can paralyse scientific thought for generations. For more than a hundred years

3. Quoted by F.F. Cartwright, *The English Pioneers of Anaesthesia* (Bristol, 1952), p. 43.

after Mayow, chemistry continued to revolve around a non-existent substance called 'phlogiston'. The term had been coined towards the end of the sixteenth century;[4] but the 'phlogiston theory' to explain changes during combustion was elaborated in full by a German chemist, Georg Ernst Stahl, towards the end of the seventeenth. Stahl suggested that all combustible materials contained 'phlogiston', a substance which he picturesquely but wrongly described as 'fire in a fixed state'. During 'active combustion' this substance was released. After combustion materials were said to be dephlogisticated. The theory was generally accepted despite the fact that several workers other than Mayow had shown that the air space around burning material tended to diminish rather than increase and that material combusted – that is, the material minus phlogiston – increased rather than decreased in weight.[5] Only towards the end of the eighteenth century did three chemists in three countries, working independently of each other, explode the concept and launch modern biochemistry.

The oldest of the three, Joseph Priestley, was a dissenting English minister born in 1733 in Leeds. Without neglecting the care of souls, his mind explored and probed widely. His published work, all concisely and elegantly phrased, ranged over electricity, the physiology of vision, colours, grammar, education, the human mind and all aspects of theology. While living in Calne, in Wiltshire he read Black's work on carbon dioxide; and 'resident in the neighbourhood of a public brewery, a little after midsummer in 1767 I was induced to start experiments with fixed air of which there is always a large body, ready formed in vats over fermenting liquor . . .'[6] He showed that fixed air extinguished candles but that water saturated with it acquired a pleasant tingle.[7] But it was his isolation of 'inflammable air' by a variety of ingenious chemical processes that proved a landmark. In 1774 he discovered that he could drive off a gas from red oxide of mercury with heat and that this gas not only supported combustion but did so more vigorously than ordinary air. Finally, with some trepidation for he suffered from asthma and was sensitive to many gases and smokes, he tried to breathe in the gas himself. Instead of finding that it precipitated an attack of choking, he was

4. From the Greek φλογιστός (*phlogistos*, inflammable). It was coined by Raphael Eglin (died in 1622, aged seventy-one) but was brought into general use by Ernst Stahl. Metals were supposed to contain it in greater or lesser degree depending on the ease with which they could be calcined.
5. This is slightly complicated when the diminished gas space is replenished by the evolution of a gas other than oxygen during the burning.
6. J. Priestley, *Researches* (London, 1775), Appendix 30.
7. He recommended that water on board ships should be carbonated to make it more durable and thirst quenching. The Admiralty thanked him for his 'valuable suggestion' and passed it on to the admirals. Nelson thought the idea ridiculous.

delighted to experience instantaneous physical and mental invigoration, the first of millions of slightly anoxic individuals to do so.

So pervasive was the phlogiston theory at the time that Priestley could not at once rid himself of the concept and called the gas 'dephlogisticated air'. It was left to Antoine Laurent Lavoisier to interpret Priestley's discovery correctly and show that combustion did not 'liberate' but 'consume' a gas; and that once that gas was consumed, the atmosphere could sustain neither combustion nor life. Fortunately for Priestley's fame (which exercised Priestley not at all) Lavoisier was already recognised as the greatest chemist of France and even Europe and had neither the inclination nor the need to claim somebody else's discovery as his own. Coming from a prosperous but not aristocratic family, he had been admitted to the Académie des sciences in 1768, at the age of twenty-five, and within a few years had changed the character and language of chemistry. Ironically, among his many excellent terminological innovations his most famous coinage, 'oxygen', was based on the mistaken idea that the element was an essential constituent of acids.[8] He also named nitrogen and hydrogen, the latter recently discovered by the reclusive English aristocrat, Henry Cavendish.

The third discoverer of oxygen was Karl Wilhelm Scheele of Stockholm, the most modest and unassuming of the three, who spent his life serving customers in his apothecary's shop and snatching the odd half-hour to experiment in his combined kitchen-laboratory at the back of the pharmacy. He published only one slender book, *On Air and Fire* (1777), and probably not enough papers to gain him tenure in a middle-ranking American university today; but into those he crammed an astonishing number of discoveries. Beside producing and characterising oxygen, he isolated and described chlorine, tungsten, molybdenum, ammonium, hydrofluoric acid, arsenic acid, lactic acid, gallic acid, pyrogallol, oxalic, citric, tartaric, malic, mucic and uric acids, glycerine and lactose. He determined the nature of borax, demonstrated that graphite was a form of carbon and synthethised a number of important

8. From the Greek ὀξύς, sharp or acid and γεννάω, I generate. Like most great discoveries, oxygen too was anticipated by almost three hundred years. In 1489 the German 'alchemist', Eck von Sulzbach, noticed that red oxide of mercury gave off a 'spirit' when heated and that metals heated in this spirit gained weight. He attributed this to the union of the spirit with the metal. Mayow called the same 'spirit' 'fire air'. Scheele in 1774 (without knowing about Priestley's work) called the gas 'empyreal air'. Lavoisier was shown the gas by Priestley and correctly described the nature of oxidation. He published his historic paper, 'Mémoire sur la nature du principe qui se combine avec let métaux pendant leur calcination' in 1775. His error about oxygen being an essential constituent of all acids was not demonstrated till 1837 when Liebig showed that it is hydrogen, not oxygen, which is essential and suggested (happily in vain) that hydrogen should be renamed oxygen, and vice versa. In 1886 Dewar first obtained solid oxygen in the form of snow.

pigments and dyes, some of which are still cherished by artists.[9] Unlike the lives of his French and English co-discoverers of oxygen, his career was entirely uneventful; and he died, perhaps from tuberculosis, at the age of forty-six.

. . .

The man who converted pneumatic chemistry into pneumatic medicine was a neglected English genius. Doctor, scientist, linguist, political radical, sociologist, anthropologist, educationalist and poet, Thomas Beddoes embodied many of the hopes and disappointments of his generation. Like most young thinkers, writers and poets in England, many of them close friends, he ecstatically welcomed the wind of change that started to blow across the Channel during the summer of 1789 – the largely symbolic fall of the Bastille, the calling of the Estates-General, the declaration of the Rights of Man, the victory of Valmy. But then, with hardly a breathing space, bright morning darkened into the night of the Terror. 'It was the best of times, it was the worst of times.' And, almost without another break, England found herself fighting for her survival against Napoleon, who claimed to be the heir of the Revolution but turned out to be not much better than any other of history's diminutive megalomaniacs.

. . .

Thomas Beddoes was born in 1760 into a modestly prosperous tradesman's family in Shifnal in Shropshire. The forebears came from Wales. His father died young; and the boy and his sister were brought up by their grandfather, another Thomas. The boy showed promise early and received a sound education at the grammar school in Bridgnorth and later from the Reverend Samuel Dickenson of Plymhill in Staffordshire. He then entered Pembroke College, Oxford, to study medicine and taught himself French, Italian, Spanish and German. Moving to Edinburgh, he was inspired by Joseph Black and William Cullen, professor of medicine. Eventually he set out for a continental tour and met Lavoisier and Spallanzani. He returned fired by the Promethean vision that medicine, like society in general, was on the threshold of a glorious revolution. He began to perceive disease and its main manifestations, pain and fear, as an enemy to be conquered, just as the imminent political upheaval would conquer oppression, poverty, ignorance, superstition and bigotry.

9. Among them Scheele's brilliant green. Many of Scheele's pigments were based on arsenic and cheap to produce; and for almost two centuries they were widely used to cover walls and large surfaces. One of them may have been responsible for Napoleon's death in St Helena from arsenical poisoning and for the nearly fatal illness of Mrs Claire Booth Luce, United States ambassador to Rome in the 1960s.

Disease would always remain to Beddoes more than an external foe. A doctor worthy of the name should see it as a malign but integral part of his patients, reflecting their entire past, born of their social background, their wealth or poverty, their class, their gender and their mind and beliefs. It also spoke about the distribution of plenty and poverty in society, of power, of oppression, of knowledge, of ignorance, of support and neglect. It had, in short, its social pathology and political dimension. It was an attitude that, independently of his discoveries, made him one of the great radicals of English history and the greatest of English medicine. Small and plump, often probably slightly comical, generally short of breath, he was too busy in an astonishing range of causes to be wholly effective in any one, but remained undaunted. He read voraciously, translated Scheele but also reviewed Goethe's *Wilhelm Meister* for the *Literary Gazette* and devoured the profusion of incendiary political pamphlets that poured from the underground presses of Paris.

Returning to Oxford, he was appointed lecturer in chemistry. It was to be a fruitful but not a happy period. Chemistry was a newcomer among academic disciplines and, compared to physics, mathematics and of course the humanities, was viewed with suspicion. It reminded the more enlightened fellows of medieval practices, of witches and cobwebbed alchemists' caves. It figuratively and literally stank. As the revolutionary events unfolded in France, Beddoes made no secret of his political sympathies, apostrophising Mr Pitt and his government as the 'most mendacious ever to govern Britain'. (The charge, probably true at the time, has now been rendered obsolete.) This was too much for conservative, Royalist Oxford. Disapproval of Beddoes' political views was aggravated by his propensity to denounce in virulent terms noxious but enshrined practices. The state of the Bodleian Library in particular aroused his ire. The books were uncared for, uncatalogued, inaccessible and regarded by the librarian as a distraction from his main interest, which was butterflies. But, while Beddoes himself was dissatisfied, his students left glowing testimonials of his numerous, ingenious and 'sometimes successful' experiments. One of them, James Melville, later wrote that 'the time of Dr Beddoes' residence in Oxford produced a taste for scientific research that bordered on enthusiasm'.[10] But when Beddoes' proposal to establish a chair in chemistry was turned down, he resigned and, after some exploratory travelling, moved to the city where pneumatic medicine, the precursor of inhalation anaesthesia and painless surgery, was to be born.

. . .

10. Quoted by Cartwright in *The English Pioneers of Anaesthesia*, p. 245.

Bristol had not been noted in the past for its scientific attainments; but as a centre of maritime commerce it had grown into England's second most prosperous city. Tall ships sailing up the Avon brought to it the riches of Spain, the Mediterranean, America and the West Indies in exchange for Bristol cloth, blankets, pottery, lead from the Mendips and wool from the Cotswolds. Most profitably, those carrying 'black ivory' from West Africa paused here before continuing their long journey to the plantations of the New World. As the century drew to a close, this profitable exchange of goods was already declining. The American colonies had been lost, the slave trade had found Liverpool more accommodating, the wool trade was shifting to London and commerce in iron and coal was edging north. But the city still drew all the talent from the West of England; and the circle which Beddoes joined was incandescent. Among his pupils were two future presidents of the Royal Society – Davis Giddy and Humphry Davy[11] – he was to marry the sister of the novelist Maria Edgeworth, he became a close friend of Robert Southey, future poet laureate, and of Wordsworth, Coleridge, Joseph Cottle and Walter Savage Landor, all in their exuberant radical youth. There were the gifted and rich Wedgwood tribe, James Sadler the ballooning aeronaut, James Watt and his family and, not far north and frequent visitors, Erasmus Darwin and other luminaries of Birmingham's Lunar Society.[12] Later Peter Mark Roget came to start his distinguished medical career under Beddoes' wing, years before, in retirement, he was to compile the best treasury of words in any language.

There was another 'growth industry' which would profoundly affect Beddoes' career. The wells of Bristol and especially those springing up under the hill on which stood the old parish church of Clifton, were proving a serious challenge to Bath. There was an influx of those of rank and wealth with their servants, retainers and fashionable ailments; but, more important and numerous, there were the victims of the terrible 'new' disease which seemed to be a camp-follower of the recent stunning advances in technology and industry. The Wattses, the Coleridges, the Cottles, the Edgeworths and the Wedgwoods all had consumptive members in their families; and they came seeking a cure or at least a merciful respite in the gentler climate of the West

11. On being elected to the Royal Society Giddy changed his name to his wife's maiden name, Gilbert.
12. Founded by Matthew Boulton in 1768, the Lunar Society was probably the most brilliant of several provincial 'dining clubs' dedicated to the promotion of science, technology, the arts and engineering. The name derived from the practice of meeting at the time of the full moon so that members who lived some distance away could return with greater ease. James Watt, Erasmus Darwin, Thomas Day, Lovell Edgeworth, Josiah Wedgwood, Samuel Galton and William Withering (of digitalis fame), many of them friends of Beddoes, were among the members. Priestley was the last member to be elected when he was appointed minister to the Presbyterian Meeting House in Birmingham in 1780; and the Society came to an end when Priestley fled to America.

Country. Consumption (as tuberculosis was called in England for much of the nineteenth century) would become the backbone of Beddoes' clinical practice; and, among his vast output of medical writing, he gave a poignant account of the new plague.

> There are few today in this city who have not beheld a person labouring under consumption of the lungs. I cannot recall seeing it as a youth. Now I am watching it constantly, from the very beginning, hardly suspected, to the last harsh breath of life . . . and I must declare that the oftener it has fallen under my observation, with the greater horror have I learnt to view it. I no more pretend to enter into all the feelings of these young consumptives than I do into the sorrow of a mother who has lost her only child; but I clearly perceive that it has miseries in common with most other disorders of the worst kind as well as disorders peculiar to itself . . . Consumption has the harrowing chills and heats of fevers in general, it has the drenching and draining sweats of agues. It has the insupportable fatigue and languor of some slow disorders. It has the oppression of asthma – indeed a worse oppression since it often grows stronger and stronger up to actual drowning by inches or suffocation from matter in the chest, which there is not strength to bring up . . . Belonging to an advanced stage, there is a scene which I have often witnessed as the consumptive expires under increasing shortness of breath. They have that sort of pain that will last for hours and even for days with some intermissions. The sufferer gasps: at times he snaps convulsively at the air; and calls with dismal shrieks upon the bystanders for breath! breath![13]

Formerly a rarity, it was suddenly everywhere.

> Go upon some of the high grounds round Bristol . . . take in its multitude of houses and people densely packed. Will you not be startled when I tell you that as many souls as are in this extent of houses and streets, so many are destined to perish in Great Britain every year under the grip of this giant malady.[14]

Soon the old graveyard around Clifton parish church would become too small and a new cemetery would have to be opened and assigned exclusively to deceased consumptive visitors. But for a generation after the sudden and terrible upsurge, there lingered the conviction that the disease was not only

13. T.L. Beddoes, *Rules of the Medical Institution for the Benefit of the Sick and Drooping Poor; with an Explanation of its Peculiar Design and various necessary Instructions* (Bristol, 1796), p. 87. This wonderful little booklet which contains far more than 'rules' was handed out free to poor patients attending the Institution.

14. Ibid., p. 56.

curable in principle but also preventable; and it inspired some of the most powerful medical writings in the language.

...

The fact that consumption in its commonest form centred on the lungs was the immediate inspiration for studying respiration and using inhalable gases in its treatment. Several were becoming available in more or less pure form; and liquids like ether were volatile enough to be warmed and inhaled. In 1797 Richard Pearson of Birmingham stated that during the previous two years he had given ether as an inhalation in phthisis on a number of occasions and obtained more hopeful results than with any other drug. He made no extravagant claims and did not suggest that the inhalations prolonged life or made the late stages of the disease less distressing; but they gave better symptomatic relief from pain than anything else.

Beddoes experimented with oxygen and carbon dioxide as well as ether, using apparatus designed and built for him first by James Sadler and later by James Watt. He looked after Watt's much-loved consumptive son Gregory; and he and Watt jointly published *Considerations on the Medicinal Use and on the Production of Factitious Airs*. Slowly his repertory expanded and he began to experiment with 'vital air' or oxygen, 'fixed air' or carbon dioxide, 'inflammable air' or nitrogen and 'hydrocarbonate' which was impure carbon monoxide. To those Watt added from time to time his 'alchemic inventions' like fumes derived from the burning of feathers or from charring meat. They noted again and again that oxygen often brought relief of breathlessness and that carbon dioxide could ease the pain in the chest; and Watt observed with amazement that 'Dr Beddoes's sheer presence often acted as a powerful sedative'.[15]

In terms of cure none of these experiments were notably successful; yet they kindled and sustained hope, the most precious commodity of the consumptive. Experimental work and a nearly overwhelming clinical practice did not stop Beddoes from pouring out poetry, political and sociological booklets and pamphlets dealing with diets, healthy living, the evils of drunkenness, education, the purpose of the universities and any other topic that attracted his attention.

But as Beddoes' professional success grew, so the political horizon darkened. Horrific tales were recounted by refugees streaming across the Channel. What could happen in France could also happen in England. The monarchy – a mad king, a dissolute Prince of Wales and a bunch of disreputable dukes and duchesses – was never more unpopular. England's landed gentry who had for a century looked askance at the absolutism of continental rulers, now saw

15. S. Smiles, *Lives of Boulton and Watt* (London, 1865), p. 80.

their own leisurely (and no less absolute) rule in Westminster threatened. Pitt's government had no difficulty in introducing a series of repressive laws to 'forestall' a revolution.[16] Joseph Priestley was among the victims of this shift in public opinion. In 1791, living in Birmingham, he addressed an open letter to Edmund Burke, describing his reactions to events in France and urging (quite mildly) the abolition of outdated privileges. The document was used by the government to incite a mob who attacked Priestley's house and, chanting 'Priestley be damned, damned, damned for ever and ever', burnt it down with all his experimental gear, notes and books. Priestley himself escaped; moved to Hackney in London and started to rebuild his laboratory. Then news came that in France the Convention had elected him 'Free Citizen of the Republic'. In the prevailing paranoid atmosphere, continuously stoked by a corrupt government, this put him in imminent danger of being arrested as a traitor. He fled to America where he was welcomed – at least initially. He purchased a farm in Pennsylvania, where he died ten years later.[17] His son, Joseph junior, became one of the circle around Beddoes.

...

By a terrible irony, the same revolution that drove Priestley into exile from England for sympathising with it, murdered his friend and much-admired colleague Lavoisier in France. The great chemist was one of the country's 'tax farmers' – or licensed tax collectors – and had married the daughter of another. The breed was hated for their greed and corruption but nobody had ever impugned the honesty of Lavoisier.[18] His real crime was to ridicule Jean-Paul Marat's vapourings on respiration. Marat, a failed doctor and scientist and one of the vilest creatures washed up by the revolutionary tide, now had his revenge. Lavoisier was arrested and sentenced to death. When his friends tried to impress on Robespierre that the condemned man was France's and perhaps the world's greatest chemist, Robespierre famously replied that the Revolution had no need of chemists. Lavoisier died on the guillotine.[19] It was a signal. In anguished disappointment most of those who had welcomed the Revolution in England now turned against it; and Beddoes and almost all of his friends would back the war against Napoleon.

16. Including the infamous 'Gagging Act' and the suspension of Habeas Corpus.
17. Quite soon Priestley came to be regarded with as much suspicion in America as in England. 'The change that has taken place is hardly credible, as I have done nothing to provoke resentment; but, being a citizen of France and a friend of the Revolution, is sufficient'.
18. These *Fermiers généraux* occupied the more or less hereditary office of royal tax collectors and many were indeed corrupt, greedy and immensely rich. Lavoisier was not among them.
19. Marat was murdered in his bath by a brave young woman, Charlotte Corday from Normandy. She died on the guillotine shortly after Lavoisier.

. . .

It was dissatisfaction with early experiments that led to the foundation of Bristol's Pneumatic Medical Institution, the first clinical research institute in England and perhaps the world. Financed mainly by Thomas Wedgwood but also by numerous individual enthusiastic admirers and patients,[20] it opened its doors on 29 March 1799 in two converted houses in 6 and 7 Downey Square. Its newly recruited 'superintendent' was a young man from Penzance who had had no formal training either in medicine or in chemistry but who, in Beddoes' eyes, made up for this by other qualities.

Humphry Davy was twenty-one and had probably been recommended to Beddoes by his assistant, Davis Giddy. Giddy in turn may have met him through Gregory Watt who had been wintering in Penzance.[21] The young superintendent spoke with a broad Cornish accent, was undersized, weedy-looking and sallow-complexioned. He had uncouth manners which he never lost, yawning cavernously as soon as he lost interest in a conversation. Although he wrote verse and thought of himself as a poet, in a circle steeped in literary matters he was both unlettered and unwilling to hide his lack of interest in 'pointless philosophising'. He loved laboratory work but was described by Watt as 'an incomparable wrecker of instruments and breaker of glassware'.[22] Both his experimental planning and execution were haphazard. He was usually too impatient to have only one experiment on the boil at a time. Conducting three or four in wobbly parallel, he worked so fast that onlookers often thought that he was still preparing a demonstration when in fact in his own estimation it had been concluded to everybody's satisfaction. And yet there must have been a quality in him which in youth made him attractive to a wide circle of people. Southey immediately took him to his heart, the wayward Coleridge became a life-long friend, the Watt family and the Wedgwoods loved him, John King who became a surgeon to the new institution found him delightful and Mrs Beddoes became a second mother. (He lodged with the family throughout his stay in Bristol.) As for Beddoes, he wrote to Erasmus Darwin that the young superintendent proved 'fully equal to my wishes and far superior to my hopes . . . he is quick and eager to learn and has a mind as sharp as steel'.[23] Of the last there could never be any doubt.

20. Including the entire Medical Faculty of the University of Edinburgh.
21. Penzance, Davy's birthplace, was a 'coinage town' of some 3,000 souls, the market-place piled up with glittering ingots of tin, the people mostly small traders, fishermen and farmers. Davy attended the local grammar school which, under Mr Coryton, had a poor reputation. Davy did not excel at school but privately devoured books on chemistry, including Lavoisier's *Elements* which he found in Giddy's library.
22. Quoted in Smiles, *Lives*, p. 96.
23. Quoted in Cartwright, *English Pioneers*, p. 126.

# Chapter 17
# LAUGHING GAS

Before Davy's arrival the most promising fields of study for the new Institution had been a matter of debate. With Davy in post, the choice evaporated. The young Cornishman's enthusiasm had been aroused by the new gases and by their seemingly infinite potential; and, whatever was true of his often incoherent arguments, his enthusiasms were irresistible. He experimented on mice, on guinea pigs, on dogs and on horses but with greatest enthusiasm on his friends, Beddoes' patients and himself. He found that inhaling ether made people pleasantly giddy and sent animals to sleep. Carbon dioxide made him drowsy. It could not sustain life, but it could be soothing. Oxygen was extraordinarily invigorating; but breathing the pure gas for a prolonged period gave him a headache. Animals were often killed by prolonged exposure. He almost killed himself by inhaling 'pure hydrocarbon' – that is carbon monoxide – despite Beddoes' and Watt's warnings not to breathe the pure gas beyond a whiff or two. He had noticed that those inhaling it even for a short time became 'delightfully pink', and felt that his own success with the ladies would be improved by a similar change.[1] He was rescued in the nick of time by the kindly Dr Robert Kinglake, another member of the Institution, and treated with pure oxygen.

But the most exciting of Davy's experiments were carried out with nitrous oxide, a gas which may have been discovered by Mayow a hundred years earlier but whose production in a variety of ingenious ways had been developed by Priestley and further improved by Davy. The immediate catalyst was a publication by Samuel Latham Mitchill, professor of chemistry, natural history and agriculture in the College of Chemistry of New York. Mitchill had been hailed in America both as 'a living encyclopaedia' and as a 'chaos of

1. Carbon monoxide kills by combining with haemoglobin in place of and more firmly than oxygen. Carboxyhaemoglobin is bright pink in colour; and those committing suicide by breathing it often look uncannily 'pink' in death.

knowledge'. His rhyming scientific treatise, *Doctrine of Septon*, was modelled on Erasmus Darwin's *Botanic Garden* and contained some of the worst poetry in the language. But it was the science which mattered. Mitchill maintained that of the two gases isolated by Priestley oxygen was the giver and nitrous oxide the destroyer of life. The latter was in fact the diabolical Septon which brought fevers into the world and was the cause of cancer, scurvy, leprosy and a surprisingly variegated list of other ills. Mitchill's opinions were widely revered but made no sense to Davy. Ignoring the warnings of his elders that inhalation of pure nitrous oxide might be dangerous and possibly fatal, on the morning of 5 May 1798 he retired to his laboratory and inhaled the purest gas he could prepare. He lived. Indeed, he had to stop inhaling because he started to giggle.

If Davy's poetry tended to have a slightly laboured ring, his historic accounts of experiments with nitrous oxide are the outpourings of a poet.

> From the nature of the language of feeling, my description must remain imperfect. I have tried to give an account of the strange effects of nitrous oxide by making use of terms which are familiar; but the familiar cannot describe the entirely new. We are at the best of times incapable of describing pleasure and pain except by means of inadequate terms which have been associated with them at the moment of experiencing them . . . Pleasures and pains connected with new experiences cannot therefore be accurately described. So with the sensation caused by nitrous oxide. The sensations to me have been similar to no others and are consequently indescribable. So with other persons. Of two paralytic patients asked to describe how they felt after breathing the gas, the first answered, 'I do not know but very queer'; and the second 'I felt like the sound of a harp'.[2]

He experimented how much gas could be breathed with impunity.

> After breathing in nine quarts for three minutes I experienced the highest degree of pleasure . . . Then gradually my sense of pressure on muscles was lost, vivid ideas began to pass rapidly through my mind and voluntary power was altogether destroyed . . . so that the mouthpiece dropped from my unclosed lips.[3]

He found that on breathing nitrous oxide respiration became prolonged and deeper (almost certainly due to the accumulation of carbon dioxide) and that the effect on some people was to induce 'stamping and laughing or even dancing and vociferating' of which later they had no recollection. He

2. H. Davy, *Researches, Chemical and Philosophical Chiefly concerning Nitrous Oxide, or Dephlogisticated Nitrous Air, and its Respiration* (London, 1800), p. 57.
3. Ibid., p. 146.

experimented with the effect of the gas on sleep, appetite, sensibility and brain power.

> In one instance I had a headache from indigestion: it was immediately relieved by a sniff or two, though it afterwards returned. In another instance my exceedingly severe pain in the gums from erupting *dentes sapientiae* (wisdom teeth) was at once relieved and was milder when I stopped breathing the gas . . . Whenever I experienced pain it was relieved.[4]

So pleasant was the effect of the gas that at one stage he almost certainly became addicted to it – another first – and took fright.

> During the past fortnight I noticed that I have breathed it very often. I ought to have observed that a desire to breathe the gas was especially awakened when I caught sight of a person breathe in deeply or even of an air bag or an air holder.[5]

But, on recovering and after a brief holiday in Cornwall, he continued the experiments. There are in fact few manifestation of nitrous oxide which he did not investigate and did not try on his friends. Dr Kinglake experienced

> a sort of intoxicating placidity and delight and then, progressively, the suspension of seeing, hearing, feeling, the power of volition itself.[6]

The ladies, Davy noted with chagrin, were 'unduly reticent' when it came to acting as 'giggling guinea pigs'; but Coleridge, Southey, Joseph Priestley junior, Kinglake, Tobin, Lovell Edgeworth, William Clayfield and that master of synonyms, Peter Mark Roget, all gave vivid descriptions of their wonderfully pleasant experiences. Roget had suffered from a boil on his neck and, fingering it under the influence of that glorious gas, it burst. All pain had gone when he recovered: 'a miracle, a wonder, a marvel, a portent and an eye-opener'. Had he started in his mind to compile his *Thesaurus*? All this blaze of glory was recorded by Davy in his book, *Researches . . . concerning Nitrous Oxide, etc.* as well as in a number of papers published jointly with Beddoes. It is in the Conclusions of the *Researches* that Davy considered the potential uses as a gas.

> The transience of the effects seems an argument against using the gas in chronic diseases like consumption. But, for the same reason, *since nitrous oxide appears capable of totally destroying physical pain, it could probably be*

4. Davy, *Researches, Chemical and Philosophical*, p. 354.
5. Ibid., p. 234.
6. Ibid., p. 237.

*used with advantage during not unduly prolonged surgical operations in which no great effusion of blood takes place.* [italics added][7]

The book was published in 1801. Not for nearly fifty years did anybody act on the suggestion.

. . .

Davy departed for London and the wider world in 1801. The Society for Bettering the Condition of the Poor which had been founded in 1796 by William Wilberforce and a group that at one time included Beddoes had established the Royal Institution of Great Britain 'to promote science, art, manufacture, especially as those improved the welfare of the poorer classes'. Davy was appointed lecturer in chemistry at the salary of 100 guineas a year. He was an instant success. He was lionised by society. He was admired by his peers. In a sense his most important discoveries were still to come, as was his brilliant career as a lecturer. In another, perhaps distorted sense, his best years were behind him. The openness and generosity of youth were gone. His marriage proved a disaster. Professionally he become jealous, suspicious and twisted. He had no need to be. Honours were showered on him, including a knighthood and the presidency of the Royal Society. He was one of the country's prides, internationally acclaimed. He died in 1829 in Rome aged forty-eight after a series of 'apoplectic attacks'. Had he lived, he might have ended his life in a lunatic asylum.

Beddoes' practice continued after Davy's departure but as a research unit the Pneumatic Institute lost its sense of direction. It closed in 1802. Comparative failure of pneumatic medicine in the field of consumption turned the doctor's thoughts in other directions, especially towards the scope for preventing rather than curing diseases and in 1802–3 he published one of his most remarkable books, *Hygeia: Essays on the Means of Avoiding Habitual Sickness and Premature Mortality.*[8] A secondary heading was 'On Causes Affecting the Personal State of our Middling and Affluent Classes'. In a clear and persuasive style, he roamed over gluttony, obesity, too little exercise, too much drinking: it could be a tract for today. Nothing was too trivial to be overlooked – the timing of breakfast, the choice of beverage, the benefits of horseback riding, the effects of smoking. He thought that pipe-smoking in moderation was an excellent calmer of the mind; and that the mind was in fact as important as the body in causing pain, disability and disease. Consumption, dropsy, palsy, insanity – all came under his critical microscope and all statements and recommendations were illustrated with telling

7. Ibid., p. 345.
8. Quoted in Cartwright, *English Pioneers*, p. 231.

examples from his own experience. As the Pneumatic Institution crumbled, it recrystallised as the Preventive Institution, once again a first, not a theoretical think-tank but an open clinic where the sick and the able-bodied could come for what today would be described as a 'health check' and be given advice and occasionally medicine. From the start payment was voluntary, and none was accepted for seeing and advising children. He himself had two sons and two daughters:[9] and, despite Mrs Beddoes' intermittent infatuation with Davis Giddy, it was a happy home.[10]

In his last years Beddoes published two more books, *On Fevers* and *A Manual of Health*, the latter especially addressed to the poor and the unschooled. They are remarkable for their humanity and modernity. Throughout the texts he is especially concerned with children and the young.

> Beware particularly when girls, formerly active, grow as you think lazy. Ten to one but you scold them without mercy when you ought to get them cured. No, it is not lazy they are, but sick – sick sometimes to their very vitals.[11]

and

> I feel it sometimes a very difficult undertaking to give instructions such as shall be really useful. But there is one very unfortunate and pernicious mistake which I am sure the poor will have the good sense to see and avoid as soon as it is pointed out to them . . . Upon examining a particular child, it has occurred to me many scores of times to ask the parent: 'Pray, have you any more children?' 'Yes . . . but they none of them complain.' 'Be so good as to bring them . . . I would wish to examine them myself . . . A child roars when suffering from a sharp pain. But when pain creeps on them by degrees, when they find themselves merely in dull pain, they will rarely complain. And yet . . . that is when I need to see them.'[12]

9. Discussed in stimulating detail in V. Finlay, *Colour. Travels Through the Paintbox* (London, 2002).
10. The elder son, Thomas, became a doctor and a considerable poet. He committed suicide at the age of thirty-five in Switzerland.

    Giddy was only seven years younger than Beddoes, but, unlike the roly-poly doctor, he was tall, impressive and handsome. Anna Beddoes threw herself at him; but Giddy remained loyal to her husband and nothing came of her advances. After Beddoes' death, Giddy, now Gilbert and safely married, helped her and the family in many ways. In addition to his presidency of the Royal Society, Gilbert became an MP and an important personage in his native Cornwall.
11. T.L. Beddoes, *A Manual of Health; or The Invalid Conducted Safely Through the Seasons* (London, 1806), p. 65.
12. Ibid., p. 75.

Nobody had written about the young in this vein before Beddoes; and in recognising the danger of contagion in many diseases, especially in tuberculosis, he was half a century ahead of medical opinion in England.

He continued to compose verse. He corresponded widely on a diversity of topics. He remained generous in life but a good hater in print whenever he detected cruelty, greed or hypocrisy.

> The danger is not just from quacks. Our colleagues too, even when bedecked with diplomas, are often as needlessly prodigal of their expensive visits as quacks are of their potions.[13]

Beddoes died at the age of forty-six on Christmas Eve 1806, almost certainly of hypertensive heart disease. He was, in the words of Coleridge, 'deeply mourned by his patients who were also his family and friends'. His medical and scientific outpourings had been extraordinarily varied. The one topic he had sedulously avoided was the unspeakable terror of surgical operations. It is time to turn to it.

13. T.L. Beddoes, *Letter to the Rt. Hon. Joseph Banks on the Causes and Removal of the Prevailing Discontents, Imperfections and Abuses in Medicine* (London, 1808), p. 27.

# Chapter 18

# THE TERROR OF THE KNIFE

Little has been said so far about preanaesthetic surgery. Over thousands of years it had changed little. With increasing knowledge of anatomy, physiology and pathology, the craft did of course slowly advance. During the last half-century before ether and chloroform men like Syme in Edinburgh, Fergusson in London, Dupuytren in Paris and Mott in New York were innovators as well as master craftsmen. They devised new operations, clarified anatomical oddities, mapped out vascular supplies and invented new instruments. Fergusson's textbook listed sixteen different ways of amputating at the hip.[1] Dupuytren discussed twelve techniques for dealing with hernias.[2] The ligation of arteries to treat aneurysms crept steadily closer to the heart.[3] In

1. William Fergusson, *A System of Practical Surgery* (London, 1846). Probably the most famous of pre-Listerian surgeons in Britain, William Fergusson, later created a baronet, was born in Scotland and educated in Edinburgh but spent most of his professional life in London as professor of surgery to King's College Hospital. Among many instruments he devised, the most famous was the appropriately named Lion Forceps for gripping the lower jaw. He became surgeon to Queen Victoria and her family and died in harness in 1877, aged sixty-nine.
2. Guillaume Dupuytren, *Petit traité de la chirurgie* (Paris, 1828). Dupuytren rose from poverty to become the leading surgeon of France, baron of the Empire and later surgeon to the last two kings of France (Louis XVIII and Charles X; Louis Philippe adopted the title of 'King of the French'). He died in 1835, aged sixty-eight. He was worshipped by his patients but disliked by his colleagues.

   He is eponymously commemorated today by the contracture of the palmar fasciae, causing a characteristic deformity of the hands in the elderly. Dupuytren observed it in his coachman and attributed it to holding the reins for long periods over many years. Since Dupuytren's contracture remains exceedingly common, whereas coachmen are exceedingly rare, somebody will have to come up with an alternative explanation.
3. This was one of the showpieces of pre-Listerian surgery. Aneurysms or dilatation of the large arteries were comparatively common, usually late complications of syphilis. If allowed to burst, they were fatal. The only treatment was tying the artery on the proximal (heart) side of the aneurysm and hoping that collateral vessels would carry enough blood to the part deprived of its main supply. The closer to the heart, the riskier the operation. Valentine Mott of New York is said to have successfully tied the innominate artery a hand's breath away from its origin, the left ventricle of the heart.

the basic sciences too there were developments. Most but not all of John Hunter's pathological researches were later shown to be erroneous.[4] Sir Charles Bell was a notable neurophysiologist as well as an eminent surgeon.[5] But in practical terms the advances were superficial. The abdomen, the chest and most of the head remained barred. Even in areas declared to be 'accessible', the results continued to be appalling. In terms of survival, men were safer on the battlefield of Waterloo than on admission to a surgical ward in any of London's teaching hospitals. Success usually meant survival and temporary relief. Lasting cures were distant hopes barely glimpsed. Nor was there, as late as the 1840s, any realistic prospect of a significant change. The reason for all this was pain.

None of the pain-killing potions, not even laudanum in life-threatening doses, significantly palliated the agony of operations. Pre- and post-operative suffering could sometimes be lightened, though never abolished. The searing pain of the knife and the saw were virtually unaffected. Almost inevitably patients passed into a state of shock on the operating table.[6] The shock was sometimes irreversible. This imposed a universal imperative. Speed was essential. Prolonged pain not only hurt. It also killed.

Unrealised at the time and often overlooked even today, pain and speed had another consequence. Anaesthesia and antiseptic or later aseptic surgery were totally interdependent. Joseph Lister is rightly acclaimed as the man

4. John Hunter was surgeon to St George's Hospital and a leading surgical teacher and investigator. He wrote on a vast range of subjects (aneurysms, pyaemia, clotting, teeth and wounds, among other topics) and collected the specimens which later became the nucleus of the museum of the Royal College of Surgeons of England. He is still celebrated in eponymous lectures and orations as a patron saint of original surgical thinking and surgical research. He died in 1793, aged sixty-five, having suffered a stroke during an acrimonious meeting of the Hospital Board.
5. Charles Bell was born in Edinburgh but later moved to London and became surgeon to the Middlesex Hospital. His famous book, *Essay on Gunshot Wounds* was based on observing the wounded at the battle of Corunna in 1809. After Waterloo he went to Brussels and reputedly operated on the wounded for three days and nights without a break. His great physiological discovery was establishing the separate functions of sensory and motor nerves. He described Bell's palsy, paralysis of one side of the face due to a lesion, usually temporary, of the facial nerve. He was knighted in 1831, returned to Edinburgh as professor of surgery in 1835 and died in 1842, aged seventy-six.
6. 'Shock' is easily recognised but difficult to define. Patients in shock look cold and clammy, their pulse is weak and racing and their blood pressure drops. Even with eyes closed, they are sharply aware of their surroundings. The condition can be a response to blood loss or caused by severe fright or pain, or a combination of all three. If it is allowed to continue beyond a critical time (depending on the severity of the shock) reversible shock becomes unresponsive to treatment, i.e. irreversible.

who, in the 1860s, started to make surgery safe.[7] His clinical series of treating compound – that is open – fractures not with immediate amputation but with reduction under antiseptic cover was the beginning of modern surgery.[8] But the series would have been impossible fifteen years earlier, that is before the introduction of general anaesthesia. For operations to be bearable, even if only just, preanaesthetic surgeons had to work with a fantastic dispatch. Showmen like Robert Liston or Dupuytren turned to their students crowding the benches of the operating room before starting an amputation and ordered: 'Time it, gentlemen.' And the students did, to the second. But speed and antisepsis were incompatible. For decades surgeons who adopted Lister's ideas would work in a haze of carbolic acid vapour. No operative step was taken before the site was doused with antiseptic.[9] Lister also emphasised the need not only to kill germs but also to leave behind no dead tissue in which germs could flourish. This meant meticulous attention to the bleeding point, and careful dissection. Neither could be done in a hurry, as it still cannot. In an anaesthetised patient the slowing down did not matter. Once the patient was insensible, shock became a rarity. It was this which was the revolutionary change. Surgical anaesthesia not only eliminated the pain of operations. Since it made antisepsis possible, it also ended the suffering and death from 'hospital disease'.[10]

. . .

7. It is fashionable but silly to decry Lister's invention of antisepsis: one commentator has even described it as a 'cul-de-sac'. If being superseded by a further development turns a medical or scientific achievement into a cul-de-sac, then the term can be applied to every advance from Pythagoras to Einstein.

   Joseph Lister was born in 1827 into an English Quaker family but moved to Edinburgh after graduating from University College Hospital, London. There he became James Syme's house surgeon and later Syme's son-in-law. In 1865, as professor of surgery in Glasgow, he became interested in the work of Pasteur on 'germs' and guessed that they were probably the cause of wound infection and 'hospital diseases'. As a countermeasure he devised 'antiseptic surgery', i.e. operations carried out under cover of a continuous carbolic acid spray. He later abandoned the spray but the recognition of bacterial infection as a cause of surgical sepsis remains a landmark. He became professor of surgery at King's College Hospital, London, and was created the first surgical peer in 1883. He died in 1912, aged eighty-five. His and Semmelweis' careers are described more fully in Dormandy, *Moments of Truth*.
8. Because open fractures and wounds involving bones almost invariably became septic, resulting in general septicaemia and pyaemia, the standard treatment until Lister was amputation of the limb as soon as possible.
9. A complex ritual of 'sprays' was devised by Lister on the assumption that most germs landed on wounds from the atmosphere. This was later shown to be wrong and antisepsis was superseded by asepsis.
10. See Chapter 19.

The impact of anaesthesia, the most important single advance in man's fight against pain, cannot be appreciated without looking at the scene which it transformed. Preanaesthetic surgery reached its peak during the first decades of the nineteenth century. This is therefore the period when that scene deserves to be looked at. It also happened to be the time when a few surgical patients began to write vivid first-hand accounts of their experience in a language comprehensible today. This fact partly at least absolves the writer from trying to recreate one of the most horrendous and difficult to write about experiences in man's cultural history.[11]

. . .

In the summer of 1810 Mme d'Arblay, née Frances (or Fanny to all her family and friends) Burney, noticed 'a small pain and heaviness' in her right breast, 'a discomfort rather than a hurt'.[12] She was fifty-eight and married to Alexandre d'Arblay, a French nobleman and army officer. Despite the fact that for most of their marriage their countries were in a state of war with each other,[13] they were a devoted couple; and, living in Paris, they moved in the highest social circles. Throughout her illness therefore, Fanny had access to the best surgical care in France which at that time probably meant the world.[14] Much worried, d'Arblay pressed Fanny to see a doctor; but for some months she prevaricated. Sometimes the pain seemed to disappear. Sighs of relief. Then it returned. Despair. Eventually she consulted their family physician, Dr Ribe, and, at his urging, the leading surgeon obstetrician of the Empire, Antoine

11. The most penetrating and detailed modern account of preanaesthetic surgery in England is Peter Stanley's *For Fear of Pain*, an indispensable and riveting source-book (Amsterdam, 2003).
12. Fanny Burney was born in 1752, the daughter of the celebrated musicologist, Dr Charles Burney. Her early diaries and letters paint a wonderfully vivid picture of the musical and literary life of London. Her first and best novel, *Evelina*, was published in 1778. She married D'Arblay in 1793 and they moved to France. The quotations from her diary are in J. Hemlow (ed.), with G.G. Falle, A. Douglas and J.A. Bourdais de Charbonnière, *The Journals and Letters of Fanny Burney* (Mme d'Arblay) (Oxford, 1975).
13. The only interlude followed the Treaty of Amiens in 1803 which neither Britain nor Bonaparte in France seriously contemplated honouring. By the end of 1804 hostilities were resumed both by land and sea. Yet, the war never interfered with private contact between the upper classes or with literary and artistic intercourse. The first Science Prize which Napoleon – he was crowned emperor in 1804 – founded in 1807 was awarded to Humphry Davy, despite the fact that England and France were at war; and in 1813 the Emperor gave Davy permission to study the volcanoes in the Auvergne. The English party was fêted by French scientists.
14. The destination of young surgeons from the United States visiting Europe provides a reasonable guide to the prestige enjoyed by various European centres. Paris remained the mecca till about 1840, when the torch passed to Edinburgh and London; it remained there until the 1870s, when it passed to Germany.

Dubois.[15] Dubois in turn called in his colleague, Dominique Larrey, the already legendary army surgeon, recently raised to the barony.[16] They examined Fanny. In grave silence they retired to the library to consult with the husband. She waited in a state of anguish, trying to divert her mind with a volume of Pascal's *Pensées.* After an hour D'Arblay returned on his own. The look on his face conveyed to Fanny 'the bitterest woe'. An operation would be necessary 'to avert the most evil consequences'.[17] Other doctors were consulted and an anxious waiting period followed.[18] The 'petit salon' was designated as the operating room.[19] On the appointed day six surgeons arrived, led by Dubois and Larrey. Their demeanour was grave but reassuring. Fanny was fortified with a tumblerful of 'wine cordial'.

> M. Dubois placed me on a mattress and spread a cambric handkerchief over my face. It was transparent, however, and I saw through it that the bedstead was surrounded by surgeons and nurses. I refused to be held; but when, bright through the cambric, I saw the glitter of the polished steel, I closed my eyes. I would not trust to convulsive fear the sight of the terrible inci-

15. Dubois, the inventor of the Dubois obstetric forceps, was professor successively of anatomy, surgery and obstetrics at the Ecole de chirurgie and eventually became dean of the Faculté de médecine in the University of Paris. He attended both the Empresses Joséphine and Marie Louise and was created a baron after safely delivering the King of Rome, Napoleon's son by his second wife, in 1811. He died in 1837, aged eighty-one.
16. Larrey was famous for his care of the wounded, as well as for his courage, on more than sixty battlefields from the Pyramids to Waterloo, performing at Borodino in Russia 200 amputations in twenty-four hours. In 1792 he invented the horse-drawn 'flying ambulances' or *ambulances volantes* which, at his insistence, collected the wounded from the battlefield irrespective of nationality. He was created baron of the Empire in 1810 and became chief surgeon to the *Grande Armée* in Russia in 1812. He was taken prisoner at Waterloo, blindfolded, and would have been shot but for the intercession of a Prussian surgeon. He died in 1842, aged seventy-six, a great man and a great doctor.
17. Mastectomies had been performed for cancerous growths of the breast for at least two hundred years both in Britain and on the Continent and were recognised as being among the most cruel but potentially life-saving operations. In his *Practical Surgery* (1837) Robert Liston wrote: 'This is a disease in which no half-measures will answer; and if the patient has made up his [sic] mind, after a severe struggle, to submit to the pain and risk of a very dreadful operation . . . in the expectation of enjoying an immunity from a terrible disease, it is not fair, from any slovenliness or carelessness to throw away a chance or in any way endanger a recurrence . . . The duration of the proceeding must not for once be considered. Many operations can be done well and quickly . . . this is not one of them. The whole of the disease structure must be extirpated.'
18. Dubois' time was largely taken up with shadowing the pregnant Empress: Napoleon would not tolerate any 'distraction'. Larrey was involved with the preparations for the Russian campaign. During major military campaigns all the best surgeons were away from Paris: Fanny was fortunate, if that expression can be used in this context, in being operated on between the war against Austria, culminating in the battle of Wagram, and the invasion of Russia.
19. Admission to a hospital for operations was never even considered for the upper and middle classes.

> sion . . . A silence most profound ensued . . . during which, I imagine, they took their orders and exchanged observations by signs . . . Oh what horrible suspension – I did not breathe . . . [Then] in a voice of solemn melancholy Dr Larrey said: 'Qui me tiendra ca sein?'[20]

After a further interval and a hectic exchange of whispered consultation between the surgeons, she tried to blank out her mind.

> Yet when the dreadful steel was plunged into my breast, cutting through skin, vein, flesh and nerves, I needed no injunction not to restrain my cries. I began a scream which lasted unremittingly throughout the whole operation – and I almost marvel that it rings not in my ear still! – so excruciating was the agony. When the wound was made and the instrument withdrawn, the pain seemed undiminished, for the air that suddenly rushed into those delicate parts felt like a mass of minute but sharp poignards that were tearing the edges of the wound – but when again I felt the instrument, describing a curve, cutting against the grain, while the flesh resisted in a manner so forcible as to oppose and tire the hand, then, indeed, I thought I must have expired . . . The instrument withdrawn this second time I thought the operation was over – but no! Presently the terrible cutting was renewed – and worse than ever, to separate the bottom, the foundation of this dreadful gland from the parts to which it is anchored . . . Again, all description would be baffled – yet all was not over. Dr Larrey rested but his own hand: then I felt the knife rackling against the breast bone, scraping it. This performed, while I remained in utterly speechless torture, I heard the voice of M. Larrey, asking if anything more remained to be done. The general voice said *rien*; but the finger of M. Dubois, which I literally felt elevated over the wound though I saw nothing, pointed to some further requisition and again started the scraping, attom after attom.[21]

The letter was started six months after the operation but the subject was still almost unbearable.

> Not for days, not for weeks or months could I speak of these terrible minutes without nearly again going through it and being violently sick. Even now it is like a nightmare as I write about it . . .When all was done and they lifted me up so that I might be put to bed my strength was so totally annihilated that I could not even sustain my arms which hung lifeless . . . and my face, Nurse has told me, was utterly colourless . . . When I opened my eyes I saw the good Dr Larrey, pale nearly as myself, his face streaked

20. Hemlow (ed.), *et al.*, *The Journals and Letters of Fanny Burney*, Vol. VI, p. 612.
21. Ibid., p. 613.

with blood, and its expression depicting grief, apprehension and almost horror.[22]

...

Fanny's last sentence raises a question that is almost as difficult to answer as how patients survived their ordeal. How did surgeons bear the strain, not once or twice in a lifetime but day after day for years? Many wrote about it but few convey a convincing explanation. James Miller, a successful and candid practitioner, listed some of the causes of the surgeon's anxiety as he stood at the operating table.[23] First, he dreaded – and continued to dread throughout his career – the pain he was inflicting. Even on the comparatively rare occasions when patients did not scream, 'the limbs stiffened through agony, the face turgid, the eyes prominent and suffused' never lost their horror.

Next came the difficulty of the operation. Without pre-operative X-rays, scans and blood tests, complications could never be foreseen. Failure could affect not only the patient's life but also the surgeon's livelihood. Senior hospital appointments were unpaid. For his living the surgeon depended on his reputation. The desire and need to finish the operation quickly created 'the most acute anxiety'. At any moment silence could erupt into pandemonium. Pandemonium could lapse into a deathly hush. Many surgeons expressed their anxiety by shouting at dressers, assistants, nurses, onlookers. The bellowing, abusive operator, occasionally smelling of whisky, was not a myth. Instruments were hurled to the floor. Buckets were kicked over. But almost no surgeon shouted at his patients. On the contrary. All preanaesthetic surgical textbooks, unlike their modern successors, give lengthy instructions on how patients should be continually soothed and reassured before and during operations. 'The most effective treatment of shock is administered through the ear,' Sir James Paget wrote. 'It is words of comfort.'[24] However stressed, five minutes would rarely pass without the operator addressing the patient in a hopefully confident tone. But the tension showed in the 'short, almost convulsive strokes' with which he sawed through bones. Many undoubtedly suffered.

John Abernethy of Bart's was one of the half-century's most admired teachers. He had a reputation for no-nonsense practicality. But a colleague

22. Hemlow (ed.), *et al.*, *The Journals and Letters of Fanny Burney*, Vol. VI, p. 617.
    Fanny returned to England after her husband's death in 1818 and herself died in 1840, aged eighty-eight. She always attributed her survival to the operation; but many modern surgeons have expressed doubt about the diagnosis. The pain experienced for a year before the operation would certainly suggest today a chronic inflammatory rather than a cancerous disease. On the other hand, Dubois and Larrey were both highly experienced surgeons and had no doubt of the cancerous nature of the growth.
23. J. Miller, *Surgical Experiences of Chloroform* (Edinburgh, 1848), p. 74.
24. See S. Paget (ed.), *Memoirs and Letters of Sir James Paget* (London, 1902), p. 243.

recorded greeting him as he was walking to an operating session: 'How are you John?' 'I feel,' Abernethy replied, 'as if I was going to a hanging.'[25] As he aged, his dislike of operating deepened. He became fidgety, 'difficult to assist'. One of his students remembered years later seeing him in the retiring room after a difficult amputation with tears streaming down his face, vomiting.[26] Sir George Bell, famed for his impassive manner, went to an operation 'as one about to face an unavoidable but inevitable evil.'[27] Charles Bell wrote to his brother after a failed operation: 'I shall regret it, I shall regret it as long as I live ... The more I do, the worse it gets.'[28] Robert Liston of University College Hospital was regarded as the most 'heroic' of London surgeons. One of his assistants described him as 'crude, brash and brutally rough'; but this was not how his patients remembered him. 'He was kindness itself to me,' a Smithfield porter recalled years later.[29]

Unstoppable bleeding was the operative complication most dreaded. Samuel Cooper recalled the moment when the surgeon is suddenly faced with the onset of a serious haemorrhage: 'pale as a corpse himself, and trembling, he beholds the jet of arterial blood.'[30] In his influential *System of Operative Surgery* Sir Charles Bell describes the patient's death on the table from loss of blood. 'Having witnessed such an event a surgeon would do all he could to avoid another ... [It was] perhaps the most awful kind of death.'[31] As the bleeding continued, the patient's face became livid before turning deathly pale. The lips turned dark and hands and feet grew cold. As surgeons and dressers desperately tried to stem the flow, the patient might repeatedly faint and revive as the pulse fluttered, faded and feebly restarted. He would be sick, gasping and muttering in a low hoarse voice, while tossing his arms, the most ominous sign.[32] Almost always the dying patient would start to raise his head, gulp for air and thrash his limbs, drawing a few long convulsive breaths before the end.

It was one of the terrors shared by surgeons wherever they worked. Valentine Mott of New York, a man of legendary skill, wrote:

> How often, when operating in some deep dark wound, along the course of some great vein, with thin walls, alternately distended and flaccid with the vital current – how often have I dreaded that some unfortunate struggle of the patient would deviate the knife a little from its proper course, and that

25. G. McIlwain, *Memoirs of John Abernethy*, 2 volumes (London, 1854), Vol. II, p. 203.
26. See Stanley, *For Fear of Pain*, p. 204.
27. G.J. Bell (ed.), *Letters of Sir George Bell* (London, 1870), p. 374.
28. See Stanley, *For Fear of Pain*, p. 206.
29. J. Miller, *Surgical Experiences of Chloroform* (Edinburgh, 1848), p. 87.
30. S. Cooper, *The First Lines of the Practice of Surgery* (London, 1813), p. 34.
31. C. Bell, *A System of Operative Surgery* (London, 1814), p. 56.
32. W. Hay, jun. *Practical Observations in Surgery* (London, 1803), p. 231.

I, who fain would be the deliverer, should become the executioner, seeing my patient perish in my hands by the most appalling form of death.[33]

Mishaps were inseparable from speed. Surgeons like Liston and Fergusson were virtuosos but had to 'practise' their craft as assiduously as would a modern concert pianist. At the height of his fame James Syme might spend several hours a day rehearsing operations on cadavers.[34] But, however fast, skilful and lucky, many admitted that continuous contact with pain could not but blunt sensibilities. 'Since surgeons cannot turn away from the horrors of their calling, they have to try to look beyond it.'[35] Most did so successfully; but even successful surgeons like Sir James Paget recalled late in life the years of wakeful torture before anaesthesia. 'They had been the worst nightmares. I can remember them still. Even now they sometimes rouse me from my sleep. I wake drenched in sweat.'[36]

But they stuck to it, or most of them did; and in a few operating, with all its pain, gore and danger, became an addiction. Sir William Blizzard, widely known as Sir Billy Fretful, became the first president of the reformed Royal College of Surgeons of England in 1800. He performed his last amputation at the hip at the age of eighty-four. Sir Astley Cooper, 'the Wellington of British Surgery', was followed round by an army of adoring students and retired full of honours to his magnificent country mansion in Norfolk aged sixty-eight. 'I had my fill of butchery,' he declared. 'What a relief.' But a few weeks later he entertained his French counterpart, Baron Dupuytren.[37] The Frenchman had become a charmer in old age and expansively admired the superb trees in the park. 'Yes, my friend,' Sir Astley commented, 'and every morning I speculate which I should choose to hang myself from.' Dupuytren made one of his celebrated spot diagnoses. 'You must at once return to London, *cher confrère*, and resume your proper calling. You are wasted in this morass.' Cooper did and died in harness ten years later.

In other European countries the scene was not significantly different. Dupuytren himself became a surgeon to the last two kings of France; and when, in 1828, Charles X was in financial straits, he offered his monarch one million francs. 'Sire; thanks in part to your benefaction, I possess three millions. One I offer to you, one I save for my daughter, and one I keep for my

33. Quoted by Robinson in *Victory over Pain*, p. 79.
34. James Syme was the most famous Edinburgh surgeon in the years before and immediately following anaesthesia and antisepsis, both of which he was among the first to adopt. He devised an amputation at the ankle and a special knife still known by his name. A genial but taciturn man, he was said never to waste a word, a drop of ink or a drop of blood. He died in 1870, aged seventy-one.
35. W. Hay, jun. *Practical Observations in Surgery*, p. 134.
36. See Paget, *Memoirs and Letters of Sir James Paget*, p. 76.
37. B. Cooper, *The Life of Sir Astley Cooper, Bart*, 2 volumes (London, 1843), Vol. II, p. 54.

retirement to which I look forward with the utmost keenness and anticipation,' he told Charles. But he never retired. 'The brigand of the *Hôtel Dieu*' as he was known not particularly affectionately to his colleagues, collapsed on a ward round and died a few days later.[38] In Vienna, at the height of the medical school's fame, young Ignác Semmelweis was deterred from surgery by the horrors he witnessed in the department of the celebrated Professor Daun. The morbid anatomist Rokitansky summed it up a few years later in English when entertaining Lister: 'Diagnosis, Daun, Death, Dissection'.

Pain did not alliterate; but of course it was the patients who suffered most. In 1841 George Wilson, professor of chemistry in Glasgow, consulted Syme about his 'mortifying foot'. Syme explained what Wilson already guessed, that amputation offered his only chance of survival.

> Waiting for the day was like being a condemned criminal waiting for the execution . . . Counting the days and hours, sleepless from fear and pain . . . The operation itself subjected me to suffering so great that it cannot be expressed in words . . . At the extremities of human experience we can observe only a kind of silence. But I can recall the black whirlwind of emotion, the horror of great darkness, the sense of desertion by God and man.[39]

The memory of such an experience lasted a lifetime – and not only for the patient. Wilson's family waited next door during the operation; and his sister, Jessie, wrote twenty years later: 'I can still hear the cries of George, like those of a wounded animal, ringing in my ear.'[40]

. . .

Other than facing a slow and perhaps even more painful death, there was usually no escape. Stones in the bladder could paralyse work and social life. Their removal, usually by way of an incision through the perineum, was one of the commonest operations performed. In skilled hands the procedure could be completed in less than five minutes. But not always. In 1819 an account by a patient, 'a gentleman, about forty years of age, who, to the advantages of a liberal education, unites an uncommon share of cheerfulness and firmness of mind', was published by Alex Marcet, a vice-president of the Medico-Chirurgical Society, forerunner of today's Royal Society of Medicine.[41]

38. J. Duns, *Memoir of Sir James Y. Simpson, Bart* (Edinburgh, 1873). The memoir contains George Wilson's account of his operation, on p. 273.
39. J.A. Wilson, *Memoir of Sir George Wilson* (Edinburgh, 1860), p. 54.
40. A. Marcet, 'A Case of Nephritis Calculosa . . . with an Account of the Operation of Lithotomy, given by the Patient Himself', *Medico-Chirurgical Transactions*, 1819 (10), 147–60.
41. Ibid.

The illness could be divided into four stages. During the first the patient noticed a 'calcerous discharge' in his urine, tinged with blood, but felt relatively little pain. This lasted about three years and was manageable without recourse to surgery. The second stage began with violent paroxysms of colic in the lumbar region, attended by uncontrollable retching. These were treated with purgatives and opium with partial success. The patient's social and professional life became increasingly restricted. He noticed with alarm that, perhaps because of his frayed temper, he was shunned by former friends and business associates. The third stage was marked by the formation of a stone in the bladder 'with all the well-known pain and suffering this entails'. The patient became incapable of riding on a horse; and even walking became difficult.

> The only position in which I could pass urine was upon my knees, with my forehead placed upon the bed or sofa and the urinal under me. All changes of posture were dreadful, especially from the erect to the recumbent position. Getting into or out of bed were the severest trial . . . When awoke by irritation and pain, I endured it as long as I could rather than move.[42]

As usual, all manner of palliatives were tried, including a visit to Brighton for a cure of tepid baths. 'I still shudder remembering the torture of the coach journey.' On arrival a doctor was summoned and he passed a sound into the bladder. It confirmed the diagnosis. Back in London the patient consulted the eminent surgeon, Henry Cline junior. Cline confirmed that an operation was necessary and urgent. It was he who eventually performed it.

> When all parties had arrived, I retired to my room for a minute or two and bent my knee in silent adoration . . . I then conducted the surgeon and his assistants to the apartment in which the preparations had been made . . . I was prepared to receive a shock of pain of extreme violence, and so . . . the first incision did not even make me wince. The forcing up of the staff prior to the introduction of the gorget gave me the first real pain, but the incision of the bladder and a rush of urine afforded a few moments of relief . . . When the forceps was introduced the pain was again very considerable; and every movement of the instrument endeavouring to find the stone, increased it. After several ineffectual attempts to grasp the stone, I heard the operator say in the lowest whisper: 'It is a little awkward . . . Give me the curved forceps.' Turning to me he said in the kindest manner: 'Be patient, sir, it will soon be over.' When the other forceps was introduced I had again to undergo the searching for the stone and heard Mr Cline say: 'I have got it.' I had by then conceived the idea that the worst was over; but when the

42. Marcet, 'A Case of Nephritis Calculosa . . .', pp. 147–60.

> necessary force was applied to withdraw the stone, the sensation was such that I cannot find words to describe it . . . The bladder seemed to grip the stone as firmly as the stone was gripped by the forceps and it seemed as if the whole organ was about to be torn apart. The duration of this stage was short: it could not have been survived had it been longer. Then I heard the words 'now, sir, it's over.'[43]

Recovery from the operation was satisfactory, though attacks of renal pain, bleeding and the occasional passage of 'sediment' continued. Nevertheless, the patient wrote five years later: 'The illness had been so dreadful that I would again submit to the same mode of relief provided I could place myself in equally capable hands.'[44]

. . .

In the big teaching hospitals of London, Paris, Vienna and New York, dominated by larger-than-life figures, preanaesthetic surgery was, among other things, grand theatre. Forthcoming capital operations were given advance notice in the press and were often attended by crowds of spectators. In 1831 a thirty-year-old Chinese labourer travelled at the expense of his employers, the East India Company, from his native village near Canton to London. He wanted and needed to have a huge tumour removed from his groin; and the Company had secured the services of the great Aston Key of Guy's Hospital to accomplish the removal. The patient, Hoo Loo, was a man of irrepressible cheerfulness, communicating his jokes and sallies through Cantonese-speaking friends. He quickly became popular with doctors, nurses and medical students; and a steady stream of lay visitors came to gasp at the gigantic growth. The operation promised to be as exciting an event as an opera performance or a public hanging. So many wanted to witness it that, at the last minute, the venue was moved from the regular operating room to the largest of the hospital's anatomy theatres.

At one o'clock on the appointed day, laughing and joking, Hoo Loo and his interpreter were ushered in. Hoo Loo was placed on the operating table and secured by straps. A handkerchief was placed over his face. By then almost 500 spectators, in addition to the fifty or so medical students, had pushed their way into the auditorium designed to hold 200. A few fainted in the heat even before the operation began. Key was followed by a retinue of dressers, assistants, porters and nurses. He briefly and 'with brilliant lucidity' outlined the course of the illness, the procedure he was proposing to follow, the difficulties he envisaged and the measures he proposed to take to overcome them. Warm applause rewarded his exposé.

43. Ibid.
44. Ibid.

With military precision his dressers and assistants took up their allotted positions. All seemed to bode well. Hoo Loo's screams were 'no more than expected'. Key spoke to him in kindly tones which, in the words of the *Times* correspondent, seemed to be 'appreciated even if not perhaps fully understood'.[45] After the initial incision Key rested for a minute or two while his assistants secured the veins emerging from the tumour. After this well-conducted manoeuvre there was again scattered applause. Key had planned to preserve the genitals but 'that would have prolonged the procedure beyond endurance'. On incising the scrotum, an 'aberrant artery', every surgeon's nightmare, appeared to be punctured. The vessel instantly retracted into the depth of the wound; but blood welled up. Within a minute or two the patient lost enough blood to make him pale and distressed. He started to moan and then to shout and croak. The interpreter duly translated: 'Unloose me! Unloose me! Water! Help! Water!' There was no question of turning back. Eventually the artery seemed to be secured but the bleeding continued. Medical students volunteered to give blood and a fit-looking athlete was selected. A blood transfusion was improvised, running blood directly from the donor's vein in front of the elbow to what Key identified as the patient's femoral vein in the groin. After two or three ounces were run in, the transfusion stopped.[46] The bleeding increased. The patient started to convulse. Brandy was forced through his lips. He vomited. The scene became chaotic. Assistants struggled to quench the bleeding while Key, 'with admirable self-control', continued to excise the tumour. The bleeding slowed to a trickle. The tumour, measuring four feet in circumference and weighing 65 pounds, was held up for inspection. The air in the theatre had by then become so foul that many more spectators, unable to make their way out of the throng, had collapsed. The patient's last comprehensible words were: 'Let it be – let it remain. Unloose me! I can bear no more.' By the time Key announced that the operation had been satisfactorily completed Hoo Loo was dead.

Next day, 11 April, the event was discussed in the press. Key was lauded for his calm but the hospital authorities were criticised. The correspondent of *The*

45. Several accounts exist of this famous operation, including a report in *The Times*, 11 April 1831; S. Wilks and G.T. Bettani, *A Biographical History of Guy's Hospital* (London, 1892), p. 209; and the *Glasgow Medical Journal*, May 1931, p. 211. They differ in detail.
46. Blood transfusions were tried in an extreme emergency long before blood groups were discovered. (The chance of two people's blood being compatible in any one location was of the order of 1 in 3.) Because anticoagulants had not yet been invented, the transfusions had to be direct: even so, some clotting almost certainly occurred in transit. No reliable statistics exist of the outcome of such procedures during the nineteenth century but the transfusions themselves may have been fatal in about half the cases.

*Times*, 'though not a medical man, [was] well versed in matters surgical'; and he attributed the patient's death not to blood loss but to the 'heated and contaminated atmosphere . . . Too many people had been admitted. I myself at times felt distinctly faint. However much benefit to science and to the education of the public, the spectators' and indeed the patient's well-being should have been paramount.'

# Chapter 19

# HOSPITAL DISEASE

Death on the operating table from blood loss or shock was the surgeon's nightmare but fortunately rare. Or not so fortunately. Most patients who died from preanaesthetic surgery died not on the operating table but from 'hospital disease'. This often followed apparently successful operations, declaring itself a day or two later. It caused more suffering and deaths than the actual cutting.

In John Brown's autobiographical story, 'Rab and His Friends', the author as a medical student in Edinburgh meets James, his wife Allie, and their big mongrel dog, Rab, in 1830.[1] Allie consults him about a hard painful growth in her left breast. He takes her to his teacher, James Syme. Syme, known for his taciturnity, explains to Allie that 'the growth will no doubt kill her and soon. It could be removed and it might not return.' 'When can you remove it, Mr Syme?' 'Tomorrow.'[2]

Next day the operating room in Syme's private hospital, Minto House, is crowded with noisy, joshing students, eager to secure a good place for the 'capital operation'. Syme perches on a stool in his operating frock-coat and reads a newspaper. Allie enters in her street clothes, her quiet dignity silencing the noise. She is followed by her husband and Rab. A porter explains that dogs are not allowed in the operating room, but Syme, a dog-lover, motions him to be quiet. Rab stays. Allie climbs on to the operating table and, holding Brown's hand, arranges her dress to expose her breast. James sits with Rab between his knees at the head of the table, silencing the dog's growling. Syme makes his first incision. Allie remains silent and composed throughout the operation. When the wound has been dressed she steps down from the table, turns to Syme and the students, curtsies and begs their pardon if she has behaved ill. Rab jumps up and wags his tail. Syme clears his throat. Brown and his fellow

1. 'Rab and His Friends', in J. Brown, *Horae subsceviae* (London, 1907), p. 265.
2. For Syme, see Chapter 18, note 34.

students cry like children. Allie is helped back to the ward and to bed. Rab settles down next to her.

At first all seems well. Then, on the second post-operative day, Rab begins to whimper. There is no way to silence him: so he has to be taken away. A few hours later Allie starts to shiver. Soon she lapses into delirium. She imagines that she is holding her long-dead baby to her bandaged chest. Only from time to time does she seem to come round and complain of 'really very bad pain'. She has developed 'hospital disease'. She dies on the fourth post-operative day.

. . .

Hospital disease, sometimes called 'hospital infection', 'hospital fever' or 'hospital sepsis', was often assumed to be several different diseases.[3] 'Pyaemia' was reckoned to be the commonest. After a spike of temperature, sweats and shivering, crops of boils started to appear all over the patient's body. Others were internal and invisible. Bony and abdominal abscesses were thought to be the most excruciating. A dental abscess before starting treatment with antibiotics is probably the nearest to that experience in civilised life today. The condition was almost totally unresponsive to drugs, even to large doses of laudanum. Death, the invariable outcome, came as a relief within two or three days.

In the late 1840s childbed or puerperal fever was recognised by the Hungarian obstetrician, Ignác Semmelweis, as a harrowing manifestation of the same pathology.[4] At times it killed up to a quarter of previously healthy young mothers after apparently normal hospital deliveries. This was in sharp contrast to the comparatively safe deliveries by midwives in the mothers' homes. Semmelweis was hounded out of Vienna for suggesting that the disease was introduced into the wombs by 'noxious matter' picked up at autopsies by doctors and medical students, not by some mysterious 'miasma' or 'lunar emanations'. Few believed him. He himself died insane from pyaemia eight years later.[5]

Septicaemia would be classified today as a slightly milder or earlier form of the same illness in which bacteria rather than clumps of pus circulated in the patient's bloodstream. It was no less fatal. Hectic fever heralded its onset. Like Allie, patients often lapsed into unconsciousness at an early stage and died in a coma before the condition could progress to pyaemia. Delirium or 'madness' was a common feature. The ravings of such patients

3. For more details see Dormandy, *Moments of Truth*, p. 315.
4. He had several clear-sighted precursors and contemporaries who knew nothing about him and who were equally ignored. Among them were Alexander Gordon of Aberdeen and doctor, poet, wit and man of letters Oliver Wendell Holmes of Boston
5. He sustained a wound during a scuffle while being admitted to a mental home for general paralysis of the insane.The wound went septic and led to pyaemia.

were the most sinister of the 'evil noises' that Victor Hugo's Jean Valjean heard at night in a hospital.

Erysipelas was like septicaemia but accompanied by a rash. The rash spread slowly like a moving burn from open wounds or surgical incisions. It was sometimes called 'The Rose' or 'St Anthony's Fire'. St Anthony was a fourth-century saint who propagated the monastic way of life and was thought to be especially efficient in interceding on behalf of sufferers. It was their best hope.

'Hospital gangrene' was in some ways the most sinister. It seems to have become common in the 1820s and reached its peak in the 1840s. What the causative organism was remains a mystery: it was probably related to the *Clostridia* which still occasionally cause gas gangrene and tetanus. The gangrene started in the wound and continued to spread after death for several hours.

. . .

The incidence and severity of these conditions were changing alarmingly during the 1830s and 1840s. Of course wound infections of various kinds had been known since accidents and operations began; but there was a general perception that 'hospital diseases' were becoming more common every year. With increased overcrowding in the wards, the perception was probably correct. It made a mockery of the much-vaunted advances in operative techniques, the new instruments, the broken speed records. The disease could occur as an isolated event even in patients' homes; but more characteristically it appeared as epidemics in large hospitals. An institution might be free from it for months or even years. Then, for no obvious reason, the scourge would strike. The only warning sign was an ominous sweet sickly smell that sometimes wafted up and down the corridors a day or two before the outbreak. It was this that Rab probably sniffed at Allie's bedside. But nothing could be done to stop the inevitable. Nobody was safe. Doctors, students, nurses and visitors were all at risk. Pricking one's finger when threading a needle could be fatal. Several Victorian charities catered for the relicts of hospital staff who died in the line of duty. A wide range of different countermeasures were advocated, tried and proved futile. Patients who could swallow were treated with laudanum, alcohol and hot broths. Leeches and cupping were usually tried but were of no avail. Hospital chaplains worked overtime.

Confronted with an outbreak, wise surgeons closed their ward and waited for the epidemic to subside. As outbreaks became more frequent, this became less and less practical. A proposal to close all hospitals in England was seriously debated in Parliament. Florence Nightingale, whose prestige after the Crimean War surpassed that of any politician or doctor, advocated hutted accommodation or at least widely spaced pavilions dotted around in wooded grounds.[6] Hyde Park and the New Forest were mentioned as possible sites. Unknown to Miss Nightingale, salvation was by then in sight.

What solved the problem was recognition that all the different hospital diseases were caused by living 'germs', later renamed bacteria, which invisibly but in unbelievable numbers and variety populated the visible world. Some were harmless inhabitants of the skin and body cavities but became killers when they gained access to open wounds. Unlike the 'miasmas' and 'cosmic emanations' of the past, germs were alive and could therefore multiply. They could be carried by the lymphatic system and the bloodstream to every part of the body. The ubiquity of germs and their extraordinary capacity to bring about chemical changes were discovered by the French chemist Louis Pasteur. Their role in surgery was the outcome of lateral thinking by a ponderous and deeply conservative English professor occupying the chair of surgery in Glasgow University.[7] The carbolic acid spray which he introduced to kill germs (and later abandoned) was a clumsy weapon; but it worked and it was the first measure that did. The later principle of antisepsis was a natural development. What made both antisepsis and asepsis possible and virtually extinguished hospital disease was general anaesthesia.

6. This was the origin of the block system of building hospitals. The corridors were originally unglazed and often arctic in the winter.
7. See Chapter 18, note 7.

# PART III

# PAINLESS SURGERY

# Chapter 20

# TO THE THRESHOLD

Few until nearly half-way through the nineteenth century seriously envisaged painless operations. Even fewer tried to put the idea into practice. Those who did were unsuccessful. The more reason perhaps for remembering them. They were an odd miscellaneous lot – an English country practitioner, a French military surgeon, a Nottinghamshire hospital doctor, a London barrister and an eminent professor of medicine hounded from his chair for his subversive ideas. They were unaware of each other. All they had in common was that they came to the threshold of *practical* painless surgery but, for one reason or another, never quite got beyond it. They bridge the gap between the idea of anaesthesia mooted by the pneumatic doctors and the acceptance of the practice fifty years later.

. . .

The idea of nerve compression may date back to the Kingdom of the Pharaohs. René Fülöp-Miller, the not always reliable first chronicler of the history of anaesthesia, mentions wall-paintings in an excavated necropolis in Saquarrah which 'clearly show' a patient exerting pressure on the nerves in the armpit while some kind of an operation is being performed on his hand.[1] The works are not actually reproduced. Ambroise Paré was certainly aware that compression by a tourniquet for a few minutes caused a degree of numbness and could sometimes ease the pain of amputations.[2] The technique was not, however, systematically investigated until the last quarter of the eighteenth century.

1. R. Fülöp-Miller, *Triumph over Pain* (New York, 1938).
2. See Chapter 12.

James Moore was the son of a Scottish doctor and man of letters and the brother of one of Wellington's most capable generals.[3] James himself served as a surgeon with the British redcoats in the American War of Independence before settling down in practice in London. Like his friend, Edward Jenner, he was both inquisitive and resourceful. In his spare time he mapped out the course of sensory nerves in the limbs in both animals and man. He then designed a tourniquet with a pressure device and described its use in a short book, *A Method of Preventing or Diminishing Pain in Several Operations in Surgery*.[4] Application of the clamp for half an hour in the right position on the arm or the thigh achieved almost complete numbness in the parts distal to it. He sent the book with a personal letter to some of the leading surgeons of the day, including John Hunter of St George's Hospital whose house surgeon Moore had been. Hunter had an original if volatile mind and, like many champions of the Enlightenment, was prepared to try most things once. He performed a below-knee amputation after applying Moore's tourniquet to the patient's thigh and noted a 'definite lessening of the pain'. Unfortunately, he also noted an increased tendency to bleed. The drawback put him off using the device again. Pain was preferable to bleeding. Benjamin Bell, the author of a standard textbook of surgery, mentioned that nerve compression as a way of easing pain 'has long been known' and that patients sometimes expressly asked for it; but that 'taking an overall view and all things into account, the effect is generally inconsiderable'.[5] The last remark sounds like a piece of textbook pomposity – Bell had probably never tried the contrivance himself – but the bleeding was a genuine drawback. It is almost impossible to compress a major sensory nerve to a limb without also compressing the accompanying vein; and, unless the artery too is compressed, this can lead to severe engorgement and alarming haemorrhage.[6] Moore returned to his practice and

3. The older John Moore was for some time surgeon to the British ambassador at Versailles and wrote an entertaining and perceptive journal of his travels as well as several novels. He died in 1802, aged seventy-three. The younger John (later Sir John), James' brother, served in many wars on the Continent and overseas and evolved a new training method for the light infantry based on that of the French *tirailleurs*. He was fatally wounded at the battle of Corunna in 1809 which ended with a hard-won victory over Soult. He was forty-eight.
4. J. Moore, *A Method of Preventing or Diminishing Pain in Several Operations in Surgery* (London, 1784).
5. B. Bell, *A System of Surgery*, 3 volumes (Edinburgh, 1801), Vol. II, p. 34.
6. The slight dilatation of veins induced by a tourniquet is still used on countless occasions when a venous blood sample is collected. (Even then the tourniquet is released before blood is actually withdrawn). The engorgement depends on the fact that pressure by the tourniquet cannot entirely obstruct the arterial inflow but can and does obstruct the venous outflow. Von Esmarch's desanguinating tourniquet which is an elastic band wound around a limb starting at the periphery and which ensures a bloodless field of operation was not invented till the 1860s.

devoted his energies to championing Jenner's vaccination. He also wrote a still readable two-volume biography of his famous brother.[7]

...

Another ancient practice of lessening pain was revived by the seventeenth century Neapolitan surgeon Marco Aurelio Severino. He used snow from Mount Vesuvius to deaden the pain when he incised abscesses.[8] A thousand years earlier Ibn Sina (or Avicenna) had noted that 'while the most powerful narcotic is opium ... among the less powerful ones is snow'.[9] Writhing in agony in his sunbaked tent outside the walls of Accra in the Holy Land, Richard the Lionheart believed himself to be at death's door. It is, as usual, difficult to make a retrospective diagnosis: it could have been any one of a dozen fevers. A caravan of camels unexpectedly arrived, bearing snow and ice from the Syrian mountains as well as fresh pears and peaches, a present from his hospitable enemy, the Sultan Saladin. The effect, according to a contemporary chronicler, was miraculous. The king was soon on his feet again, ready to slaughter many more infidels.[10]

Larrey, Napoleon's great surgeon-general, also noticed the anaesthetic effect of freezing. After the battle of Borodino, in the midst of the terrible Russian winter, he could amputate hundreds of frozen limbs without inflicting pain.[11] Sadly, most of those successfully operated on had to be left behind as the *Grande Armée* continued its catastrophic retreat. Most froze to death or were hacked to pieces by the Cossacks.

In civilian life freezing small areas of the skin with a spray of a fast-evaporating organic solvent seems to have come into use some time during the second quarter of the nineteenth century. Who first had the clever idea remains uncertain; but the apparatus, clearly modelled on sprays of perfume, began to appear in the catalogues of surgical instrument makers around 1810.[12] Lister is

7. He married an heiress and became director of National Vaccine Establishment. He died in 1860, aged ninety-eight.
8. He is not known to have published anything, but his 'invention' was recorded by his Dutch pupil, Thomas Bartolinus (died 1680, aged sixty-four): 'Before employing the cautery he applied snow to dull sensation . . . To avoid gangrene he applied it in narrow parallel lines. After a quarter of an hour all feeling would be deadened.'
9. See Chapter 8.
10. He inflicted two defeats, neither of them decisive, on Saladin.
11. Larrey's experience at Borodino was described in his classic *Mémoires de la chirurgie militaire* (Paris, 1812–14). The book was favourably reviewed, despite the war between the two countries, in the *Edinburgh Medical Journal* in April of that year, only a few months after its publication in France.
12. The earliest mention the present writer has been able to trace is in the catalogue of Bremner und Söhne, of Dresden and Leipzig, dated 1801, in the Semmelweis Medical Historical Museum in Budapest. The inventor is not mentioned. By the third decade of the century the freezing spray was a standard part of the surgeon's equipment in England. A variety of organic solvents was used.

known to have used a freezing spray in addition to his antiseptic carbolic acid spray when, as surgeon to the queen in Scotland, he incised an abscess in Queen Victoria's armpit while she was in Balmoral.[13] He later frequently recalled – one of his many increasingly ponderous jokes – that he was the only person allowed 'to plunge the knife into her Majesty's sacred flesh'. Thereafter refrigeration as a means of local anaesthesia remained stagnant until fitfully revived during the Russo-Finnish War of 1939–40. Anaesthetics were scarce on both sides: snow was plentiful.[14]

. . .

Unlike Moore and the pioneers of freezing, Henry Hill Hickman has perhaps received an excess rather than an insufficiency of posthumous recognition. His claim as a pioneer of inhalation anaesthesia was first espoused by Dr Thomas Dudley of Kingswinford in 1847, sixteen years after Hickman's death, partly as a gesture of piety but also as a counterblast to the Yankee claim that general anaesthesia was an American invention. The counterblast convinced nobody; but in 1870 Sir James Young Simpson, already famous as the inventor of chloroform anaesthesia, published an article in support of Dudley. More effectively, during the first decade of the twentieth century Sir Henry Wellcome, founder of the Wellcome Medical Historical Museum and Library in London, began to collect Hickman memorabilia. Eventually the Section of Anaesthesia of the Royal Society of Medicine was galvanised into providing sufficient funds (£118 13s. 9d.) to unveil a plaque on the centenary date of Hickman's burial, 5 April 1930, in Broomfield Parish Church, a few miles west of Ludlow in Shropshire. The Section also instituted a lecture and a medal in memory of the 'the pioneer of anaesthesia' and 'for the encouragement of anaesthetic research'.[15] In fact, what little is known about Hickman's short life and work does not always fully support the more extravagant claims made on his behalf.

About his background, it would be impossible to improve on F.F. Cartwright's sympathetic if somewhat generalised encomium.

> The Hickmans were one of those old and undistinguished families which are the true backbone of England. For centuries they had lived at Old Swinford on the border of Shropshire and Worcestershire . . . Living their own quiet lives amid the ordered routine of seed time and harvest, politics, wars, the rise and fall of dynasties passed over their heads to be a subject of

13. The spray was wielded enthusiastically but 'very clumsily' by Sir William Jenner, physician to the Queen.
14. On the Russian side any organic solvent tended to be purloined to be made into vodka.
15. The plaque was unveiled by Sir St Clair Thomson. He describes Hickman as 'the earliest known pioneer of Anaesthesia by Inhalation'.

> conversation in dining room and inn, subjects of far less importance than the state of the crops or the prospects of the lambing season. Such a family, never knowing the extremes of wealth or poverty, kept upon its own even middle path; now rising by a hardly sensible gradation to the ranks of the magistracy and the smaller squierarchy, now relapsing, only perceptible to the eyes of the village community, to the place of a tenant farmer or to the village professional class. Such a stock is the most enduring in our land . . . and, from its very mediocrity, finds it equally easy to obey or to command.[16]

Cartwright also reproduced a charmingly romantic portrait preserved by the Hickman family, showing the young doctor gazing into the world a little wistfully, resting his right elbow on a book and holding in his left hand that perennial symbol of mortality and medicine, a skull.

Hickman qualified in Edinburgh, married Eliza Hannah Gardner of Leigh Court near Worcester and settled down to general practice in his native county. A card which hung outside his surgery has survived. It is inscribed: 'at home every Tuesday. from 10 o'clock until 4, for the purpose of giving advice, gratis to the poor and the labouring classes'. He was undoubtedly a man of sensibility and kindness, representing one face of an elegant society whose other face often displayed extraordinary brutality. At the age of twenty-three he wrote:

> I have frequently lamented when performing my own duties as a Surgeon that something has not been thought of whereby the fears and pains of surgery may be tranquillised and its suffering relieved.[17]

To remedy this state of affairs he embarked on a series of experiments inspired, it seems, by numerous reports of 'suspended animation' (some exceedingly ghoulish involving hanged but surviving criminals) in the lay as well as in the scientific literature. Such a state could apparently be 'induced by breathing carbonic acid', today known as carbon dioxide.[18] Essentially what he was aiming at was insensibility through 'controlled asphyxia', that is the effect of strangulation just short of death from lack of oxygen. He proposed to achieve this – and indeed did achieve it in a comparatively small number of animal experiments – by the inhalation either of pure carbon dioxide or of deoxygenated air. He did not even aim at anaesthesia in the modern sense or

16. Cartwright, *English Pioneers*, p. 270. This extended essay is the only monograph on Hickman.
17. From the *Souvenir Booklet, Henry Hill Hickman Centenary Exhibition* at the Wellcome Historical Medical Exhibition (London, 1930), p. 65.
18. In a lightly coded language, asphyxia with a plastic bag is the manner of suicide recommended by voluntary euthanasia societies today.

insensible sleep as could sometimes be achieved even in his day with drugs or hypnosis. In a typical experiment:

> I took an adult dog and exposed him to carbonic acid gas prepared in large quantity. Life appeared to be extinct in about 12 seconds. Animation was suspended for 17 minutes, allowing respiration to intervene occasionally by the application of inflating instruments, allowing me to amputate a leg without the slightest appearance of pain to the animal. There was no haemorrhage from the smaller vessels. The ligature that secured the main artery came away on the fourth day and the [three-legged] dog recovered without expressing any material discomfort.[19]

The procedure was not significantly different from that used in the Grotto del Cane near Naples for centuries where it did not entail mutilation of the animal.[20] It is impossible not to sympathise with Hickman's dedication to the cause of abolishing pain at a time when the possibility was widely derided. It is equally impossible not to sympathise with those who viewed his 'controlled asphyxia' with misgiving. To be effective, asphyxia induced by whatever means must be so deep that it becomes life-threatening. A window between consciousness and irreversible anoxia exists but it is too narrow for safety. The evidence that Hickman tried any more effective gases (like ether) is at best inconclusive.

In August 1824 Hickman published an open letter on suspended animation, addressed to T.A. Knight of Downton Castle, a local grandee, amateur horticulturist and fellow of the Royal Society. Knight was a kindly man and a lover of roses who was moved by Hickman's philanthropic enthusiasm. But, though he probably listened with sympathy, he would not actually do anything to propagate the young doctor's dubious ideas. (By an odd coincidence the open letter was printed in Ironbridge, a village only a few miles from Shifnal, Thomas Beddoes' birthplace.)[21] The response was almost total silence, the only contemporary reference being a snide note in the *Gentleman's Magazine* under the title of 'Intriguing Surgical Experiments'. The note-writer concluded that the method proposed to render operations painless was probably more distressing than the operations.

In April 1828 Hickman travelled to Paris partly at least to interest the French medical and scientific establishment. He stayed for three months during which time he addressed a lengthy memorandum to Charles X, the last of France's Bourbon kings. He may have had some influential friends at court

19. Quoted by Cartwright in *English Pioneers*, p. 276.
20. See Introduction.
21. Despite the geographical proximity, there is no evidence that Hickman was directly influenced by the practitioners of pneumatic medicine.

because the memorandum was forwarded to the Section of Medicine of the Académie royale. There it was discussed three months later. Hickman's claim of 'prolonged and reversible insensibility' was received with incredulity; but a subcommittee was appointed to investigate the discovery. There is no record that the subcommittee ever met or reported back; but this was not quite the end. Eighteen years later Horace Wells addressed a letter to the Académie claiming to have discovered anaesthesia by means of nitrous oxide. Jean-Philippe Gérardin who had originally tabled Hickman's letter then recalled the invention by the English doctor of some kind of an 'inhalable gas'. He also recalled that Baron Larrey was the only academician at that meeting who had thought that the idea merited further consideration.

Hickman returned to England and resumed practice in Tenbury on the Worcestershire side of the county border. No documents or letters survive about his remaining two years, but the house where he practised in Teme Street, now a chemist's shop, still stands. In 1830 he died at the age of thirty, almost certainly of tuberculosis, leaving behind a widow and two children.

. . .

The doctors who in England came closest to achieving surgical anaesthesia before the breakthrough with ether and chloroform were the mesmerists or, as some preferred to call themselves, the 'zoists'. Their leader, John Elliotson, was a noble figure. Though of diminutive stature and lame in one leg from a riding accident, he was blessed with ferocious energy, an incisive mind and strong and admirable principles. Born in Southwark, the son of an apothecary, he took his MD degree at Jesus College, Cambridge and was then appointed physician to St Thomas's Hospital. There he acquired the reputation of an ingenious physician and inspiring teacher. He wrote voluminously on a wide range of topics and was chosen to be both Lumleyan and Goulstonian lecturer of the Royal College of Physicians of London. He also acquired one of the largest practices in the city: in the words of *The Times*, 'during the season Conduit Street [where Elliotson lived] was daily filled with carriages almost as thickly as St James's Street on a levée day'.[22] In 1832 he was appointed first professor of the theory and practice of medicine at the recently opened University College Hospital in Gower Street. A little ironically in the light of later events, he praised the institution 'for its toleration and forbearance towards those of other creeds and doctrines'.[23] Five years later he attended a demonstration by Baron Dupotet, the French mesmerist, and began to experiment in his own wards. Soon he achieved a few 'excellent cures'

22. In a premature obituary published in *The Times* on 14 April 1868.
23. It was the first university college in England (though not in Scotland) which accepted students and awarded degrees to graduates without stipulating a religious test.

and 'striking results', mainly in diminishing or curing pain. He also became convinced that the method could and should be used in surgery.

Though Elliotson could be irascible, he was a man of great kindness. His generosity and obvious concern for his patients won him the friendship of Dickens. Thackeray based the character of Dr Goodenough in *Pendennis* on him. Mrs Frances Trollope thought well of him.[24] His irascibility, on the other hand, and even more the sensational success of his public demonstrations, won him the enmity of his colleagues. At a stormy meeting of the University College Hospital Board on 16 December 1840 he was forbidden to practise mesmerism or 'any related humbug' in the wards or the lecture theatre. He was also admonished by the College Mentor, Lord Brougham, to stop indulging in 'this undignified tomfoolery'. The student body came out resoundingly in his support, passing a unanimous vote of confidence in him; but, uncompromising in his beliefs, Elliotson resigned from both his chair and the hospital.

But it was not Elliotson who provided the *cause célèbre* of mesmerism in England. Mr William Squire Ward, a Nottinghamshire surgeon, was preparing to operate on a forty-two-year-old farm labourer, James Womball, when he was persuaded by his friend, William Topham, a London barrister, to do so under 'mesmeric sleep'. The patient needed to have his leg amputated for gangrene but was 'hysterical with fear'. Ward testified that he himself, 'a Bart's graduate and conservative by temperament and outlook', was 'by no means sanguine' that mesmerism would work. But Topham was an old friend, and Ward was prepared to give the method a try. The patient 'went under easily', and, after gazing at his quietly sleeping figure for a few minutes, Ward plunged in the knife. Womball slept throughout the operation 'like a child, occasionally moaning softly'. Neither Ward nor any of several onlookers had any doubt that the operation had been painless and Womball later swore that he had felt nothing.

The case caused pandemonium in the Royal Medical and Chirurgical Society. Several leading London surgeons expressed frank disbelief in this 'provincial farrago'. Others felt that it was 'no great shakes'. Rutherford Allcock had seen many 'unhypnotised but brave patients, including some plucky little women' who had undergone operations without flinching. Elliotson who was present replied that he had trained under Astley Cooper and that he too had seen such cases. In particular, he remembered a sailor who had once astonished Cooper by not uttering a sound during a particularly prolonged amputation of the leg. But the patient's compressed lips, deathly pallor and general demeanour left no doubt that he was in pain almost beyond endurance.

24. In addition to Dickens and Thackeray, several royals, though not the Prince Consort, regularly attended his lectures and became fans.

Womball, by contrast, had slept peacefully throughout the amputation. Thomas Wakley, belligerent editor of the *Lancet* and at one time a friend and ally of Elliotsons, now described hypnotism as 'gross deception' and insisted to his colleagues that 'the imposters should be hooted out of the profession'. Robert Liston, at the height of his power and fame, wondered if Messrs Ward and Topham could also make their hypnotised patients 'read with the backs of their necks or talk with their bellies'. Many felt that such incandescent wit settled the matter for good. In an attempt to calm tempers the president, the emollient Sir Benjamin Brodie, remarked wisely but irrelevantly that some people did not seem to be as susceptible to pain as others. Everybody had a story to tell and, as happens on such occasions, everybody insisted on telling it. Some timidly congratulated Ward on his 'boldness'; but even they took the view that if operations could be rendered painless, this might be a 'grave disservice' to patients and surgeons alike, for 'It is pain which alerts the surgeon that he must proceed with circumspection or change course.' Mr Hector Mansfield suggested that 'pain was a necessary evil: we must endure for the sake of our own safety'. Several members of the Society chimed in with similar sentiments. 'Pain was essential for healing.' 'Without pain we would be lost.' The usually mild-mannered Dr Samuel Copeland expressed the opinion that 'Dr Elliotson and his henchmen should be incarcerated . . . Pain was a wise provision of nature . . . patients ought to suffer pain . . . they were the better for it'.

. . .

Although Topham and Ward were censured and 'forbidden to publish their case',[25] this was not the end of mesmerism in Britain. James Esdaile, a Scottish surgeon working at the Native Hospital in Hooghly near Calcutta, used hypnosis in 950 operations, including amputations of limbs, the breast, the penis and 200 'huge scrotal tumours'. His results were endorsed by witnesses who included such unimpeachable characters as a British army officer, the wife of a missionary and even a colonial bishop. Major Corfield of the 20th Bengal Infantry wrote in the *Englishman*, a Calcutta newspaper, that 'Dr Esdaile is destined to be a blessing to mankind and an honour to his profession'. Sadly, in the aftermath of the Womball case, Esdaile was vilified in his own country; and, after retiring to Scotland in 1851, he found few prepared to undergo operations in a mesmeric state. He also found that his countrymen and countrywomen were 'significantly less susceptible' to hypnosis than his

25. 'No man ever had a more complete beating,' Elliotson recorded despondently. The case was nevertheless published, Ward resigning from the Society in disgust. Details are vividly described in Peter Stanley's *For Fear of Pain*.

Indian patients. He attributed this to the excessive consumption of haggis and whisky.

A sensible variation on mesmerism was practised by James Braid, another Edinburgh graduate, who spent his professional life in Manchester. He became interested in 'somnambulism' in 1841 but objected to the 'nonsense about animal magnetism'. In his book, *Neurypnology, or the Rationale of Nervous Sleep*, he attributed mesmeric phenomena to suggestion and substituted the term 'hypnosis' for 'mesmerism'.[26] Despite his interest, Braid used the method sparingly, confining it to dental extractions and the incision of abscesses. For sixteen years, however, the journal of mesmerism, the *Zoist*, founded by Elliotson, continued to publish case reports from provincial centres and an interesting series of five cases from Jamaica.[27]

. . .

Mesmerism also had a late flowering on the Continent. The objection that 'pain was necessary, even essential for healing' was discussed in Paris where Hyppolite Cloquet reported the case of a young woman on whom he had himself reluctantly performed a mastectomy for a malignant growth. The hypnotised patient had slept throughout quite insensibly and later testified that she had felt nothing. Her recovery, however, was unduly prolonged. The Académie took the report seriously enough to appoint four subcommittees. The number of subcommittees on such occasions provided a star rating of the presumed importance of the topic. (Hickman's letter had merited only one.) More than three, however, tended to be counterproductive, effectively burying any contentious issue. Discussions rumbled on for years. Since the argument that pain was useful both during an operation and even more for subsequent healing could neither be proved nor disproved, no clear conclusion ever emerged.

In contrast to such uncertainty, in Milan Cardinal Berlusconi, perhaps a kinsman of Italy's twenth-first-century prime minister, delivered a much-quoted sermon in which he castigated those who sought to abolish the pain of surgery, 'one of the Almighty's most merciful provisions'.[28] He did not specify how the provision operated but his views were shared by many. It is, in fact,

26. James Braid, *Neurypnology, or the Rationale of Nervous Sleep Considered in Relation to Animal Magnetism* (Manchester, 1867).
27. After the cessation of publication of The *Zoist* in 1856 (with the 13th volume) little was heard of Elliotson, though he continued to contribute articles to the *Medical Times and Gazette*. He warmly welcomed ether and chloroform as 'achieving more easily what mesmerism has endeavoured to do'. He became deeply religious. His finances failed in 1865 and he retired to the house of his pupil, E.S. Syme, in Davies Street in Mayfair. There he died of 'natural decay' and was buried in Kensal Green cemetery in 1868, aged seventy-seven.
28. Quoted by Father Johann Franz, in *Unsere Schmerzen* (Vienna, 1868), p. 232.

unfair to judge the frustrated efforts of hypnotists and other pioneers without looking at the cultural background of their battles.

...

In the Europe of the early nineteenth century physical pain still stalked life from cradle to grave. To most people surgical operations were a terrible ordeal but neither outrageous nor exceptional. Suffering was part of the human condition. It was not confined to those labouring in the new factories and mines or the wretches in workhouses, prisons, lunatic asylums or hospitals. Boys from aristocratic homes, rioting in vicious freedom in their public schools, were birched unmercifully by high-minded classical scholars. The injuries inflicted by these often reverend gentlemen would land many in prison today. Some, however, might successfully plead self-defence. In the exclusive 'seminaries' for young ladies, those culpable of minor misdemeanours were made to kiss the whip before being chastised till they collapsed in front of the school. These practices were mild compared to those prevailing in the services. In the army and navy sentences of up to 300 lashes were commonplace. Some of those sentenced took weeks to recover. Some never did.

Civilian life was not much kinder. Pretty sixteen-year-old Amy Lyon from Scotland started her London career as a nursemaid to a doctor's family. She became pregnant and was thrown out onto the street. Her only asset was her splendid teeth. Like hair, these could fetch a fair price when the supply from cadavers was short. She was on her way to have all but her molars pulled out in a single session without an anaesthetic when, fortunately for her, near the dentist's surgery she was picked up by Captain Payne of the Royal Navy. He offered her and her baby an alternative and less painful means of livelihood. Eventually, by a circuitous route, she ended up with perfect teeth as the wife of Sir William Hamilton, Britain's ambassador to Naples, and lover of Britain's greatest sailor. Others were less lucky.[29]

In parts of the country scolding women were regularly bridled with metal hooks piercing their tongues or ducked in rivers until they were half drowned. Felons were branded with red hot iron. Lesser criminals went to the whipping post or the stocks. Prostitutes were flogged in the Bridewell. Men and boys of ten were sent to the gallows for stealing sheep or transported for damaging a bridge. Executions were public and drew cheering crowds. The bodies were

29. The price of 'fresh teeth' was not high: live donors had to compete with the haul of the resurrectionists, a flourishing trade despite the threat of the gallows. The battlefield of Waterloo was scoured for days by 'the tooth collectors', the thousands of young bodies offering a rich harvest of 'Waterloo teeth'. In the American Civil War dental pirates were still snapping at the heels of both armies. Most of their loot was exported to England, where American teeth could be purchased by mail order.

left to rot on the gallows and the sites were visited as part of instructive Sunday outings by Quaker families like the Listers. When no executions or public floggings were on offer, cock-fighting, bear-baiting and bare-knuckle fist-fights were favourite pastimes.

Among the country's rulers the few who dared to advocate a regular police force were ridiculed. A gentleman armed with a shotgun was expected to face and outface an angry mob. Duelling was fashionable and deadly: an estimated 500 men were killed in this way every year in Britain. In France and Germany the figures were higher. In such a society men like Elliotson with his hatred of capital punishment, loathing of torture, dislike of duelling and objection to unnecessary pain were regarded with suspicion or pity. Pain was not something that could or even should be avoided. It was normal. Since it was normal, it had to have a purpose. Indeed, its purpose was self-evident. It maintained discipline in schools and the armed services. It provided the firm foundation of the social order.

...

Then came an extraordinary change. By the time the eighteen-year-old Princess Victoria ascended the throne, the tide was turning.[30] Why such turnings come about remains a mystery. Yet they are irresistible. Dr Thomas Arnold's headmastership of Rugby changed the face of education in the public schools.[31] His regime was rigorous and even brutal by modern standards; but the worst barbarities were put beyond the pale. The Factory Act of 1833 and the Mines Act 1842 outlawed some of the abominations of female and child labour. The legislation was a modest beginning; but the trend could not be reversed. By the mid-century the duel in England had virtually disappeared. (On the Continent it lasted longer.) The excesses of service discipline were being tempered. Cock-fights, bear-baiting and the brutal prize ring went underground. Slavery was abolished throughout the British Empire. In darkest Africa the Bible began to follow the flag. At home the death penalty was no longer enforced except for treason and murder. Within a decade executions would cease to be public.

In 1804, had either Davy or Beddoes pressed for 'laughing gas' or ether to be used routinely in surgery, they would have been laughed at. They themselves would probably have regarded the idea of painless operations with scepticism. Even Hickman only partly believed in his own invention: he simply hoped that others might improve on his method of partial strangulation.

30. In 1837. She reigned for sixty-four years.
31. Thomas Arnold, one of Lytton Strachey's four 'Eminent Victorians', became headmaster of Rugby School in 1828 and remained there for nine years. He died in 1842, aged forty-seven.

Forty years later physical pain was no longer universally seen as an absolute, let alone a salutary, necessity. Those who thought otherwise included a significant number of doctors. To some pain had become an evil to be defeated at all cost. In 1842, the year of his death, Sir Charles Bell wrote: 'When pain will be taken out of surgery, the earth and the lives of all who dwell on it will have changed'.[32] He did not write 'if' but 'when'. But he would probably not have guessed that the good news would come from Britain's still uncouth former colonies across the Atlantic.

32. Sir Charles Bell, *Practical Essays* (Edinburgh, 1841), p. 67.

## Chapter 21

# A GENTLEMAN FROM THE SOUTH

The man who discovered ether anaesthesia, the first general anaesthetic to abolish the pain of major operations, was never hailed as one of the benefactors of mankind. He was an upright, dignified figure, honourable and hard-working. He contrasted sharply with those who later staked their claim. The mantle of a medical pioneer would have sat well on his shoulders. Personality and circumstances conspired against it.

At the time of his momentous discovery Crawford Williamson Long was the town physician of Jefferson, a small agricultural community in the State of Georgia. His Scottish-Irish grandfather had arrived in Carlisle, Pennsylvania, about 1760, fought in the War of Independence and settled in Madison County, Georgia. His father, James, became a successful and wealthy planter, clerk of the State Superior Court and a state senator. The Longs, in short, became part of the old South's cotton aristocracy.

Crawford was born in Danielsville on 1 November 1815. Numerous more or less true-sounding anecdotes testify to his intelligence and resourcefulness as a child. At Franklin College in Athens, the future University of Georgia, his room-mate and best friend was Alexander Stephens, later a leading figure of the Confederacy. According to Stephens, Long was 'undoubtedly the brightest star of our year', though 'not notable for sustained uncongenial effort'. On his returning home, the young man expressed the wish to become a doctor. The elder Long felt that a 'few years' reflection as a teacher' might make him choose a more gentlemanly profession. The post of principal at the local academy, maintained largely by the elder Long, happened to be vacant. The son was duly installed. But, despite his reputed indolence, even during that time he busied himself acting as unpaid assistant to the town's physician, George R. Grant. Eventually he embarked on the long track to Kentucky to enter Transylvania Medical College at Lexington, a famous institution of learning at the time. Benjamin Winslow Dudley, the fastest man in the Union to cut the bladder for stones, was professor of surgery, and Ephraim McDowell, celebrated as the first man to remove an ovarian tumour, practised in nearby Danville and lectured occasionally.[1] In 1837 Long transferred to and obtained a medical degree at the University of Pennsylvania, then spent two more years attending medical and surgical courses in New York. None of this was essential or even customary. Four out of five doctors who catered for the needs of small-town and rural communities and many of those practising in large cities had qualified by apprenticeship. 'Doctor' designated a trade like 'Baker' Jones or 'Carpenter' Smith. But a degree from one of the new colleges was beginning to be valued. The institutions provided a grounding in the basic sciences like anatomy and botany – the latter important since most doctors acted as their own pharmacists – but no clinical teaching. Clinical skills and the craft of surgery had to be learnt in general hospitals in the big urban centres like New York, Philadelphia or Boston.

The crop failure of 1841 put an end to young Long's leisurely travels. A projected study tour to Europe had to be cancelled. At twenty-six he returned to Jefferson in his native county and purchased a practice. He married a Southern belle, Mary Caroline Swain, sixteen-year-old niece of Governor David Lowry of North Carolina. Six of their twelve children would survive into adult life.[2] He then settled into the life of a respected country practitioner.

. . .

At what stage Long made the acquaintance of 'gas frolics' is uncertain. By the 1830s showmen visiting country fairs and giving exhibitions of 'laughing gas' were popular attractions. (Serious lectures too on chemical topics by bearded academics were starting to be well attended.) Among the most successful fairground hustlers was a young man from Ware, Massachusetts, by the name of Sam Colt but calling himself Professor (or sometimes merely Doctor) Colt of New York, London and Calcutta. Having acquired the skill of preparing nitrous oxide, he set out with his portable apparatus to make money. He had no interest in science or medicine but he needed funds to patent his clever six-barrelled rotating automatic shooter which would eventually make his

1. The first occasion almost ended in him being lynched. Yielding to the entreaties of the patient who was in desperate pain with a huge ovarian tumour, he agreed to operate on her. Such an operation had never been performed in the United States. The rumour spread in the town that he was terminating a pregnancy and killing the patient. The sheriff went to inspect while an excited crowd waited outside. By the time the sheriff entered, the operation had been completed. The patient, Jane Crawford was alive – just. She eventually recovered. McDowell performed two more similar operations and became a celebrity and a medical hero. He died in 1830, aged fifty-five.
2. Two became doctors and one married a doctor. One daughter, Frances, wrote a biography of her father: F.L. Taylor, *Crawford W. Long* (New York, 1889).

fortune.[3] For a time he settled in Cincinnati and on one occasion almost got into deep water by putting six hired American Indians so fast and so deeply to sleep that they could be aroused only with difficulty. In the process they were savagely kicked and pummelled but later remembered nothing. It was Colt on one of his travels who introduced the frolics to Georgia. In a land where the puritan ethic was still honoured and competing party entertainments were few, gas frolics became the rage among young people in search of more or less innocent fun. At twenty-seven Long no longer counted as young; but he had four teenage apprentices and other young friends whom he was ready to oblige. As town physician he was also town pharmacist and prepared and sold the medications he prescribed. He later wrote:

> In early 1842 I found myself without the material or apparatus to prepare the gas. I told my young friends that I did have another medicine, sulphuric ether, which would produce equally exhilarating effects; that I had inhaled it myself and considered it as safe as nitrous oxide. One of the company said that he had tried it at school and was willing to try again . . . I gave it first to the young gentleman who had said that he had inhaled it before, then I inhaled it myself and then I gave it to the rest of the company. They were so much pleased with the effect that they afterwards inhaled it frequently and induced others to do so. It soon became fashionable in the county and extended from this place through several counties in this part of Georgia.[4]

Many of the 'ether frolics' were held in the Long's house and some bizarre side-effects set the doctor thinking. Ether sometimes made him furiously excited with loss of control of his movements. When he emerged from his 'state of intoxication' (as he began to call it) his arms and hands were often covered in bruises. Yet he could not recall sustaining any painful injuries. His friends too sometimes fell about under the influence and seemed to hurt themselves. Yet all agreed that they had felt nothing during the sessions.

A wealthy young resident of the town, James M. Venables, was among Long's patients. He had two unsightly but painless tumours on the back of his neck. Long wrote that Venables frequently consulted him about them.

> He wanted to have the tumours removed but would always postpone the operation for dread of pain. At length I mentioned to him the fact of my receiving bruises while under the influence of the vapour of ether without

3. His was a typical early American success story. Entirely self-taught, with his earnings from demonstrating laughing gas and selling the paraphernalia needed for its production he founded the Patent Arms Company in New Jersey in 1835 and in 1852 built the huge Colt Firearms Manufactory in Hartford. He died in 1862, aged forty-eight.
4. Quoted by H.H. Young in 'Long, the Discoverer of Anaesthesia, a Presentation of His Original Documents', *Johns Hopkins Hospital Bulletin,* 1897 (8), 104.

suffering, and as I knew him to be fond of and accustomed to ether inhalation, I suggested the probability that the operation might be performed without pain. . . . He eventually consented to have one tumour removed, and the operation was performed the same evening. The ether was given to Mr Venables on a towel; and, while under the influence, I extirpated the tumour. It was encysted and about half an inch in diameter. The procedure took seven and a half minutes. The patient continued to inhale ether during the whole of the operation; and when informed it was over seemed incredulous until the tumour was shown to him.[5]

The contemporary entry in Long's ledger reads simply: 'James Venables, 30 March 1842. Ether and excising tumour, $2.00'. Healing was uneventful. With hindsight it was an epoch-making event – or could have been. It was later suggested that Long was unaware of its implications. This was not, however, how Long recalled his thinking.

I was anxious before making my publication, to try etherisation in a sufficient number of cases to fully satisfy my mind that anaesthesia was produced by the ether and was not the effect of the imagination, suggestion, or some peculiar insusceptibility to pain of the person experimented upon . . . I determined to wait also . . . to see whether any other surgeon would present a claim to having used ether by inhalation in surgical operations prior to the time it was used by me.[6]

The attitude may have been over-cautious but remains familiar. The more important and beneficial a new observation, the more reluctant many doctors are to announce their discovery. Some of course have an unquenchable thirst to announce momentous breakthroughs at the drop of a hat. But even today they are probably a minority. Most give a passing thought to the false hopes and unnecessary anguish premature claims might engender. Long was clearly of that breed. He was intelligent enough to know that painless operations would change the face of surgery. He did not publish his findings for three critical years despite the fact that he used ether again on several occasions, several times for operations lasting more than twenty minutes. He was well enough read in the literature to know that his discovery was revolutionary. It was a significant advance on 'nitrous oxide' as it was then administered.[7]

5. Quoted ibid., 121.
6. Quoted by Taylor in *Crawford W. Long*, p. 67.
7. Nitrous oxide as an anaesthetic only became safe in major operations after the introduction by Edmund Andrews of a nitrous oxide–oxygen instead of a nitrous oxide–air mixture in 1868 (Andrews, 'The Oxygen Mixture', *Chicago Medical Examiner*, 1868 (9), 656). Combined with oxygen, the gas could be administered in relatively high concentrations without running the risk of severe anoxia and its consequences.

Nitrous oxide was useful for short operations like tooth extractions; but then the associated anoxia became dangerous. Ether seemed safe for longer anaesthetics since it did not deprive patients of oxygen. Long later regretted his reticence. He then began to claim in private conversation and correspondence to have been the first to anaesthetise a patient effectively for major operations, a claim that was almost certainly true. Nobody took much notice. His discovery and subsequent labours had no influence on the historical development of anaesthesia. Had his achievement been a work of art or literature, its late rediscovery might have been sensational, but medical discoveries do not keep. By the time he started to claim priority – and then, by contemporary American standards, not very aggressively – the assertion provoked little more than mild commiseration.[8]

. . .

In 1850 Long moved to Athens, Georgia, because the town offered better educational facilities for his children. He became a successful surgeon. A few surviving photographs of his home show a stately pile surrounded by a large and well-tended garden. He continued to use ether for operations and difficult deliveries. His discovery was locally known but did not raise distant echoes. Medical conferences were coming into fashion in Europe at the time, a happy (or otherwise) effect of the railways. They were grand occasions marked by frock-coated and be-medalled professors parading in procession to the strains of national anthems. In the cosy world of the antebellum South such 'futile and godless gatherings' were unknown. Of Long's travels, if he did any travelling, nothing is known.

At the beginning of the Civil War he, like most of his family and friends, volunteered for the Confederate Army and was put in charge of the military hospital installed at the University of Athens. In his flight before the advancing Union forces he is said to have carried with him a roll of paper, 'my proof of the discovery of ether anaesthesia'. Nobody asked him to produce it. He returned to a burnt-out Athens a year later. The surgeon general in Washington put him in charge of both the Union and the Confederate wounded. He was later praised for his 'selfless and excellent service'; but the peacetime army had no permanent post for him. The economic collapse of the 'Reconstruction' years that hit the South in the wake of the military defeat ruined him and his class. Long had his skills and reputation; but there could be no question of a graceful retirement. He continued in practice to make

8. There were enough claimants fighting over priority already. Long owed his final recognition to the testimony of the renowned obstetrician, James Marion Sims ('Discovery of Anaesthesia', *Virginia Medical Monthly*, 1877, p. 81).

ends meet. Most of his patients were as poor as he was himself; and making ends meet was as far as it went.

Ten years later and in indifferent health, he was still in harness. He used ether regularly and continued to keep meticulous notes. These are impressive, anticipating some later technical discoveries. He may have circulated papers privately but he published nothing. No recognition came from the outside world. On 16 June 1878 Long attended a woman in labour. The delivery proved prolonged, painful and difficult. He successfully administered ether and delivered the baby. As he handed it to a nurse he suffered a stroke. He died a few hours later.[9]

9. On 30 March 1926 the state of Georgia presented a marble bust of him to the Statuary Hall of the US National Capitol in Washington, DC. The engraving is a quotation from Long: 'My profession is to me a ministry of God.' There are other memorials to him in Danielsville, Philadelphia and Athens and his house in Jefferson is now a charming small museum. He was commemorated by the US Postal Service on a two-cent stamp in 1940.

# Chapter 22
# 'THIS YANKEE DODGE'

## *I*

'This Yankee dodge sure beats Mesmerism hollow' was Robert Liston's memorable comment at the end of the first anaesthetic administered at University College Hospital, London.[1] 'Dodge' hints at something faintly fishy; and the origins of general anaesthesia were certainly fishy enough to populate an aquarium. Rarely has one of the great advances in medicine been accomplished by an unholier crew. Even more rarely has greed and paranoia played so dominant a part. To find one's way in a maze of conflicting claims, it may be useful to start with a cast list and their provenance.

*Gardner Quincy Colton* (not to be confused with Sam Colt) was a New England boy, tenth son of Walter and Thankful Colton, born on 7 February 1814. At ten he was apprenticed to a chair-maker at $10 a year, a trade which brought him to the metropolis of New York. He made cane chairs until an older brother offered to finance his medical studies under Dr Willard Parker. Funds ran out long before he could qualify as a doctor; but during his apprenticeship Gardner learnt all that was to be known at the time about nitrous oxide. A sharp-witted youth, he persuaded a group of friends to hire a hall where, as Professor Colton, he could give an illustrated lecture on the marvels of the recent advances of medicine. A large audience paid an amazing total of $535 to listen to him. Colton was twenty-eight. Fame and fortune beckoned.

*Dr Horace Wells* was born on 21 January 1815 in Hartford, Vermont, into a comfortably off middle-class family. He was a handsome and bright boy,

1. Contrary to many statements in the literature, the first surgical operation under ether anaesthesia in Britain (as distinct from London or England) was William Scott's operation two days earlier at the Dumfries and Galloway Royal Infirmary. But news spread from London to the rest of the country and to Europe.

indulged by his parents. He received a good education at Amherst, Massachusetts and Walpole, New Hampshire. Passing through a deeply religious phase, he toyed with the idea of entering the ministry but eventually decided against it. He did, however, remain a devout Christian, the only one of the present cast who seems to have been troubled by ethical scruples or indeed scruples of any kind. At the age of nineteen he embarked on dental studies in Boston and received a sound practical training from some of the best dentists in the city. Within a few years of qualifying he himself became a sought-after practitioner. He married Elizabeth Dainton, who came from a well-to-do family like his own, and started to accept students. Among them were John Mankey Riggs and William Thomas Green Morton.

*William Thomas Green Morton* was born in Charlton on 9 August 1819, the son of James Morton, a Massachusetts farmer, and his wife, Rebecca. He seems to have had a troubled time at his local school, which perhaps contributed to his later delusions of persecution. On one occasion, according to his commissioned biographer, he was 'falsely accused of a misdeed by another student and outrageously punished'. His health was affected for months. His formal education ended when he was seventeen and he then tried his hand at various jobs, mostly related to publishing. He certainly acquired a facility in self-advertisement, composing circulars and commissioning ghosts. His biographer (who had much difficulty collecting his fee) refers to these ventures as 'adverse circumstances' and concludes:

> Duped by his unscrupulous partners who were older, more shrewd and better versed in business, his mercantile career terminated in decided disaster and its abandonment forever.[2]

From business Morton turned to dentistry and later claimed to have studied at the College of Dentistry in Baltimore. There is no documentary evidence of this. But he certainly became Horace Wells' pupil and assistant and the two men went into a somewhat uneasy partnership in 1842.

*Charles Thomas Jackson* was perhaps the most bizarre and in some respects the most brilliant of the characters involved in the birth of anaesthesia. He was of old New England stock, born on 21 June 1805 in Plymouth, Massachusetts, a descendant of Abraham and Remember Morton who celebrated their wedding in Plymouth in 1657. Charles' entry to the Harvard Medical School was sponsored by Walter Channing, later the first professor of obstetrics and gynaecology and a pioneer of anaesthesia in childbirth, and by James Jackson,

2. N.P. Rice, *Trials of a Public Benefactor* (New York, 1859; 1995 edn), p. 54.

professor of the theory and practice of physic; and, throughout his career, he had access to and usually the behind-the-scenes support of some of the most influential people in academe. His interests soon turned from medicine to chemistry and geology, and he made two useful geological trips to Nova Scotia. He nevertheless qualified as a doctor in 1829 and departed to Europe for postgraduate studies at the Sorbonne.

*John Collins Warren* was sixty-eight in 1846 and one of the most famous surgeons in the United States. As a young man he had travelled widely in Europe, studying with leading surgeons like Dubois and Dupuytren in Paris and Astley Cooper in London. Returning to Boston he became professor of surgery and anatomy at Harvard and chief of surgery at the Massachusetts General Hospital on its opening in 1820. He was sarcastic and intimidating, feared rather than liked, an incisive lecturer and pitiless examiner. But he was said to be kind to patients and was widely acknowledged as the most astute diagnostician and fastest operator in the business, 'wholly devoid of debilitating sentiment'. His operating suite was under a glass dome on top of the Bullfinch Building, deliberately out of earshot of the rest of the hospital. He refused to operate anywhere else; and never operated in patients' homes. He was to preside both over the first formal attempt at general anaesthesia which ended in a farce and over the second, which proved to be a turning point in the history of mankind. He died in 1856.

*Henry Jacob Bigelow*, the son and grandson of Boston physicians, was born in 1818. Suspected of tuberculosis, he spent some years in Cuba before returning to Boston and qualifying as MD at Harvard. He became Warren's first assistant and later successor at the Massachusetts General Hospital and was instrumental in persuading his chief to give Morton a chance to demonstrate his 'invention', ether anaesthesia. He himself later popularised the lithotrite, an instrument now named after him, which enabled surgeons to crush large stones in the bladder and extract the fragments.[3] He died in 1890.

. . .

Unlike Colton, Wells and Morton, whose claim to fame rests entirely on their part in discovering general anaesthesia, Charles Jackson had a varied career before his involvement with ether. His European tour was a notable success. Money, of course, and letters of recommendation from home helped: rich Americans were already popular, if heavily patronised, in old Europe. But Jackson also had a winning personality and quick wit. He learnt to speak French and German fluently. In Paris he made friends with eminent geologists like

3. This useful instrument was in fact invented by the French surgeon, Jean Civiale, in 1824.

Elie de Beaumont as well as enjoying the favours of beautiful tuberculous *demi-mondaines* like Jeanne de Flourence. He went on a walking tour of Italy and climbed both Mount Vesuvius and Etna. In Vienna he lodged with Professor Hofrat Johann Kuhlmauss, physician to several archdukes and archduchesses, stayed for the 1831 cholera epidemic and did post-mortem examinations on some of the victims. After the outbreak he read a paper to the Gesellschaft für Medizin which was acclaimed as a useful contribution to the aetiology of the disease (which, however, remained totally obscure). His wide contacts would later help to propagate his claim in Europe as the discoverer of anaesthesia. Before departing, he visited the famous establishment of Pinder et Fils in Paris and purchased equipment for further physical and electrical experiments.

On the return journey his curious personality first became manifest (at least to posterity). The lengthy crossing was enlivened for him by discussions with a fellow passenger, the engineer and physicist Samuel Morse. Morse had already conceived the idea of using electricity for the transmission of messages and was delighted to explain his findings and ideas to a bright young listener.[4] The bright young listener may have made a few useful suggestions; but the consequences later came as a shock to Morse. By the time the boat reached New York Jackson had decided that Morse's invention was in fact entirely his. How far he believed these fantasies is uncertain. There can be no doubt that, apart from a few unimportant details, telegraphy was Morse's brainchild; but, however flimsy the foundations, Jackson's extravagant notions always had a superficial plausibility.

In 1832 Jackson started to practise medicine and married Susan Bridge of Charlton. Over the next ten years she would present him with two daughters and three sons. Soon after his marriage he also started to take an interest in the researches of a military surgeon, William Beaumont. The interest would blossom into another fantasy invention.

In 1833 and 1834 Beaumont was touring medical societies with his French-Canadian patient, Alexis St Martin. A gunshot accident had left St Martin with a fistula or tract from his skin into his stomach. Beaumont had originally saved St Martin's life and the patient was devoted to him. The fistula now enabled Beaumont to sample gastric juices and investigate the process of digestion. His observations laid the foundations of modern gastric physiology as well as paving the way for Pavlov's historic studies on psychogenic stimuli eighty years later.[5] Unsuspectingly, Beaumont gave Jackson samples of St Martin's gastric juice 'for chemical analyses'. Perhaps the army surgeon was

4. Born in 1791, Morse was fifteen years older.
5. Pavlov created a gastric fistula in dogs and became an icon in Stalin's Russia. (See Chapter 39.) Many of Pavlov's observations were anticipated by Beaumont.

dazzled by the interest shown in his work by an eminent academic. Beaumont also described his current researches and preliminary conclusions. Jackson immediately published these under his own name, making only passing reference to Dr Beaumont's 'useful model'.

Even before publication Jackson learnt with dismay that Beaumont was being posted to the frontier and would take St Martin with him. This suited Jackson not at all. Without consulting Beaumont he drew up a petition and had it signed by 200 members of Congress. Addressed to Edward Everett, Secretary for War, this extraordinary document recounts how the brilliant chemist and physician, Dr Charles T. Jackson, is 'currently prosecuting fundamental research on gastric secretion using a young Canadian patient of a military surgeon, a certain Dr Beaumont'. This had already furnished Dr Jackson with an opportunity 'to make fundamental observations whose supremely important fruits would be lost forever to our countrymen if Dr Beaumont and his patient were allowed to depart'.[6] Fortunately for Beaumont, he had already departed; and Everett considered the menace posed by the Sioux of more immediate concern than the whimsical flow of the gastric juices. Fortunately also, Jackson's attention was turning to other matters, including publishing a claim in the *Boston Post* that 'the electromagnetic telegraph which has been claimed by Mr S.F.B. Morse of New York is entirely due to our fellow citizen, Dr Charles T. Jackson of Boston who provided Morse with all the necessary information on board the packet ship *Sully* on their return voyage from Europe'.[7] About the same time – Jackson's imagination was nothing if not versatile – he began to claim that he was the rightful begetter of the recently announced discovery of guncotton by C.F. Schönlein.

Jackson's fantasies are the more intriguing since he was making a good living and earning a reputation as an expert in geology. (He had given up his medical practice after three years.) It was an auspicious moment for this kind of expertise. The notion was beginning to grip the minds of both politicians and private entrepreneurs that the world's and especially the United States' wealth lay hidden just under the ground. Jackson was appointed to conduct geological surveys of Maine, Massachusetts and Rhode Island and later of Lake Superior, as well as acting as consultant and investor under various aliases to a number of private conglomerates and somewhat shadowy 'capital ventures'. Unsurprisingly, there were murmurs of 'impropriety' and complaints of 'inattention to duty, drunkenness and inefficiency'. Though never properly investigated, Jackson was forced to resign from his official positions. He proclaimed himself the victim of a conspiracy and continued to enjoy favour in academic circles. Overshadowing his other pre-

6. Quoted in Robinson, *Victory over Pain*, p. 249.
7. Quoted ibid., p. 252.

occupation, he was also getting embroiled with an 'ignorant dentist called Wells' and an 'impertinent and lying charlatan' called William Thomas Green Morton.

. . .

While Jackson was claiming as his own the discoveries of a succession of inventors, Horace Wells prospered in a less spectacular way in the city of Hartford. Yet he was also frustrated. The usual procedure for fitting false teeth at the time was to fix them to a gold plate and set the plate upon old roots which were seldom removed. A gold solder was used for the fixation. The solder had to be softer than the plate since any heat great enough to melt the solder would fuse the plate. Invariably the colour of the solder tended to change, leaving an unsightly line on the margin of each tooth. The gap between the plate and the solder also tended to widen, leading to an unpleasant taste as well as causing distressingly foul breath.

The first idea of a new kind of solder was probably neither Wells' nor Morton's but that of John Mankey Riggs. Wells and Morton first tried to persuade Riggs to let them exploit the invention; and, when Riggs proved difficult, went ahead anyway. The idea was to let an 18-carat solder flow upon an 18-carat gold plate. This, they rightly supposed, would make better-looking dentures and attract thousands to their surgery. In their second supposition they were mistaken. Few beat a path to their door. The flaw in the method was that to fit the plates perfectly, the old roots had to be removed. This was a torture few prospective patients were prepared to contemplate. The disappointment probably planted – or at least reinforced – the idea of painless extraction in the minds of both Wells and Morton. By coincidence, it also led to their first contact with Jackson. Jackson was recommended to Morton as an eminent academic who could be engaged (or persuaded) to write a fulsome certificate of the 'unique properties' of the new solder. Jackson duly obliged; but the endorsement did not overcome patients' reluctance to suffer excruciating pain even for a more pleasing image and breath.

The paths of the three men then temporarily diverged. Jackson resumed his geological-cum-business pursuits. Morton enrolled at the Harvard Medical School. This was largely at the insistence of the parents of his long-time betrothed, Elisabeth Whitman of Farmington, Connecticut, who would approve of the marriage only if their prospective son-in-law first qualified as a doctor. Morton never actually qualified but he established useful contacts at the Massachusetts General Hospital. A son was born to the couple in 1845.[8]

8. Physician to a mining company in Cape Town, William James Morton was also at various times prospector, big-game hunter and psychiatrist. In March 1913 he was convicted of using the mails to defraud by promoting the stock of a bogus Canadian

Wells went back to his practice in Hartford but continued to ponder the idea of painless tooth extraction.

. . .

On 10 December 1844 Gardner Colton repeated his wildly successful New York lecture demonstration in Hartford. He later recorded that, after a brief introduction about the properties of nitrous oxide

> I invited a dozen or fifteen gentlemen to come upon the stage who would like to inhale it. Among those who came forward was Dr Horace Wells, a dentist, and a young man by the name of Cooley. Cooley inhaled the gas and, while under the influence, ran into some wooden settees and badly bruised his leg. Taking his seat next to Dr Wells, the doctor said to him: 'You must have hurt yourself'. 'No, not at all'. Then he began to feel some pain and was astonished to note that his leg was covered in blood.
>
> At the close of the exhibition Dr Wells came to me and said: 'Why cannot a man have a tooth extracted under the gas and not feel it?' I replied that I did not know. Dr Wells then said that it could be done and would try it on himself . . . The next day I went to his office carrying a bag of gas.[9]

Colton would always give credit to Wells for suggesting the first practical use of nitrous oxide and he became the most convincing, steadfast and eventually successful witness for Mrs Wells in her later fight against Morton and Jackson.

Wells' life was transformed by his meeting with Colton. He was reluctant to try his idea on one of his patients; but he himself had a troublesome wisdom tooth. The morning after the demonstration he seated himself in his own dental chair, took the tube into his mouth and began to inhale the gas. He had persuaded Riggs to perform the extraction while he was under the influence. As soon as Wells was 'asleep', Riggs grasped the root of the offending molar, rocked it free and pulled. The tooth and root came out easily. Wells did not stir for some minutes. He then came to and demanded why Riggs had failed to perform the operation as promised. He would not believe the truth until the tooth was shown to him.

. . .

Colton now fades from the picture; but since his subsequent career was the only happy one among the inventors of general anaesthesia, it merits a brief

mining company. He was paroled after serving only a few months of his prison sentence and was later restored to the medical register but never prospered. He died of heart disease in poverty in 1920, the year in which his father was posthumously elected to the Hall of Fame of the United States.

9. Quoted in Robinson, *Victory over Pain*, p. 256.

follow-up. After instructing Wells how to prepare nitrous oxide, he went off on his 'exhibition business'. Later he teamed up with Samuel Morse and lectured on the electric telegraph. Then the newly discovered gold-fields of California beckoned. His luck at prospecting for gold was unremarkable, but he practised successfully as a doctor and made himself popular as a dispenser of drugs. Perhaps there were not many regular but good guys about, for after two years he was appointed Justice of the Peace of San Francisco by Governor Riley. The job was no sinecure; but it did help him to amass a small fortune. On returning to the East he lost his capital in bad investments and for a while supported himself by reporting Sunday sermons for the *Boston Transcript*. On the outbreak of the Civil War in 1860 he tried his luck in publishing a series of 'improvised war maps'; but by 1863 he was back lecturing on nitrous oxide. In surgery it had by then been supplanted by ether and chloroform; but in dentistry 'laughing gas' now mixed with oxygen still had much to offer. Colton teamed up with a dentist, J.H. Smith, and in twenty-three days extracted 1,785 teeth entirely painlessly. He was home and dry. With another dentist, John Allen, he established the Colton Dental Association in New York with the sole objective of pulling teeth painlessly. The Association prospered, spawning branches in Philadelphia, Baltimore, St Louis, Cincinnati, Boston and Brooklyn. He himself estimated that before his retirement at the age of eighty-one he had extracted 25,000 teeth without a single mishap. (If true, this was a uniquely successful series even for a comparatively safe surgical procedure.) He then retired and set out with his young third wife on a European tour, still keeping his eyes open for a possible scope for new branches in the Old World. He died at the age of eighty-four after a banquet given in his honour by the Municipality of Rotterdam in Holland.

. . .

Wells' widow later declared that the discovery of nitrous oxide anaesthesia had been to her family 'an unspeakable evil'. It ruined her handsome, red-haired, lively and lovable husband and transformed their happy home into a domestic battleground. It did not seem so at the time. Wells was exultant and set out to draw the world's attention to his momentous discovery. One of his first ports of call was his former pupil, William T.G. Morton, then still ostensibly a medical student in Boston. Together they went to consult the eminent Professor Jackson. Unlike Morton, who knew Wells well enough not to underestimate him, Jackson was scathing: the idea of painless operations was preposterous. But Wells had contacts at the Massachusetts General Hospital and Warren was persuaded to give the dentist's discovery a half-hearted trial.

On 10 February 1845, at the close of his lecture Warren addressed his audience:

> 'There is a gentleman here who pretends that he has got something here which will destroy pain in surgical operations. He wants to address you. If any of you would like to hear him you can do so'.[10]

It was hardly an overwhelming commendation; but Warren himself and most students stayed to watch. But Wells got flustered. He briefly explained what he was about to do and asked for volunteers. Since he did not expect any he had brought with him a beefy young man who had consulted him earlier and who needed a tooth extracted. Wells later admitted that in the deathly hush he probably did not administer the gas for long enough. As soon as the patient seemed to fall asleep, he started on the extraction. As he gripped the tooth, the patient emitted a howl. (He later testified that he felt or could remember feeling no pain.) The extraction was aborted. The students jeered. There was no need for Warren to comment. Wells picked up his bag and, deathly pale, rushed out of the theatre and hospital.

## *II*

Morton, who was present at Wells' humiliation, may or may not have been among the jeerers; but he drew his own conclusions. Wells was a bungler. He had never learnt to handle nitrous oxide properly. He had chosen the wrong patient. He did not administer the gas for long enough. But, though Morton avoided Wells after the fiasco, he was more convinced than ever that gas anaesthesia was not only possible but that it would bring him, Morton, fame and untold wealth. He was confirmed in this belief by meeting several of Wells' patients from Hartford who spoke in glowing terms of their painless tooth extraction. Without any chemical and with little medical training, Morton was at a disadvantage; but, if Wells had been dogged by ill-luck, fortune seemed to smile on Morton.

He too experimented with nitrous oxide both on himself and on a few patients. When he ran out of the gas he went to see Jackson again. Jackson had heard of the Wells débâcle at the Massachusetts General Hospital and spoke about it with glee. It was exactly as he had foretold. The whole idea of surgical anaesthesia was preposterous. Operations were not party frolics. Why on earth did Morton want to press on? In any case, he had run out of 'laughing gas'. He was a busy man. But if Morton insisted, why did he not try ether? It was just as useless but the purified product (used for coughs, difficulties in breathing, and chest pain) could be obtained at Burnett's pharmacy in Price Street.

There is little doubt – though Morton would always deny it – that the idea of ether was planted in Morton's mind by Jackson. It is indeed likely that for

10. Quoted in J.M. Warren, *Etherisation, with Surgical Remarks* (Boston, 1848), p. 234.

once Jackson was telling the truth when he maintained that Morton, 'the ignorant little trickster', had never heard of ether before and that the chemical properties of the liquid had to be explained to him 'in pitiful detail'. Jackson gave one other crucial piece of advice. He told Morton that, if disasters were to be avoided, the ether inhaled must be 'corrected', that is purified. (Long had prepared his own, the probable reason for his success.) Many had found cheap commercial preparations unreliable and even dangerous. The anaesthetic effect of the impure liquid also tended to wear off. Jackson had himself certified the purity of the preparation sold by Burnett's. But he had no more faith in the surgical use of 'ether frolics' than he had in nitrous oxide; and his later assertions and depositions under oath that Morton was no more than his 'messenger boy' acting under his instructions as an 'unskilled agent' were lies. Equally, Morton's testimony in years to come that he had informed Jackson merely as an act of courtesy of his impending use of ether was a fabrication. It is difficult to decide which of the two was the greater scoundrel.

But Morton's extraordinary luck held. That same evening, 30 September 1846, the back door of his Boston lodgings was opened to a late and distressed caller by John Haydon, the older of Morton's two apprentices . The man at the door was Eben H. Frost, a young estate agent. He was in severe pain. He had never had toothache before, but he had read somewhere that teeth could be extracted by a painless method called mesmerism. He had brought all his savings and was ready to part with them if Dr Morton could remove his tooth painlessly. It was unprofessional to see patients out of hours and Haydon was on the point of turning him away. Then he changed his mind and passed the buck to his junior, Glen Tenney. Tenney was ready to have a go at the offending tooth himself but when Frost insisted that he should be put into a mesmeric trance first Tenney decided to risk his chief's ire and inform him of the caller. To his surprise Morton, though relaxing with a newspaper and a nightcap, jumped from his armchair in delight, picked up a bottle from the shelf and rushed into the surgery. Frost repeated his request for mesmerism but Morton assured him that he had something much better to offer.

One reason for the ban on late visitors was the poor light on winter evenings. Morton ordered Tenney to hold up a candle while he pressed a piece of cloth soaked in ether to Frost's nose. Among Morton's strokes of luck was the failure of the ether to ignite. When Frost suddenly relaxed in mid-sentence and seemed to fall back in the dental chair in a faint, all three dentists were too taken aback to move. But Morton recovered in time, opened the patient's mouth and swiftly removed the inflamed canine. Frost woke up a minute or two later. What is happening? When will the doctor start? he asked. Morton showed him the extracted tooth. Frost was delighted. Delight was not enough. Having relieved Frost of his savings, Morton penned a declaration for Frost to sign.

> I hereby certify that this evening at nine o' clock, I visited Dr Morton at 42 Prince Street with a terrible ache in my right lower canine tooth. I certify that Dr Morton made me inhale his specially prepared fluid applied to my nose with his handkerchief. Within a minute I lapsed into a deep sleep. I did not come round until my tooth had been extracted and was lying on the floor. I had felt no pain whatever. I stayed for another twenty minutes. There were no untoward after-effects. My toothache had disappeared.[11]

Frost's testimony was countersigned by Haydon and Tenney as witnesses. That same evening Morton visited the offices of the *Boston Daily Journal*, perhaps accompanied by Frost. Next day an unsigned news item appeared in the paper informing the readership of an 'epoch-making event' that had taken place the night before: the painless extraction of a dangerously inflamed tooth under the influence of 'a pain-killing liquid specially prepared by the well-known dental surgeon, Dr Morton'.

. . .

There was now no holding Morton back. He went to see Bigelow (see p. 210) and begged him to persuade Warren to give inhalation another trial. Bigelow was at first noncommittal: he would let Morton know. By now Morton was coping with an onrush of patients who had read the newspaper announcement. Etherisation was successful in several cases, but not always. On one occasion Morton and his assistants had some difficulty in resuscitating the patient. Morton's landlord and colleague, Marshal Gould, suggested that some kind of a rebreathing apparatus instead of a handkerchief could avoid dilution of the ether vapour with expired air. It was probably Gould who sketched out the glass balloon with a valve opening which allowed the escape of expired air but concentrated the ether vapour. If not, it was Morton's one and only original contribution to anaesthesia. He at once commissioned the implement; but next day he received an unexpected summons to present himself with his 'gear' in Professor Warren's operating theatre on 16 October. The balloon now became a rush job collected on the way to the hospital. Morton also picked up Eben H. Frost as a back-up. This made him late.

At the hospital the scene was set. The auditorium was packed with students and staff. Gould was present, as well as Bigelow, C.F. Heywood and S.D. Townsend. Warren had concluded his lecture: he was now ready to perform the operation. The patient, Gilbert Abbot, was a consumptive young man with a vascular tumour at the angle of his jaw. His poor physique would work in Morton's favour unlike Wells' robust young patient, who had been relatively resistant to nitrous oxide. But for fifteen minutes there was a pause. Warren

11. Quoted in Robinson, *Victory over Pain*, p. 265.

was not used to being kept waiting. He was assured that Morton had been given precise instructions about the time when he should appear. With ill-concealed irritation he prepared to start the operation without the pain-killing gas. He turned to his audience.

'As Dr Morton has not arrived, I presume that he is engaged on more urgent business,' he announced. At that moment Morton burst in. The guffaws stopped. Warren nodded his acquiescence to a further *short* delay. Morton retired to a room behind the seats of the amphitheatre. He needed a few seconds not to prepare anything or even for prayer but to camouflage the odour of ether with an essence of herbs and add a pink cooking dye he had borrowed from Mrs Gould's kitchen. He then entered again and spoke to Abbot, who was already strapped to the dark purple plush operating chair. Morton was a big hearty man with upswept moustaches and clad in a fancy waistcoat. Unlike Wells a year earlier, he appeared wholly unflustered and at once gained Abbot's confidence. He also introduced Frost, 'an old patient of mine', who assured his successor that all would be child's play: he would feel nothing. Morton then asked Abbot: 'Are you afraid, laddie?' Despite his pallor and poor physique, 'laddie ' (about whose life, past and future, nothing else is known) was obviously no weakling. 'No,' he replied, 'I feel confident and will do exactly as you tell me.'

Morton put the tube leading from the globe to the patient's lip and told him to breathe in and out quietly through his mouth. At first Abbot's face seemed unnaturally flushed and his arms and legs moved spasmodically; but Morton held the globe steadily in one hand and continued to press the tube into Abbot's mouth. There was total silence for what seemed to many an interminable length of time but was in fact no more than three or four minutes. Then Morton turned to Warren and uttered the words that would echo down the centuries in uncounted operating theatres the world over:

'Sir, your patient is ready.'

Warren seized the engorged veins of the tumour with his left hand and with his right made the first incision. Years of hearing the anguished outcry in response to the first impact of the knife had made him steel himself for the inevitable scream. There was always a barely perceptible pause before the next step. Patients occasionally lapsed into a state of shock. Some had been known literally to die of fright. This time no sound came. With the controlled swift movements for which he was famous, Warren completed the excision. Because of the proximity of important nerves, the dissection took fourteen and a half minutes. Then the bleeding points were tied, the skin sutured, the face washed of blood. With the tube removed from his mouth Abbot gradually regained consciousness. Morton was the first to break the silence.

'Did you feel anything, Mr Abbot?'

'It was good,' Abbot mumbled with a sigh.

Warren turned to the students, and remembering perhaps the scathing words he had uttered after Wells' misfired exhibition a year earlier, he remarked:

'Gentlemen, this is no humbug.'

It was the high point of Morton's life and career.

## *III*

The news of Morton's success spread extraordinarily rapidly, and the historical focus must now shift to Britain and Europe; but the end of the personal saga of Morton, Jackson and the Wells family cannot be left unsung. (There is a Wagnerian quality about their tangled lives but at least there is no need to recount previous events in the second act of each part.)

Though his humiliation and failure seemed complete, Wells did not give up. Returning to Hartford, he continued to use nitrous oxide experimentally on himself and on his patients. Later forty of them, including his two daughters, gave depositions that in the course of 1846 he had extracted their teeth painlessly. Then news of Morton's successful session with ether at the Massachusetts General Hospital was relayed to him. He expected to be hailed as the pioneer if not a co-discoverer. He trusted Morton. Morton made a half-hearted attempt to forestall any claim by Wells by inviting him to become a junior partner in the 'commercial exploitation of my discovery'. Wells declined. After that, so far as Morton was concerned, Wells had never existed. There was no mention of Wells or of nitrous oxide in Bigelow's historic letter in the *Boston Medical and Surgical Journal*. It was this letter which started general anaesthesia on its triumphant tour.

When Wells realised that he was to be completely ignored, he sank into despair punctuated by attacks of rage. Early in 1847 he was persuaded by friends to try a change of scene. He also convinced himself that recognition might come to him in Europe so he set sail for France. Unexpectedly, his unrealistic hopes seemed to be realised.

In Paris he met a fashionable American dentist, C. Starr Brewster, who was quickly converted to the merits of nitrous oxide. Brewster also proved an honest friend, possessing the creative outrage of many of his countrymen at the time and since. When news broke that the Institut had precipitately granted Jackson and Morton the princely sum of 2,500 francs in recognition of their discovery of the 'principle of anaesthesia', Brewster orchestrated an outcry of protest. The details are still murky; but the award to Jackson and Morton had divided the award committee and the announcement had been pushed through without the necessary majority. In a letter to *Galignani's Messenger*, Wells stated that priority should be his. It was a well-argued case, slightly marred by the claim, perhaps inserted by

Brewster, that Wells had used ether as well as nitrous oxide. This is possible – Wells had never lied before – but there is no evidence of it. Nevertheless, Wells was lionised at parties given in his honour in fashionable salons. It was his apotheosis.

Success tended to go to Wells' head just as setbacks cast him into the deepest despair. He was enchanted by his reception in France. The pleasure was mutual. A little naïve but modest and easily pleased – 'a breath of fresh air from across the Ocean' – he charmed his hosts. He was complimented not only on his achievements with anaesthesia but also on his flair with antique *objets d'art*, *bijoux* and paintings. The latter praise was especially fulsome from those who had antique *objets d'art*, *bijoux* and paintings to sell. Wells invested the remains of his money in treasures which he convinced himself could be sold at a profit back home. When his money ran out, he took the first boat for the United States.

Back in Hartford it did not take him long to realise that his investments were fakes and imitations. Worse, Morton and Jackson, embarking on their war of mutual vituperation and lawsuits, were united only in wishing to exclude Wells from any credit for 'their' discovery. Worst of all, experimenting on himself with nitrous oxide, ether and, most recently, chloroform had become an obsession. Wells travelled to New York to lobby his remaining contacts at the Massachusetts General Hospital. He persuaded Bigelow, soon to be Warren's successor, to give nitrous oxide another trial. Bigelow had used it again without Wells' assistance for a difficult mastectomy on an emaciated young woman. The patient almost died under the influence of the gas and the operation had to be finished in a hurry. Bigelow concluded that nitrous oxide was dangerous in any major procedure. Wells was by now a quivering wreck, partly at least as a result of his daily experiments on himself – if they were indeed experiments rather than a form of addiction. The last act of his tragedy is still surrounded by mystery.

In the evening of 22 January 1848 he was caught 'red-handed' and in a state of mental confusion, having squirted a strong acid at two street-walkers in Broadway. This was wholly out of character: he was probably framed by a notorious pimp who had visited him the day before. On the other hand, he was clearly not in control of his actions and could remember nothing after being taken into custody. He was transported to the Tombs Prison. The place had recently been visited by Charles Dickens who described it as a most horrible abode, 'not fit to house beasts let alone humans'. Appearing before a judge next morning, Wells pleaded amnesia. He was obviously in a bad state and, not unkindly, the judge adjourned the hearing. At that stage – the details have never been clarified – the prisoner was allowed to return to his rooms under guard to collect a few 'personal belongings'. What he collected was a bottle of chloroform and a razor. Left alone in his cell, he penned a few lines to his wife.

> I am losing my mind. I cannot go on living without control over my actions. God will forgive me. Please you too forgive me. Your loving H.[12]

He then anaesthetised himself with chloroform for a last time. Timing the cut superbly, just before slipping into unconsciousness, he slashed his femoral artery and vein. He must have died within minutes but was not found in his cell till next morning.

. . .

Twelve days earlier a letter was posted in Paris from Brewster to Wells. It said:

> My dear Wells:
>
> I have just returned from a meeting of the Paris Medical Society, where they have voted that to Horace Wells of Hartford, Connecticut, United States of America, is due all honour of having successfully discovered and successfully applied the use of vapours of gases whereby surgical operations could be performed without pain. They have done even more, for they have elected you honorary member of their Society. This was the third meeting that the Society had deliberated on the subject. On two previous occasions Mr Warren, the agent of Dr Morton, was present and endeavoured to show that to his client was due the honour but he, having completely failed, did not attend the last meeting. The use of ether took the place of nitrous oxide gas, but chloroform has now supplanted both, yet the first person who discovered and performed surgical operations without pain, was Horace Wells, and to the last day of time must suffering humanity bless his name.
>
> Your diploma and the vote of the Paris Medical Society shall be forwarded to you.
>
> Believe me ever truly yours, Brewster.[13]

By the time the letter arrived, Wells was dead.

. . .

It is now necessary to return to Boston. Some time after the first operation under ether anaesthesia Warren wrote about the discovery in language that seems somewhat remote from his unbending personality:

> As philanthropists we surgeons may rejoice that we have a means, however slight, to confer a precious gift on poor suffering humanity . . .

12. Quoted in Robinson, *Victory over Pain*, p. 234.
13. Quoted in Robinson, *Victory over Pain*, p. 324.

Unrestrained and free as God's own sunshine, it will go forth to cheer and gladden the earth.[14]

But whatever Warren may have envisaged, for ether anaesthesia to go forth 'unrestrained and free as God's sunshine' was not what Morton had in mind. From the first he was determined to make a fortune from his 'invention'. Within a few days of the first operation he filed for a patent and started to draw up elaborate schemes for marketing 'his liquid' throughout the United States and selling it to the government. The day after the operation on Abbot he successfully anaesthetised another patient for the removal of a tumour from her shoulder; but, to the consternation of Bigelow and his colleagues, he refused to divulge what his secret 'anodyne mixture' was. The hospital or the City of Boston or the State of Massachusetts must first pay for using the 'patent' and continue to pay 'royalties' on every future occasion. He later tried to sell the invention to the United States Government for use in the war with Mexico. The dispute eventually led to an ultimatum which threatened Morton with a permanent ban from the hospital.

Morton at last realised the power of the professional establishment acting in unison and reluctantly admitted that the colour and smell of his fluid were unnecessary artefacts and that he was using rectified ether. The patent application had to be modified to cover the method of administration only. It was never granted, the judge comparing Morton's 'invention' to the discovery of 'celestial electricity', a fact of nature, by Benjamin Franklin. The fight between Morton and the hospital was punctuated by further cases. A woman who needed a mastectomy provided the first public 'failure' of the method: she refused to 'go under' and went 'wild'. The operation had to be abandoned. But next day an amputation by Bigelow on an elderly man went off without a hitch. Bigelow published the discovery, giving credit to Morton, in the *Boston Medical and Surgical Journal* on 18 November 1846.

About this time several members of Warren's department decided that the ether method of rendering patients insensible to pain merited a name of its own. 'Narcosis', 'insensibility', 'faint' and other terms had been applied to similar states before. Something newer and more portentous was wanted. There was no question who in Boston was the man to turn to. Oliver Wendell Holmes, doctor, wit, writer, classical scholar and commentator on the medical scene had recently been appointed to the 'combined chair of anatomy and physiology' at Harvard: 'more a settee than a chair', the new occupant quipped. The original letter addressed to him has been lost; but the response survives. Holmes learnedly discoursed on several possibilities, eventually coming down in favour of 'anaesthesia' and 'anaesthetic' (but not originally anaesthetist)

14. Warren, *Etherisation*, p. 34.

from the Greek 'an' (privative) and 'aisthesis' (sensation) and suffix. Holmes' letter was addressed to Morton but was widely circulated. For good or ill, the name caught on.[15]

. . .

At the end of November Jackson stepped in. His first move was to stop Morton's fame spreading abroad. Giving a detailed story of his 'invention' and a wholly mendacious account of its reception in the United States, he wrote to his French friend and colleague, Elie de Beaumont, asking him to inform the Académie of the happy event. A few days later he arranged a meeting with the American Academy of the Arts and Sciences in New York at which he claimed undivided credit for the discovery of ether anaesthesia. He had, he stated, experimented with the method on animals and on himself for *years*; and only 'in response to Morton's repeated importunities' and 'to ensure that no harm came from the unskilled use of the method' did he give the dentist instructions how to use it.

Both Morton and Jackson set about mobilising newspapers, lawyers, judges and politicians and indeed any public figure who could be persuaded or, more often, bribed to take an interest. Large sums of money were collected by both parties. Contests of priority are rarely refined arguments carried out in hushed voices; but the ensuing legal fight was one of the dirtiest on record. Extraordinary accusations of theft, embezzlement, fraud, bribery and even murder were bandied about. Both Morton and Jackson had acquired wealthy backers who were prepared to invest millions but wanted a return on their investment. The fight went on for years, marked by Congressional hearings, lawsuits, judgments overturned, appeals and supplementary actions for slander, libel and infringement of patents. The one point on which Morton and Jackson were united – the exclusion of Wells' widow from any possible benefit – eventually proved their downfall.

Lizzie Wells had been a long-suffering wife and mother and in financial resources or resourcefulness no match for Morton and Jackson. But probably her very helplessness prompted the junior Senator for Connecticut, the Honourable Truman Smith, to embark on a crusade. Smith might have stepped out of a thirties Hollywood movie starring James Stewart. He was fearless, incorruptible and single-minded. Never accepting a fee from the Wells family, he demonstrated with a pile of documents and an unanswerable array of personal depositions that the idea of an inhalable anaesthetic for which both Morton and Jackson had been awarded $100,000 by Congress was

15. Holmes was thirty-seven in 1846. He published his delightful *Autocrat of the Breakfast Table* in the *Atlantic Monthly* ten years later; and several similar works, mostly literary but also of medical interest, followed. He died in 1894, aged eighty-five.

in fact the brainchild of Horace Wells. The reward should be made to his destitute widow. Congress withdrew the grant from Morton and Jackson but an appeal with powerful backing was at once lodged. Smith showed that the appellant had himself invested a large sum in Morton's campaign, as did several congressmen who supported him. Further appeals were lost in the turmoil of the Civil War. Ether was used on the battlefields by both sides. Its value was no longer in doubt. To the wounded, the identity of the inventor mattered not at all.

...

Yet Morton and Jackson kept at it. Both showed increasing signs of mental instability. In July 1868 Morton returned to New York from one of his lobbying trips to Washington. His final breakdown seems to have been precipitated by a *New York Times* article which favoured Jackson's claim. Doctors Sayre and Yale were called to his New York lodging on 15 July. They pronounced his condition critical, ordered a roster of nurses to remain by his bedside and prescribed leeches and ice-packs. Hardly were they out of the house than Morton gave his guardians the slip, jumped into his buggy and headed for 'out of town'. He drove furiously up Broadway and through Central Park. At the north end of the park he leapt from the buggy, ran to the lake and plunged his head into the water. Bystanders persuaded him to return to his buggy. He drove on for a short distance, leapt out again and began to run. Soon he collapsed. He was taken to St Luke's Hospital conscious but incoherent. He died a few hours later.

...

Jackson's end was not far off. He too had spent years in lobbying and writing articles to promote his claim and to discredit Morton. He may also have continued with laboratory work. It is impossible to say how much truth there was in the researches he claimed to have pursued. In 1861 he published a book, *Manual of Etherisation*, at his own expense. In this he claimed to have conducted experiments at the McLean Asylum, the branch of the Massachusetts General Hospital for the care of the incurably insane. He described in detail several cases of raving inmates calmed by the administration of ether. The descriptions are extremely vivid: Jackson had retained the gift of the pen. The experiments could have taken place; and ether may have had a calming effect. But there is no record of these proceedings in the hospital archives.

Twelve years later Jackson himself was admitted to the Asylum, this time as a patient. There is no record that any of his eminent friends ever visited him. He died 'from terminal decline . . . totally insane' on 28 August 1880.

...

The argument about the respective parts played by Morton, Jackson and Wells in the discovery of general anaesthesia was to rumble on. The world needs heroes; and Bigelow and friends decided to honour Morton. He had no criminal record and he was safely dead. An imposing memorial stone was erected over his grave in Mount Auburn cemetery in Boston. Bigelow composed the inscription:

> Before Whom, in All Time,
> Surgery was agony;
> By whom, Pain in Surgery was averted,
> Since Whom, Science has Control of Pain.

Bigelow was an excellent surgeon but not much of a poet. But then Morton was not much of a hero either.

# Chapter 23

# IN GOWER STREET

Dr Francis Boott of Gower Street in London had many friends on both sides of the Atlantic. His father was English, his mother Scottish and he himself had been born in Boston. He was a graduate of Harvard and Edinburgh. On 18 December 1846 he received a letter from his father's old friend, Dr Jacob Bigelow, professor of materia medica at Harvard.

> My dear Boott: I send you an account of a new anodyne lately introduced here, which promises to be one of the important discoveries of the present age. It has rendered many patients insensible to pain during surgical operations . . . Limbs and breasts have been amputated, arteries tied, tumours extirpated and many hundreds of teeth extracted without any consciousness of the least pain on the part of patients . . . Last week I took Mary [his daughter] to Dr Morton's rooms to have a tooth extracted. She inhaled the ether for about one minute and fell asleep instantly in the chair. The tooth was then extracted without the slightest movement of a muscle or fibre. In another minute she awoke, smiled and said that she had felt no pain nor had the slightest knowledge of the extraction . . .[1]

Boott realised that the news should be brought to the attention of Robert Liston, surgeon to University College Hospital, less than a hundred yards up the road. But he hesitated. Sensational claims had been made before. Americans, though not, as a rule, Bostonians, were prone to premature enthusiasms. He knew that Bigelow was an experienced physician, not easily duped. But he was getting on in years. Boott had no wish to be at the receiving end of Liston's barbed comments. Or to witness the great man flying into one of his terrible silences. Rather than rush off to the new hospital, he went to see a dentist friend, James Robinson, who practised in Bedford Square. Could Robinson visit him with a patient who needed a tooth extracted? No problem.

1. Quoted by V. Robinson in *Victory over Pain*, p. 138.

The dentist and a patient would come to Boott's surgery the next morning. Boott was something of a kitchen chemist, as were many doctors at the time, and he had a bottle of pure sulphuric ether in his private laboratory. His wife had a perfume vaporiser not unlike the apparatus described in the article sent to him by Bigelow. It was easily adapted. All he needed to buy was a fresh roll of gauze. Next morning, 19 December, Robinson arrived with a young woman, a Miss Lonsdale. After a sleepless night she was in a bad way. She was willing to try almost anything to get rid of the pain. She sat in Boott's improvised dental chair. Boott pressed the gauze to her face. After a few whiffs of the ether squirted on to the gauze she relaxed. Robinson extracted the tooth. Miss Lonsdale came round within a few seconds. She had felt nothing.[2]

Boott's task as a messenger was still daunting. Liston might have taken exception to receiving important surgical intelligence from a mere general practitioner, if quite a fashionable one. Fortunately Liston himself had received that very morning a letter from Henry Jacob Bigelow, the son of Boott's correspondent. The younger Bigelow was as enthusiastic as his father. Of course Liston declared himself deeply sceptical. Yes, he might try the new gimmick some time. He would let Boott know when he did. In fact, he at once drove to Peter Squire's famous pharmacy in Oxford Street and the two men went into a huddle. Squire promised to rig up a vaporiser similar to that described in Bigelow's letter and fill a suitable container with pure concentrated sulphuric ether. He would then send the package to University College Hospital using his nephew as the messenger. The strictest secrecy would be observed.

Despite the secrecy the news that something unusual was afoot leaked out. An unusual number of medical students and local practitioners assembled on the benches in the operating theatre two days later. Among the former was a Quaker youth, Joseph Lister, still in his preliminary year at University College.[3] He was one of those who would later relate the events of the day. It was a cold day outside; but a roaring fire was burning in the fireplace in the operating theatre. After a short and unusual delay – Liston prided himself on his punctuality and tended to eject late-comers – Liston's assistant, William Cadge, entered. He was accompanied by William Squire, carrying a mysterious-looking parcel. With barely controlled excitement Cadge explained that

2. A blue plaque now marks the house in Gower Street, surely the most uninformative in London. While it mentions that the first general anaesthetic in England was administered in the building, it gives the name of neither the anaesthetist nor the dentist, nor even that of the patient.
3. It is surprising that he should have been present; and on the painting in the possession of University College Hospital he looks a little like an afterthought and a later addition; but he was not the kind of person to make up such things. (See Dormandy, *Moments of Truth.*)

Professor Liston was about to try a new American invention that was purported to dull pain almost to the point of abolishing it entirely. Cadge himself had made a few preliminary trials the evening before. He had asked his cousin to stick some pins into his arm after breathing in a few whiffs of the 'anodyne vapour'. The experiment was successful. He showed the marks on his arm. Some looked quite violent, yet he had felt nothing. Would some of the audience like to try it for themselves? There was no rush. Cadge then summoned a male attendant, a mountainous ex-sergeant called Sheldrake. Reluctantly, Sheldrake agreed to breathe in the vapour. Hardly had he taken a sniff or two when, with a bellow, he jumped up, pushed Cadge out of his way and charged with flailing arms at the audience. Panic. Fortunately, Sheldrake had hardly taken a few steps when he collapsed in a stupor.

In the midst of the confusion Liston entered with a surgical colleague, John Ransome, and his chief dresser, John Palmer. Nurses and porters followed. Cadge explained what had happened. Sheldrake was still unconscious. Cadge suggested that the experiment should be postponed: one could not have patients jumping up and charging around in mid-operation. Liston inspected Sheldrake, snoring happily on the floor, and decided otherwise. The method would be tried at once. Let the first patient be brought in. He was Frederick Churchill, a footman in the service of Lord Aberdeen. He had sustained an injury to his left thigh some weeks earlier and the wound had gone septic. Liston had drained the abscess but this failed to clear up the infection. The patient had developed a swinging temperature. Pyaemia was imminent. A mid-thigh amputation offered the only hope of survival.

Churchill was carried in on a stretcher and laid on the operating table. He seemed moribund but was whimpering that he would rather die than lose his leg. Liston ignored his words, as he had done hundreds of similar last-minute pleas in the past. He motioned Cadge to put the tube attached to the vaporiser to the patient's mouth and put a clamp on the patient's nose. Churchill had an immediate fit of coughing and Cadge once again suggested that the experiment be abandoned. Liston motioned him to continue. Suddenly Churchill went quiet. Was he dead? No, he was breathing. He seemed to have passed out; to be in some kind of a coma. It was at this point that Liston himself hesitated for a moment. But then he took off his frock-coat, rolled up his shirt-sleeves and picked up the knife from the instrument tray. 'Keep your hand on the pulse,' he instructed Palmer. Then he turned to the audience. 'Time it, gentlemen.' Some of the students had their watches in their palms already. At lightning speed Liston performed two U-shaped incisions and fashioned two skin flaps. Palmer handed him the saw. After five or six to-and-fro movements the detached lower limb was dropped into the basket of sawdust under the operating table. 'Two minutes twenty-eight seconds,' a student announced. Liston looked at Churchill. Churchill was still fast asleep.

In deathly silence but no longer in a rush Liston tied the bleeding points, stitched up the flaps. 'Four and a half minutes,' the timekeeper whispered.

For once there was no applause. Every eye was on the sleeping patient. Cadge had removed the tube. Would Churchill start breathing? He did. A long two minutes passed in total silence. Then Churchill opened his mouth. Once again he started to whimper. 'When will you start? No, no, I will not have my leg taken away. I will rather die.' Liston motioned one of the porters to lift the amputated limb from the basket of sawdust and show it to Churchill. Churchill closed his eyes in fright, mumbling incoherently. There was another pause. All eyes now turned on Liston. 'It was incredible,' an eyewitness later reported. 'The big man stood motionless but tears were rolling down his cheeks.' Nobody stirred. Then Liston uttered the famous words: 'Well, gentlemen, this Yankee dodge sure beats mesmerism hollow.'

# Chapter 24

## AND BEYOND

Liston himself was dead within a year;[1] but over the next few months similar scenes were enacted in operating theatres in Britain, Europe and beyond. In the new age of steam the news spread extraordinarily quickly. The *Neue Deutsche Zeitung* printed the story of the operation in London within a week. It created intense excitement. Sulphuric ether was available everywhere. The method was childishly simple. On 6 March the *Wiener Journal* carried a sensational news item.

> The entire personnel of the Master of the Imperial Horse and the Imperial School of Veterinary Science, as well as numerous noble, aristocratic and important personages, forgathered in the Imperial Stables in the presence of His Excellency, the Lord High Master of the Horse, Count Julius von Hellingshausen, to watch several stallions being castrated under the influence of ether vapour. The stallions stood firmly and motionless throughout the administration of the vapour and during the operations which were performed by Professor Hofrat Adalbert Schwarz. They showed none of the usual resistance to the procedure and needed virtually no restraint. It was generally agreed, not without surprise, that the animals were experiencing no pain. They were uncommonly lively after the operation and enjoyed the fodder set before them, showing none of the listlessness usually observed after surgery for at least six to twelve hours. They were ready to be put into harness the very same day.[2]

It was, the *Wiener Journal* proudly claimed, the first time the new invention had been tried on the Continent of Europe. The claim had to be withdrawn. It transpired that by then twenty-three dental extractions had been performed

1. He died suddenly of a ruptured aortic aneurysm. By far the commonest cause of this at the time was tertiary syphilis. He was fifty-three.
2. *Wiener Journal*, 1847 (3), 67.

under the influence of sulphuric ether not on horses but on humans at the Surgical Clinic of the University of Erlangen.[3] A little belatedly – but what else would one expect from Bavarians? – the Royal Government in Munich issued a statement:

> The experiments had been entirely successful and achieved almost total painlessness. No accidents, etc. had occurred. One operating session had been graced by the presence of His Highness the Prince Eduard von Sachsen-Altenburg. His Highness expressed his pleasure and congratulated Professor Schonhausen and his staff on their skill and their rapid adoption of an important medical advance.[4]

In Vienna what was good enough for the stallions of the imperial *Reitschule* was good enough for loyal citizens; but Franz Schuh, professor of surgery, proceeded cautiously. He was, with Skoda and Rokitansky, one of the medical avant-garde battling against a torpid establishment; but he performed painstaking experiments on animals before trying ether on humans. Even after trying it, he felt that

> in all likelihood the results obtained with sulphuric ether vapour have been accorded excessive and premature confidence which more prolonged and dispassionate examination may question.[5]

Despite his forward-looking reputation, he was concerned, in traditional Viennese fashion, that ether might open the door to the 'nefarious activities of unqualified practitioners . . . Great vigilance is needed from the authorities to prevent such a development.' In a city where, for all its reputation for *Gemütlichkeit*, every third adult was a part-time police informer, vigilance meant vigilance.

. . .

But times were changing. In Pest – not yet united with Buda – second capital of the Habsburg Empire, the administrator of the university arranged for the new discovery to be demonstrated in a series of open lectures in the Aula. 'It will give the public insight into medical advances and *progress in a wider field* [my italics] which they so greatly desire . . . A modest entrance fee would be charged to be devoted to charitable and patriotic purposes.'[6] The Aula was

3. The term 'Clinic' or 'Clinique' or 'Klinik' (from the Greek κλίνη bed) came into general use in the late eighteenth century. In contrast to the modern usage of the term in English-speaking countries, where it usually refers to outpatient clinics, in Germany and countries east it meant and still means a university medical or surgical department.
4. *Bayrischer Rundschau*, Munich, 1847 (5), 168.
5. Quoted by Robinson in *Victory over Pain*, p. 154.
6. *Pester Lloyd*, 1847 (13), 214.

crammed for the first demonstration; but what was this business of 'progress in a wider field'? It had a sinister ring. The audience included Baron Miklos Wesselényi, one of the country's reformist magnates, who wrote to his wife the same day:

> It is unbelievable but the poor lass [who had to have her foot amputated following a crush injury] remained fast asleep throughout the cruel operation and assured us afterwards that the cutting and the sawing through the bones were wholly painless! Mankind is progressing at an unprecedented pace. We must not be left behind. [7]

But six professors (out of a total of eight, all appointees of Vienna) drew up a petition protesting against the 'scandalous practice' of turning a procedure that 'belongs within the sacred halls of the science of healing' into a public spectacle. The Palatine, the Archduke Stephen, had no wish to quarrel with the Faculty and the demonstrations were discontinued; but soon a notice appeared in the city's German-language newspaper advertising private demonstrations by 'an eminent medical man'.[8] It is not known who the mystery performer was or how the venture fared; but the date, 21 October 1847, was significant. A breeze, heralding the storm of 1848, was blowing across Europe, nowhere more noticeably than over the lands of the Habsburgs. Cautious, complacent and conservative Herr Bieder Meyer, exemplary citizen who had sedulously kept out of politics and other nefarious activities for the past forty years, was in retreat.[9] Anything new in science, literature, politics or medicine was suddenly received with enthusiasm. And it was no longer just students and starving poets. Mature citizens were ready for novelty, perhaps even for a brief and not too violent revolution. Even in Prague, generally regarded as the most loyal of Habsburg cities, the idea of painless surgery became, for a brief period, an emblem of political change. It was tried by Franz von Pitha, professor of surgery and reputedly a friend of the constitutional movement.

> When we consider how many unfortunates must have for years suffered the most acute anguish in fearful anticipation of an operation even when a cure is promised . . . we realise the priceless value of this discovery for suffering humanity . . . But it is a great a gift also to the operating surgeon who will no longer be the dread figure of the past . . . and will dare the most painful

7. M. Wesselényi Jun. (ed.), *The Letters and Journals of Baron Miklos Wesselényi* (Budapest, 1893), p. 369.
8. *Pester Lloyd*, 1847 (13), 386.
9. Herr Bieder Meyer who gave his name to the homely and comfortable Biedermeier style, was an invented literary character, the figment of the imagination of two German poets, Ludwig Eichrodt and Adolf Kusmaul.

> operations with composure and tranquillity without seriously disturbing emotions.[10]

Did his declaration have political undertones? He was certainly cheered by his students.

...

But it was in the numerous small universities of Germany that developments were most varied and promising. The Royal Saxon Ministry of Education reported that at the Royal Hospital in Dresden in a series of sixty tooth extractions ether inhalation failed to bring about total insensibility in only one subject, 'an ignorant Polish woman who behaved very stupidly, failed to inhale and brought her suffering on herself ... Most subjects, awakening from narcosis, did not realise that one or several teeth had been drawn until they became aware of the blood trickling from their mouths.'[11]

In Berlin a still unknown first assistant in the Department of Surgery, Dr Rudolf Virchow, wrote a characteristically sober report on the experience of his chief, Professor Johann Christian Jungken.

> In the case of refractory or anxious patients great caution is necessary in order that they may not be thrown into a state of intoxication as a result of improper or too prolonged inhalation ... This can have grave consequences even during the operation but more specifically after. We must gain further experience about the effect of multiple inhalations ... It is well to note that, despite excellent results so far, the method should not be used by inexperienced persons.[12]

Virchow was also among the first to comment:

> It is of the utmost importance that during the administration of the gas an atmosphere of silence and calm should prevail.[13]

The young assistant was right, as he would disconcertingly remain on most topics throughout his long career; but an atmosphere of silence and calm was easier to achieve in the Berlin of the Hohenzollerns than in many less well disciplined surgical centres. In most operating theatres, even in reputedly phlegmatic England, the atmosphere during public surgical sessions more closely approximated the calm of a cock-fight.

In Erlangen Johann Ferdinand Heyfelder published the first monograph on ether inhalation based on eighty cases, mostly minor operations and tooth

10. Quoted by N. Menczer, *A Pragai orvosi iskola* (Budapest, 1954), p. 17.
11. I. Krausen, *Medizin in Dresden* (Dresden, 1893), p. 245.
12. Quoted in A. Binder, *Rudolf Virchov* (Leipzig, 1939), p. 50.
13. Quoted ibid., p. 56.

extractions without a single failure or mishap. He also advocated muriatic (hydrochloric) rather than sulphuric ether as both less irritating and faster in action. The claim was probably true but the exorbitant cost of muriatic ether always ruled it out for routine use.

Konrad Johann Langenbeck of Göttingen was the German Liston: he was said to amputate an arm at the shoulder in less time than it took other men to inhale a pinch of snuff. Painlessness would soon render such virtuosity obsolete, even reprehensible: nevertheless, he adopted the new discovery with enthusiasm. Though he did not disguise his surprise that such a refinement should originate in America, a land inhabited mostly by savages.

> This splendid discovery that has so strangely come to us from the New World will prove a blessed gift to many . . . It can be predicted with certainty that, unlike many other acclaimed innovations which prove two-day wonders, this will prove a lasting blessing.[14]

. . .

Inevitably there were dissenting voices. Even Langenbeck warned that 'ether anaesthesia might prove dangerous unless all relevant circumstances were considered'. He noted in particular that ether might cause inflammation of the respiratory tract, 'especially in consumptive patients who were coughing up blood'. What concerned him was the aggravation of the patient's illness. The possibility that the disease might also be transmitted by using the same breathing apparatus in several patients in succession did not occur to him. In most countries north of the Alps tuberculosis was still regarded as the outcome of an inborn familial predisposition, the terrible 'diathesis': another forty years would pass before Robert Koch established the infective nature of the illness.[15]

Geheimrat (Privy Councillor) Johann Friedrich Dieffenbach, head of the Frankfurt Surgical Clinic, was troubled by other risks. 'Who,' he asked rhetorically (a grammatical device more favoured in France than in Germany, but Frankfurt was not far from the Rhine), 'has the power to determine the quantity of ether vapour which will cause not anaesthesia but instant asphyxiation through paralysing the lungs and brain?' But his concerns were wider.

> Hitherto we have regarded the injuries suffered by intoxicated persons who felt little or no pain as far more dangerous than injuries sustained by the sober. Should we not look upon doctors who deliberately operate on unconscious patients as potentially dangerous?[16]

14. Quoted ibid., p. 68.
15. Discussed in more detail in Dormandy, *The White Death.*
16. Quoted by E. Lesky in *Die Wiener Medizinische Schule im 19. Jahrhundert* (Graz, 1965), p. 256.

He was right. The manual skill and anatomical knowledge of preanaesthetic surgeons would soon be a thing of the past; and some other restraints might go as well. He nevertheless decided, 'though not without grave and prolonged hesitation', to try the new technique. Eventually he became a fervent advocate.

Motivated by similar concerns to those which exercised Dieffenbach, the Kingdom of Hanover had already promulgated decrees prohibiting 'barber surgeons and all dentists, accoucheurs and midwives without medical degrees' from administering ether. It was a necessary precaution perhaps but an odd one in the history of an invention that was almost entirely the brainchild of dentists. But both in Germany and in Austria the better class of dentists had medical qualifications; and Vienna's famous professor of dentistry, Moritz Heider, was both delighted and mortified by the implications of painlessness.

> Not only can operations under narcosis entail many difficulties but it might tempt imprudent and unconscientious dentists to operate with still less care than they would otherwise ... Teeth might be broken or unnecessarily removed, the soft tissues might be injured . . . and indeed, patients might be assaulted![17]

The last generated genuine unease, especially as stories began to circulate about the erotic nature of ether dreams. The prince-archbishop of Passau, Paschal von Fallersweisen, felt it necessary to draw attention to the danger of 'sinful delusions' and 'immoral yearnings' while under the influence of 'stupefying gases'. Such aberrations were to be avoided at all cost. (He did not explain how this might be achieved.) In fact, Langenbeck rightly denied that ether dreams were particularly sinful.

> On the contrary, they depend on the personality of the dreamer. Ether will set the pious praying, the bully to draw his dagger and the loafer to carouse in the tavern. The dreamer may hear music, look at a pleasing landscape, sit at a well-laid table or cuddle a pretty maiden.[18]

Yet the prince-archbishop expressed a fear – or at least an unease – that many felt. The young Hungarian poet, János Arany wrote: 'Pain is the most universal of human experiences. To try to create a painless world this side of Heaven is not only doomed. It is wrong.' But, surprisingly perhaps, the first three years of ether anaesthesia in Central Europe passed without serious ill effects: at least, none were reported in the medical press.

...

17. Quoted by E. Lesky in *Die Wiener Medizinische Schule im 19. Jahrhundert* (Graz, 1965), p. 289.
18. Quoted ibid., p. 239.

But there was much heated argument. Quarrels over priority were not a New World monopoly. In Europe they centred on technical refinements rather than on the discovery and choice of the gases used. Albert Heinrich Kreuser, assistant physician to the Surgical Clinic in Tübingen, saw the limitations of administering inhalable gases by mouth, especially for operations on the face and in the mouth itself, and concluded that 'the invention of a safe method of inhaling through the organ designed for that purpose by Nature would be most welcome'. Surprise, surprise, such a device was just up Kreuser's sleeve. Within a month, he published a design for a wire-mesh mask to fit on the nose, in shape not unlike the masks used today. But a similar idea had also occurred to a Berlin chemist, Adalbert Wolff, from whom, if Wolff was to be believed, the eminent surgeon, Josef Bergson, filched it.

> Shortly before ether intoxication seized the minds of European surgeons, Dr Bergson paid me a visit; and, under cover of recommending a patient, carefully examined the apparatus [for breathing gas through the nose] I had just designed. I readily explained to him the ingenuity of the invention and gave him full details of how to make it. After ether inhalation became widely acclaimed, Dr Bergson could not resist publishing papers in medical and political newspapers proclaiming my invention as his.[19]

Such diatribes enlived the medical press, not otherwise full of laughs, for some years.

...

In France ether was widely but not universally welcomed. As early as January 1847 the acclaimed biographer of Ambroise Paré, Joseph-François Malgaigne, reported twenty operations under ether anaesthesia to the Académie des sciences. He did so with a proper Gallic sense of occasion.

> The first case was a workman who had fallen under a railway car and had his leg crushed. My assistant, Doctor Charlerois, under my direction, administered ether to the unfortunate and I speedily severed the crushed limb below the knee. On awakening, the patient declared that he had suffered no more pain than if he had been scratched with a needle. I realised that on that day I was not writing surgical history but making it.[20]

In 1840 the head of surgery at the Charité in Paris, the great Alfred Velpeau, had declared that '*Eviter la douleur par des moyens artificiels est une chimère*.' Seven years later he announced that his own and his colleagues' experiments had removed all doubts from his mind and that ether inhalation

19. *Deutsche medizinische Zeitschrift*, 1852 (34), 532.
20. Quoted in R. Rey, *L'Histoire de la douleur* (Paris, 2000), p. 342.

was one of the most important discoveries ever made, 'of incalculable benefit not only for surgery but also for medicine, physiology, chemistry, dentistry, veterinary science and even psychology'.[21] He could have added revolutions to the list.

In February 1848 it was Paris as usual which gave the signal for a wave of uprisings across Europe. Karl Marx' 'beautiful little Revolution' against King Louis-Philippe's boring regime lasted only a few days and was relatively bloodless;[22] but, before the king hastily departed for England, medical students – those at least who felt that they could be spared from the barricades – established first-aid stations in strategically placed public buildings. Julien Marchaux, later a staid professor of anatomy in Lyons, recalled that as a hot-headed youth he helped to man a surgical post at the *hôtel de ville*. The twenty-two-year-old Julien Delorme, by the 1870s a famous ophthalmologist, had acquired a generous supply of sulphuric ether from the Salpêtrière and had used it liberally when dressing the wounded and even amputating a finger. Also present was the budding writer and poet, Charles Beaudelaire, a mature twenty-seven, who perhaps held the ether bottle. It was the first though not the last time that a general anaesthetic was administered against a background of gunfire.

France (rather than Edinburgh) also led the field in using ether in obstetrics. Paul Dubois, surgeon-obstetrician to the Hôpital Beaujon and nephew of the surgeon who had operated on Mme d'Arblay, described the case of a fifteen-year-old girl bearing her first child. Labour was prolonged and painful and it looked as if the infant might have to be sacrificed to save the mother. However, after a few whiffs of ether had sent the mother to sleep, forceps were applied and the baby was safely delivered. 'The lusty sound of the infant crying brought the mother back to consciousness and she joyfully assured us that she had felt nothing.' Dubois was gratified to note that in his first eight cases ether did not seem to interfere with uterine contractions. Sadly, two of the eight mothers died from childbed fever within a few days of giving birth: the epidemic was raging at the Maternité and Semmelweis had not yet been heard of.[23] The post-mortem findings did not suggest that the ether had

21. 'To eliminate pain artificially is a delusion.' Quoted ibid., p. 167.
22. For once, it seemed, the property-owning bourgeoisie, the favoured but always insufficiently favoured class under Louis-Philippe, the working class and the destitutes of Paris were in some kind of an alliance. The National Guard joined the rebels. The king's departure seemed like a victory. But the February Revolution settled nothing. The next revolution in July was a far bloodier affair, with thousands killed, executed or transported. It ushered in the reign of Napoleon III.
23. The Hungarian obstetrician, Ignac Semmelweis, had in fact demonstrated two years earlier that puerperal fever was essentially caused by unwashed hands; but his message took years to percolate through to Western Europe. For a more detailed discussion see Dormandy, *Moments of Truth*.

played any role in the infections and deaths; but Dubois expressed himself with some reserve. 'Perhaps, after all, Nature should be allowed, whenever possible, to take its course.'

Not everybody was enthusiastic. The great physiologist, François Magendie, delivered a diatribe in the Académie on 7 February 1847. He suggested that ether inhalation and insensitivity to pain would make surgical operations 'doubly dangerous' and might 'lead to the most distressing dreams'. He was congratulated by the president on his 'sublime oratory'; but, pressed to give examples, he had to admit that on the spur of the moment he could think of none.

With patchy success French doctors also explored ether inhalation in some non-surgical conditions. Trials in a range of mental disorders were abandoned when the oracular Felicien Failleret reported that 'all hopes of useful application in this field were disappointed'. In some cases the mental state of patients actually deteriorated under and after anaesthesia. Comas induced deliberately as a therapeutic measure had to wait another eighty years to stage a comeback and another hundred to be finally (one hopes) abandoned. More hopefully, ether anaesthesia was reported to have been useful during the acute phase of meningitis in an outbreak in Algiers.

In 1850 the method had its first trial in forensic work. In Rennes ether inhalation was used in two conscripts suspected of seeking to evade military service. In the first the joint deformity shown by the patient remained unaffected by anaesthesia; but in the second the large hump on his back disappeared under the 'beneficial influence of ether'. The unfortunate later confessed to malingering and had to start his seven-year military service with a six-month stretch in jail.

. . .

Russia was the unlikely setting for another early advance. In 1847, under the stern rule of Our Little Father, the Tsar Nicholas I, this was still a country where to teach the alphabet to moujiks or their offspring was a criminal offence and where a succession of imperial ukases forbade the use of any language other than Russian in any official document in any part of the vast colonial Empire. In that year the surgeon, Nikolai Ivanovich Pirogoff, appointed at the age of twenty-eight to the chair of surgery at Dorpat (today's Tallinn) in Estonia, introduced the successful administration of ether vapour by the rectum. He published his results in a monograph, *Recherches pratiques et physiologiques sur l'étherisation*. Though the book became a European bestseller, the rectal route never caught on in Western Europe. Yet it had many advantages over inhalation and proved valuable in treating Russian wounded in the Crimea. What Florence Nightingale and her troupe achieved on the British side, Pirogoff and the

Grand Duchess Helena Pavlovna and her nurses accomplished in the Russian army.[24]

. . .

Back in Britain ether anaesthesia had become for a year or two a social as well as a medical sensation. Within a week of Liston's operation at University College Hospital, Prince Jérôme Bonaparte visited Westminster Ophthalmic Hospital to watch the surgeon, George Guthrie, remove a bladder stone. (Why at an ophthalmic hospital remains a mystery.) He was then driven straight to St George's Hospital to watch an amputation. 'It was amazing,' he reported back to the Princesse Mathilde. 'The patients experienced absolutely no pain. I am seriously considering having my own haemorrhoids treated.' The list of spectators at an operating session at the Westminster Hospital the following week, featuring the painless excision of a tumour from the buttock of a young woman, could have been extracted from the Social Register of the *Morning Post*. They included Lord Walsingham, Lord Merton, Viscount Falkland, Sir Henry Mildmay, Sir George Wombwell, Admiral Sir Charles Napier and General Sir George Baker, as well as 'many distinguished foreigners', among them the ubiquitous Prince Jérôme.

The new technique rapidly spread outside London. On 23 December, the day after Liston's operation, James Miller read out a hastily scribbled note from Liston to his audience of students in the lecture room of the University of Edinburgh. The note announced that 'a new light has burst on Surgery'. The news 'excited, surprised and charmed' not only the medical profession but Edinburgh society as well. Next week a number of prominent citizens and citizenesses, 'not a few titled', attended an operating session at the Royal Infirmary for the first time.

> No doubt they felt apprehensive but also curious. During the operation, the removal of a foot, though some doubtless felt a little queasy at the bloody operation, they remained composed, feeling little since the patient felt nothing at all.[25]

In January 1847 in Bristol ether, 'specially prepared by an eminent chemist', was used in an above-knee amputation for a man with a 'white knee', almost certainly tuberculous. From Poole in Dorset came a report that, with a ten-day interval between the two operations, both feet of a seventeen-year-old Scottish sailor had to be amputated for frost-bite. The ether was administered by Mr Swansea, 'our excellent dentist', and the second opera-

24. Pirogoff was also among the first on the Continent to introduce Lister's antiseptic methods. A great and humane surgeon, he died in 1881, aged seventy-one.
25. J. Miller, *Surgical Experiences with Chloroform* (Edinburgh, 1850), p. 46.

tion was 'not without its light diversions'. After inhaling for three minutes the patient rose from the operating table, shouting 'wild, incoherent and rude expressions' before losing consciousness. Asked how he felt after coming round, he replied: 'How can a drunken mon tel?'[26] In an animal-loving country no less excitement was caused by a report from Liverpool that Mr Langridge, a veterinary surgeon, had painlessly removed a tumour from the neck of a Newfoundland dog.[27]

Some surgeons tried to use ether only for the most painful part of an operation. Frederic Skey of St Bartholomew's Hospital in London performed a Caesarean section on a twenty-seven-year-old unmarried woman with a severely deformed pelvis. 'The unfortunate patient, a dressmaker, had become pregnant when, under a temporary excitement, she had connection with her lodger, and did not discover her condition till the seventh month of pregnancy.'[28] Her contracted pelvis made an attempted vaginal delivery hazardous. Skey used ether only for the eleven-inch initial abdominal incision: he then delivered a healthy infant in four minutes without further inhalation. It was, so far as is known, the first Caesarean section performed partly at least under general anaesthesia.

In November 1847 Henry Johnson of St George's Hospital excised 'a huge and difficult fungoid tumour' from the face of a nineteen-year-old man, 'adding to the ease and comfort of the patient during the hour-long operation by intermittent inhalation of ether vapour'. At their next meeting the Council of the College of Surgeons specially commended Johnson for his skill and humanity.

. . .

By the summer of 1847 the news had encircled the world. The first public demonstration of ether anaesthesia in Australia was made on the same day, 5 June, in Sydney and Launceston. Both were minor procedures; but a week later, 12 June, in Van Diemen's Land, today's Tasmania, Surgeon Captain Bruce Sweetnam amputated the crushed leg of a sailor, John Wellbegone. The first operation under ether in New Zealand was performed by Mr Alfred Butler in Wellington for the extraction of bladder stones on 27 September. All procedures were successful.

. . .

26. Quoted by Stanley in *For Fear of Pain*, p. 309.
27. *The Times*, 27 March 1849.
28. *London Medical Gazette*, 17 January 1848.

Yet ether had many disadvantages; and they were almost as quickly apparent as its near-miraculous benefits. A reaction was bound to set in. Before opponents could muster their ammunition, a new method of anaesthesia appeared on the scene. Less than a year after Liston's historic performance in Gower Street, chloroform made its début at a dinner party in the Georgian town house of Professor James Young Simpson of Edinburgh.

# Chapter 25
# CHLOROFORM

James Young Simpson was the eighth child and seventh son of David Simpson, village baker of Bathgate in Linlithgowshire, today's West Lothian, in Scotland. David's wife Mary Jarvie, was forty-two when she conceived, old in those days; and from the moment of Jamie's birth the family scrimped to give him a superior education. They remained close and watched his progress with pride. At fourteen he enrolled in the arts classes in the University of Edinburgh, 'very young, very solitary, very poor and very friendless'. But he did not stay friendless for long. He was, by all accounts, an attractive and gifted boy; and, after a difficult start, he would never be short of admirers. At twenty-one he graduated Doctor of Medicine.

Edinburgh people liked to call their city the Athens of the North and with some justice. The medical faculty of the university in particular, though the youngest in Scotland, had a colourful history and was still peopled by brilliant eccentrics.[1] Too brilliant and eccentric in fact to be contained in one small

1. Among the most remarkable in the Faculty's history were Robert Sibbald, first professor of medicine and founder of the Royal College of Physicians, Andrew Balfour, 'the 'Morning Star of Science in Scotland', Archibald Pitcairn, of such renown that he was called to Leyden where he taught Boerhaave, three generations of the Monro dynasty of anatomists (of sharply declining merit), Robert Whytt, pioneer of neurology who first described tuberculous meningitis in children, William Cullen, scholar and the most influential teacher of pharmacology of his generation, John Brown whose generalised system of diseases divided Europe into Brunonians and anti-Brunonians, the great surgical brothers Bell, the great chemist Joseph Black, John Hughes Bennett who described leukaemia, the first blood disease to be recognised, and uncounted Gregories and Hamiltons of varying merit.

The terrible days of the body-snatchers and of the murderers and body-sellers Burke and Hare were not far in the past; and their eminent and possibly conniving customer, the great anatomist, John Knox, was still remembered.

Still very active, often to Simpson's chagrin, was the famous surgeon, James Syme: their feud would provide the city with a long-running soap opera. Less well known is the fact that Simpson frequently consulted Syme when he or his family needed sound surgical advice.

city. Some of the best had always migrated south and had made their fortune in England.[2] Simpson too toyed with the idea but stayed. He never explained what made him embark on obstetrics.[3] 'Male midwives' were still a highly controversial species. As recently as 1846 a pamphlet published in Boston but enjoying a wide circulation in Britain fulminated against a practice that was 'grossly indelicate, pagan, obscene and a temptation to immorality'.[4] In Edinburgh the Faculty was divided; and division implied more than an exchange of polite unpleasantries. The professor of materia medica, James Gregory, a founder of the Edinburgh College of Physicians, gave James Hamilton, first professor of midwifery, such a thrashing with his walking stick in the university quad that policemen had to intervene. In court Gregory was deprived of some hard cash. He paid up, roaring that he would gladly repeat the performance under the same conditions. The judge remarked that 'next time the professor would be incarcerated'. But Hamilton was no innocent either: the kindly John Christison described him as a 'snarling, foul-mouthed boozer'.[5] Perhaps that was not altogether surprising. Hamilton had battled all his life to have his subject properly recognised; and though, according to Simpson, he was the most 'soporific lecturer in the university', Simpson decided as soon as he had qualified that he would eventually be Hamilton's successor.

In the event the election that followed Hamilton's untimely death became the most closely contested in the history of the university. Edinburgh was known for admitting students without religious qualifications and for keeping its doors open to crofters' sons, but no university anywhere in the United Kingdom was a bastion of social equality. The twenty-nine-year-old village baker's son was gifted and could be delightful company; but he was still a village baker's son. All the professors were in favour of Emory Kennedy from Dublin, already master of the Rotunda and the scion of a distinguished family. Fortunately for Simpson, the seventeen representatives of the City Council on the selection committee were unanimous in their determination to frustrate the professors and, by some ancient by-law, they were in a majority – just. Even so, fortunes (by local standards) were expended on posters, pamphlets and testimonials, as well as on bets. Rival gangs of students canvassed members of the committee and their womenfolk and intimidated ordinary

2. The two leading professors of surgery in London, Robert Liston of University College Hospital and William Fergusson of King's College Hospital, were both migrants from Scotland and Edinburgh graduates.
3. The ordeal of surgery horrified him and medicine 'bored him to tears'as a medical student. On the other hand, he was always good at handling women across a broad spectrum of situations.
4. Quoted by J. Duns, *Memoir of Sir James Young Simpson, Bart.* (Edinburgh, 1873). p. 67.
5. Quoted ibid., p. 56.

Rest, the oldest of pain-killers. *Girl with her Arm in a Sling*, Roman (Vatican Museum). Resting the injured or diseased part until the pain subsides is still the most universally applicable of pain killers. Animals do it instinctively; humans have always used various aids – slings, splints, plaster casts. The last reached its apogee in the mid-twentieth century, before the introduction of internal fixation. Its apostle, Sir Reginald Watson-Jones, allegedly declared: 'Put two billiard balls into a well-moulded plaster cast for long enough and they will unite.'

Jean Antoine Watteau (1684–1721), *L'Indifférent* (Louvre, Paris). Contrary to operatic lore, pulmonary tuberculosis in its last stages, especially when complicated by laryngitis, was a terribly painful illness. Dying at the age of thirty-six, Watteau worked almost to the end in a haze of laudanum. The drug did not so much abolish pain as make patients almost indifferent to their suffering. This was probably the inspiration of this enigmatic masterpiece.

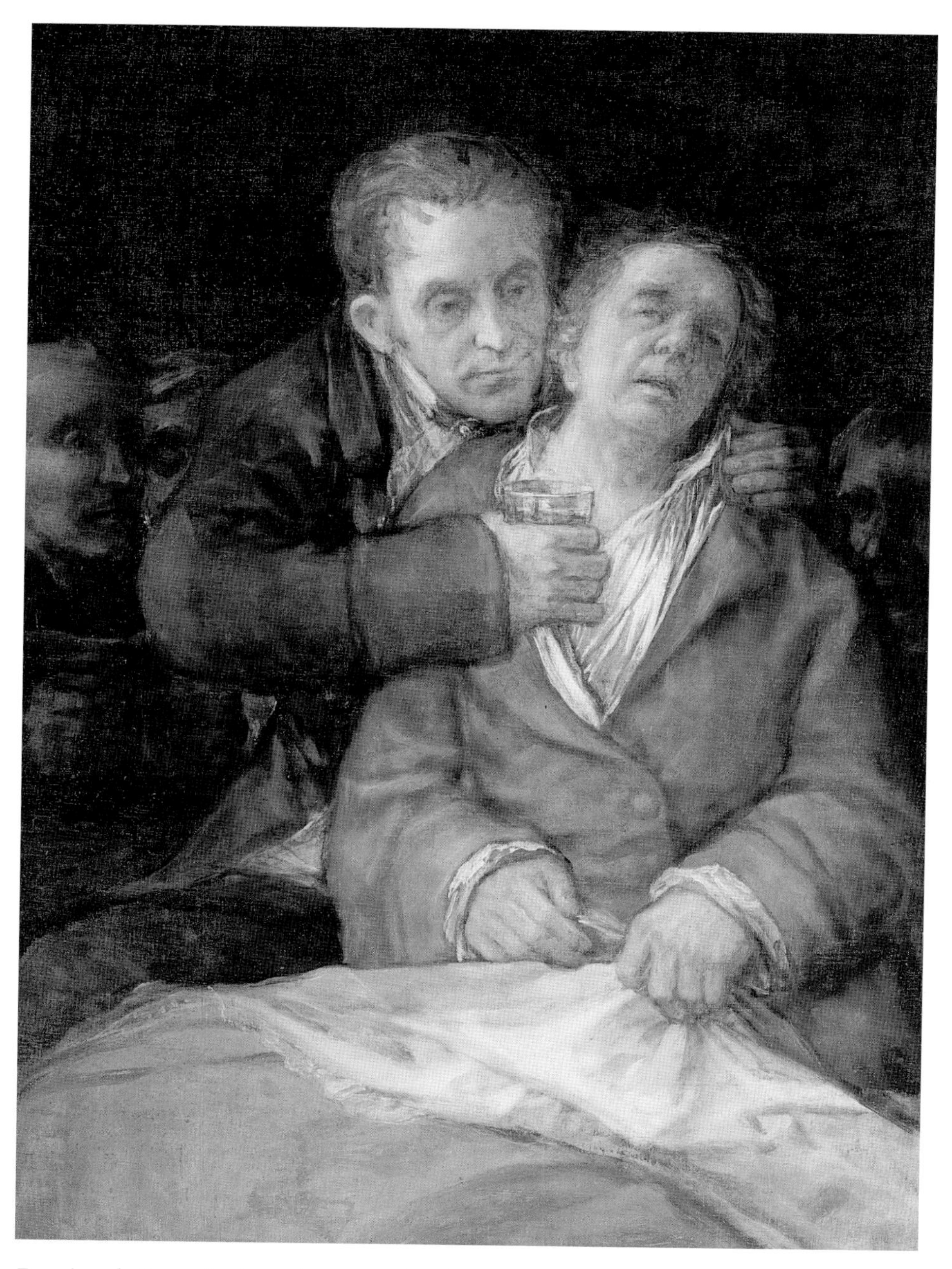

Francisco de Goya y Lucientes (1747–1828), *Self-portrait with Dr Arrieta* (Minneapolis Institute of Art). Arguably the greatest medical documentary in art. The artist is in the grip of an attack of cardiac asthma. The posture, the tinge of the skin, the fingers clutching the bedclothes, the flared nostrils, the open mouth are all diagnostic. The image of the caring doctor (to whom the painting was a present) shows the profession at its best. He is probably offering a concoction of laudanum, the only remedy to relieve the distress at the time. The patient survived for another twelve years.

The Laocoön, Hellenistic (first century) (Vatican Museum, Rome). Probably the greatest sculpted image of the fight against physical pain. It inspired Michelangelo (see Chapter 11).

Matthias Grünewald (*c.*1478–*c.*1531), *Crucifixion* (Unter den Linden Museum, Colmar, France). Is this the greatest representation of physical suffering in painting (see Chapter 11)?

Images of terminal disease:

Top
Ferdinand Hodler (1853–1916), *Valentine Godé-Darel im Krankenbett* (Kunstmuseum, Solothurn, Switzerland). This portrait of his dying wife is unlike the artist's better known historic tableaux and symbolic paintings. The facies of terminal cancer and low-grade but unrelenting pain must be familiar to every doctor and most people past middle age today. It is this suffering which modern palliative care must address.

Opposite
Ejnar Nielsen (1872–1956), *The Sick Girl* (Statens Museum for Kunst, Copenhagen, Denmark). The painting by the twenty-three-year-old artist is probably that of his sister in the last stages of tuberculosis. Its realism is in striking contrast to Edvard Munch's more famous image of his tuberculous sister, Sophie; but both are true. In an interview in 1927 Nielsen observed: 'Death serves life and life serves death. I cannot understand why people cannot view death with its majestic beauty as they do life.' But there is nothing majestic about the terminal suffering portrayed in this painting.

Preanaesthetic dentistry. Honoré Daumier (1808–79), '*Voyons . . . ouvrons la bouche!*' Nobody has surpassed Daumier in the biting depiction of the medical and dental mores of his day.

A pioneer. *Henry Hill Hickman* [1800–30] *Performing Experimental Surgery on an Animal Under 'Suspended Animation' Induced by Carbon Dioxide*, c. *1826* (Wellcome Library, London).

The ecstasy of pain. Gianlorenzo Bernini (1598–1680), *The Vision of St Theresa* (Cornaro Chapel, Santa Maria della Vittoria, Rome). No single photograph can do this work justice. It is part sculpture, part painting, part architecture, part stage set. It has repelled many – 'it is impossible to sink lower than Bernini', John Ruskin pontificated (though he later repented) – but no other work of art so brilliantly conveys the ecstasy of pain. St Theresa herself spoke of the experience of the angel piercing her heart as an indescribable mixture of torture and joy.

Hysteria. Pierre André Brouillet, *Une leçon de Charcot à la Salpetrière, 1887* (Faculty of Medicine, Lyon). Charcot demonstrating the famous 'arc-en-ciel' attitude adopted by a 'grande hystérique'. She would fall and injure herself if she were not caught by an assistant, on this occasion by the famous-to-be neurologist, Josef Babinski. Subconsciously, Charcot would explain, she is aware of this.

Jean-François Millet (1814–75), *L'Angélus* (Louvre, Paris). Better than any photographic image of the shrine itself, the painting illustrates the simple faith of the French countryside which provided the spiritual soil of the miracles of Lourdes.

Horace Wells (1815–48), the Hartford dentist who first systematically used nitrous oxide anaesthesia in dental extraction.

John Snow (1813–58). No other doctor advanced two unrelated specialities – epidemiology and anaesthesia – while working as a general practitioner.

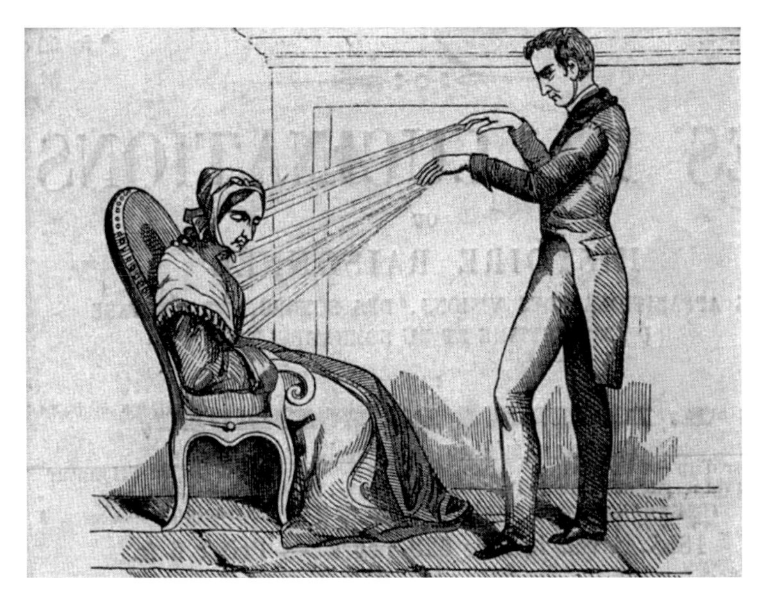

Animal Magnetism. By the time of Mesmer's death in 1815 several features of his original procedure – the bucket of water filled with iron objects, the bodily contact or hand-holding had disappeared. Animal magnetism had become what is recognisably modern hypnosis, though the term was not coined till 1863.

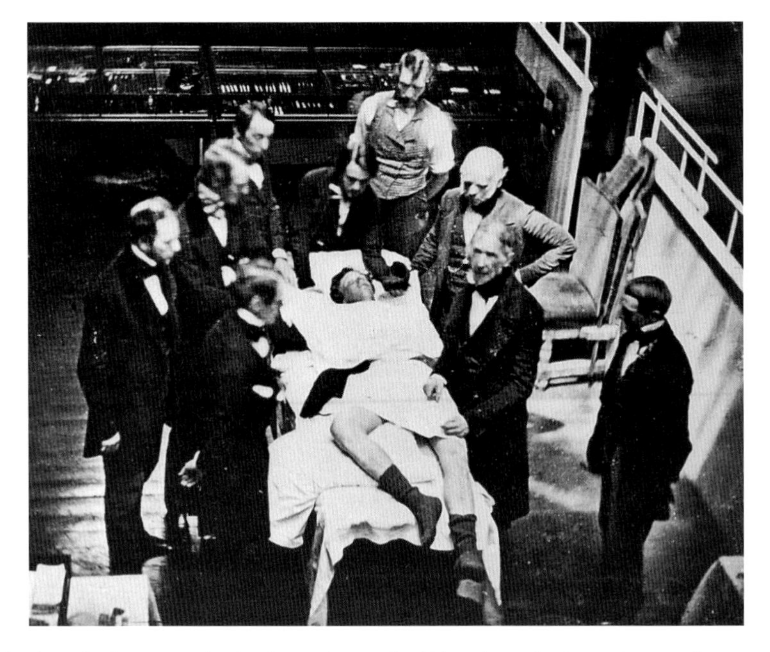

The beginning of ether anaesthesia. *Operation under Ether Anaesthesia,* c.*1847, at the Massachusetts General Hospital* (daguerrotype attributed to Southwood and Hawes). The surgeon in the foreground is John Collins Warren.

Sir James Young Simpson (1811–70), professor of obstetrics in Edinburgh, who in 1847 introduced chloroform anaesthesia into obstetrics.

Cicely Saunders (1918–2005), pioneer of the modern hospice movement.

Sir Luke Fildes (1844–1927), *The Visit* (Tate Gallery, London). The sick child was the artist's daughter. Doctors before the schism of the mid-twentieth century (see Chapter 47) could do little in the way of curing; but perhaps they were better at caring than many of their scientifically well equipped successors.

citizens. In the midst of the electioneering Simpson married Jessie Grindley of Liverpool, who immediately prepared a detailed catalogue of her husband's private museum both of antiquities and of curiosa that would, if he were elected, become the pride (or one of the prides) of the city. 'As the finish approaches, both Kennedy and I are digging our spurs deeper into our horses' sides,' Jamie, a racing man, reported to his father-in-law. And three days later to his mother:

> My dear Mother, Jessie's honeymoon and mine can begin today. I was elected Professor yesterday by one, I repeat one, vote. Hurrah!!! Your affectionate son, J.[6]

With the townspeople – but not with his colleagues – he was immensely popular from the start. 'Thrifty hotel keepers wept tears of gratitude over their ledgers,' one disgruntled colleague wrote, 'for soon they would be rich beyond the dreams of Hamilton.' He was referring to the female clientele who would be flocking to the new professor. They would come not only from Scotland but from south of the border as well. He was right. Christison soon reported that

> when I called on Simpson, his two reception rooms were crowded with patients as usual. Female faces stared at me in vacant expectancy and a titled lady attended by a maid and a dog was ringing the door bell.[7]

To see duchesses the new professor was prepared to travel south. Ladies of a lesser rank had to journey north. Oddly perhaps, he was regarded with as much affection by crusty colonels and gouty peers as he was adored by their ladies. Even more remarkable, a torrent of learned papers on an astonishing variety of topics began to flow from his pen; and distinguished medical and scientific visitors from the Continent started to travel to Edinburgh to meet him in person. They were as charmed by his erudition as they were by Jessie's cuisine.

Simpson could be vain and overbearing but he had a fertile mind and great determination. The problem which exercised him from the moment of his appointment, perhaps even as a medical student, was painless childbirth. He tested ether within a few months of Liston's historic operation in London; but was not entirely happy with it. The inhalation undoubtedly eased labour pains but tended to make the mothers (and some of the onlookers) sick. He and his closest friends, George Keith and Matthews Duncan, spent regular sessions after formal dinners inhaling an assortment of vapours, volatile oils and

6. Quoted by H. L. Gordon, *Sir James Young Simpson and Chloroform* (London, 1897), p. 214.
7. Quoted ibid., p. 87.

organic solvents to see if they would have a stupefying effect. His neighbour in Queen Street, the surgeon James Miller, would call in the mornings on his way to the Infirmary to assure himself that the crackpot professor was still among the living.

. . .

On one such occasion, the date was 4 November 1847, the three obstetric conspirators had already run through an assortment of liquids, including mixtures containing acetone, iodoform, ethyl oxide, benzene and some nondescript oils. After remaining disappointingly sober Simpson remembered that a Liverpool chemist, David Waldie, had promised to prepare for him a liquid which was said to induce a kind of 'dopiness'. Always impatient, Simpson had procured some of the material from an Edinburgh pharmacy – and then forgot about it. He remembered it now and began to search for it. His desk was chaotic as always. Just about to give up, he discovered the bottle under a sheaf of printers' proofs. He took the stopper out and poured the contents into three tumblers. The experimenters pressed their noses to the brim.

When the ladies returned, they found the scene transformed. The usually slightly morose Keith was waltzing round the room. Duncan, his eyes shining brightly, was guffawing heartily for no discernible reason. Their host was on his feet but not for long. Within seconds the three men reassembled on the floor. Soon Duncan was snoring. Keith's legs were trying to overturn the table. Simpson was giggling inanely. To the relief of Jessie and the other wives, all three soon came round. The first words Simpson uttered were 'But this is better than ether!' One of the ladies, Miss Petrie, insisted on proving that she was as brave as the men. She picked up one of the tumblers and inhaled deeply. Almost at once she dropped the glass, fell on the floor, stretched out her arms and started carolling: 'I am a fairy, I am a fairy!'

. . .

Simpson began to use chloroform at once. He persuaded the mother of the first child born under the influence to call the baby Anaesthesia.[8] A photograph of the pretty lass, aged seventeen, would later hang over his desk. A month after the memorable dinner party he read a paper before the Edinburgh Medico-Chirurgical Society. He had by then extracted the information that his magic potion had been discovered by three men more or less synchronously and independently of each other, one in the United States, one

8. He was inspired by the story that the first child in Russia to be vaccinated had been christened Vaccinov.

in Germany and one in France.[9] None of them had noticed its anaesthetic qualities: indeed, even its chemical properties were not to be fully described till a characteristically thorough and exhaustive paper by the great French chemist Jean Baptiste Dumas ten years later. But Simpson was neither surprised nor particularly disconcerted when his first publication aroused an outcry. The noise was polyphonic.

Comparatively mild objections were raised by a few obstetricians who worried that chloroform (like ether) might interfere with uterine contractions and delay or complicate labour. Some concern was also expressed over the possibility that the drug or one of its metabolites might cross the placental barrier and adversely affect the foetus. Such possibilities today would lead to prolonged and carefully controlled trials. Few would blithely accept Simpson's assurance that in his experience – as many as two dozen cases, perhaps – he had observed no complications. In the mid-nineteenth century few quibbled.

More vocal were objections raised on moral and religious grounds. Many maintained that under the influence of anaesthetics 'the holy pangs of labour' would metamorphose into unseemly exhibitions of sexual passion. Indeed, the ill-effects of anaesthesia would be felt long after labour, for no child brought into the world without pain could later expect to enjoy the blessings of maternal love. In some mysterious way even fathers might evince a certain indifference to painless offspring.

But even these concerns were as nothing compared to the deep religious misgivings expressed both by ministers of the Kirk and by devout layman. In Genesis 3: 16 God explicitly predicts and ordains that 'I will greatly multiply thy sorrow and thy conception; in sorrow thou shalt bring forth children'. Simpson was prepared for the onslaught and had his answer ready. Far from God potentially disapproving of the new development, He might justly be described as the first obstetric anaesthetist. For does not Genesis 2: 21 state that 'the Lord God caused a deep sleep to fall upon Adam, and he slept: and He took one of his ribs, and closed up the flesh instead thereof'? Today this might settle the dispute. It did not do so in the Edinburgh of the late 1840s. Hardly had Simpson put forward his argument (which, many felt, bordered

9. In America it had been Samuel Guthrie, a descendant of James Guthrie, the Scottish Covenanter and martyr, who prepared chloroform in his self-built laboratory while working as a country doctor in Sackett's Harbor in Connecticut. He called it a spirituous solution of Chloric Ether and described his discovery in a letter in 1830 to the famous chemist, Benjamin Silliman. Silliman published it with his commendation in the second volume of his *Elements of Chemistry* in 1832. About the same time the same substance was described by Eugène Soubeiran in France under the name 'ether bichlorique' and by Justus Liebig in Germany as 'carbon trichloride'. It was Jean Baptiste Dumas who, in his article in the *Annales de chimie*, named it 'chloroform'. Its anaesthetic effect in animals was noted by Flourens a few months before Simpson's experiments on himself and friends in 1847.

on the blasphemous) when more than one biblical scholar turned on him in fury. The passage in Genesis quoted by the godless professor referred to the blessed period *before* the Fall. It could not be adduced as an argument in favour of anaesthesia *after* that event. But Simpson was not easily beaten. He pointed out that 'sorrow' is not 'pain'. He also contended that he had studied the biblical passage not in translation, as had most of his adversaries, but in the original Hebrew. These studies had led him to the philological conclusion that a word had been mistranslated for centuries. The word 'sorrow' should in fact be 'effort'. Such Scottish chutzpah was difficult to counter; but the argument continued. In fact, the dispute was finally settled neither by Simpson, nor by Hebrew scholars but by Queen Victoria.[10] But before commemorating the royal intervention, chronology ordains a brief review of the fast changing drug scene.

10. See Chapter 26.

## Chapter 26

# THE SHAPE OF DREAMS

For the first two decades of the nineteenth century opium, mainly in the form of laudanum, was fashionable among the affluent and the literati; but it was not yet a popular remedy. Factory workers, farm labourers, miners, soldiers and sailors still sought solace and oblivion in gin, beer and other forms of cheap alcohol. But by the mid century the drug had permeated every class of society.[1] Between 1825 and 1850 imports to Britain rose from 28,300 kilos to 125,000 kilos per year. About a third of this was re-exported, mostly to the United States; but the rise in home consumption was still a steady 3–4 per cent a year. Most of the drug now came from Turkey. Payment was simple. Britain had enjoyed commercial links with the Ottoman Empire since the foundation of the Levant Company in 1581; and Turkey was still a lucrative market for British goods. By 1860 even the best Persian and Egyptian opium, the former including the produce of today's Afghanistan, was repackaged and made to look Turkish in Constantinople. Later analyses showed that genuine Turkish opium had a morphine content of 10–13 per cent in contrast to the 4–6 per cent of the repackaged Indian product. Traditional routes had also changed. Venice was in decline. Marseilles had become the main reception centre on the Continent and would remain so till the 1970s. Amsterdam and Rotterdam had also developed into opium ports, as they still are. But by the 1850s most of the opium was shipped directly to Britain in British vessels. Dover, Liverpool and Bristol were all important destinations; but the hub of the trade was in London.

The mechanics of the business had evolved gradually. With the demise of the Levant Company in 1825 the cosy price-fixing cartel of the eighteenth century had disintegrated. New aggressive wholesale importers had moved in, purchasing opium by competing deals and auction. More than 90 per cent of

1. All the old herbal remedies like mandrake had fallen into disuse. Opium was reliable, cheap and generally believed to be harmless.

the transactions were now settled within a hundred yards of Mincing Lane in the City, a significant proportion in Garraway's Coffee House near the Royal Exchange. The fortnightly auctions were conducted by a system known as 'buying by the candle.' A small candle was lit at the start of auctioning each lot and the highest bid made before the wick burnt away was accepted. The custom was endearingly quaint. Trading in opium was not. It required cut-throat expertise. Mincing Lane dealers seldom worked under a 50 per cent profit margin. This could rise to 200 per cent but could also fall to nothing or end in a staggering loss. Risks were great. Profits could be greater. The operation of market forces led to a steady fall in retail prices. The monthly publication of current price lists allowed chemists to shop around for the best buy. Among the buyers were now wholesale houses like Allen and Hanbury's and the Apothecaries Company. There was also a stepwise reduction in import duty. From 9 shillings per pound in the early 1820s it fell to 4 shillings in 1826 and to 1 shilling in 1836. It was finally abolished by the free-trade agreement of 1860. But the most important reason for the cheapness of the drug was availability. Even when in 1865 the SS *Crimea* ran aground off the coast near Smyrna, today's Izmir, with a loss of several tons of her cargo (as well as most of the crew), the price in the City rose by barely 2 per cent for a week.

The Turkish connection irked many patriots, who felt that so valuable a plant should be cultivated within the Empire or, even better, at home in Britain, not bought from 'the rascally Turk'. Home growth was not an idle dream. Since 1763 the Society of Arts had been promoting the study of medicinal plants, offering prizes and medals for new discoveries. By the end of the century their interest had turned from rhubarb to the poppy. Some varieties had been indigenous in Britain at least since the seventeenth century. Dr Alston, professor of materia medica in the University of Edinburgh, had described a way of producing opium from the garden plant as early as 1735. The method was cumbersome but could be simplified. In 1789 a 50-guinea prize was awarded to John Ball of Plymouth for the 'excellent quality' of his home-grown opium. A few years later Thomas Jones set five acres of ground near Enfield north of London with white poppies and won a gold medal. In 1820 John Young, an Edinburgh surgeon, won another for showing that opium could be harvested even in a cold damp climate. By the 1830s poppy was grown in most southern counties in England, including the market gardens around London. It was mainly for home consumption: no family with six or eight children was ever without pains and aches or diarrhoea. The surplus could be sold to the local chemist more cheaply (but still at a profit) than the market price. The morphine content of 'good English opium' was reckoned to be around 4–5 per cent. Although it was inferior to Turkish, its purchase, like that of some inferior English wines today, lifted the purchaser's heart.

While poppy-growing was never going to be profitable in most parts of Britain, the Fens, the low-lying marshlands of Lincolnshire, Cambridgeshire and Norfolk, had become 'the Kingdom of the Poppy'. In the cathedral city of Ely, in the heart of the region, there was no need to ask for a stick of opium or an ounce of laudanum in the shops. A coin laid on the counter meant only one thing. Dr John Hawkins described a King's Lynn farmer entering a chemist's shop, buying one and a half ounces of laudanum and drinking it there and then. He would then return twice the same day for similar draughts and then buy half a pint to take home for the evening. A chemist in Spalding told Hawkins that he had sold more laudanum in his first year in the town than he had sold over twenty years in Surrey. A Holbeach grocer took about £800 a year for laudanum from the people of this poor working-class parish. Charles Kingsley in *Alton Locke* (1850) makes a Fenman explain to a visitor:

> Oh! ho! ho! – you goo into the druggist's shop o' market dayt, into Cambridge, and you'll see the little boxes, dozens and dozens, a'ready on the counter, and never a venman's wife go by, but what cals in for her pennord o' elevation, to last her out a week. Oh! Ho! ho! Well it keps womanfolk quiet it do; and it's mortal good agin the pains.

One Ely brewer added opium to his ale at source and, despite the slight increase in price, did well from the innovation.

The reasons for such a regional addiction were never entirely clear. In the Fens the bleak winters and the damp climate promoted rheumatism – as they still do. 'Aigues', fevers and malaria remained common even after the introduction of quinine at the end of the seventeenth century. Life in the marshlands could be hard. Opium softened it.

. . .

By the 1830s opium was by far the most commonly used and most effective painkiller both in Britain and in Europe. The Auvergne was the French Fens, for reasons as obscure or as complex as in England. It had some of the most luscious farmland in France. Poppies were grown on a large scale in the fertile Po Valley in Italy and in Andalusia in Spain. In Transdanubia in Hungary the acres of cultivated poppy were the pride of the agricultural reformer, Count György Festetich. But even in warm Mediterranean countries the home-grown plant could never satisfy the demand of the cities. Most of the drug continued to come from Turkey via Marseilles and London. Several reasons boosted its sales.

Opium was uniquely effective not only against aches and pains but also against diarrhoea. In most of Europe the worst kind of diarrhoea, cholera, was a killer. In Britain the epidemics of 1830–31 and 1848–49 were the most powerful propagators of the drug. Although nothing was known about salt and water balance, it had long been realised that people died from the disease because of the massive loss of intestinal fluid.[2] If the loss could be reduced, there was a chance of survival. But treatment had to be early and heroic; those who survived owed their lives to massive doses of opium. Babies and children died from the milder 'summer diarrhoea' or 'dysentery' too. A few were saved by opium.

Another promotional point was poor people's reluctance to visit a doctor. Even the cheapest consultation could cost a third of a skilled workman's weekly wages. The chemist's advice, on the other hand, was part of his sales pitch. Almost always it ended with a prescription for some form of opium. When the Sale of Poisons Bill was first discussed in England in 1857 John Brande of the Royal Institution submitted to Parliament that the use of opium-based medicines was so general that any restriction was unenforceable. How could a chemist keep opium under lock and key when he dispensed it a hundred times a day? On the Continent doctors were always cheaper but destitution could be more extreme too. In the United States doctor and chemist were often the same person. Pain was universal. Opium was the universal standby.

For those who disliked asking for advice from either doctor or chemist, patent medicines, freely available in most grocers' or village shops, offered a simple way of self-medication. Laudanum and Dover's powder were still popular but they were competing with newcomers. Dr J. Collis Browne's Chlorodyne, invented as a cholera remedy by an Indian army doctor, saved lives. Of course, it did more. It was also an escape from the unending drudgery of the loom, the coal-face and the plough. Women took it more than men. It eased the despair of poverty, the hurt with which they struggled to raise a family and survive. In the 1850s it made the makers an annual profit of about £50,000, more than the income of a small German principality.

. . .

In the United States non-specific 'pain killers', all based on opium, were the commercial discovery of the mid-century. Of course patent and proprietary nostrums had long flourished in the New World: a characteristic brand of folksy quackery was part of the American scene. But in the past most magical brews had been marketed as magical brews against *diseases*. There had been drops against 'rheumatism', concoctions against 'consumption', plasters against 'dropsy'. Perry Davis' 'Celebrated Pain Killer' broke new ground. It did not aim at any particular ailment: it was both a universal panacea and a

2. It was intravenous saline treatment, especially intravenous potassium, introduced in the 1950s, rather than antibiotics which made cholera a non-lethal disease.

specific remedy against all pains. From the start American advertisers showed a zest and lack of inhibition unknown in Europe. No pain was a match for 'Samuel's Herculean Embrocation'. 'Pond's Extract, the Universal Pain Extractor' extracted all pains. Most cultivated a populist, anti-professional image. Because Pond's pain extractor was of 'purely vegetable origin' it was safe. It did not need the ministrations of a doctor or a druggist. It was, as the advertisement proclaimed, almost always showing children at play, 'The People's Remedy'.

...

The age range continued to expand. By the mid-century most people in Britain were introduced to opium as soon as they left their mother's breast: some before. A bewildering variety of 'baby-calmers' was on the market. The best-known was Godfrey's Cordial. A good draught of the old 'cordie' would stop the worst colic. At least it would stop the squealing, and it disguised the bitter taste of the drug by drowning it in thick syrup. A Nottingham chemist reported selling 600 pints a year. In Long Sutton, Lincolnshire, a chemist testified that he sold 2 gallons a month to a population of 6,000. Among the competition Mrs Winslow's Soothing Syrup was almost as popular. Street's Infant's Quieter came next. Atkinson's Baby Preservative in its attractive shiny silver wrapper was aimed at upmarket customers and was bought by nannies to the aristocracy. But most baby preparations were bought by the poor, not only by mothers but also by the proliferating breed of baby-minders. In the industrial cities many were in charge of dozens of infants and children during the long working day and usually had a second home job as well. They charged about a fifth of the average wage packet; and their objective was to keep their flock semi-comatose. Mothers, returning from a day's work and in desperate need of a few hours' sleep, continued the dosing.

A useful side-effect of opium was that it suppressed appetite – or, in tens of thousands of households, hunger. A side-effect of the side-effect was to make children undernourished and sickly. By the age of three or four many 'had shrunk up like old men, wizened like little yellow monkeys'.[3] Few could ever benefit from any form of free education, even when that became available. If and when they grew up, they would provide the next generation of the workforce, illiterate, poor, short-lived and dependent on opium.

Inevitably, the accidental or deliberate poisoning of children at home and in the baby-farms was widespread. This was known, yet charges, when brought, could rarely be proved. Juries were mostly common folk who knew what was going on. But the loss of a sickly child, while sad, could be a merciful dispensation. When found guilty of negligence, the mother was usually

3. Victor Hugo (who else?) in *Les Misérables.*

cautioned. Fathers were rarely in evidence. Opium was also used by dockland prostitutes. Clients might be drugged before being robbed. Smaller doses would counteract the stiffness and pain of a long night's drudgery and the symptoms of venereal disease. The greater the poverty, the greater the benefits. Eventually opium might provide the last painless escape.

. . .

Opium addiction provoked little public debate until the last quarter of the century. It was often pointed out that many regular users, mainly of course among the wealthy, lived to a ripe old age. Some were in their eighties or nineties. Their yellow complexion gave them away; but their only serious trouble was constipation. That could usually be remedied by gargantuan doses of laxatives. When in 1828 the Earl of Mar died at the age of sixty-five, the inquest found that he had been eating opium for fifty years, consuming 49 grains of the solid and an ounce of laudanum a day. The insurers refused to honour his life policy, won their case in the lower court but were forced to pay up on appeal. Professor Christison's testimony that the opium habit 'undoubtedly shortened life' was rejected. On the contrary, the judge ruled, for individuals suffering from such painful conditions as the late earl's gout, only opium made life bearable. Household publications like Mrs Beeton's in Britain and Mme Bonnefoix' in France contained detailed instructions on how to deal with overdoses.

. . .

But by the mid-century the monopoly of crude opium as a painkiller, pacifier and escape route was beginning to be undermined. The hunt for the 'basic principle' – or, in the French literature, *l'essence* – of opium had begun in the 1780s. In 1786 John Leigh of Virginia in the United States published his *Experimental Inquiry into the Properties of Opium.* He suggested that getting rid of 70 or even 80 per cent of waste would ensure that the effectiveness and reliability of the drug could be 'multiplied manifold'. Seventeen years later, Jean-Pierre Derosne, in France, succeeded in preparing a salt from opium which in animal experiments was three times more powerful than the parent substance. With commendable modesty he named it *Sel Derosne.* What he had in fact discovered was one of opium's constituent parts now called *noscarpine*, formerly known as *narcotine.* In 1804 Armand Seguin read a paper to the Institut, describing a method of extracting active substances from poppy juice. He was on the threshold of discovering morphine; but, before his work could be published, he was arrested for diluting medical supplies destined for the army. The charges were never proved; but in the year of Napoleon's coronation he was lucky to escape with his life.

French writers tend to emphasise their countrymen's contribution – which was considerable – but there can be little doubt that morphine was actually discovered by a twenty-one-year-old German pharmacist's assistant in the small town of Paderborn in Westphalia. Friedrich Wilhelm Adan Sertürner, born in the village of Neuhaus on 7 June 1783, had no scientific training and never acquired academic qualifications. What he had was the inborn chemist's 'touch'. His laboratory equipment was mostly self-made and his book knowledge rudimentary. His paid employment left him little spare time, but his patience was prodigious and his curiosity about opium inexhaustible. First in Paderborn and later in Einbeck in Hanover, still earning a modest living as a pharmacist, he performed hundreds of experiments purifying and characterising active ingredients and trying their effect on himself. In 1806 he published fifty-seven experiments in the *Journal der pharmazie.* Twenty described the properties of meconic acid, which he termed poppy acid. He then discovered that poppy acid had a significantly different effect on blue plant pigment than the parent opium. He rightly deduced that this was due to something non-acidic in his material. This might strike the modern reader as unremarkable. It was revolutionary in its day. Most organic compounds, especially those with a biological action, were firmly believed to be acids. Their acidity was in fact the key to their effect. Sertürner was bluntly told by the famous Professor von Roon of Cologne that his dedicated analytical work was commendable but Narrisch.

In private Sertürner was a shy, retiring and weedy-looking individual; but when it came to chemical experiments he possessed divine self-confidence. When he eventually came round to studying the chlorate precipitates of infusions of opium he hit gold – or better. He isolated a new and almost pure compound which was not only alkaline but also at least ten times more potent than opium. He first named his discovery 'principium somniferum', or sleep-inducing substance. Then, succumbing to the lure of classical Greek which the great Johann Joachim Winckelmann and Goethe had recently unleashed on their countrymen, he renamed it 'the shape of dreams' or, more concisely 'shapes'.[4] The word, slightly corrupted, is known in the English-speaking world today as morphine.

...

For years Sertürner was ignored. He had no professional standing and he was a windy writer. He was not perturbed. He continued to experiment on

4. Contrary to what is stated in almost every textbook of anaesthesia, pharmacology and medicine, he did not derive the name from Morpheus, a problematic god of dreams, but from μορφη shapes (as in morphology). The official name in the British pharmacopoeia was changed to morphina in 1885.

himself. He discovered most of the important effects of morphine, from euphoria to depression, from nausea to constipation, and, most strikingly, its capacity to dull pain. He even realised that morphine did not so much suppress pain as make the sufferer 'beautifully indifferent to it'. It was an extraordinary insight, worthy of Watteau's superb representation of the same effect in art.[5] Sertürner almost certainly became addicted and was aware of the dangers. It was the risk which partly at least impelled him to continue to press his discovery on the scientific establishment. In 1812 he wrote: 'I consider it my duty to attract attention to the terrible effects of this new substance, called morphium, in order that calamity may be averted'. In 1816 he wrote that 'chemists and physicists will find that my observations largely explain the remarkable activity of opium and that I have enriched chemistry with a new acid, meconic, and a new substance which is distinctly alkaline and which I have called morphium.'[6] Oddly perhaps, he did not seem to envisage that his discovery might have medical repercussions as well.

In 1817 Sertürner's publications at last attracted the attention of Joseph Louis Gay-Lussac, professor of chemistry at the Ecole polytechnique in Paris. Gay-Lussac was already famous for his balloon ascents to study the properties of the atmosphere as well as for a string of important chemical papers. In collaboration with the German naturalist and traveller, Baron Alexander von Humboldt, he had recently discovered that water was the product of oxygen and hydrogen in the ratio of 1 : 2.[7] Only five years older than Sertürner, Gay-Lussac was outraged that the discoveries of 'an unschooled but exceptionally gifted individual' had been ignored by his 'blinkered' colleagues. Even then it took another fourteen years before the Institut awarded the German pharmacist a 2,000-franc prize for 'having opened the way to important medical discoveries by his isolation of morphine and the exposition of its alkaline character'.[8]

Sertürner's discovery sparked a frantic search for other 'alkaloids': naturally occurring biologically active substances of an alkaline nature.[9] 'The spirit of Lavoisier is alive in France again,' the chemist and politician Jean Marchais

5. To the present writer *L'Indifférent* in the Louvre is the perfect representation of the indifference to pain induced by opium and later by morphine. Watteau suffered from tuberculosis and was in constant pain (probably from tuberculous laryngitis) in the last years of his tragically short life. The London physician Mead treated him with opium, and he went on painting till his last days. *L'Indifférent* was one of his last works.
6. Quoted by M. Booth in *Opium*, p. 58.
7. Gay-Lussac later moved to the Sorbonne and then to the Jardin des Plantes. His chemical investigations brought him into direct rivalry with Davy and both were stimulated by it. Gay-Lussac died in 1850, aged seventy-two.
8. It was one of the lesser prizes but the only one going to a non-French chemist.
9. Another awful hybrid from the Arabic 'alkali' and the Greek εἶδος *eidos*, meaning 'like', alleged to have been coined by Liebig in the 1820s.

exulted in 1825. It was a chemical flowering that would lead directly to Pasteur. Strychnine was isolated in 1817, caffeine in 1820, nicotine in 1828, aconitine in 1830, atropine in 1832. The excitement eventually touched the medical profession, or at least a new elite of scientific physicians. François Magendie was one. His *Formulaire pour la préparation et l'emploi de plusieurs nouveau médicaments* was brilliant and practical.[10]

Though official recognition came late, Sertürner's work was also expanded.[11] Two French chemists, Joseph Caventou and Pierre-Joseph Pelletier, developed a comparatively cheap and simple process for producing pure morphine. Magendie published an account of the first case where the new substance was applied therapeutically. His patient was a young woman with an aortic aneurysm 'who, for some years, had sought relief from her constant and gnawing pain in vain from charlatans, pharmacists, magnetisers and herbalists'. Aneurysms eroding bone could be diabolical. Since she claimed that opium in any form made her sick, Magendie gave her two kinds of morphine pills, one stronger containing the sulphate and one weaker containing the acetate. Both calmed her symptoms and 'made her life tolerable, even happy, even though her condition remained incurable and she died nine months later'. Magendie went on to prove the palliative potential of the new drug in other patients. It was a historic moment.

Yet it is possible that for some years after Magendie morphine was used as widely for suicide and homicide as for relieving pain. After a sensational case of multiple murders in Lyons – a whole family was exterminated by a greedy heir – Balzac introduced the drug into literature. In *Comédie du diable* the Devil claims that it had caused a sudden increase in the population of Hell. This was no deterrent above ground. By the 1830s morphine was available commercially. A translation of Magendie's work into English in 1828 gave it an entrée to Britain and the United States. For the first time doctors could prescribe more or less accurate doses of a painkiller that was at least ten times more potent than opium. Other opium alkaloids were soon isolated. Robiquet identified codeine in 1832 and Thioboumery and Pelletier discovered thebaine two years later.

. . .

Yet morphine did not entirely replace opium for several decades. Jonathan Pereira's British textbook on therapeutics published in 1854 still described

10. François Magendie made numerous contributions to both physiology and medicine, most notably perhaps completing the work of Charles Bell, identifying and mapping out the pathways of the sensory and motor nerves. He died in 1855, aged seventy-two.
11. He never sought or gained much academic recognition. He died in 1841, aged fifty-eight.

opium as 'by far the most valuable remedy available against pain: its effect is immediate, direct and obvious and its operation unattended by severe complications'. A few doctors and pioneering ladies of the temperance movement warned against the danger of addiction. They were ignored.

...

In the meantime the search for administering morphine by some route other than by mouth continued. 'Clysters' or enemas had already been tried with opium; now morphine suppositories coated with wax or animal fats enjoyed a passing popularity. Inhalation was too time-consuming and tended to induce nausea. Skin patches were tried, as well as the direct application to raw flesh. But morphine, even more than opium, could cause blistering. Something else was needed; and something else turned up.

It was the coming together of two strands of ingenuity. First, the invention of the syringe (or some implement like it) has long been attributed to Hero of Alexandria. He was the mechanical genius of Antiquity, inventor of the steam engine and the water clock.[12] Why not the syringe? Second, in 1830 a Dr Hyacinth Lafarge of the little town of Saint-Emilion suggested a way of introducing morphine directly into the tissues. He dipped a pointed lancet into a morphine solution, pushed it under the skin and left it there for a few minutes. He later suggested that tiny pills could be pushed into the subcutaneous tissues. In 1832 Dr Isaac Taylor in America started to introduce drugs through an incision by means of a blunt-nozzled syringe. Dr Francis Rynd of Dublin injected drugs with a gravity-fed bottle connected to a hollow needle. The classic glass syringe with metal plunger attached to a sharp pointed hollow needle finally assumed its modern form in the hands of a French surgeon, Dr Charles-Gabriel Pravaz in 1851. Though (like Dr Guillotine) a mild-mannered and philanthropic practitioner, his splendidly sonorous and faintly sinister-sounding surname name became a synonym of the morphine habit on the Continent. In Britain the independent inventor was Dr Alexander Wood of Edinburgh. In 1851 he ordered a glass syringe made to his specification from the famous instrument makers, Ferguson's of Southwark. Wood's design was modified and improved by the addition of a locking device a year later by Dr Charles Hunter of St George's Hospital, London. Hunter also became an ardent advocate of what he described his 'ipodermic method'. In 1856 Dr Fordyce Barker of New York was giving the first hypodermic injections in the United States.

Both Wood and Hunter assumed that morphine by the hypodermic route would not only act more quickly and eliminate the unpleasant gastric side-

12. The steam engine was to drive a fountain. He also discovered the formula which expresses the area of a triangle in terms of its sides.

effects but would also prevent the 'considerable inconvenience' of an 'appetite' developing. The term 'appetite' hints at the belief, widely shared at the time, that craving for the new drug was no different from hunger or, at worst, a 'sharp need' for alcohol. 'Remove the act of swallowing,' Wood maintained, 'and the hunger would be assuaged.' Firm in this belief, Wood himself and others following his assurance began to use morphine injections often and at frequent intervals to treat every kind of pain, from 'rheumatism' to inflamed eyes and from dysmenorrhoea to delirium tremens. Even a mild hangover or passing headache provided an indication. Disastrously, some doctors taught their patients to inject themselves. Many probably needed little tuition. The manoeuvre was simple and virtually painless. Small syringes, some with prettily engraved silver plungers, could be bought from a range of shops, including a famous firm of Regent Street jewellers. Morphine itself began to be marketed in convenient glass phials. The gear could be accommodated in the handle of an elegant walking stick. Many of the elegant walkers became addicts. The condition soon became widely known as 'morphinism'. In 1854 Mrs Wood became one of the first victims of an accidental overdose.

. . .

From London, Paris and New York morphinism spread across Europe and the United States. Like opium, it remained for some years a habit of the well-to-do. It was slightly more expensive than opium. There also existed a not so slight difference in effect. Compared to opium, morphine acquired the reputation of being a superior painkiller. In truth the effect depended on dosage and route of administration. The doses of morphine injected were by and large significantly higher than those consumed by old-fashioned opium addicts. An ounce of laudanum contained an average of 1 grain of morphine; so a laudanum addict dosing himself on 2 ounces a day (and Coleridge even in an advanced stage of his addiction took only about two pints a week) took the equivalent of 2 grains. Those developing morphinism not infrequently injected 6–10 grains and some up to 40 grains daily.[13] Laudanum was also often poorly absorbed, especially in addicts with gastritis. Hypodermic morphine bypassed the stomach and the gut. No doubt the new drug was more powerful against pain. It was also more lethal.

. . .

Historical developments tend to find their characteristic and sustaining drug. New heights in morphinism were reached in those fighting in the savage and

13. Approximate equivalents: 1 grain = 60 milligrams; 1 fluid ounce = 29.5 millilitres. 2 pints = 1 litre.

prolonged wars of the second half of the century. Morphine addiction became an epidemic in the Crimea. It was widely described as a blessing. Perhaps it was. The drug was extensively used in both armies fighting in the Civil War in the United States. The wars eventually ended, the habits often did not. It was the first surfacing of an ongoing tragedy.

# Chapter 27
# MR ANAESTHETIST

There is a temptation both to exaggerate and to disbelieve the achievement of the eminent Victorians. As writers of fiction, statesmen, scientists, social reformers or African explorers, they laboured in many fields. Even allowing for the blessed absence of search engines, computer-generated bibliographies, peer reviews and other modern encumbrances, the bulk of what they accomplished is intimidating. In medicine, accomplishment is difficult to measure; but few today could leave an indelible mark on two specialties while working in a third. John Snow was, in more than one sense, the first anaesthetist. He was also a pioneer of epidemiology. In anaesthesia he became a role model. His work on the spread of cholera helped to save thousands, perhaps millions of lives. To him the two fields were of equal importance. While pursuing them, he was also a busy general practitioner. Even in a book on the relief of pain, to ignore his achievements outside anaesthesia would be to diminish him. He deserves better.

## *I*

John Snow was born on 15 March 1813 into a poor family of Yorkshire craftsmen and had to struggle to obtain a medical education.[1] He never tried to disguise his humble origins. He was plain in speech, dress and habits. He never married and, so far as is known, had no romantic love affairs. His friends were few. He was admired for his brains, hard work and integrity; but he had little charisma. He died at the age of forty-five after a stroke, probably secondary to hypertensive kidney failure; but the post-mortem also showed

1. John's father, William, was a carter who later became a farmer near York. His mother, Frances, was the illegitimate daughter of Mary Ashkham who later married the girl's father, Charles Empson. Empson was a wealthy man and a leading figure in Bath society who guided Snow on several occasions and helped the family financially. Snow was devoted to him.

evidence of tuberculosis. The last was the great romantic killer of the age, but if Snow coughed, he coughed in private. He died as he lived, without any drama.[2] Not surprisingly perhaps, he never attracted multi-volume biographers as did many other Victorian worthies.[3]

He was not a Quaker but he may have been influenced by the slightly austere life-style and reverence for learning of the Quaker school which he attended.[4] But after leaving school he was too poor to take the high road to medicine – a degree offered by one of the ancient English universities, a university in Scotland or abroad.[5] At fifteen he was apprenticed to a busy general practitioner, Dr William Hardcastle, of Newcastle. His years there provided him with his first experience of cholera. Unlike York, an ancient and wealthy cathedral city, Newcastle was a busy port and industrial sprawl, invariably hard hit by pestilences imported from distant lands. The cholera epidemic of 1831–32 struck here first in Britain, killing 178 people out of the 320 who fell ill with it during the first week. Eventually nearly 22,000 deaths were to be reported from England and Wales. There was no treatment and nobody had any idea how the disease spread or how the spread might be prevented, contained or cured. Snow was horrified by the ravages among the miners and their families who were the majority of his patients.[6]

2. He was unwell on occasions during the winter of 1857 and spring of 1858 but ignored the symptoms (mainly gastrointestinal, so far as one can tell from his correspondence). On 9 June 1858 he developed severe headache and his housekeeper called in two colleagues, Drs Budd and Murchison. They diagnosed a cerebral haemorrhage. He died on 16 June. He was buried in Brompton cemetery.
3. The eminent Victorian physician, Sir Benjamin Ward Richardson, wrote two memoirs of Snow as well as editing Snow's unpublished work after Snow's death. Neither memoir is free from error, apart from being written by an unreserved admirer and friend. The first modern biography is that by David A.E. Shephard. What will undoubtedly be regarded as the 'definitive' life and work for many years by Vinten-Johansen and colleagues was published recently: *Cholera, Chloroform and the Science of Medicine* (Oxford, 2003).
4. His later preference for abstinence and vegetarianism, as well as his studiousness and industry, may also have been nurtured by the school. He remained an evangelical Christian all his life.
5. University College, including University College Hospital, London, was the first university college in England which did not require students or staff to declare their acceptance of the Thirty-Nine Articles, the doctrinal foundation of the Church of England. It did not start accepting medical students till the 1840s. Outstanding Quaker physicians like John Coakley Lettsom and Thomas Hogdkin qualified in Edinburgh or in Leyden in Holland.
6. To many cholera, a new disease in Europe, struck at the core of Victorian faith in progress. To others it was a 'visitation from God', its cause not poverty and poor sanitation but, in the words of J.C. Ryle, punishment for 'our national sins of drunkenness, Sabbath breaking and infidelity' (quoted in L. Bending, *The Representation of Bodily Pain in Late Nineteenth-Century English Culture* (Oxford, 2000), p. 17). Despite Snow's evangelical faith, some attacks on him would always come from the extreme evangelical wing of the Church.

Teetotal, vegetarian and a serious Christian, young Snow worked hard. As his admiring biographer, Benjamin Ward Richardson, later put it, 'already his habits puzzled his landladies, shocked the cooks and astonished the children . . . but his kindness won them over and his culinary peculiarities were always accepted'. Samuel Smiles, a near-contemporary and a doctor like Snow, wrote in his famous *Self-Help:*

> Nothing credible can be accepted without application and diligence . . . the student must not be daunted by difficulties but conquer them by patience and perseverance – and, above all, he must seek elevation of character without which achievement is worthless and worldly success is naught.[7]

It was a precept Snow would have endorsed.

. . .

In the summer of 1836 he walked the 200 miles from York to London (just as his tuberculous predecessor in medicine, Laennec, walked about the same distance from Rouen to Paris at the outset of his career). On the path to medicine but with meagre funds, the licentiateship of the Society of Apothecaries, without which a doctor could not prescribe or dispense drugs, and membership of the Royal College of Surgeons were the next steps. He may also have read the editorial admonition printed in the *Lancet* of 24 September 1836:

> In its results, the study of the science of medicine yields to none, whether viewed in relation to its enabling conceptions or the well-being of society in the everyday walks of life. It includes that most useful of researches, the study of mankind; whence the mind is led throughout an ever-varying and ever-instructing path to the great beneficent cause of Nature. It teaches us not only to discover but also to promulgate truth and it humanises our feelings by prompting us, on all occasions, to direct our knowledge to the alleviation of human suffering.[8]

Medical leader-writers did not yet shrink from the beautiful phrase or the uplifting sentiment. And this too Snow may have taken to heart.

He took lodgings in Soho, a working-class district at the time, where he would stay and where his practice would remain for the rest of his life. With his savings he entered and attended the famous Hunterian medical school at 16 Great Windmill Street.[9] He also carried out his first research, investigating

7. S. Smiles, *Self-Help* (1857; revised and enlarged edn New York, 1943), p. vi.
8. *Lancet*, 1836 (2), 7.
9. The school had been founded by William Hunter, the famous anatomist and obstetrician (brother of John) in 1746. It remained a private institution named after him. See D'A. Power, 'The Medical Institutions of London. The Rise and Fall of the Private Medical Schools', *British Medical Journal*, 1895 (1), 388.

the mechanism of the toxicity of arsenic. He then enrolled as a student at Westminster Hospital; and, after qualifying, set up in practice in 54 Frith Street. (In 1853 he moved to slightly more upmarket 18 Sackville Street.) He graduated MB London in 1843 and a year later passed the difficult MD examination. While busy looking after patients, he also experimented with the effects of carbon dioxide. He became a member of the Westminster Medical Society, later joined to the Medical Society of London, and his research findings on a diversity of topics were mostly presented at their meetings.

...

A comparatively mild form of cholera (if it was that), referred to in Britain as 'English cholera', had been endemic in Europe for centuries. The much fiercer 'Asiatic' variety, which had long been known and feared in India, first reached Europe in around 1817. Moving slowly westwards, it affected millions and killed about a quarter of its victims. In Hungary alone, one of the first countries to be hit, 100,000 succumbed out of about a quarter of a million affected. As with most other epidemics, ideas about its origin and mode of spread proliferated. None were helpful. By the 1830s speculation had largely given way to fatalistic acceptance. The best that could be tried was to get out of the way of the disease by guessing its route and moving at speed. The debate between 'contagionists' who believed in the importance of personal contact and the 'environmentalists' who postulated an invisible 'miasma', 'putrid emanations', 'lunar influences' or 'cosmic radiations' flared up from time to time. No clear winners emerged. Living microbes had already been demonstrated by amateur microscopists; and their wriggling in pond water and skin scrapings had been watched with delight. Nobody as yet had connected them with disease. There was no effective treatment. This meant that remedies proliferated. Blood-letting had been the great therapeutic nostrum for centuries in every form of illness; and in cholera patients whose blood quickly became 'tarry thick, dark and sticky' it seemed to many eminently sensible. But as the disease almost visibly 'dried up' the bodies of its victims, there was soon nothing left to bleed. A surgeon by the name of Holmes left a picture of a patient called William Sproat, possibly the first victim of the great 1831 outbreak:

> The pulse scarcely beating under the fingers, countenance shrunk, eyes sunk, lips dark blue as well as the skin of the lower extremities, the nails livid; too weak to talk . . . Yet the eyes wide open, staring, patient conscious, aghast and terrified.[10]

10. Quoted in N. Longmate, *King Cholera* (London, 1966), p. 54.

It was the heart-rending image of extreme dehydration, what was sometimes referred to as 'the blue death'. Opium in massive doses and administered early did help sometimes; but it could also hasten death. The same could be said of an astonishing variety of domestic remedies – hot bricks, linseed, mustard and bran poultices, plasters of hayseed, warm sand and salt, peppermint and, inevitably, gin and brandy. Some as early as the 1830s realised that fluid loss was the immediate cause of death and advocated its replacement; but nobody knew how to set about it.[11]

...

Assessments of Snow's contribution to the understanding of the disease tend to focus on the last scene, the Broad Street outbreak of 1854. It remains one of the most satisfying tableaux in the history of medicine. Too satisfying, perhaps. Recent research has cast doubt on many of the details and has provided overdue corrections.[12] But it was certainly the climax of many years of painstaking labour. While working as a general practitioner and as an anaesthetist, Snow was collecting and analysing statistics, publishing papers and writing letters and memoranda to ministries, committees, working parties and medical associations. He was also lecturing to learned societies about the disease which he came to regard as his life's mission to understand and to defeat. Cholera visited London again in 1848–49 and it was the analysis of the beginning and spread of the illness which became the foundation of Snow's ideas.

On 22 September 1848 the steamer *Elbe* from Hamburg docked in London. A seaman, John Harnold, left the ship and took lodgings in a district known as Horsleydown. Next day he fell ill with cholera. Three days later he was dead. A man named Blenkinsop came to lodge in the same room two days later. He developed cholera eight days after Harnold and was dead within four days. A Dr Russell who had treated both patients reported on ten subsequent cases in Thomas Street, a slum area 'a little way removed from the two preceding deaths'. This intrigued Snow.

> Though there was no apparent connection, a closer study [which meant exploring the area house by house] showed that an open sewer, up which the tide flows, runs past both places, and the sewage from the houses of the

11. The suggestion was first made by William O'Shaughnessy in 1831 and an infusion of salt solution into a vein was tried by Thomas Latta in 1832 and reported in the *Lancet*, 1832 (2), 274. It was fraught with difficulties and abandoned until the early twentieth century.
12. It has been provided by H. Brody, P. Vinten-Johansen, N. Paneth and P. Rachman, 'Map-making and Myth-making in Broad Street: The London Cholera Epidemic of 1854', *Lancet*, 2000 (356), 64. See also J.M. Eyler, 'The Changing Assessment of John Snow and William Farr's Cholera Studies', *Sozial und Präventivmedizin*, 2001 (46), 225.

> first neighbourhood [where Harnold and Blenkinsop died] is carried past all houses of the second outbreak.[13]

It was probably the first time that cholera – or indeed any disease – was seriously suspected of being spread by water. Water was not of course the only means. The causative agent, whatever it was, could also be spread by 'fomites' – bedlinen, towels, articles of clothing – provided they were moist. But from now on contaminated water was the focus of Snow's thinking.

Snow's *Mode of Communication of Cholera* was published on 29 August 1849 and in an embryonic form it contained all that needed to be known about the spread of the illness. The booklet went beyond aetiology.

> Cholera might be checked and kept at bay by sanitary measures that would not interfere with social or commercial intercourse: the enemy would be shorn of its chief terrors. It is contaminated water which most often carries the causative agent . . .[14]

This was written twenty years before Pasteur established the potential pathogeny of micro-organisms and eighty years before the discovery of the cholera vibrio by Koch.

While 'the causative agent' would eventually have to be identified, cleanliness was crucial. Though in a different context, Snow made his discovery only a year after Semmelweis in Vienna realised the rôle of unwashed hands in the causation of puerperal fever. In ignorance of that far-away event, Snow wrote in the first edition of his monograph:

13. J. Snow, 'On the Prevention of Cholera', *London Medical Gazette*, 1853 (3), 559, p. 32.
It was during the 1848–49 outbreak that *The Times* published a letter from an address not far from Snow's home. The letter was, at the time, widely attributed to a social reformer pretending to be semi-literate; but the newspaper checked the details and attested to its veracity. It is more eloquent than any later reconstruction:

> Sur, – May we beg and beseech your protecshion and power. We are Sur, as it may be, livin in a Wilderness, so far as the rest of London knows anything of us, or as the rich and great people care about. We live in muck and filth. We aint got no priviz, no dust bins, no drains, no water suplies and no drain or suer in the hole place . . . The stench of a Gulley-hole is disgusting. Some gentlemen comed yesterday, and we thought they was comishioners from the Suer Company, but they was only complaining of the noosance and stenche of our lanes and corts was to them in New Oxforde Street. They was much surprised to see the sellar in No 12 Carrier Str, in our lane, where a child was dyin of fever and would not believe that Sixty persons sleep in it every night. This here seller you couldent swing a cat in, and the rent is five shillings a week . . . We all of us suffer, and numbers are ill, and if the Cholera comes Lord help us . . . (5 July 1949)

The cholera did come and Snow's work was perhaps one manifestation of the Lord's help.

14. J. Snow, *The Mode of Communication of Cholera* (London, 1849), p. 67.

> For all persons attending or waiting on the patients it is essential to wash their hands carefully and frequently, never omitting to do so before touching food.[15]

He recognised the role of poverty and overcrowding; but it was the sanitary consequences which determined the spread of the disease.

> In Thomas Street, Horsleydown, there are two courts close together. The South side is occupied by Truscott Court, the north side by Surrey Buildings. Both are occupied by poor people, mostly sleeping four to eight in a room . . . The courts are separated by a small intervening space in which are situated the privies of both courts. They communicate with the same drain and there is an open sewer which passes the further end of both courts . . . After a girl and then an elderly woman contracted cholera in the Surrey Building the fateful damage was done. Evacuations were passed into the bed and the water in which the foul linen would be washed would be emptied into a channel which communicated with the water people in Surrey Buildings drank . . . I found the top of the overflow-drain from the cesspool belonging to these Buildings only a few inches above the water tank . . . The overflow drain had no bottom and it crossed at right angles the earthenware pipe of the water tank the joints of which were leaky . . . In Surrey Buildings the disease caused fearful devastation . . . killing 76 out of 110 inhabitants, 52 out of the 61 children under ten . . . In adjoining Truscott Court where the outflow of the cesspool did not cross a water pipe the deaths numbered only 8 out of 92.[16]

Even such a small sample may convey the extraordinary – in his day almost certainly unique – style in which Snow carried out his investigations. He worked single-handed, ferreting around under difficult and depressing conditions. He was surrounded by squalor and must often have met hostility. Nobody paid his expenses. On his published pamphlets he invariably made a loss. Yet no detail escaped his attention; nothing difficult was fudged. In addition to looking at the physical evidence, water bills and receipts were scrutinised and the multiplicity of water companies that provided drinking water even in the same district were identified.

Altogether he studied eight outbreaks in and around London. Statistical analysis of the figures must have taken him hours of concentrated labour. Yet first results were frequently reviewed, reconsidered and checked. From 1850 onwards he also performed chemical analyses, mainly measuring the sodium concentration of the water samples, to help identify their source.

15. J. Snow, *On Cholera*, p. 46.
16. Ibid., p. 124.

He usually communicated his findings and conclusions to the Westminster and later to the London Medical Society. He was listened to with respect – it was impossible not to be impressed by his industry – and was eventually elected orator and then president of the Society. But he was not a spell-binding communicator, and his message was less than universally understood. What was needed for the importance of his findings to become accepted was a piece of drama. It was provided by the Broad Street outbreak of 1854.

. . .

In 1854 the cholera struck barely a hundred yards from Snow's home. It involved only a small area but it was of exceptional virulence. Five hundred victims were carried off within the first ten days. Inhabitants of the area fled in panic, 'much as in medieval plagues', *The Times* commented.[17] The credit for tracing the source to 40 Broad Street must go to the local curate, the Reverend Henry Whitehead. He wrote in his later report that he had started his investigation 'deeply sceptical of Dr Snow's seemingly somewhat arid numerical approach to disease, almost wishing to disprove his ideas'. In the event, the investigations provided the 'first unequivocal proof that these ideas were entirely sound'.

Most localised outbreaks tended to abate after three or four days. This one was still increasing in violence a week after its onset. One vestryman of St James's described the atmosphere in the council chamber on 7 September as that of mortal sinners awaiting Doomsday. Something had to be done – but nobody had any idea what that something should be. Then a local doctor asked for a hearing. As one vestryman later wrote, 'we would have admitted Satan himself if he had offered us any hope'.[18] Snow was humble but firm. If the infection was to be arrested, the handle of the water pump in Broad Street had to be put out of action. A vestryman objected that to do so would deprive numerous houses and workshops of their drinking water. That was precisely the reason why Snow wanted its handle removed. He explained that a cartridge factory which drew tubs of drinking water from the pump had been ravaged. A brewery next door whose workmen drank malt liquor instead of water had been spared. An almshouse surrounded by houses which had been hard hit had its own well. It was spared. Two out-of-London visitors had been traced who had died of cholera having drunk a single tumblerful of water from the pump. There was more evidence, all pointing in the same direction. The conclusion to be drawn was clear. At least it was clear to Snow. Most vestrymen confessed that they were sceptical. Fortunately, they had no choice.

17. *Times*, 6 September 1854.
18. It was Sir Harry Peachey, owner of the pharmaceutical business which marketed Peachey's drops against dysentery and cholera.

On 8 September at dawn the pump handle was removed. The epidemic started to abate.[19]

## II

The man who dealt with pump handles also laid the scientific foundations of modern anaesthesia. He brought to the second task the same surprising rigour as he did to the first. Among those surprised was Sir James Young Simpson. He did not entirely approve of dwelling on the possible complications of chloroform.[20] But he was big-hearted enough to give Snow a mention in the article he wrote for the *Encyclopaedia Britannica*.[21]

Snow's interest in gases in fact antedated his preoccupation with epidemics. In his first published article in the *London Medical Gazette*, 'Asphyxia and the Resuscitation of the Newborn', he suggested that when applying artificial respiration to neonates a double air pump should be used, one to draw off and expel expired air and the other to pump fresh air or oxygen into the lungs. He emphasised the need for oxygen in the neonatal lung as he later did in anaesthesia.

Yet how exactly he became involved with anaesthetic practice is uncertain. He never wrote the memoirs or reminiscences so popular with eminent Victorians. He probably regarded private landmarks as of no general interest. For the historical record, one biographer states that Snow was present at the first ether anaesthetic administered by Boott. This is unlikely. On the other hand, James Robinson, the dentist, became an enthusiast and extracted at least fifty teeth under ether anaesthesia over the next few months. He was not shy in talking about his successes both to prospective patients and to colleagues and his first case reports were incorporated in a short monograph within six months.[22] Snow almost certainly attended one of the early sessions. His devout biographer, Benjamin Ward Richardson, on the other hand, relates that, emerging from his home one winter morning in 1847, Snow

> met a druggist of his acquaintance and wished him his usual friendly 'Good morning'. Holding a large ether inhaler under his arm, the druggist replied, 'Good morning to you, doctor! But don't detain me. I am wanted to give

19. Providence helped. The epidemic was already on the wane.
20. Nor did he think much of the new science of statistics. A 'double-blind' trial would probably have meant to him keeping *both* eyes closed while sniffing a new anaesthetic agent.
21. He was asked to write the entry on 'Anaesthesia' but managed to do so without mentioning ether or nitrous oxide once.
22. J.A. Robinson, *A Treatise on the Inhalation of the Vapour of Ether, for the Prevention of Pain in Surgical Operations* (London, 1847).

ether urgently. Indeed, I am wanted to give ether here, there and everywhere, all urgently. I am quite getting into the ether practice. Good morning doctor.' An ether practice! Snow had devised ether inhalers for years for his experiments with animals. The possible practical applications had not occurred to him. This was something he had to explore.[23]

Whatever started him off, as early as January 1847 he asked for permission to administer ether to outpatients needing tooth extractions at St George's Hospital at Hyde Park Corner. The surgeon, John Fuller, stood by, watched and was impressed. He mentioned Snow and his improved inhaler to his colleague, Thomas Cutler, who had had an unfortunate experience with ether the day before – the patient had had a fit of coughing and both the anaesthetic and the operation had to be aborted – and he was planning a major operation without ether the next day. But he invited Snow to try his hand and was won over. From then on Snow was regularly called in to administer ether on operating days at St George's. Little of importance in the surgical world escaped the notice of Robert Liston of University College Hospital. He too invited Snow to anaesthetise for him. Quickly the two became a team, and remained one until Liston's premature death nine months later. Snow's soft-spoken methodical approach proved the perfect foil for the great surgeon's flamboyance. Within a year Snow published his first monograph. *On Ether* was a landmark, reviewing in full detail 52 anaesthetics administered at St George's Hospital and 23 at University College Hospital.[24] Most operations were superficial; but some were major amputations. Six of the patients died but none under the anaesthetic. All were followed up, until their death or discharge from hospital. The book confirmed Snow's position as an authority. Soon news came from Edinburgh of the first operations performed under chloroform. Snow at once gave it a trial, and concluded that in most cases it was to be preferred to ether. He communicated his experiences, both clinical and experimental, in several papers; but his *On Chloroform*, a comprehensive treatise running to 900 pages, was not published until 1858, after his death.[25]

. . .

Within a year Snow was the busiest anaesthetist in London. This was not simply the result of his publications. Circumstances and character traits combined to bring it about.

Unlike the first experimenters with ether in the United States and on the Continent, Snow's approach was from the start dispassionate and scientific.

23. B.W. Richardson, 'The Life of John Snow, M.D.' in *J.Snow, On Chloroform and Other Anaesthetics: Their Action and Administration* (London, 1858; reprinted Park Ridge, Illinois, 1988).
24. J. Snow, *On Ether* (London, 1847).
25. The final sentences to the book were penned two days before Snow's death.

Many of his surgical colleagues probably had only a hazy grasp of the gas concentrations and dead spaces which Snow was discoursing about; but few failed to be impressed. That 'harmless little man' (as Liston once referred to him) was clearly not in the business of promoting a new invention. He was not after riches and was not thirsting for glory. He had no bees in his bonnet – or only one. If anaesthesia was to be more than a passing and dangerous fad, a scientific approach was essential. The questions which Snow set out to answer and often actually answered make a formidable list, and one still worth rehearsing.

He compared and defined the physical, chemical and pharmacological characteristics of a group of volatile anaesthetic agents. He proposed and established the inverse relationship between their anaesthetic potency and their solubility in blood.[26]

He showed that the concentration of the anaesthetic agents in blood determined the depth of the anaesthesia. Depth in turn determined not only the effectiveness of the agent as a painkiller but also the risks involved. The main risk, in Snow's dry prose, was the 'irreversibility' of the anaesthetic effect.

He demonstrated the importance of the ambient temperature. Since the solubility of gases in liquids is temperature dependent, this might have seemed obvious. Yet it had been ignored. The ambient temperature also influenced the rate of vaporisation of the anaesthetic liquids. This too had been neglected.

Vaporisation was also affected by mechanical factors: hence the importance of designing efficient apparatus. In particular, Snow demonstrated the potential influence of the 'dead space' both in the anaesthetic apparatus and in the respiratory tract. Nobody before him appears to have given this a thought. The effective elimination of carbon dioxide from the anaesthetic circuit as well as from the body could be critical.

He identified the main physiological characteristics of a number of anaesthetic gases, including those of ether and chloroform. He made no bones about the fact that each and every one was potentially lethal. Complications and fatalities would inevitably occur. If the dangers were to be minimised, every death should be investigated. Less dramatic complications should be documented. His insistence on the high quality of the anaesthetic liquids had an unforeseen effect. Ether and chloroform were the first drugs where purity

26. This is now recognised as only one of several important variables.

not only mattered but was, according to Snow, critical. Perhaps it was critical in other drugs too.

Many of those who administer anaesthetics the world over today – and world-wide only a small minority actually specialise in the field – perform their task without being aware of Snow's discoveries, let alone his name. Yet the safety of their practice depends on knowingly or unknowingly following his advice and taking note of his injunctions.

...

If Snow was not in the business of promoting a new invention, he was even less in the business of promoting himself. In and out of the operating theatres surgeons tend to be the lead actors in a drama that is centred on the usually passive figure of the patient. The anaesthetist can play a useful supporting role. The fact that Snow was eagerly sought – indeed fought over – by the leading surgeons of London suggests that he filled this role well. It was almost certainly not one which he consciously assumed. For anaesthetists it rarely is. The aptitude was part of his nature. He became, in other words, the first professional anaesthetist not only because of his expertise but also because of his character.

But while Snow was recognisably a modern anaesthetist, the circumstances under which he laboured were primitive. Before the operation, he never saw the patients whom he was to anaesthetise; and there was neither time nor opportunity to prepare them for the anaesthetic. Many were far advanced in their disease and would today be considered grave risks. He had no knowledge of their pre-operative blood pressure, their heart rate and rhythm, the oxygen and carbon dioxide concentration in their blood, the presence of a multitude of medical conditions which might adversely affect their response – alcoholism, liver failure, epilepsy, insanity, congenital heart disease, bronchitis, kidney disease or addiction to drugs, to list just a few. No X-ray or electrocardiography was available either before or during the operation. No blood tests. The level of anaesthetic gases in the blood (or of oxygen or carbon dioxide) could only be guessed. Tracheal intubation was unheard of. In short, like fighter pilots in the First World War, he flew solo a ramshackle biplane, much of the time being shot at. Yet even modern anaesthetists will recognise the almost unbearable tension generated by critical moments and drily but vividly recorded in his case-books.

In 1853 he anaesthetised a six-year-old child in whom an orbital tumour 'necessitated the removal of the greater part of an eyeball'.

> The usual inhaler was employed and when the child seemed sufficiently insensible withdrawn. Mr Travers [one of the assistant surgeons] introduced a needle through the globe of the eye to draw it forward with a liga-

> ture while Mr Reynolds [the operating surgeon] divided the eye. As the needle was introduced the child cried out and I poured more chloroform a little hastily without measuring it on a rather large sponge and placed it over the nostrils and mouth. It was pressed closer to the nose than I had intended or would have wished by Mr Travers's hand, but it was removed after the child gasped once or twice. The operation was quickly concluded without any further sign of feeling . . . but at the conclusion the boy's face was pale, his lips were turning blue and the limbs seemed inert. I tried to feel the pulse at the wrist but there was none. I did not immediately fear the worst since the boy was breathing and the chloroform had been left off for some time. But the pallor and blueness continued and the breathing became slow and embarrassed and then appeared to cease altogether, the pulse being still absent. The windows were quickly opened and cold water was dashed in the child's face. He made a few gasping inspirations now and then. It was not clear whether the cold water was good or bad since the inspirations did not immediately follow each application . . . Then the gasps became more frequent till the breathing was slowly re-established and colour began to return to the face and lips . . . Eventually the pulse restarted.[27]

Snow was undemonstrative by nature and often emphasised the need to remain calm; but on one occasion he was asked by an eminent and elderly surgeon: 'Anaesthetist . . . anaesthetist . . . that's our new speciality, isn't it? What's it like?' Snow replied: 'I would not recommend it, Sir. It's the speciality where you die with your patient.'

. . .

His self-control but also his commitment were among the qualities surgeons appreciated. In general he himself liked those who had been trained in the 'preanaesthetic' age. Even when the patient was insensible they operated with great speed. 'No need to replace the theatre clock with a calendar,' one old-timer quipped, on hearing about the marathon sessions of one of his younger colleagues. The formidable Sir William Fergusson of King's College Hospital (and of the aptly named 'Fergusson's lion forceps') was Snow's favourite.

27. J. Snow, *Case Books*, 3 vols (manuscript held by the Royal College of Physicians, London), Vol. II, p. 56. The rare but severe and occasionally fatal complications of chloroform anaesthesia, especially massive irreversible liver failure, were not recognised till the last decade of the nineteenth century and not fully described till the first decade of the twentieth. Had these delayed complications been recognised at the time the anaesthetic was introduced, its opponents would have had a field day; and it is doubtful if even Simpson's advocacy and Snow's support could have ensured its survival.

. . .

After the first wave of enthusiasm, surgical anaesthesia became a hard fought-over issue everywhere. One of Snow's cases decisively influenced the outcome, at least in the parts of the globe painted red on the map.

On 3 April 1853 he was summoned to Buckingham Palace for an audience with the Prince Consort. This required the donning of a court dress, including a ceremonial sword.[28] A little girl out with her mother stopped him as he left his house. 'Oh, doesn't the soldier look pretty?' she said. It was one of Snow's few recorded vanities that he regaled his friends with the story. By now Snow had many aristocratic patients. His case-books are records of a large working-class and middle-class practice peppered with dukes, marchionesses and the daughters of archbishops. The former he treated as their general practitioner; the latter he saw in his capacity as anaesthetist to the most fashionable surgeons in the capital. But their ailments and anaesthetics are recorded in the same flat prose.

At the palace Snow was surprised at how conversant the prince was with recent developments. His Royal Highness had even read some of Snow's papers. Snow gathered that the queen's obstetricians, Drs Locock and Fergusson, had suggested that Snow should be asked to anaesthetise her or at least to stand by. The queen's senior physician, Sir James Clark, was less enthusiastic. Clark was a walking medical calamity, but the queen would not act against his advice. In the light of conflicting counsel the prince wanted to check out the proposed anaesthetist himself. Of course there was no question of Snow seeing the queen before the confinement. But he might be called on the day. He was, at nine o'clock in the morning, just as he was setting out on his first round. He made detailed notes, as he did of all his cases.

> Administered chloroform to the Queen in her confinement this morning. Slight pains had been experienced since Sunday . . . Dr Locock was sent for about nine o'clock and he found that the *os uteri* had commenced to dilate a very little . . . I had remained in an apartment near that of the Queen along with Sir J. Clark who was amiable. At twenty past twelve by the clock in the Queen's apartment [which did not agree with the clock in the room where Snow had waited] I commenced to give a little chloroform with each pain, by pouring about 15 minims on a folded handkerchief. [He deliberately did not use a mask since he did not want to induce total loss of consciousness.] Her Majesty expressed great relief from the application, the pain being very trifling during the uterine contractions, while in between contractions there was complete ease . . . Dr Locock thought that the

28. It is not known whether the apparel was hired, borrowed or bought.

> chloroform prolonged the intervals between pains and retarded the labour somewhat. The infant was born after chloroform had been applied intermittently over a period of 53 minutes. The Queen appeared cheerful.[29]

She was in fact more than that. In her diary that evening she referred to 'that blessed chloroform' and described the effect as 'soothing, quieting and delightful beyond measure'.[30] When she saw that it was good, everybody – or everybody who mattered in the Empire upon which the sun never set – saw that it was good. Perhaps royalty have their uses.

29. Snow, *Case Books*, Vol. III, p. 58.
30. Quoted by C. Woodham-Smith, *Queen Victoria: Her Life and Times* (London, 1962), p. 342.

This was the birth of Prince Leopold, later Duke of Albany. Two years later Snow was asked again to anaesthetise the queen at the birth of Princess Beatrice. It was fortunate for the future of chloroform and of general anaesthesia that it was not known for some years that these were the two children of the queen in whom the haemophilia gene first became manifest (as a carrier gene in the case of the princess). Anaesthesia had of course nothing to do with this but non-relationships are difficult to disprove.

## Chapter 28

# CONFLICTING VIEWS

Anaesthesia did not abolish the pain of operations: it only made the pain of operations optional. To some and for some time the preferred option remained in doubt.

The benefits were obvious. Only a few days after Simpson's memorable dinner party James Miller, Liston's friend, performed the first operation under chloroform at Edinburgh's Royal Infirmary. His patient was a five-year-old Gaelic-speaking boy who needed a diseased 'white elbow' excised. Not understanding what was happening around him, the child was hysterical until he began to inhale the anaesthetic. He then quickly went to sleep and slept peacefully throughout the operation. It lasted only about ten minutes. When he woke up, he remembered nothing. 'Oh the dear boy! What a relief it's over,' Sheena McDonald, one of the nurses, sighed.[1] A few days later his parents took him home to their Highland croft and obscurity.

Some disadvantages of ether – its occasional ineffectiveness, the accompanying sickness, the chest complications – were already apparent. Chloroform seemed to remove many of these drawbacks. But there are few advances in medicine without complications. The first was unexpected and emerged within a few months. George Keith, one of Simpson's dinner guests, became addicted to the 'delightful' vapour. 'I had such a craving,' he later confessed, 'that I sometimes broke through all the rules of propriety in attempts to get possession of it.'[2] He eventually recovered, but remained 'neurotic and difficult to live with' for some time. Simpson regarded such lapses as revealing a flaw in the addict's character rather than a risk of the anaesthetic. It certainly

1. One of the significant side-effects of anaesthesia was to open the way for women to become members of the operating theatre team. This had been possible in an emergency or war; but preanaesthetic butchery in peacetime had been incompatible with Victorian womanhood. The way was now open even for woman surgeons. One of the first, Miss Emily Blackwell, sister of Elizabeth Blackwell, chose Sir James Simpson as her precentor.
2. G. Keith, *Fads of an Old Physician* (London, 1897), p. 200.

did not, in Simpson's view, amount to a contraindication. But soon there were graver effects to cause concern.

In January 1848 in Newcastle Hannah Greener, 'a fifteen-year-old bastard child . . . much thrust about by her family', needed to have her in-growing toenail removed.[3] She had already had one such operation, probably under ether; and this had left her with a 'heaviness in her head'. Her stepfather urged her that 'she had better suffer a bit of pain for a moment' rather than have 'her head made bad again'; but Hannah was adamant. She would not have the operation without being put to sleep. Dr Tom Meggison, their doctor and a kindly man, came to their home and agreed to give her chloroform. In front of a friendly fire she was put into a straight-back chair, sitting upright. Her stepfather held her foot. A teaspoonful of chloroform was poured on to a handkerchief and Meggison motioned the surgeon, a Mr Lloyd, to start operating. Lloyd made an incision but Hannah gave a ferocious kick. Thinking that the girl had been given too little anaesthetic, Meggison poured out more. Then he noticed that the patient's lips were turning white and then blue. There was frothing at her mouth. A few seconds later she stopped breathing. Then her pulse disappeared. The operation was abandoned. Nobody knew what to do. Hannah was stretched out on the floor. Meggison tried pummelling her chest. Then, in an attempt to bleed her, an arm vein and then the jugular vein in the neck were opened. No blood flowed from either. Brandy was poured into her mouth. There was a gurgle (as, in the absence of the swallowing reflex, the brandy presumably flowed straight into her lungs)[4] and she was dead. As happened with most early fatalities of anaesthesia, it was impossible to say what exactly the patient died from; but 'the trifling operation undoubtedly cost her life'.

Such events were rare, certainly rarer than death from shock or haemorrhage on the operating table before anaesthesia; but, then as now, deaths that could be attributed to treatment, especially to what many regarded as non-essential treatment, were harder to accept. The issue of the *Edinburgh Medical and Surgical Journal* which had carried a detailed account of the Boston operation, the first successful application of ether anaesthesia, had also published

3. The case was recounted in the London *Times* on 24 March 1847, and the *London Medical Gazette* on 11 February 1848.
4. Pouring fluid down the throat of unconscious patients was for long a comparatively common cause of anaesthetic death. Normally as soon as fluid or food enters the throat the trachea or windpipe is closed off by the epiglottis (one of the cartilages of the larynx), directing the fluid or food into the oesophagus or gullet. In deeply unconscious patients this reflex may not be functioning and fluid and food can then pass into the trachea and then the lungs, often with fatal consequences. This is also the reason why vomiting in anaesthetised patients can be so dangerous: the vomit emerging from the oesophagus can pass straight into the open trachea and hence the lungs.

an unsigned leading article expressing doubts about the value of the invention. 'Pain is not by far the worst thing in an operation,' the writer claimed. He had often observed that 'those who suffered severe pain often made a good recovery . . . while those who were able to suppress external indications of pain died soon after.'[5] He did not specify how the suppression of pain had been achieved; but he cautioned against a 'mania of etherism'. In fact, the 'mania' had hardly begun. The warning was an early manifestation of what James Miller would later apostrophise as 'The Rights of Pain'.[6]

. . .

As often happens after the introduction of a beneficial novelty, it is easy to find evidence of its welcome. Painless surgery was so spectacular, even near-miraculous, that expressions of joy were loud and ubiquitous. Many hailed the invention as 'the greatest discovery of the century'. Justified jubilation is always pleasant to record, but it would be misleading to portray the reception as uniformly or even overwhelmingly favourable. For some years at least the outcome hung in the balance.

At a meeting of the South London Medical Society a few months after Simpson's original report on chloroform Dr William Gull, a fashionable doctor known for his sharp tongue, maintained that it was 'a dangerous folly to try to abolish pain'. Even if it were desirable, 'ether was a well-known poison'.[7] Most of those present agreed. Benjamin Travers, later an ardent advocate of anaesthesia, described both ether and chloroform as 'among the most dangerous remedies'. A Dr Cole called them 'pernicious'. A Dr Nunn 'failed to see how surgeons could get on without pain'. One member had brought good news from France where the eminent Dr François Magendie had denounced anaesthesia in scathing terms. He had finished his impassioned oration with the words: *La douleur toujours*! Dr Radford, summing up, concluded that 'there was nothing but evil' in the new invention.

Some lay newspapers too printed forebodings. One or two quoted 'recognised medical authorities'. Dr James Pickford, a respected practitioner in Brighton, described the 'sudden, grave and violent effects of ether . . . producing black vitiated blood, similar to that found in putrid and malignant fevers'.[8] Experienced doctors claimed that healing by first intention – that is, without infection and pus formation – would be prevented by anaesthesia. More gravely, ether would inevitably cause 'asphyxia, coma, the spitting of blood, convulsions, pneumonia and, most certainly and inevitably, explosions,

5. *Edinburgh Medical and Surgical Journal*, 1847 (68), 256.
6. J. Miller, *Surgical Experiences with Chloroform* (Edinburgh, 1856), p. 408.
7. Quoted in Stanley, *For Fear of Pain*, p. 302.
8. Quoted ibid., p. 306.

killing the surgeons and his team as well as the patients'. It was not clear which of the two disasters was more to be feared. A Dr Meigs expressed the view that anaesthesia 'abrogated feeling pain, one of the most ancient and general rights of man'; and a London vicar wrote to say that it was not so much the alleviation of pain which was the problem as the 'abandonment of consciousness'.

Simpson, by now recognised as the champion of progress, was never at a loss for an answer. Riding and railways also abrogated one of the most ancient and general rights of man, he pointed out, namely walking; and everybody abandoned their consciousness every night when they went to sleep. He also embarked on a sustained campaign based on figures and facts collected from many hospitals in different parts of the country. Even over a comparatively short period for most surgical procedures mortality everywhere seemed to be significantly higher when performed without anaesthesia. For amputations of the thigh, the most formidable standard operation, anaesthesia cut the mortality from one in two to one in four. Infection and other complications were less common after anaesthesia. By 1850 surgeons at St Bartholomew's Hospital had operated with chloroform on several thousand patients 'without notable adverse effects'.

Nobody in their senses believed everything Simpson said;[9] but almost everybody believed Snow. And Snow's message was similar. Anaesthesia with either ether or with chloroform was not without dangers and even the occasional fatality; but, as new methods went, it was both extraordinarily safe and undoubtedly beneficial. Paradoxically his most important contribution was his careful search for all cases of anaesthetic deaths reported in the medical literature world-wide. The first fifty included a housewife in Cincinnati, a young woman in Boulogne, a young English woman in Hyderabad, two young men in the *hôtel-Dieu* in Lyons, a young Australian visiting his family in Scotland, an Irish seaman in the United States Navy, a labourer in Westminster, an artilleryman on board ship off Mauritius, a patient in the Seraphim Hospital in Stockholm and a boy from the Highlands admitted to the Glasgow Royal Infirmary. In most cases the death was unexpected, tragic

9. Which did not prevent Simpson from becoming Scotland's leading surgeon and Edinburgh's favourite son (after the death of the author of the Waverley novels), surgeon to the queen in Scotland and a baronet. He adopted for his coat of arms the rod of Aesculapius and the motto *Victo dolore*. He died on 6 May 1870 in the arms of his older brother Alexander, the beloved Sandy, who had watched him grow from a barefoot boy in Bathgate to become Scotland's most famous doctor. Westminster Abbey was open to receive his body but Edinburgh would not give him up. On the day he was laid to rest in Warriston cemetery, where five of his nine children had already preceded him and where Lady Simpson was soon to follow, flags flew at half-mast, schools and the university were closed, business throughout the city ceased and the sound of sobbing everywhere was punctuated by the tolling of the bells of St Giles.

and perhaps unnecessary; but, seen in perspective, the total was small. By the early 1850s leading surgeons in Britain like James Syme in Edinburgh and John Erichsen, Liston's successor at University College Hospital, used anaesthesia 'almost invariably'. Even William Gull was reported to have allowed an anaesthetic to be administered to 'one of his august but frail patients' who needed a hernia repaired.

...

But even in Britain, where Simpson, Snow, Queen Victoria and a climate of faith in technological progress all worked in favour of anaesthesia, it took decades for the innovation to be universally accepted. Medicine has always been a conservative profession. Any elderly doctor can regale his hearers – and often does – with examples of premature enthusiasms which later prove useless or positively harmful. Sadly, the value attributed to knowledge in any field is proportional not to its content of truth but to the labour involved in its acquisition. Much false wisdom in medicine had been acquired at great cost.

After the first few years the delaying force was inertia rather than hostility. The 'literature' can be misleading. At no time did more than a minority of doctors read medical journals or attend professional meetings. Many eminent members of the profession ignored anaesthesia because they had never heard of it. In the early 1850s no medical men could get more eminent than Sir James Clark and Sir Benjamin Brodie, both baronets. In their youth they had been innovators. Sir James had acquired his first stethoscope from Laennec in Paris. Sir Benjamin had shed new light on bony abscesses. But that was a long time ago. In the eyes of the queen Sir James, grave and bewhiskered, was still the incarnation of medical wisdom. She would never hear of replacing him. Sir Benjamin was the first surgeon-president of the Royal Society. He had declined a peerage because he had felt that on his relatively modest income he could not maintain a life-style appropriate to such an exalted station. Sir James had more than once confided in his colleagues that he had never adopted a new treatment until it had been in general use for at least ten years. It was a sentiment with which Sir Benjamin would have concurred. And they were right – up to a point. Over the years their attitude to innovations had almost certainly saved many patients disappointment and aggravation. But occasionally it turned them into lethal incompetents.

On 29 June 1850 Sir Robert Peel delivered a passionate speech in defence of Greek independence to the House of Commons. He was loudly cheered even by some of his political enemies. Dazzled perhaps by such a reception, riding up Constitution Hill a few hours later, he was thrown from his horse and injured his left shoulder. He screamed at any attempt to move him. A local doctor was summoned or appeared unsummoned and offered to anaesthetise him there and then. He could then be securely bandaged before being lifted. It was an eminently sensible suggestion. Disastrously, the news of the fall reached Buckingham Palace, where the queen happened to be in the company of her favourite physician. She begged Sir James to hurry to the scene. For Sir James to make the wrong decision was always a matter of seconds. Ignoring the patient's protests and the offer of anaesthesia – never heard of such a thing, especially not in the middle of a busy London thoroughfare – he had Sir Robert placed in a carriage and ordered the carriage to convey him to his home in Whitehall Gardens. It must have been apparent that Sir Robert had broken his collarbone and several ribs beneath the shoulder blade. The bone splinters were eroding blood vessels and causing bleeding and swelling as well as severe pain. The evidence was ignored. The jogging of the carriage behind the galloping horses caused Sir Robert to faint. It may have led to the erosion of a major vein.

On arrival in Whitehall Gardens two leading surgeons, Sir Benjamin Brodie and Caesar Hawkins, were summoned by Sir James. Attempts at applying a bandage to immobilise the patient's shoulder 'caused such intense pain that the plan was abandoned'. Neither of the surgeons seems to have considered an anaesthetic. Despite 'huge doses of opium' the patient survived for two days in 'excruciating pain'. 'It was,' as Sir James recorded, 'most distressing. I hardly left the poor man's bedside. Death came as a merciful release. A sad story.'

...

The decisive shift of opinion probably came from patients as much as from surgeons. From 1849 onwards hospital notes frequently refer to new admissions 'even of the vagrant classes' actually 'demanding to be put to sleep'. Relatives too were becoming obstreperous. Catherine Dickens' obstetrician, a fierce opponent of anaesthesia, was forced to use chloroform for her eighth confinement in 1849. 'Charles was his usual obnoxious self and insisted on it.'

## Chapter 29

# THE RIGHTS OF PAIN

Ironically perhaps, it was in the United States where surgical anaesthesia was invented that resistance to it was strongest and most prolonged. Over the ten-year period of 1853 to 1862, during which few major operations were performed without an anaesthetic in London, Paris or Edinburgh, about a third of limb amputations for compound fractures at the Pennsylvania Hospital were still carried out on conscious patients. Even at the Massachusetts General Hospital, cradle of ether anaesthesia, one in every three potentially painful operations was performed without etherisation. It was not a question of wealth: private surgery followed a similar pattern.[1] The respected *New York Journal of Medicine* warned practitioners that 'ether is not safe even when administered in the most skilful manner . . . Serious and almost fatal consequences have followed the inhalation of it.'[2] Army surgeon John B. Porter, a veteran of the Mexican War declared that

> by the inhalation of ether even in the most cautious manner but in a sufficient quantity to produce insensibility to pain, the blood is poisoned, the nervous influence and muscular contractility is destroyed and the wound is put in an unfavourable state for recovery.[3]

The Committee of Surgery of the American Medical Association listed the most important complications 'which could be expected'. They were:

> convulsions, more or less severe and protracted, prolonged stupor, high cerebral excitement, alarming and long-continued depression of the vital

1. The American scene is admirably described by M.S. Pernick, *A Calculus of Suffering* (New York, 1985).
2. *New York Journal of Medicine and the Collateral Sciences*, 1847 (8), 122.
3. John B. Porter 'Medical and Surgical Notes', *American Journal of Medical Science*, 1853 (23), 33.

> powers and asphyxia . . . and as secondary effects . . . bronchitis, pneumonia and inflammation of the brain.[4]

Stories of sudden, unexplained deaths under anaesthesia featured prominently in the popular press. Should innocent members of the public be exposed to such perils? According to John P. Harrison, vice-president of the American Medical Association, pain was curative.

> The actions of life are maintained by it. Were it not for the stimulation induced by pain, most surgical operations would be followed by dissolution.[5]

In most countries such views were held with especial conviction by the armed forces. The Confederate Virginia Manual of Military Surgery quoted Sir John Hall, inspector-general of British hospitals in the Crimea, with approval. Sir John was known to disparage chloroform and to laud 'the lusty bawling of the wounded from the smart of the knife'.[6] Pain was the most powerful and valuable stimulant. Many a brave soldier seemingly dead was brought back from the brink by the feel of the surgeon's steel. Illustrative examples were numerous, vivid and patently invented.

But it was not only grizzled veterans who sang the blessings of pain. Benjamin Hill, surgeon to the religious Eclectic Sect, advised patients to submit to the thorough cauterisation of all cancers not only to obtain a cure but also for the 'moral regeneration' provided by the experience.

> I have not infrequently had patients, after submitting, perhaps for an hour, to this burning alive without flinching or groaning, open their mouth for the first time to express their fears that the operation had not been carried far enough because they had felt it so much less than I had given them reason to expect . . . I had told them beforehand that unless they had fortitude enough to . . . hold out like Christian youths of old while they were burnt . . . the final 'moral cause' of cancer, to develop heroism in them, would be frustrated.[7]

The fashionable New York gynaecologist, Augustus Gardner, was far from being a fundamentalist Christian but he too could rhapsodise about the suffering of his female patients:

> I feel not only that their suffering strengthens all their other faculties. Their baptism of pain regenerates them as individuals to become little lower than the angels.[8]

4. *Transactions of the American Medical Association*, 1848 (1), 189.
5. Quoted in Pernick, *A Calculus of Suffering*, p. 125.
6. A.D. Farr, 'Early Opposition to Obstetric Anaesthesia', *Anaesthesia*, 1980 (46), 901.
7. B. Hill, *Lectures on the American Eclectic System of Surgery* (Cincinnati, 1850), p. 68.
8. A.K. Gardner, 'Physical Decline of American Women', *Knickerbocker*, 1860 (55), 37.

The influential Philadelphia *Presbyterian* warned: 'Let everybody who values freedom beware of the slavery of etherisation.'[9]

It was not just doctors, hardened perhaps by their terrible profession. Poets and writers would sometimes express the 'rights of pain' even more poignantly than doctors or preachers. Ralph Waldo Emerson declared that all bodily pains carried with them their own spiritual reward; and Emily Dickinson wrote:

> Power is only Pain —
> Give balm to Giants
> And they'll wilt
> Like men.[10]

. . .

Pain was more than a question of rights or of morality. Ordinary people had good reason to believe that pain was an essential part of life. In Nature loss of pain usually indicated the approaching end. Even today many look back on their first anaesthetic as an uncanny experience. Nineteenth-century man and woman had seen such insensibility only in comas, severe strokes, brain injuries or fatal poisonings. These were all portents of impending dissolution. Pain, by contrast, clearly required sound and healthy organs. The greater the pain, the greater could be one's confidence in one's 'natural fitness', even in being alive. It was easy to believe that anaesthesia itself was an assault on vital functions. Even that 'anaesthesia is death'.[11]

. . .

More even than the Old World, the United States was alive with quacks. They were of an astonishing variety, ingenuity and pretension. Anaesthesia in their hands was a prospect to frighten the boldest.

9. *The Presbyterian* (Philadelphia) 1847 (17), 20.
10. *The Poems of Emily Dickinson* (1861), ed. Thomas H. Johnson (New York, 1963), Vol. I, p. 181.

It was more than a century later that another writer expressed the 'rights of pain' even more uncompromisingly. Aldous Huxley wrote in *Brave New World* (1932):

> 'But I don't want comfort. I want God, I want poetry, I want real danger, I want freedom, I want goodness, I want sin.'
> 'In fact', said Mustapha Mond, 'you're claiming the right to be unhappy.'
> 'All right then', Savage said defiantly, 'I am claiming the right to be unhappy.'
> 'Not to mention the right to grow old and ugly and impotent; the right to have syphilis and cancer; the right to have too little to eat; the right to be lousy; the right to live in constant apprehension of what may happen tomorrow; the right to catch typhoid; the right to be tortured by unspeakable pains.' There was a long silence.
> 'Yes, I claim them all', Savage said at last.

(Vintage Future Classics Edition, London, 2005), p. 211.

11. *American Journal of Medical Sciences*, 1849 (18), 182.

> Every mountebank who digs out a corn and dignifies himself with the title of chiropodist; every itinerant dentist who gouges out a tooth or fills a cavity with some home-made amalgam; or anything that can creep, crawl or sneak into any of the unguarded sanctuaries of medicine can arm himself with an inhaling apparatus and a bottle of an anaesthetic with which he expects to prey on the public.[12]

The nightmare conjured up by the American Society of Dental Surgeons was horrific. Think of the prospect of patients slipping out of the benign if painful control of properly qualified individuals! Think of them being tempted to have their tooth extracted painlessly by a charlatan! It was hair-raising. Fortunately, 'most patients actually prefer painful remedies because they can feel them working'.[13]

Some warnings against anaesthesia had an apocalyptic ring. Dr William Henry Atkinson, first president of the American Dental Association, expressed his conviction that

> anaesthesia is of the devil and I cannot give my sanction to any Satanic influence which deprives man of the capacity to recognise the law! I wish there were no such thing as anaesthesia! I do not think man should be prevented from passing through what God intended them to endure.[14]

It was widely felt that 'anaesthetic intoxication' posed a special threat to women. Not only were they liable to be raped – in the hands of unqualified dentists a virtual certainty – but, under the influence of the drug genteel ladies might attempt to seduce their anaesthetist. It was 'well known' that, drunk on ether or chloroform, respectable women sometimes did the most shocking things; and, even when they did not actually *do* things, they might experience distressing fantasies. The *American Journal of the Medical Sciences* quoted an 'internationally renowned obstetrician' who 'insisted on the unmitigated impropriety of etherisation, leading to sexual orgasm instead of the natural throes of parturition'.[15]

Some saw not only respectable women but the entire nation as in jeopardy. Professors of history pointed tremulous fingers at 'the disastrous effects of inhaling intoxicants on the ancient civilisation of China'. Even wholesome cities like Boston might soon be sprouting opium (or ether) dens.

...

12. Quoted in Pernick, *A Calculus of Suffering*, p. 45.
13. *American Journal of Dental Science*, quoted by Pernick in *A Calculus of Suffering*, p. 67.
14. Quoted ibid., p. 131.
15. Quoted ibid., p. 80.

It was in America that the gruesome spectre of the apparently anaesthetised but only paralysed patient first raised its head. Did anaesthesia really remove pain or did it merely induce paralysis followed by amnesia?[16] It was a vision worthy of Edgar Allan Poe; and it was canvassed by influential medical journals.

> Can it be that because patients on awakening express no recollection of that suffering, it is mistakenly assumed that there was no painful impression conveyed to the sensorium at the time?[17]

The spectre – and the reality – would re-emerge more than a century later with the introduction of curare-like muscle relaxants.[18]

...

Many of the more extreme opinions reflected the way in which American medicine had evolved and was still evolving. Like the country itself, the profession was younger, more varied, more heterogeneous and more experimental than in most European countries. It was also acquiring self-conscious gravitas. In the 1830s and 1840s many doctors were still in thrall to Dr Benjamin Rush of Philadelphia, a signatory of the Declaration of Independence and founder of America's first medical school.[19] Rush had firmly believed that most, perhaps all diseases were the result of 'increased tension' and would yield to 'depletive' remedies such as blood-letting, cupping, scarifying, blistering and purgation. He had also believed that the more dangerous the disease, the more extreme treatment had to be. Even epidemics like yellow fever and patients already reduced to ghastly yellow skeletons would respond to drastic measures like blood-letting and laxatives provided these were undertaken on a sufficiently bold scale. There was no such thing as an optimal dose. The dose of anything had to be increased until the patient either recovered or died. It was usually the latter; but then they would have died anyway.

It was partly as a reaction to this radical – or 'Republican' – trend that a variety of 'natural' healing schools gained adherents. In the 1840s and 1850s the most popular was homoeopathy. The invention of the German physician Samuel Hahnemann, it was always more highly esteemed in the United States and later in Britain than in the inventor's own country.[20] Hahnemann taught that a drug which produced a particular symptom in a healthy person

16. The question was asked by Robert Barnes, in the *American Journal of Medical Sciences*, 1851 (21), 214.
17. Leading article, ibid., p. 260.
18. See Chapter 50.
19. Rush died in 1813, aged sixty-eight.
20. Hahnemann died in 1843, aged eighty-five.

would cure a similar symptom in a sick person; and that the potency of most drugs increased in proportion to their dilution. An impressive number of 'homoeopathic' remedies containing – or allegedly containing – minute amounts of much-feared poisons sold in large quantities and were claimed to be curative.

Hydrotherapy, another product of the German Romantic imagination, was also acclaimed. Hydrotherapists deplored all drugs and surgery as meddlings with the beneficial workings of Nature. They used only water, steam and ice to assist healing, supported by judicious exercise, rest, temperance and hygiene. They believed that physical pain could be wholly vanquished by adopting a natural way of life. The doctrine influenced vegetarian health reformers like the wheat-cracker pioneer, Sylvester Graham, and the promoter of flaked cereal, Henry Kellogg. In their hands it not only worked: it could also be hugely profitable. That could not be a bad thing. But hydrotherapy also acquired adherents among the proliferating millenarian sects like the Millerites and the Adventists. To those groups the encouragement of Nature was the only reliable way of treating illnesses. To them the pains both of labour and of operations were 'righteous chastisements'; and Nature, not God, was wielding the lash. It was indeed 'an insult to Providence', the hydropath T.L. Nicholls declared, to suggest that God intended mankind and even more womankind to suffer.

> This world is the work of infinite power and benevolence ... Perfect freedom from pain is both possible and intended ... In a state of health no natural process is painful ... There is no more certain fact in physiology than that the nerves of organic life in a healthy condition are not susceptible to pain.[21]

Pain, in short, was unnatural, abnormal, pathological, a manifestation of misguided human free will battling against the laws of Nature. Happily, it could be totally eradicated by following those laws, that is by adopting a life based on exercise, fresh air, cleanliness, virtue and temperance. Anaesthesia was not merely sinful. It was also unnecessary.

21. T.L. Nicholls, 'The Curse Removed: A Statement of Facts Respecting the Efficacy of Water Cures in the Treatment of Uterine Diseases and the Removal of the Pains and Perils of Pregnancy and Childbirth', quoted by Pernick in *A Calculus of Suffering*, p. 54.

## Chapter 30

# WHO NEEDS AN ANAESTHETIC?

In a self-consciously moderating attempt between extremes many leading American physicians and surgeons evolved what became known as 'conservative' medicine. To practitioners like Austin Flint, Oliver Wendell Holmes, Worthington Hooker and the surgeon Frank H. Hamilton, all leaders of the profession, this implied more than a 'well balanced mix'. The duty of a 'conservative' doctor was to select whatever measure was appropriate to any particular case. It drew upon the admirable ethical precepts of putting individual patients at the centre of medical care.

On the face of it, 'conservative' medicine made sense. By rejecting universal treatments for all diseases, the legacy of Benjamin Rush, doctors would take into account every aspect of the patient's personality, way of life, habits, background, hopes and aspirations. Inevitably – but this was not always and immediately apparent – they would also take into account their own personality, background, ethical preconceptions and prejudices. The two could not always be separated. Tailoring treatment to individual patients also meant tailoring treatment to individual doctors.

'Individual selection' was soon part of a wider trend. Teachers too were urged to 'accommodate themselves to the peculiar nature, habits, background and stage of development of their pupils'.[1] This also meant accommodating schools to what the teaching profession regarded as the widely different needs of widely different classes of children. The principle was even applied to religion, ministers being encouraged to 'mould' their ministry to their flock.[2] Not everybody saw the potential conflict between individuality and equality, two ideals equally cherished. Walt Whitman did and expressed it admirably, as he expressed many of his country's teasing moral dilemmas. After listing at

1. *New York Teacher*, 1852 (1), 70.
2. C.G. Finney, *Lectures on Revivals of Religions*, ed. W.G. McLoughlin (1835; reprinted Cambridge, Massachusetts, 1960), p. 199.

length the ways in which human individuals differed – physically, morally, intellectually and racially – he asked:

> Have you ever loved the body of a woman?
> Have you ever loved the body of a man?
> Do you not see that they are exactly the same to all?
> In all nations and times all over the earth?[3]

. . .

Theoretically and often in practice, conservative medicine gave doctors enormous power. This was a new development. For all his limitations, Benjamin Rush had entertained no delusions of grandeur. Like his friend Thomas Jefferson, he believed that medicine was basically simple common sense. Of course there was room for wisdom and the application of experience. Even a certain amount of book-learning could be useful. But, by and large, any gentleman of moderate intelligence could be his own and his family's medical adviser. In the hands of the new conservative physicians medicine became dauntingly abstruse and complex. The requirement of every patient needed to be individually weighed and assessed. This did not mean that general rules disappeared. The very reverse. They became more numerous and complicated. The American Medical Association looked forward to the time when 'our precepts will clearly indicate to what class of patients any treatment is applicable'.[4] This was especially relevant to new and untried remedies. To none did it seem to apply with greater force than to anaesthesia. In itself the new invention was neither good nor bad. It simply – or not so simply – had to be used with refinement, insight and discretion. Most obviously, the value of a painkiller depended on how much suffering individual patients might experience without it. This, according to the beliefs of conservative medicine, varied widely.

. . .

Gender was the most obvious divide. Few doubted that it profoundly influenced the perception of pain. 'The nerves themselves are smaller in women than in men and of a more delicate structure . . . Women's senses in general are more acute . . . Women are more delicate in their whole economy.'[5] Poets and men of letters reinforced professional opinion and what passed for

3. W. Whitman, 'Salut au Monde', in *Leaves of Grass*, New American Library Edition (New York, 1958), p. 127.
4. *Journal of the American Medical Association*, 1852 (23), 449. Less ambitiously, the editor of the *British Medical Journal* thought that this was neither desirable nor practicable.
5. R. Welter, 'Cult of Womanhood', in *American Writings on Popular Education of the Nineteenth Century* (Indianapolis, 1971), p. 149.

experimental evidence.[6] Oliver Wendell Holmes, never at a loss for the telling phrase, marvelled at how 'much more fertile [women are] in their capacity of suffering pain. They have so many varieties of headaches! Quite astonishing.'[7] How different were men, and most particularly the 'brawny ploughman' or the 'virile soldier or sailor'. Especially the last two and especially in the heat of battle. 'Heroic manly fortitude, heightened by the exhilaration of a good fight, makes soldiers almost insensitive to the pain of almost any operation.'[8] There was no doubt that broadly speaking – and subject to all kinds of reservation – women were infinitely more vulnerable and needed anaesthesia more than men.

Not all women agreed. To Christabel Mondale, writer and intrepid explorer,

> women who expect to go to bed at every menstrual period . . . constantly complaining of their nerves, urged to consider all their minor discomforts by well-intentioned but misguided male advisors, pretty soon become nothing but a bundle of nerves.[9]

And Elizabeth Cady Stanton blamed a 'man-made culture' and the urging of the 'male medical profession' for making women 'unnaturally sensitive'.[10] To be called vulnerable was acceptable. To be patronised was not. But it was not always easy to draw the line.

. . .

Age was another important variable. It was also fraught. Most experts agreed that the old – the 'elderly' had not yet been invented – had a diminished sensitivity to pain. It was also beginning to be realised that old people were particularly prone to the complications of anaesthesia. Chloroform especially seemed to affect many adversely. On the other hand, frail patriarchs were liable to pass into a state of irreversible shock when subjected even to minor surgery. They might be spared such a fate by having their operations under anaesthesia. It was a fine balance, needing individual consideration. In most cases it was best to let Nature take its course and not to operate either with or without an anaesthetic.

To legislate about children could also be hard. According to Abel Pierson of Massachusetts, infants properly handled could sleep peacefully during any operation.[11] Henry J. Bigelow argued that sensitivity to pain was closely

6. R.W. Emerson, 'Women', in *The Complete Writings of Ralph Waldo Emerson*, ed. T.H. Johnson (Cambridge, Massachusetts, 1963), p. 1178.
7. O. W. Holmes, in the *American Journal of Medical Sciences*, 1852 (23), 509.
8. J.B. Porter, 'Medical and Surgical Notes of Campaigns in the War with Mexico', *American Journal of the Medical Sciences*, 1852 (23), 13.
9. C. Mondale, *Women, Their Nature and Their Diseases* (Chicago, 1878), p. 56.
10. E.C. Stanton, *The American Woman and Womanhood, 1820–1920* (New York, 1972), p. 45.
11. A. Pierson, 'On Operations in Infants and Children', *American Journal of the Medical Sciences*, 1852 (24), 546.

related to experience, intelligence, memory and reasoning.[12] This meant that, 'like lower animals', small children lacked the mental capacity to suffer, however loudly they bellowed or pitifully they whimpered. Others disagreed. 'The constitution of infants in terms of debility and irritability approaches that of the female.'[13] Bigelow's lumping together of babies and 'lower animals' offended animal lovers. Babies might not suffer pain but animals most certainly did. Bigelow hastily explained that he had specified *lower* animals. Like worms. Or fish. His own hound 'would certainly equal or surpass a Bushman or an Indian in suffering during surgery'.[14]

In Britain, despite bear-baiting, cock-fighting, the slaughter of birds and other popular pastimes, the pro-animal and pro-baby lobbies were increasingly vocal. Jeremy Bentham expressed their views succinctly as usual: 'The question is not, Can they *reason*? Nor, Can they *talk*? But Can they *Suffe*r?'[15] As the century progressed, the tide of public opinion moved in their favour. In the Romantic world-view women, children and pets tended to join hands/paws/hoofs.

. . .

In a country where the equality of men was the bedrock of the Constitution and of every homily delivered from the pulpit, social class was bound to become divisive. A hundred years earlier Alexis de Tocqueville had contrasted the European with the American pauper. In Europe, De Tocqueville believed, paupers had become so inured to perpetual misery that they were hardly aware of it. In America, by contrast, there were no 'permanently degraded' people (except of course the slaves). Everybody shared equally in the 'extreme hypersensitivity' of the middle classes.

> America's constantly fluctuating fortunes prevent anyone from becoming accustomed to the physical pains of poverty and expose everyone to the risk of having to undergo them.[16]

But by the mid-nineteenth century and to the Americans themselves this was far from being obvious. The Connecticut novelist, John W. De Forest felt that

12. H.J. Bigelow, 'Sensitivity to Pain and Intelligence', *Transactions of the American Medical Society*, 1848 (1), 221.
13. H.M. Lyman, *Artificial Anaesthesia and Anaesthetics* (New York, 1881), quoted in S. Rothman, *Woman's Proper Place: A History of Changing Ideals and Practices, 1870 to the Present* (New York, 1978), p. 56.
14. J.H. Bigelow, 'Inaugural Lecture delivered to the Course on Surgery', Harvard University Medical School, Boston 1849, reprinted in *Surgical Anaesthesia: Addresses and Other Papers* (Boston, 1900), p. 322.
15. J. Bentham, *Animals and Man in Historical Perspective*, ed. J. and B. Klaitis (London, 1971), p. 151; italics in the original.
16. A. de Tocqueville, *Democracy in America*, 2 vols (New York, 1945), Vol. II, p. 214.

> we waste far too much unnecessary sympathy on poor people. A man is not necessarily wretched because he is cold, hungry, unsheltered or in pain. Provided such circumstances usually attend him, he gets along very well with them: they are annoyances but not torments.[17]

His books sold well, presumably mostly to the well-heeled.

Alcohol further inured many of the poor to pain, though only during their happy state of drunkenness. 'I love being a-drunk for then all my pains are gone,' went music hall artist Ben Mort's popular ditty. It was a sad irony that chronic alcoholics deprived of their drink were – like criminals – often hypersensitive. They lacked fortitude. Or pride. But then how much sympathy did they deserve? If they drank beyond their means, they had brought their sufferings on themselves.

Recent immigrants were often different from poor but well-rooted Americans. Of course one could not generalise. But everybody did. The hydropath Thomas Nicholls had no doubt that 'our newly-arrived Teutonic friends' harder strung nerves and blunter sensation make them less exposed to suffering pain than are our own finer grained humanity'.[18] It was nevertheless undeniable that it was the Teutonic Brothers Grimm who created the prototype of the hypersensitive female. Their fairy princess who was kept awake by a pea hidden under mountainous layers of mattresses was the perfect representative of the 'finer grain'.[19] Such precious creatures were often too sensitive even to carry out more or less normal bodily functions. According to the respected gynaecologist, Dr Gunning S. Bedford, menstruation in 'all women of the better classes' caused them to suffer 'pathological levels of pain'. By contrast, 'the nervous system of women of the poorer classes, fortified by constant exercise in the open air and strengthened by frugal habits, was not subject to such a morbid reaction'.[20]

...

When it came to suffering pain, cultural refinement was a double-edged weapon. 'For in much wisdom is much grief,' Koheleth, the Preacher in Ecclesiastes warned, 'and he that increaseth knowledge increaseth sorrow.'[21]

17. J. W. De Forest, quoted by Pernick in *A Calculus of Suffering*, p. 138.
18. T.L. Nicholls, 'Natural versus Acquired Habits', *Massachusetts Teacher*, 1949 (2), 303.
19. The story of 'The Princess and the Pea' was first published by the Grimm brothers in their collection of fairy tales in 1812 but the idea has its roots in ancient Indian (and probably other) folklore. The Grimms' collection was not published in the United States until 1842 but the stories were published in England in 1823–6 and were probably known in America before publication there.
20. G.S. Bedford, *Clinical Lectures on the Diseases of Women and Children*, 8th edn (New York, 1867), p. 2.
21. Ecclesiastes, 1: 18.

Many on both sides of the Atlantic took their Ecclesiastes literally. But one did not have to be religious to be worried. Eve and the apple was only one example. Look at Athene and her restless mind. Or Pandora and her box! There could be no doubt that knowledge increased suffering. And it was no accident that those who ignored such wisdom were usually women. Education as such was not perhaps bad; but any broadening of the mind had to be compensated for by vigorous physical training. Since women were already more sensitive than men, the education of girls in particular had to be supplemented by bodily discipline.[22] Or perhaps it was safest to dispense with both education and gymnastics.

Civilisation in general, like education, was a mixed blessing when it came to experiencing pain. Medical sages contrasted the 'savage life' which virtually conferred freedom from pain to the 'general effeminacy' of the pampered yet always suffering town dweller. This was once again particularly true of women. Dr Duwees of Philadelphia, a leading obstetrician, compared the prolonged and painful labour of 'women of the upper walks of life' to the quick, painless experience of their 'more savage sisters'.[23] The difference even seemed to apply to the animal kingdom, separating domestic breeds from those living in the wild. Dr Henry M. Lyman, professor of obstetrics at Chicago's Rush and Woman's Medical College expressed the view that

> really normal labour is not a painful process . . . But in civilised society the majority of mankind are living under quite abnormal conditions . . . Hence in civilised society, it is the rule rather than the exception to find parturition attended by suffering.[24]

Even feminists like Elizabeth Cady Stanton felt that 'refined civilised women suffered inordinate pain in childbearing'. She exulted in her own 'wholly painless' labour: 'Am I now as good as a savage?'[25]

. . .

Darwinian ideas of evolution generated grave concern. If Darwin was right – admittedly a big if – freedom from pain seemed to give the uncivilised and the uneducated an evolutionary advantage.

22. V. Bullough and M. Voigt, 'Women, Menstruation and Nineteenth-Century Medicine', *Bulletin of the History of Medicine*, 1973 (44), 66.
23. W.P. Duwees, *An Essay on the Means of Lessening the Pain in Certain Cases of Difficult Parturition* (Philadelphia, 1806), p. 165.
24. H.M. Lyman, *Artificial Anaesthesia and Anaesthetics* (New York, 1881), p. 57.
25. E.C. Stanton, quoted by Pernick in *A Calculus of Suffering*, p. 167.

> In our process of becoming civilised, we have won, I suspect, an intensified capacity to suffer pain. The savage does not feel pain as we do and will therefore always survive us.[26]

This was an ancient conundrum. In various guises it had troubled Plato, the elder Cato, Pliny and before them probably many of the more contemplative of the pharaohs. But the emergence of anaesthesia, potentially abolishing some of the worst physical torments of civilisation, gave the problem a new edge. Painlessness was a gift long prayed for but almost overwhelming in its implications.

In America the advantage 'physical toughness' could give to more 'primitive races' caused apprehension about the future. There the 'primitive races' were on everybody's doorstep. Happily they still lacked the killing technology of their white brethren. They were also exceptionally susceptible to the diseases and afflictions of Europeans, including the depredations of alcohol. But they were a ubiquitous reminder. Many believed that the Indian was virtually incapable of experiencing pain (or indeed any sharp feeling). Benjamin Rush, apostle of equality, had explained that 'Indian men could often inure themselves to pain by burning parts of their bodies with fire or cutting themselves with sharp instruments from early boyhood'.[27] Samuel Stanhope Smith, foremost race theorist of enlightened America lectured that

> among your red Indian and other uncivilised tribes the parturient female does not suffer the same amount of pain during labour as the female of the White race . . . Nor does physical pain generally torment them or the males nearly to the same degree.[28]

This had clear implications for selecting patients for anaesthesia, a procedure neither without risk nor without expense to the community. Fortunately the dilemma did not often arise. Most Indians still preferred their own brand of doctoring: few were ever admitted to the Massachusetts General Hospital.

. . .

Black people were widely believed to run Indians a close second – or even surpass them – in insensitivity to pain. The belief apparently antedated slavery. 'Early European travellers to Africa' (whoever they were) had noted

26. T.E. Clegg, 'Some of the Ailments of Women due to Her Higher Development in the Scale of Evolution', *Texas Health Journal*, 1890 (3), 57.
27. B. Rush, *Medicine among the Indians of North America: A Discussion* (1774) in *Selected Writings*, ed. D.D. Runes (New York, 1947), p. 259.
28. S. Stanhope Smith, *An Essay on the Causes and Varieties of Complexion and Figure of the Human Species* (1787; ed. W. D. Jordan, reprinted Cambridge, Massachusetts, 1965), p. 72.

African imperviousness to physical suffering. Medical explorers recounted amputating 'countless black limbs' and meeting with no expression of pain.[29] At least Dr Charles White, expert on race, cited 'many such accounts from numberless sources' to support his contention that black skin was 'excessively thick . . . and as a result insensitive'.[30] None of the numberless sources was ever named, but who cared?

The most comprehensive account of black insensitivity to pain came from the New Orleans physician Samuel A. Cartwright, the discoverer of 'dysaesthesia Aetiopis' or the non-sensitivity of the Aetiops, that is blacks.

> This dysaesthesia is a widespread hereditary disease of blacks which causes such obtuse sensibility of the body that its victims are insensitive to pain when subjected to punishment.[31]

While many doctors probably felt that Cartwright was a mite exaggerating, the relative insensibility to pain of blacks went virtually unchallenged.

> Nervous action in the Negro is comparatively sluggish. In contrast to their acuity of hearing, seeing and smelling, their sense of touch and taste are obtuse. They require less sleep and submit and bear the infliction of the rod with a surprising degree of resignation and even cheerfulness.[32]

Marion Sims, grand old man of obstetrics, explained that he had carried out lengthy and agonising experiments on slave women not because he could force slaves to submit but because white women were 'too sensitive to pain'.[33] A Virginia physician of a scientific bent anticipated Nazi practices in the twentieth century by pouring 'near boiling water over the backs of Negro subjects . . . The treatment seemed to arouse their sensibilities somewhat.'[34] A Georgian doctor, trying a similar experiment, was surprised when 'the patient leaped up and appeared to be in great agony'.[35] Some attributed the 'morbid insensitivity of Negroes to pain' to congenital leprosy. This was also held to account for their body odour.

. . .

29. T. Gossett, *Race: The History of an Idea in America* (New York, 1963), p. 34.
30. C. White, quoted by Pernick in *A Calculus of Suffering*, p. 168.
31. S.A. Cartwright, 'Report on the Diseases and Physical Peculiarities of the Negro Race', *New Orleans Medical and Surgical Journal*, 1851 (7), 34.
32. A.P. Merrill, 'Distinctive Peculiarities of the Negro Race', *Memphis Medical Recorder*, 1855 (4), 1.
33. B.M. Harris, *The Life Story of J. Marion Sims* (New York, 1959), p. 109.
34. F.W. Jones, 'On the Utility of the Applications of Hot Water to the Spine in the Treatment of Typhoid Pneumonia', *Virginia Medical and Surgical Journal*, 1854 (3), 108.
35. T.L. Savitt, 'The Use of Blacks for Medical Experimentation and Demonstration in the Old South', *Journal of Southern History*, 1982 (48), 341.

Many philanthropic Christians saw in black insensitivity the working of God's benevolence. The happy demeanour of slaves exposed to the lash was to the young Abraham Lincoln clear evidence of God's compassion. The abolitionist Lydia Maria Child also praised 'the merciful arrangement of Divine Providence . . . whereby in Negroes the acuteness of sensibility is lessened when it becomes merely the source of suffering'.[36] Though De Tocqueville had inclined to the same view, as a spiritual child of Voltaire and Enlightenment, he was aware of its moral ambiguity.

> Am I to call it a proof of God's mercy . . . that a man in certain states appears to be insensible to pain? . . . Am I to praise God because the Negro, plunged into this abyss of evils scarcely feels his own calamitous position?[37]

There were no easy answers.

. . .

While few seriously disputed that different sexes, ages and social and racial groups required different degrees of anaesthesia if they required it at all, there was much argument about why this should be so. Reformers and abolitionists vehemently denied that blacks and Indians were insensible to pain by nature. The alternative explanation was 'brutalisation'. White plantation owners too might grow 'insensitive to pain if they were subjected to repeated flogging, degradation and privations'.[38] Like De Tocqueville, Lydia Maria Child and men like the mature Lincoln saw in black insensitivity a form of numbness. Freeing slaves would in time restore their normal – and God-given – sensibility. Many slaves agreed. One runaway, subjected to much brutality reported that 'I began to grow less feeling, even indifferent to my own punishment.'[39] On the other hand, most mid-century Americans (like Europeans) accepted as a matter of course that acquired characteristics could be inherited. This meant that brutalisation over many generations would eventually lead to hereditary numbness. Southern slave holders and frontiersmen might agree that environment, not race, caused the differences in susceptibility to pain; but they did not believe that setting slaves free and kindness to the Indians could reverse this process within the foreseeable future. How foreseeable? Not for many generations certainly, not perhaps for thousands of years.

36. Quoted in W. D. Jordan, *White over Black: American Attitudes toward the Negro* (New York, 1969), p. 508.
37. De Tocqueville, *Democracy in America*, Vol.I, p. 345.
38. H. Mann, *Lectures in Education* (Boston, 1855), p. 313.
39. Henry Watson, quoted in Pernick, *A Calculus of Suffering*, p. 158. This is of course universal experience reported by survivors of wars, concentration camps and natural disasters.

In 1605 Shakespeare wrote:

The poor beetle, that we tread upon,
In corporal sufferance finds a pang as great
As when a giant dies.[40]

In pretty, poetic language the Bard usually expressed what most of his contemporaries felt. Two and a half centuries later few would have agreed with him.

. . .

An odd semantic consequence of different responses to pain was the distinction in literature – and then in 'cultivated intercourse' – between 'endurance' and 'insensitivity'. Insensitivity, that is the inability to feel pain, was broadly speaking 'bad', a trait of ruffians, savages and the inferior races. Endurance, by contrast, that is the capacity to perceive but to withstand pain, was 'good', a sign of nobility and moral fibre. The difference might not be obvious to ordinary folk in everyday life; but to writers of romantic fiction the signs were unmistakable. And successful practitioners of the genre made sure that their readers were left in no doubt which category a character belonged to. The moral significance of the two responses was utterly distinct. Only characters of whom the writer approved ever 'endured'. The rest were insensitive. Even level-headed doctors and thinkers like Silas Weir Mitchell, poet and father of American neurology, convinced themselves that they could tell the difference.

The distinction was eventually adopted even by the 'insensitive races'. To hide one's pain from the oppressor became an act of endurance as well as of defiance.

Nobody knows the trouble I see. Nobody knows but Jesus.
Dear God! I can feel the torture now, the terrible excruciating agony.
I did not scream. I was too proud to let my tormentor know
What I was suffering

as one former slave remembered.[41] Some white anti-slavery campaigners like Harriet Beecher Stowe also attributed the black man's apparent insensitivity to 'Christian endurance'. The slave-owning Thomas Jefferson, stung by the French naturalist Buffon's description of the American 'fauna' as 'puny and degenerate', even had a stab at exalting the Indians:

40. Shakespeare, *Measure for Measure*, Act III, scene i, lines 77–9.
41. J.W. Blassingame, *The Slave Community: Plantation Life in the Antebellum South* (Oxford, 1972), pp. 54, 124.

> The noble native American endures torture with a firmness unknown to us . . . yet his sensibility is keen though in general they endeavour to appear superior to human events.[42]

. . .

Jefferson's high-minded image of the 'Red Indian' made a deeper impression in Europe than in his own country. On the Continent especially over the next hundred years the Apache warrior became a heroic figure who, in the vast and wonderful yarns written for the young by Karl May, pre-eminently endured.[43] In two dozen languages May inspired generations of cosseted middle-class towny boys[44] who could not bear to watch the killing of a chicken and had not the faintest conception of what torture really meant.

. . .

In America such hero worship was reserved for the white frontiersman. Most nineteenth-century Americans, from doctors writing in learned journals to Walt Whitman, accepted that the Frontier environment produced actual physical changes that made the pioneers – and by some kind of a long-distance osmosis even urban East Coast Americans – a tougher and fitter breed than were most Europeans. They were brave, noble and taciturn; but, most importantly, they endured.

Women too tended to endure. 'What angelic serenity! White endurance!' Dr Gardner of New York gushed over his suffering female patients. 'What beautiful gentleness and love! What a mild radiance pervading the whole being of one so severely afflicted.'[45] The combination of hypersensitivity and saintly suffering (spiced with male guilt) put medical men under a double obligation to protect women from pain. It sometimes compelled even those who vociferously disapproved of anaesthesia in labour, to make surreptitious use of it. The secretly administered whiff of chloroform in stately mansions overlooking New York's Central Park became Sir James Young Simpson's ultimate triumph.

42. T. Jefferson, 'Notes on Virginia', in *The Life and Selected Writings of Thomas Jefferson*, ed. A. Koch and W. Peden (Philadelphia, 1994), p. 45.
43. K. May, *Winnetou*, 3 vols (Leipzig, 1907).
44. And even privileged out-of-townies like the present writer.
45. A.K. Gardner, 'Physical Decline of American Women', *Knickerbocker*, 1860 (55), 37.

# PART IV

# THE BEGINNING OF THE MODERN

# Chapter 31

# THE NEW PHYSIOLOGY

Whatever the blessings and dangers of anaesthesia, the fact that pain could be abolished showed that it existed. Of course in common parlance pain had existed since time immemorial. No human experience except perhaps love has been so much spoken and written about. But in most people's minds it had always been part of a diseased, injured or imperfectly functioning body part, like the redness and the swelling of a sprained ankle or the tenderness and throbbing of an inflamed tooth. It was always *something* that hurt; and the ankle, the tooth, the head and the stomach (and even, if you believed in it, the soul) all hurt differently. This was not surprising, since they themselves were different.

Some thinkers had suggested that pain was no more than an exaggeration of other sensations to the point where they began to cause distress. This too was an attractive notion – and still is. A smack is only a violently administered caress; and even a smack is not always and necessarily painful. The difference between the luxury of a warm bath and the suffering of a burn can be represented as one of degree. But in the light of anaesthesia, most scientists and doctors began to think of pain as an experience like no other, a function that existed in its own right. It was a recognition that raised interesting questions, almost a new science.

. . .

The foundations had of course been laid long before anaesthesia. In 1811 the surgeon physiologist Charles Bell published a modest pamphlet for the entertainment of his friends, entitled *An Idea of a New Anatomy of the Brain.*[1] In this he addressed questions about nerve and brain function which had not been aired since the anatomical studies of Thomas Willis in the seventeenth

1. See C. Bell, *An Idea of a New Anatomy of the Brain* (Edinburgh, 1811).

century.[2] He showed that it was possible to separate two kinds of nerve fibre. One kind carried commands *from* the brain to muscles and to other peripheral organs. The other kind carried information from the periphery – mainly but not exclusively from the skin – *to* the brain. He demonstrated that in peripheral nerves the two kinds of fibre were often mixed but that they entered or left the spinal cord separately, as the anterior and posterior nerve roots. In the spinal cord, he guessed, the sensory and motor tracts remained separate on their way to the brain (in the case of sensory fibres) or from the brain (in the case of motor fibres). In France François Magendie aroused misgivings by his experiments on live animals but came to more or less the same conclusions. The question of priority did not exercise either scientist though their followers exchanged more or less polite insults.

On the Continent the man whose influence dominated neurophysiological research for half a century was Johannes Müller, professor of physiology first in Bonn and then in Berlin. As Justus von Liebig did for chemistry, Müller pulled together many strands of investigation; and, in his textbook, *Handbuch der Physiologie des Menschen*, formulated a series of general laws. He had no doubt that pain was a sensation in its own right, usually but not always accompanied by other sensations; that it was conveyed from the periphery and internal organs to the brain first by special nerve fibres and then in special nerve tracts. It was processed and interpreted in the brain. His electrical experiments led him to the concept of specific nerve energies. This meant that every sensory nerve fibre, in whatever way it was stimulated, generated a single specific sensation. Stimulation of the optic nerve, for example, whether by light, heat, pressure or pain, always produced a visual sensation. This meant that the human mind did not perceive the events of the external world directly but only as 'interpreted' by a variety of 'sensors'. This had to be true of pain as well. It suggested the existence of special pain receptors widely distributed both on the surface of the body and in interior organs. It was a departure from Descartes' simpler scheme of a bell and a bell-pull.[3]

Müller was an inspiring teacher. His pupils – Helmholtz, Du Bois-Reymond, Brücke, Ludwig, Köllicker and many others – bestrode physiological research

2. Thomas Willis graduated from Christ Church, Oxford, in 1642 and became a member of the 'Philosophical Clubbe' for the discussion of questions of science. Moving to London at the suggestioin of Dr Sheldon, Archbishop of Canterbury, 'he became so noted and so infinitely resorted to that never any physician went beyond him or got more money'. He still had time to write *Cerebri anatome*, his greatest work, published in 1664. Sir Christopher Wren drew some of the illustrations. In the book, which is a little verbose, Willis clearly envisaged different parts of the brain as having distinct functions. He died in 1675, aged fifty-five, and was buried in Westminster Abbey. See C. Zimmer, *Soul Made Flesh* (London, 2005).
3. Müller died in 1858, aged fifty-seven.

in German-speaking lands. A group of them, led by Helmholtz, published a *Physiological Manifesto* in 1847. It announced the aim of the life sciences to be the explanation of all vital functions in terms of physico-chemical laws. It was an act of defiance since Müller, the master, still clung to the ancient but somewhat woolly idea of 'vitalism'. The manifesto became influential and more immediately fruitful than Karl Marx' *Communist Manifesto* published a few months later.

Hermann von Helmholtz, the most original member of the group, was, on his mother's side, a proud descendant of William Penn, founder of Pennsylvania. He too would bequeath his name, if not to a state, at least to a useful physical instrument. But his beginnings were not particularly promising. He had planned to become a physicist but poor career prospects in academe in the decades following the Napoleonic Wars led him to become a doctor and then an army surgeon. The last profession in peacetime was far from demanding and left him enough leisure time to pursue his private researches. His seminal paper on the law of the conservation of energy, published in Müller's *Archiv*, led to his appointment to the chair of physiology in Königsberg. Twenty years earlier Müller had stated in his *Handbuch* that, while much indirect evidence pointed to the electrical nature of nerve impulses, it would never be possible to measure their actual speed. This was the kind of encyclical that was a challenge to the new breed of physiologists. Using instruments he designed himself, Helmholtz showed that the speed was 'exactly' 20 metres per second. The same genius for instrumentation led him to the invention of the ophthalmoscope. With it, he was the first man to see the human retina – that is the back of the eye – and the entry (or, pedantically, the exit) of the optic nerve.[4]

Emil du Bois-Reymond was a native of Neuchâtel, now in Switzerland but then part of the Kingdom of Prussia. He went to study with Müller in Berlin where he stayed for the rest of his long and productive life.[5] Like Helmholtz, he was a prolific designer of new apparatus and a lover of anything electrical. Fortunately, nerve impulses fell into this category. As a medical student, while other youths wasted their time on wine, women and song, he explored the electrical organ of the torpedo fish. As a professor of physiology he designed the 'algometer', essentially a generator of galvanic current, the first device for measuring sensitivity to pain. He used it to test the effect of various drugs on skin sensitivity. Like many other physiologists, he was sometimes more

4. After holding the chair of anatomy in Königsberg Helmholtz moved to the chair of pathology in Bonn, then the chair of physiology in Heidelberg and finally to the chair of physics in Berlin. His *Handbook of Physiological Optics* (1867) was a landmark, as was his *Sensation of Tones* twelve years later. He died in 1894, aged seventy-three.
5. Du Bois-Reymond died in 1896, aged eighty.

interested in designing instruments than in using them; but the algometer later became the favourite tool of Havelock Ellis and the Italian anthropologist Cesare Lombroso. They demonstrated 'beyond doubt' the grossly different sensitivity to pain of upright citizens and of various groups of criminals and mental defectives.[6] The device is still occasionally seen in medical school laboratories though it is no longer used to assess criminality.

Ernst Brücke migrated from Berlin to Vienna where his interests embraced physiology, biochemistry and histology but where his main research clarified the mechanism of neuromuscular transmission. Like most of Müller's disciples, he was fascinated by the mechanism of pain and experimented with cocaine without realising its potential as a local analgesic. He showed that it blocked the transmission of motor commands from nerve to muscle, the first compound shown to do so. One of his admiring young students, Sigmund Freud, would later carry the cocaine story further.[7]

Karl Ludwig became the most influential member of Müller's kindergarten and the most determined to topple such romantic notions as 'vitalism' and 'the spirit'. After spells in Marburg, Zurich and Vienna he was appointed director of the Physiological Institute in Leipzig, one of the grandest of the new temples to the sciences erected by Wilhelmine Germany. In his view – self-consciously different from French teaching based on pathology – there could be no effective medicine without physiology since all illnesses were essentially normal functions gone wrong. Among his inventions was the kymograph, the ancestor of all modern visual display units, for recording continuous changes in pressure and other measurable characteristics. He accepted the idea that pain was an electrical impulse transmitted by specific nerve fibres but emphasised the importance of the brain in interpreting it.

6. Havelock Ellis, *The Criminal* (London, 1889, p. 112). In this terrible book skin sensitivity is inversely related to the gravity of the crime, murderers being less sensitive than petty thieves. Ellis' discoveries were 'amply borne out' by tattooing habits. 'The extent to which tattooing is carried out among criminals in prison, sometimes not sparing the parts so sensitive as the sexual organs which are rarely touched even in extensive tattooing among barbarous races, serves to show the deficient sensibility of criminals to pain. The physical insensibility in criminals has indeed been observed by every one familiar with prisons . . . The instinctive criminal resembles the idiot to whom, as Galton remarks, pain is a welcome surprise. [Francis Galton never quite said such an idiocy.] He may even be compared with many lower races, such as Maoris, who do not hesitate to chop off a toe or two in order to wear European boots.'

Using Du Bois-Reymond's algometer Cesare Lombroso reported that sensitivity to pain was 'more obtuse' in murderers and 'incendiaries' (arsonists) than in normal citizens. The difference was less marked in non-violent criminals like embezzlers and swindlers. G. Lombroso-Ferrero, *Criminal Man according to the Classification of Cesare Lombroso, briefly summarised by his Daughter* (New York, 1911), p. 115.

7. Brücke died in 1893, aged seventy-three. Freud regarded him as a mentor and 'the greatest physiologist of his day'.

This interpretation, in his view, 'undoubtedly involved the production of specific chemical agents'. It was a prophetic concept and he felt frustrated that the chemistry of his day was not advanced enough to identify 'his' agents. He regularly chided his chemical colleagues for wasting their time on less important problems (like discovering new drugs). 'The brain secretes sensations and thoughts as the stomach secretes gastric juice or the liver secretes bile. Pain is a substance. It is waiting to be identified,' he wrote. Many of his wilder though usually inspired notions, like that of warring algesics and analgesics produced by brain cells, were more or less vindicated a hundred years after his death. Leipzig drew young physiologists and scientifically inclined doctors from all countries and they spread Ludwig's ideas. Among them were Pavlov and Sechenov from Russia; Gaskell, Lauder and Brunton from Britain; and Bowditch and Welch from America.[8]

...

The earliest studies of nerve cells were associated with yet another of Müller's pupils, Theodor Schwann.[9] Starting with plants, Schwann gradually came to identify the cell as the basic unit of all living things. But he was a versatile man, discovering, besides the Schwann cells, pepsin, the main digestive enzyme of the stomach, and suggesting that microscopic organisms might play a part in putrefaction. He also distinguished for the first time the conducting nerve filaments from their covering and insulating 'sheath', now known as myelin, and speculated about the possible consequences of losing this 'protective coating'. The concept of demyelinating diseases (like disseminated sclerosis) can be traced to his researches. Rudolf Virchow, future 'pope of pathology', also acquired his taste for laboratory work under Müller and carried Schwann's principles further. Forced to leave Berlin after his support for the 1848 uprising, he laid the foundations of cellular pathology – that is modern pathology – during his 'exile' as professor in Wurzburg. Political migration was one of the fructifying features of German science, ensuring the rise of previously insignificant backwaters. In the same way Jewish workers excluded from prestigious posts in Berlin and Heidelberg established the fame of 'second-class' provincial universities like Breslau. Virchow showed, among many other discoveries, that pus cells were wandering white blood cells from

8. Ludwig also discovered the gas pump to extract gases from fluids like blood and the 'Stromuhr' to measure blood flow. Many of his discoveries were published by his famous pupils such as Pavlov and Meyer who were devoted to him. (He was related to the Friedrich Karl Ludwig who described the inflammation in the throat now known as Ludwig's angina.) He died in 1895, aged seventy-nine. The quotations are translations from Ludwig's *Lehrbuch der Physiologie* (Leipzig, 1852–6), Vol. 3, p. 345.
9. After his apprenticeship with Müller, his most productive years, Schwann became professor of anatomy in Louvain and later in Liège in Belgium. He died in 1882, aged seventy-two.

the bloodstream, about the same time as John Hughes Bennett of Edinburgh described a new disease named 'leukocythaemia', later abbreviated to leukaemia.[10]

...

What in Germany were a dozen or so inventive minds scattered over the country's numerous seats of learning merged in France into a one-man band called Claude Bernard. Descended from generations of Burgundian wine growers, he was born in 1813 in one of the country's famous but in post-Napoleonic days impoverished regions. By the beginning of the nineteenth century the vineyards around the confluence of the Rhône and the Saône had been divided too often to support large families; and the Bernards were prolific. Claude, the youngest of four brothers, would have to leave, but the farm in Saint-Julien-en-Beaujolais would always remain home to him. Every year till his last illness he would return there for the *vendange*.[11] Bright and studious, he attracted the attention of the village *curé*, the learned abbé Donnet.[12] Donnet recommended him to the Jesuit college in the nearby industrial town of Villefranche and later to the college at Thoissey. He was a promising pupil; but at eighteen the family could support his studies no longer. As he showed an aptitude for science, he was apprenticed to a pharmacist in Lyons.[13] He then tried his hand at journalism and had some success in Lyons with an atrocious romantic drama called *Rose de Rhône*. It was enough to earn him a letter of recommendation to Marc Girardin, professor of literature at the Sorbonne. Girardin, as Bernard later recounted, was 'kindness itself but valued literature too highly to encourage me in my puny efforts. Did I really expect to compete successfully with young lions like Hugo? A visit to *Les Huguenots* convinced me that I did not.' Instead, the professor persuaded him to sit for a medical scholarship. Bernard never regretted the change of direction, though he would never practise as a doctor.

10. Virchow called it *weisses Blut*. In middle age Virchow once again branched into politics, becoming leader of the liberal opposition to Bismarck in the Reichstag, one of the few parliamentarians to strike fear in the heart of the Iron Chancellor. ('The trouble with the Herr Geheimrat,' Bismarck remarked, 'is that he regards all political problems of Europe as the result of defective sanitation capable of being solved by improved drainage.')
11. The house still stands with a spacious view over the valley of the Saône and a glimpse of Mont Blanc in the distance. It serves the same purpose as it did in the days of Claude Bernard.
12. Later Archbishop of Bordeaux and a generous patron of scientists.
13. His published lectures – *Leçons* – are collected in 17 volumes, easily dippable into. Few physiological and biochemical topics entirely eluded his creative mind. The *Introduction to the Study of Experimental Medicine* (1927) is his only work translated into English.

As a medical student, a little older than the rest, Bernard became Magendie's prize pupil and would eventually succeed his master at the Collège de France. His writing experience stood him in good stead. Unlike most nineteenth-century scientific literature, his flamboyant publications are still captivating reads, conveying the thrill of new insights as well as often startling factual information. They also convey his pleasing contempt for benighted colleagues who disagreed with him. His *chameaux* were the scientific equivalents of Monet's artistic *pompiers* fifty years later. Bernard developed his own, largely biochemical methodology, a legacy perhaps of his pharmacological training, and the range and scope of his experiments were extraordinarily wide. Yet he was often troubled by ill health, some kind of digestive trouble, perhaps a peptic ulcer, which forced him to take extended leaves of absence. He never went anywhere except Saint-Julien, maintaining that fashionable spas like Vichy and Baden-Baden deliberately poisoned their customers to make rich invalids prolong their sojourn indefinitely. During one of his illnesses, fearing the worst, his friend Louis Pasteur published a eulogy in *Le Moniteur universel*: 'Bernard is not a physiologist but French physiology personified,' he wrote. There was some truth in this. In the French tradition Bernard did not believe in a sharp divide between the chemistry of the living and the non-living, nor between the sciences and the humanities. 'With increasing knowledge our poets and our physiologists will come to understand each other perfectly.' Some of his own boldest physiological statements even scanned.

At various times in his life Bernard studied the anaesthetic properties of carbon dioxide and clarified many of the observations made by Henry Hill Hickman half a century earlier. In a series of experiments with frogs he showed how general anaesthetics like ether and chloroform are distributed throughout the body by the bloodstream but that they can affect the brain only by way of the spinal cord. In other words, anaesthetics did not act on the peripheral nerves directly, nor could peripheral nerves conduct their anaesthetic action. These findings led him to one of his few mistaken generalisations, that 'local anaesthesia does not exist'. 'The occasional error is a hallmark of genius,' Emile Roux said of Pasteur. The observation can also be applied to Bernard. Dead and dying amphibians populating their home eventually proved too much for Mme Bernard, a dedicated anti-vivisectionist, and she left him in 1870. Happily by then her considerable dowry had been spent.

Bernard's interest was also aroused by a South American arrow poison known to the natives as 'curare' (or something which must have sounded like it). Brought back to Europe by the English explorer Charles Robertson and described in his *History of America* in 1775, the poison paralysed but did not kill its victims. Bernard's initial interest was in the effect of the substance on sugar metabolism. (It causes a massive breakdown of sugar reserves in the liver.) But he also demonstrated that the poison acted on motor but not on sensory nerves, the second such selective action to be demonstrated. The first, also studied by Bernard, was the action of strychnine, the exact opposite of curare since it excited motor nerves, causing convulsions and death from painful muscular spasms, but had no effect on the sensory system. Of particular interest was the observation that the muscular paralysis induced by curare did not affect involuntary muscles like the heart or the muscular coat of the intestines or blood vessels, only voluntary or 'skeletal' ones.[14] Victims of poisoning eventually died from paralysis of the respiratory muscles and the diaphragm, which belong to the second category.

The two different kinds of muscle led Bernard to what many regard as his most important achievement, the recognition of the 'autonomic' nervous system. The action of autonomic nerves is not consciously registered, nor are they under voluntary control; but they regulate blood flow, the movement of the bowels, the renal excretory apparatus, sweating, the heart rate and other vital functions. Bernard recognised that, though the action of these nerves was unconscious, they could also convey pain of a special kind. Angina pectoris, 'the fierce sensation of constriction of the chest' due to insufficient blood supply to the heart, together with the 'terrifying sense of impending dissolution' which sometimes accompanies it, was such a pain. Intestinal and renal colic were others. Bernard's pupil, Paul Chambord, added the choking caused by bronchial spasm and urinary retention. It was the opening of a new chapter in the science of pain.[15]

Bernard also experimented with the recently discovered chloral hydrate, showing that it was a hypnotic, that is, a sleep-inducer, but not an anaesthetic, a painkiller;[16] and he demonstrated the synergism or mutual reinforcement of chloroform and morphine. More fundamental was his identification of the body's overriding physiological objective as the maintenance of the constancy of its own 'internal milieu'. The activity was later called homoeostasis.[17] This brilliant insight was as obvious as are all brilliant insights into Nature (like gravity or genetic inheritance); but nobody had grasped it before. Every second, Bernard pointed out, billions of unrelated chemical reactions take place in the body. Each one of those either consumes or generates a certain amount of heat. No known man-made machinery under these circumstances could maintain itself at a constant temperature. Yet the body does. And the same is true of a small number of other essential body constants. Without their constancy life would cease. This, he speculated, suggested the underlying

14. See Chapter 45.
15. See Chapter 46.
16. See Chapter 45.
17. The term was coined in 1910 by the Harvard physiologist Walter Cannon.

purpose of pain. It was in its origin the alarm signal which warned against any threatened disturbances of the *milieu intérieur.* But Bernard disliked – or professed to dislike – speculation: 'facts are so much more amazing'. (The preference, according to his detractors, sometimes led him to turn speculations into fact.) He ended his career as professor at the Sorbonne, president of the Académie des sciences and Senator of the Third Republic with a clutch of gifted pupils to carry on his work. He was the first scientist in France to be granted a state funeral and the only one to have his bust removed from the Pantheon by the Nazis in 1940.[18]

. . .

Reflex action, the body's first line of defence against injury, had puzzled doctors, naturalists and philosophers since Aristotle and Descartes.[19] Another pioneer, the Reverend Stephen Hales of blood pressure fame, showed around 1730 that contractions could still be demonstrated in decapitated frogs by pricking the skin of their legs but only if their spinal cord was left intact. His work was confirmed by Robert Whytt, an Edinburgh professor, who found that many such responses did not even require the whole of the spinal cord but only a segment. Activities like blinking were 'unconscious' because they did not involve the higher brain centres. He identified the constriction and dilatation of the pupil provoked by light and darkness as well as by focusing on close and distant objects as 'unconscious but beneficial reactions'.[20] Their role in animals and originally perhaps in man had to be 'defensive'.[21]

Earlier work was brought together and expanded by a Nottingham doctor, Marshall Hall, who in the English tradition never obtained an academic post or official patronage but conducted his fundamental researches as a hobby, financed by his private practice. He showed that after transection of the spinal cord in the thoracic region the front legs of a frog could still move voluntarily but that the back legs were drawn up in a motionless spasm. But when a stimulus was applied to the back legs, they twitched violently – indeed more violently than before – but only once per stimulus. He concluded that the

18. He was sixty-five when he died in 1878. His bust was removed in 1940, presumably by a zealous *Obersturmführer* who thought that both the name and the image suggested a Jew.
19. The term is derived from *reflectere*, to bend or turn back.
20. In 'An Essay on the Vital and Other Involuntary Motions of Animals' published in 1760.
21. The pupillary response was for a time referred to as Whytt's reaction. Whytt died in 1766, aged fifty-four.

    Hales' and Whytt's ideas were elaborated by Georg Prochaska, professor of anatomy in Vienna (died in 1820, aged seventy-one), who stated that 'the impressions coming through the nerve to the brain are *reflected* into the motor nerves'. This may have been the first use of the term 'reflex'.

nervous system was made up of a series of segmental reflexes and that these segments were coordinated by ascending and descending nerve tracts to and from the brain. It was the complexes as a whole which established both patterns of movements and complicated sensations. The spinal cord therefore was more than a conduit: it had a life of its own like the brain and could 'perceive' pain and other sensations. Indeed, in Hall's view it was superior to the brain, since 'it needed no sleep but was always ready to function'. This was fortunate for, besides blinking, coughing and all kinds of other protective responses, spinal reflexes were also responsible for most actions of the newborn. 'Without spinal reflexes no neonate could survive.' Some of Hall's ideas sound like a dress rehearsal for the 'Gate Control' theory of pain developed in the mid-twentieth century.[22]

After Hall, reflexes became a prime preoccupation of neurophysiologists as well as a prop in the theory of evolution. Wilhelm Heinrich Erb, professor of physiology in Heidelberg, gave the first complete description of the knee jerk, the most famous of the diagnostic reflexes. He showed not only that it was abolished when the critical segment in the spinal cord was destroyed but that it could also become abnormally brisk. This seemed to happen in some diseases of the brain.[23] He concluded – prophetically at the time – that the brain 'sometimes exercised a restraining influence' on centres in the spinal cord.[24] Soon the patella hammer would become as numinous a repository of medical wisdom as the stethoscope.

. . .

Steam travel by land and sea, international journals and medical conferences were abolishing separate national schools in medicine. Augustus Waller, an Englishman, studied in France and Bonn and presented his discoveries to the French Académie des sciences in 1852. Peripheral nerves had already been shown to be extensions of nerve cells in the spinal cord or in two chains of ganglions on either side of the spine. Waller demonstrated that cutting such nerves led to the irreversible degeneration of the distal but not of the proximal fibres; and that this depended on the proximal fibres remaining connected to the parent cell. Experimentally, 'Wallerian degeneration' helped to map out areas supplied by individual nerves since the cutting of a nerve resulted in a

22. Marshall Hall eventually moved to London and died there in 1857, aged sixty-seven. The gate control concept is discussed more fully in Chapter 48.
23. Erb never retired. He died in 1921, aged eighty-five. He was feared for his gruff manner but was a great music lover, dying from a stroke during a performance of Brahms' Fourth Symphony.
24. This accounts for the 'paradoxical' combination of a limb incapable of being moved voluntarily but showing abnormally violent reflex responses This is seen in some cases of stroke and other abnormalities involving the motor cortex.

circumscribed area of numbness and the paralysis of a characteristic group of muscles.[25]

Charles Edouard Brown-Séquard was a native of Mauritius, studied medicine in France, taught and practised in England, became professor of neurology at Harvard and eventually returned to France to succeed Claude Bernard at the Collège de France. He correlated clinical syndromes, especially changes after strokes, with the course of nerves and nerve tracts in the central nervous system. Among other observations, he showed that fibres from the left side of the brain crossed to the right of the spinal cord and vice versa, an anatomical arrangement so bizarre that it was for some time disbelieved.[26] 'Nature is never unnecessarily complicated' was one of the grand but demonstrably untrue axioms pronounced by Victorian sages.

. . .

Cerebral localisation grew out of the pseudo-science of phrenology. Towards the end of the eighteenth century Franz Joseph Gall of Vienna proposed that different mental functions were localised in different parts of the brain and that these areas could actually be detected by palpating the head for lumps and bumps. This remained for almost a century a popular pastime, with phrenological maps and busts – the latter often occupying places of honour on top of bookcases between Homer and Cicero – showing thirty-odd 'character areas' the seats of aggression, amiability, holiness, liberality, combativeness, ambition and many others. Gall had also postulated a special region of the brain concerned with pain but its whereabouts escaped his agile fingers.[27] At a more scientific level the idea that certain cerebral regions served specific functions started to be investigated in the 1830s. In the 1850s the French anatomist Marie Jean-Pierre Flourens used pigeons to show that removal of both cerebral hemispheres made the birds blind whereas removal of one hemisphere made them blind only in the opposite eye. A pigeon whose cerebellum or hindbrain had been removed could see and hear but its balance and muscular coordination was destroyed. He concluded that the cerebrum was the organ of thought, sensation, will and conscious pain whereas the cerebellum governed 'purposeful movement and hidden feelings'. But, contrary to phrenological ideas, in his view all parts of the nervous system 'concurred, conspired and

25. Waller died in England in 1870, aged fifty-four.
26. Brown-Séquard died in France, having published over 500 papers, in 1894, aged seventy-seven.
27. Gall was eventually expelled from Vienna and died in Paris in 1828, aged seventy. Goethe was among his admirers.

consented' in directing essential functions, and character was the product of the whole, not of parts.[28]

Clinical observations and physiological research began to complement each other. Flourens' friend, the surgeon Jean Legallois, noted that a patient whose hindbrain had been damaged in a duel developed progressive respiratory difficulties. Legallois' localisation of the respiratory centre in the medulla was almost entirely accurate.[29] Philip Gage, a Connecticut quarryman, had a crowbar passed through his skull in an accident. He recovered and lived for another ten years, though it was possible to poke a finger through the front part of his brain. His only 'disability' was a welcome change in character. His former attacks of violent temper ceased, as did his unquenchable thirst for alcohol.[30]

The localisation of functions in special areas of the brain received a further impetus from the work of Eduard Hitzig, a military surgeon in the Prussian army. He and his assistant, Gustav Theodor Fritsch, took an interest in soldiers who had had parts of their skulls blown away and survived. The injuries made areas of their brain surface – or the membranes covering the cortical surface – accessible to electrical stimulation. Hitzig and Fritsch were thrilled when they could produce involuntary eye movements and then other facial contractions by applying low-voltage electrodes to certain areas. They then generated involuntary sounds – 'some almost mimicking coherent speech' (German, of course) – and finally screams without any distress. Hitzig was anxious to find the 'pain area' in one soldier whose injuries had left him with a severe burning sensation in his amputated limb, perhaps with the idea of selectively destroying it. His search was unsuccessful; but by increasing the voltage of the current he could make painless movements painful.[31] The two investigators then went on to animal experiments and found that the convolution immediately in front of a prominent indentation known to anatomists as the central sulcus seemed to control most movements. In London, at King's College Medical School, David Ferrier repeated many of their animal experiments and named the region the 'motor cortex'.[32]

Sensations were more difficult but not impossible to localise. Using rabbits and monkeys the Liverpool physician Robert Caton reversed the

28. Flourens became professor of comparative anatomy in Paris in 1825 and died in 1867, aged seventy-three. He demonstrated the anaesthetic effect of chloroform in dogs and published his findings a year before Simpson's experiments on himself and his friends. Simpson was unaware of Flourens' work.
29. Legallois died in 1860, aged fifty.
30. His can be regarded as the first successful case of a frontal lobotomy.
31. Eduard Hitzig became a major-general and later professor of mental diseases in Zurich. He died in 1907, aged seventy. Fritsch died in 1891, aged fifty.
32. Ferrier died in 1928, aged eighty-three.

electrical experiments of Ferrier, applying a galvanometer to the brain and recording electrical currents set up by peripheral stimulation.[33] The first successful demonstration elicited a current in the visual cortex when a light was shone into the animal's eye. Peripheral pain provoked diffuse and uncoordinated activity, often difficult to repeat or to reproduce. No 'pain centre' could be identified. The spikes of current provoked by different kinds of pain – heat, pressure, electrical current – showed no specificity. All were usually but not always attenuated and often changed but not abolished by general anaesthesia. That was odd. If anaesthesia did not abolish electrical activity associated with pain in the brain, how did it work? The mystery deepened. It remains a mystery. Later a sensitive galvanometer could pick up electrical activity even when applied to the intact skull, the beginning of modern electroencephalography.

Ferrier became a virtuoso at reproducing clinical syndromes in animals by removing or irritating certain parts of their brains. His book, *The Functions of the Brain*, published in 1876, became a bestseller. Two years later, at the International Medical Congress in Paris, he showed some of his ataxic monkeys to a spellbound audience. So astonishingly 'human' was their behaviour that Jean-Martin Charcot of the hôpital Salpêtrière interrupted the demonstration exclaiming: 'Mais ce n'est pas un singe! C'est un malade.' (But this is not a monkey! It is a patient.)

But patients and Charcot belong to the next chapter.

33. Caton died in 1926, aged eighty-four.

# Chapter 32

# THE NEW PATHOLOGY

In an inspired moment to which he was prone Helmholtz declared that every disease was a physiological experiment and every physiological experiment was the model of a disease. To prove him right, during the second half of the nineteenth century physiology (the study of normal life) and pathology (the study of disease) intertwined. It was a productive intertwining which makes the separation of the 'new physiology' from the 'new pathology' convenient rather than real.

## *I*

In pathology as in physiology, there had been notable forerunners. As long ago as 1680 Johann Jakob Wepfer, a Swiss anatomist, showed that strokes were usually associated with blood clots inside the skull and suggested that bleeds in different regions might account for different patterns of pain and paralysis. Two years after Waterloo an English physician, James Parkinson, published his *Essay on the Shaking Palsy*, describing the characteristic tremor and stiffness still diagnostic of the disease.[1] He noted that the trembling was different from that seen in most old people: voluntary effort to control it made it worse. (The term 'intention tremor' was probably coined by Hughlings Jackson sixty years later.) Parkinson commented on the 'surprising fact' that the disability could almost totally disorganise 'muscle function' without affecting 'sensibility' and speculated that the seat of the disease might be in some specialised area of the brain where the action of muscles was 'harmonised'.

. . .

The new pathology did not fully emerge till the mid-nineteenth century. When it did, the causes and mechanisms of pain were one of its main pre-

1. Parkinson died in 1824, aged seventy-nine.

occupations. Of course pain had always been a cardinal manifestation of disease. But some of the pain syndromes commanding the interest of the new pathologists seemed to be new themselves. If they had existed before, they were getting more common. They had nothing to do with injury or acute infections like boils. They were not the result of gangrene. They were not even associated with cancers eating into normal flesh. Yet they were terrible.

Among the early pioneers, Moritz Romberg became director of the Berlin Neurological Clinic in 1840. He was to devote his life to bringing order to an extraordinary array of symptoms and signs which, he suspected, were caused by a single progressive disease.[2] Worst were the attacks of 'lightning pain', usually but not always in the lower limbs, striking for no obvious reason. Sufferers who had known festering wounds and had undergone operations without anaesthesia described them as worse than anything they had experienced before. Morphine barely touched them. Romberg's description has never been bettered. The pain seemed to pierce the calves, heels or thighs almost always in a transverse direction and was sometimes described as 'like a hot knife plunged into the muscles and then turned'. For some reason still unexplained the attacks often got worse or more frequent in cold and wet weather, 'a feature', Romberg warned, 'which must not mislead you into diagnosing rheumatism'. The attacks were often followed by pins and needles in the toes and a 'girdle' sensation of constriction around the trunk. Gradually the attacks would become associated with a variety of sensory signs. Feeling might be lost over the chest, the insensitive area being distributed 'like a cuirasse', over a 'butterfly area' on the bridge of the nose spreading out on either side (later known as the *masque tabétique*), on the soles of the feet and in other characteristically circumscribed patches. As months or years passed, muscular weakness and then paralysis of certain groups of muscles might develop. Mental symptoms were initially ascribed to the almost unbearable suspense of impending attacks. They were later recognised as manifestations of another late complication of the underlying disease, GPI or general paralysis of the insane.[3]

2. Romberg died in 1873, aged seventy-eight.
3. Tabes and general paralysis of the insane (GPI) combined were later called taboparesis. The congenital forms, the causative spirochete crossing the placenta from mother to foetus, could occur in young people, declaring itself in their teens or early twenties. The acquired disease most commonly started to cause neurological and mental symptoms in the late thirties and early forties, about twenty years after the infection.

Among numerous famous sufferers from painful tabetic crises were Friedrich Nietzsche and the composer Delius. GPI was by far the commonest cause of 'mental breakdown' and madness in middle age. Victims included Schumann, Hugo Wolf, Manet, Maupassant, the poet Lenau, Smetana and Semmelweis among many, many others.

What emerged from Romberg's studies was the nearly complete clinical picture of tabes dorsalis. He correctly related the symptoms and signs to areas of degeneration in the spinal cord without realising (though, it seems, strongly suspecting) that the process was a manifestation of syphilis. The disease usually started in the lower limbs because it was the lumbar region of the spinal cord which was most commonly affected. Nobody knew the reason for this then: nobody knows the reason for it now. Romberg's *Lehrbuch der Nervenkrankheiten des Menschens* (1857) was ground-breaking. His name survives in 'Romberg's sign', still one of the set pieces of neurological diagnosis. Patients with a 'positive Romberg' can stand upright with no difficulty with their eyes open but begin to sway and fall as soon as they close their eyes. The reason is that normal vision can inform individuals of their position in space; but with the eyes closed that knowledge depends entirely on 'joint sense', that is on signals from the ligaments, tendons and capsules of the joints. This 'joint sense', in fact a combination of sensations from joints and muscles, is lost early in tabes. Ironically, Romberg warned that his sign was not diagnostic of tabes but could also be positive in nervous individuals as well in some other neurological conditions.[4] On the other hand, the characteristic 'tabetic gait', wide-based, the heels being brought down jerkily with a stamp and the feet lifted high in stepping, all due to the faulty projection of the position of the limb, 'sometimes allows the diagnosis to be made as soon as the patient enters your consulting room'.

Tabes was finally identified as a late stage of syphilis by Guillaume Duchenne, the son of a Boulogne sea-captain.[5] He accurately located the lesion in the nerve fibres ascending in the posterior columns of the spinal cord. His description of the clinical features and pathology, published in 1858, was so complete that in France the illness became known as Duchenne's Disease. It was a triumph of the new pathology. As was true of many of these

In fiction Ibsen provided the most memorable portrayals. Dr Rank in *A Doll's House* is a puzzling case. Obviously a victim of congenital syphilis, he is waiting for tests which will confirm (or contradict) the diagnosis. But what were the tests which prove positive in the last act? The first diagnostic test for syphilis, Wassermann's reaction, was not described until twenty-five years after the play was written (1879). In another Ibsen play, *Ghosts* (1881), congenital syphilis is at the centre of the plot, the hero, Oswald, going mad on the stage in the last act.

4. The second and more reliable diagnostic sign of tabes was described six years after the publication of Romberg's book by the Scottish ophthalmologist, Douglas Argyll-Robertson (died in 1909, aged seventy-two). Argyll Robertson, a pioneer of ophthalmology in Britain, showed that in tabes the pupils react more or less normally to changing from near to distant vision but do not react by constriction when a light is shone into the eye. The exact reason for this characteristic dissociation is still not entirely clear.
5. Known in France as Duchenne de Boulogne, he died in 1865, aged fifty-nine.

triumphs, it was purely pathological. There would be no effective treatment for almost a hundred years.[6]

. . .

A younger colleague of Duchenne, Jules Déjérine, focused on another, different kind of pain. This was caused not by changes in the spinal cord but by degeneration of the peripheral nerves. The symptoms and signs accompanying the pain – tenderness, altered sensations and muscle weakness were the main ones – were almost always symmetrical, spreading from the periphery centrally. Déjérine described them as showing a 'glove and stocking distribution', a term still used.[7] In Déjérine's days the commonest underlying cause was chronic alcoholism. Later the abnormality was shown to be the result of Vitamin B deficiency such as occurred in many other diseases as well. The condition could also complicate diabetes, lead and other heavy metal poisonings, diphtheria, a number of more or less common toxic states and inborn biochemical abnormalities. The pain was sometimes mild but could also be excruciating. When the nerves affected were those of the autonomic nervous system (described about the same time by Claude Bernard) the attacks could simulate almost any internal disease, including appendicitis and intestinal obstruction. In such cases emergency operations not only failed to demonstrate any abnormality visible to the naked eye but could also make the condition worse.

. . .

Another new pain-related syndrome that emerged became associated with the name of Jean-Martin Charcot. Like Napoleon who eclipsed his peacock marshals with the studied simplicity of his attire, Charcot, 'the Napoleon of neuroses', eclipsed leading French clinicians and neuropathologists by his understated superiority.[8] He was one of three sons of a poor Paris coachbuilder, and his parents were unable to finance the higher education of all their male offspring. They conducted an improvised intelligence test. Jean-Martin, the youngest, emerged as the winner. Given a free choice, he elected to study medicine. With stiff entrance examinations this was at the time the longest and most arduous of the university courses; and the young man was

6. Many treatments were tried in syphilis, some (like mercury) as lethal as the disease. The illness did not become effectively treatable until the advent of penicillin in the late 1940s.
7. Déjérine died in 1917, aged sixty-eight. His wife, one of the three gifted and self-assured American Klumpke sisters, was the first woman to qualify as a doctor in France and became a distinguished neuro-anatomist and neurologist herself.
8. Short and a little plump, he was said to show a passing resemblance to the first emperor. This did not displease him. He died in 1893, aged sixty-eight, from coronary disease.

twenty-eight by the time he qualified. But then, even as a young assistant, he earned the reputation of being the most astute diagnostician in Paris. He was said to be particularly good at spotting impending mental disease. This led to a large and lucrative practice that enabled him to repay his parents and to support his brothers in their careers. They remained a close family. In 1862, at the age of thirty-six he was appointed physician to the hôpital Salpêtrière. This was where in 1799 the great Philippe Pinel first removed the chains from lunatics. It was famous but, like many old Paris hospitals, vast and understaffed. Part acute hospital, it was also workhouse, lunatic asylum and a shelter for vagrants. Charcot and his friend, Edmé Felix Alfred Vulpian, were responsible for over 500 patients or 'souls'. They had the help of one junior doctor and a clutch of medical students. In this setting – often daunting but never short of interesting 'clinical material' – Charcot established within a few years the greatest neurological clinic of all time.

In the realm of nervous diseases Charcot's range of interests was all-embracing. He shed new light on the neuralgias, on epilepsy, spastic conditions, muscular degenerations, sensory syndromes and every aspect of neurosyphilis. Eponymous fame links him today with a clinically disabling disease associated not with pain but with the abnormal *absence* of it. It became and has remained the 'clinching argument' of those who see pain as intrinsically useful. It was never intended by Charcot to be seen in this way. In tabes dorsalis and in other conditions where 'deep' sensation is lost, the wear and tear of movements, unprotected by the restraints imposed by pain – or rather, by the feeling of stress that precedes pain – leads to the progressive disorganisation of the joints. Swollen, unstable and grossly deformed they become painless but useless appendages. Charcot, a friend of England – despite Waterloo which, he assured Lawson Tait, he was prepared to forgive – presented some of the first specimens of such joints preserved in pots of formaldehyde (complete with details of clinical history in two languages) to pathological museums in London and Edinburgh. It was British surgeons and pathologists who coined the name 'Charcot joint'.

His Sunday morning demonstrations of interesting clinical cases became one of the 'musts' for visitors to Paris, easily eclipsing the Louvre and almost rivalling the Folies Bergère. Admission was by ticket or 'influence'. Charcot liked to see 'a good audience'. The painter Edvard Munch from Norway rubbed shoulders in the front row with Dr Freud from Vienna; and Verlaine exchanged pleasantries with English milords and seducible American millionaires.

Like all great showmen, Charcot prepared his demonstrations with meticulous care, examining his exhibits well in advance, probing their past histories, rehearsing special 'effects'. Each of his assistants – by the 1870s he had dozens of unpaid acolytes from many countries of Europe – had a carefully assigned role. Apart from his mastery of the subject, he was well read in ancient and modern languages – his English and Latin were French sounding but fluent – and he had a gift for the apt quotation. His greatest fame today rests – a little unfairly – on his study of hysteria and on demonstrating the essential features which separate the hysterical syndrome, especially hysterical pain, from the clinical manifestations of 'organic' lesions.[9] Contrary to many of his followers, he never decried hysterical symptoms as 'faked' and disliked the term 'simulated'. The simulation was unconscious and required in his view as much ingenuity and understanding from the doctors as 'genuine' diseases.

In Charcot's view the essential feature of hysterical manifestations was their imitative character. Patients always unconsciously reproduced, sometimes with uncanny accuracy, the symptoms and signs they had seen in individuals who had suffered from an organic disease. The task of the doctor was to demonstrate that, despite the resemblance, the objective clinical signs (like the pattern of disturbed reflexes or the areas of insensitivity to pain) could not be explained in terms of an anatomical lesion. Even with a thorough knowledge of neuropathology, this was often difficult. One of Charcot's celebrated cases was the wife of a doctor who had studied the symptomatology of cerebellar lesions in her husband's textbook (not with any conscious intention to deceive) and mimicked every improbable detail.

A famous painting by Pierre-Andre Brouillet shows Charcot in 1887 demonstrating a 'hysterical fit' in a woman, the patient assuming a posture which he likened to a rainbow or *arc-en-ciel*. She would certainly have fallen with perhaps catastrophic consequences had she not been caught by assistants. In the painting the bearded character who supports her shoulders is the soon to be famous Polish neurologist Joseph Babinski.[10] 'She knew that she would be caught but not consciously,' Charcot would assure his spellbound audience. 'Unlike real epileptics, hysterical patients rarely if ever injure themselves.' He would then demonstrate another cardinal hysterical feature: the patient's non-response to pain. 'She behaves like a saint but again without knowing it.'

Both the painting and the description conjure up an unattractive individual. He is using a patient as a party prop. But all clinicians in his day did that. In fact, despite his imperious manner, Charcot was a kindly man. He was, as Freud remarked many years later, rather pleased with himself; 'but why not? He had every right to be.' From the cramped living quarters over his father's workshop in the rue Lafitte to the palatial mansion in the boulevard

9. In Charcot's day the commonest hysterical 'showpiece' was the epileptic fit because epilepsy was common, uncontrolled and spectacular. Even in the late 1940s one could not travel around in Paris or London for more than a day or two without witnessing a fit. Today epilepsy is no less common but far better controlled and fits, especially in public places, are rare. Hysterical fits have declined in parallel.
10. He was to describe the important plantar reflex. When the sole of the foot is scraped, the toes normally turn down. In cortical disease they may turn up.

Saint-Germain had been a long journey; and the professor never forgot his origins. 'The grounds of his house are infested with vagrants and drunks whom he had picked up in the course of his work,' Madame de Cros complained. 'You cannot approach the front gate without being accosted.' His wife and their children, Jeanne and Jean, adored him and he them. He also kept an assortment of pets, including two cosseted and by all accounts exceedingly destructive monkeys.

...

Whatever clouds gathered on the political horizon – and they seemed far from menacing during the 1860s – the Paris of the Second Empire was the gayest, most artistic and most frivolous city in Europe. The Charcots were regular attenders at Charles Garnier's splendid new Opéra though the professor's musical tastes did not extend beyond Offenbach whose marvellous tunes he whistled ceaselessly. But, like his friend and patient Louis Pasteur, he was an accomplished artist and a generous patron of artists. He was especially fond of Daumier's savage caricatures of doctors at work. Of course, he explained to guests, the images were the depictions of his colleagues (like the poor ineffectual Potain), not of 'real doctors' like himself. Mme Charcot too and both children were artistic. Camille Corot, kindly, pipe-smoking painter of wistful landscapes and another grateful patient, gave them lessons. He resolutely refused to accept a fee. Jean was to become a famous Arctic explorer.

When eventually Offenbach's General Boum (French militarism) received his come-uppance at the hands of Prince Puck (Bismarck), Mme Charcot and the children fled to Dieppe and then to Upper Phillimore Gardens in London.[11] The head of the family stayed behind. His long letters to his wife, written on the thinnest paper in minute handwriting since the correspondence had to be smuggled out of the city by carrier pigeons, were peppered with schoolboy jokes. Yet life in besieged and starving Paris and later in the Paris of the Commune was no children's party. Charcot maintained strict neutrality, making his way to the Salpêtrière every day with a Red Cross armband. The organisation was less than fifteen years old but the symbol was already known. At least nobody shot him. Jeanne would wear her father's band in 1914 as an army nurse.

...

11. Napoleon III abdicated in 1871 after France's disastrous defeat by Prussia. He himself had been taken prisoner at Sedan. Most people regarded the Third Republic as a temporary, makeshift regime which would soon give way to a third Bourbon Restoration. This foundered largely on the refusal of the Bourbon pretender, the comte de Chambord, to accept the *tricolore* as his kingdom's flag.

After the war Charcot became one of the prides of the Republic. He acquired a taste for travelling. He was fêted in Dublin (and offered a chair at Trinity College) and visited Russia on several occasions. The last involved a circuitous journey since, despite his claim to be entirely free from 'petty provincial prejudice', he would not set foot in Germany after 1871.

Contrary to legend, he was not an atheist. He may have timed his clinical demonstrations to clash with Sunday High Mass in Notre Dame (which he regarded as a pagan rite), but he himself liked to attend a short mass at Saint-Sulpice at dawn 'to collect my thoughts'. When pressed he would say: 'Of course I believe in God, but he is far, far away.' But he laboured in an age when France became bitterly divided into a Catholic, monarchist and often anti-Semitic Right who hated the Republic, and a republican, anti-clerical and anti-religious Left who hated the Church. It was in his capacity as a representative of the Left as well as the greatest living authority on hysteria that he was called upon to express an opinion, indeed to pass a verdict, on an event which convulsed the country during the last years of the Empire and for decades after the downfall of the 'little Napoleon'. The event inaugurated a way of dealing with pain that was unheard of in hospital wards, unresearched in university laboratories and undiscussed in learned academies. It was both new and ancient. It was timeless.

## *II*

On 11 February 1858 in one of the most impoverished and remote corners of France, the foothills of the Pyrenees, a fourteen-year-old girl by the name of Bernadette Soubirous, small and sickly even for a region where childhood disease and undernourishment were endemic, had a curious experience. She had only recently returned to her native Lourdes from a neighbouring village where she had been farmed out as a shepherdess. Her father, François Soubirous, was a failed miller who now tried to earn a living as a casual labourer. But casual labour was scarce. Perhaps he also drank too much. His wife, Louise, née Casterot, had to take in laundry as well as work in the field to scrape together enough to support their five children.

On the memorable winter's day Bernadette had been sent by her mother with her sister Toinette and a friend to collect driftwood for cooking. Bernadette had joined the older girls though her weak and often painful chest – usually described as her 'asthma' – made it unlikely that she would be of much use. The threesome had to avoid places where they might be accused of stealing so they headed for the common land. The area was called Messabielle after a nearby rock formation. It had a cave at its base. While the two older girls waded into a stream Bernadette sat down near the entrance to take off her stockings. (She was the only one allowed such a luxury because of her poor health.) First she merely heard the wind making an unusual sound; but then she beheld a soft light coming from a niche in the rocks. From that moment on the cave would become the grander-sounding Grotto. Out of the Grotto stepped a beautiful young woman, hardly more than a girl of Bernadette's own age, smiling. She beckoned to Bernadette to come closer. Startled, Bernadette instinctively reached for her rosary; but in her confusion she was unable to pick it up. The girl/lady stepped forward and handed it to her with a smile. The apparition then made the sign of the Cross and disappeared. No words had been spoken. As on all subsequent occasions, nobody other than Bernadette had heard, seen or felt anything unusual.

Bernadette begged her sister to keep quiet about the apparition; but that was more than Toinette could bear to do. Both girls received a thorough thrashing and Bernadette was forbidden to visit the area again. She did her best; but three days later an irresistible force drew her back. This time she was with about a dozen friends. The girl/lady in white appeared again and Bernadette begged her to leave unless she came from God. To make sure, she sprinkled holy water on her. To Bernadette's relief, the beautiful apparition merely inclined her head and smiled again. Then one of the girls threw a big stone in the direction where Bernadette was pointing. All but Bernadette jumped with fright. She remained pale and seemingly paralysed, so her friends ran for help. By the time Bernadette emerged from her trance she was surrounded by women amazed by her general happiness.

Once again Louise Soubirous forbade her daughter to go near the Grotto. Once again she was drawn back. Following instructions from the local *curé*, the abbé Dominique Péyramale, she held out a pencil and paper for the apparition to write down her name; but the vision in white only laughed gently. But she addressed Bernadette for the first time with great politeness. Speaking in the local patois, she said 'Would you be kind enough to come back here for fifteen days?' Bernadette refused to guess the apparition's identity. She called her simply *Aquero*, the patois for 'this one'. It referred to someone not quite human but not necessarily divine either. The event now began to make waves and somebody, it was never clear who, suggested that the apparition might be the Virgin Mary. A growing number of local women and even a few men began to accompany Bernadette on her daily visits to the Grotto. After the sixth visit she was interviewed by the local gendarme, a kindly man concerned that the concourse of idle folk might trigger off a disturbance. He was sure that he was dealing with a childish prank. The region was teeming with ghosts, resurrected dead or *revenants* and other supernatural beings. But Jacomet, the officer, was the first to be confronted with an assurance that was to confound hundreds of officials over the next ten years, both lay and clerical. Putting Bernadette's words into writing, he was also the first to give a written account of the apparition, of the white robe drawn together with a blue sash, the white veil over the head and the yellow roses on each of the bare feet. (The fact that under her beauty the girl/lady had no cloven feet would prove an important detail.) Once again Bernadette was threatened with prison if she returned to the Grotto. Once again there was no holding her back. The abbé Péyramale was much feared for his temper but also revered for his charity. Initially deeply sceptical, he became the first 'official personage' to become Bernadette's protector.

On 24 February *Aquero* asked for penitence and prayers and instructed Bernadette to kiss 'this holy ground'. Bernadette scratched the earth and discovered the stream or fountain which, to this day, remains the focal point of all pilgrimages to the Grotto. The eleventh and twelfth apparitions saw only more gestures and enjoiners to prayer and penitence; but on the thirteenth another message came from the lady/girl: 'Go and tell the priests to come here in procession tomorrow and then build a chapel.' Péyramale was furious. How could he, even if he were that way inclined, arrange a procession for the next day? But next day was market day in Lourdes and a procession did form. The assembled crowd was disappointed. *Aquero* did not announce her name, nor did she make the rosebush at the mouth of the Grotto blossom. Three weeks passed while Bernadette did not go near the place. Then, on 25 March, she was drawn back at dawn. It was only five o'clock but already there were women waiting, as well as Jacomet, the gendarme. The girl/lady in white appeared and Bernadette pressed her to say who she was. After four requests, *Aquero* finally put her hands together and turned her eyes heavenwards. Then she spoke the words that would catapult Lourdes to fame and Bernadette Soubirous to sainthood. 'Que soy era Immaculada Counceptiou' or 'I am the Immaculate Conception'.

. . .

For centuries the Virgin had been addressed as *Mère immaculée* by the faithful but the Immaculate Conception was a new dogma, promulgated by Pius IX only four years earlier. Partly a response to the godless revolutionary upheavals of 1848, it would confront and confound the seemingly irresistible onward march of secularism. Its declaration, which made Mary the first and only woman in creation to be free from original sin, was deliberately provocative. It was received with fury by Protestants, deists, freethinkers, radicals and even by many 'progressive' Catholics. But it forged an unlikely alliance between rural, medieval piety, the Vatican, the deeply devout section of the bourgeoisie and the old Gallican aristocracy. None of this could have been known to Bernadette. To her there were two more apparitions. On 7 April Bernadette remained unburnt when she touched the flame of a candle. Finally, on 16 July, she saw *Aquero* for the last time, no longer in the Grotto which the authorities had boarded up to 'prevent disturbances', but on the other side of the stream, waving her farewell.

. . .

That was all. Bernadette never claimed to have witnessed a miracle but what she had seen and heard she repeated a thousand times. Her story never changed. It never became embellished. It never contradicted earlier accounts. Her simplicity was her strength. She was eventually moved to a strict convent in Nevers in central France, was forbidden to respond to contacts from outside or to revisit 'her' shrine. She died of tuberculosis in 1879.[12] Her body was said to be miraculously preserved, the ground for her beatification and canonisation by Pope Pius XI in 1921. By then hundreds of thousands were visiting the shrine every year, rich and poor, men and women, old and young, all convinced on leaving as well as on arrival of the transcendent significance of her vision. Even among those whose health did not benefit or who actually died at the shrine – as thousand did – none complained that they had been let down.

Yet there was nothing 'inevitable' about Lourdes' rise to fame. Pious children seeing visions were two a penny in France. Miraculous grottoes were scattered over its wild, hilly regions. Brittany alone had – and still has – more than 200 holy sites of pilgrimage to cope with every conceivable ailment. Material coincidences marginally contributed. The opening of the railway line in 1864 made Lourdes relatively accessible. Soon special 'white' or 'white and gold' trains would transport the sick and dying from all corners of Europe. During the early years the courting of the Catholic vote by Napoleon III discouraged sceptical but ambitious bureaucrats. Above all, in a rampantly secular and materialistic age, a mass of people were clearly longing for purity and sanctity, remorse and redemption, the protection of the Immaculate Virgin. And to crown it all there were the medical miracles, above all the cases of miraculous relief from pain.

This is what ultimately made Lourdes unique. From the start it was dedicated not only to the sick but also to the carers of the sick. The pilgrimage was a celebration of service which recognised no geographical or social barriers. Aristocratic ladies from turreted châteaux held the same stretchers as did poor working-class women. Men and women mingled on easy terms. Physical pain and relief from physical pain through prayer were at the centre of the cult as they still are, omnipresent and often inarticulate but transcending language and the mind–body divide.[13] 'I came and prayed and my pain disappeared:

12. It was inevitable that the 'white death', the death of the innocents, would kill the two most charismatic female saints of the nineteenth century, Bernadette Soubirous and Thérèse Martin of Lisieux. But, unlike Sainte Thérèse, Bernadette never wrote down anything for posterity. Her handwriting was childish and unformed.

13. Lourdes has been fortunate in finding outstanding modern chroniclers: see R. Harris, *Lourdes* (London, 1999) and F. Werfel, *The Song of Bernadette* (London, 1942).

thank you Virgin Mary' is the commonest inscription on the hundreds of thousand of votive tablets which now line the approach to the Grotto. Or 'I was in pain. I am no longer. Thank you, Mary.'

...

But if Lourdes was an unexpected success story to the faithful, it was also, to the other half of France (and of Europe) a horrible provocation. The miracles happened at a moment when medicine seemed at last to have rid itself of the superstitions of centuries, when medical scientists were beginning to understand the nature of pain and hoped soon to eliminate it. Outraged rationalists turned to men like Charcot to express their revulsion and castigate the folly of the pilgrims. And of course Charcot was quite prepared to become the voice of science. He had on many occasions referred to the visions of saints like Saint Teresa of Avila as 'hysterical manifestations' and cures at their shrines as 'fantasies'. But he disappointed his disciples by refusing to relegate 'hysteria', whether religious or secular, to a quasi-criminal category. 'Fantasy' was not the same as a lie. Or even a figment. The 'deception' was not the work of confidence tricksters or indeed of the Devil. To Charcot hysterical pain was pain of a certain kind, no less real or bad than 'organic pain'. Nor did he have any doubt that 'faith can heal'. And, as the 'miracles of Lourdes' multiplied, he was prepared to extend hysterical manifestations to include ulcers, tumours, unhealing wounds and even stubbornly un-united fractures.[14]

But the profession was generally hostile. In 1874 the Parisian physician Jean de Bonnefon dispatched 11,000 questionnaires to his colleagues, asking them to express their views. Most of the responders deplored the unhygienic water and the long and arduous train journey. Their concern with hygiene reflected

14. Perhaps the most remarkable was the healing of a Belgian labourer, Pierre de Rudder, who visited not the shrine in France but a replica of the shrine built in Oostaker near Gand in Belgium. When he was helping two men to clear a tree in 1898, the trunk had fallen on his leg, shattering it. The compound fractures were followed by gangrene and the extrusion of bony fragments. The destruction was so painful and complete that all doctors agreed that amputation held out the only hope of survival. This was a case where the medical details were fully documented by several non-Catholic surgeons. After De Rudder had been carried to the shrine he felt himself shaken and fell to his knees in prayer. He then got up and walked for the first time since the accident, making his way without difficulty to the statue of the Virgin. His wife fainted when she saw him walking. The wound disappeared, the bones reunited, there was no loss of length of the limb. De Rudder still walked on his slightly thinner but perfectly adequate left leg years later. A careful post-mortem examination ten years after the cure showed perfectly united fractures and only a thin healed scar on the skin.

There have been half a dozen or so similar cases besides the thousands where the medical evidence was persuasive but not fully documented. Of course medical sceptics would maintain that a handful of unexplained and unexpected cures in a large sick population is par for the course for medical diagnosis and prognosis. There will never be an answer to convince all.

the recent Pasteurian revolution establishing the microbial causes of disease.[15] Few categorically denied the possible benefits to 'credulous and misguided women', but to the shrill minority the shrine was a 'charnel house', a place of disorder and death that should be burnt down. They had no doubt that under the pious veneer unspeakable crimes were being committed and concealed. The place was a hotbed of cholera, 'un bouillon de la tuberculose'. 'If I wished you harm, I would advise you to take a bath in this foul, purulent, stagnant puddle,' said Jean de Bonnefon.[16] But this was not Charcot's view. In his astonishingly temperate article, 'La Foi qui guérit', he noted cures by religious belief in an extraordinary variety of disorders never previously labelled hysterical. They were no doubt 'functional' rather than 'organic'; 'but the borderline between the two could become blurred'.

Catholic believers, many of them no less intolerant than their opponents, saw in Charcot's pacific assessment a more subtle and therefore a more dangerous menace. It denied the miraculous nature of the cures as surely as did the incredulity of the atheists. Hippolyte Bernheim, a self-proclaimed agnostic and a professor at the University of Nancy, argued that no symptom was beyond the power of 'suggestion': the mind could cure even such conditions as the pain of chronic arthritis.

> Suggestion does not kill microbes, it does not eradicate tubercles and *perhaps* it does not heal chronic indurated ulcers of the stomach . . . yet faith moves mountains and faith performs miracles because faith is blind, because faith does not reason and because faith acts directly on the imagination without moderating second thoughts. Above all, faith can nullify pain.[17]

Remarkable in themselves, many of the miracles were even more persuasive (or seductive) when contrasted with the results – or lack of results – of orthodox medicine or even of 'scientific hypnotism'. Surveying the chanting crowds, the ecstasy of the swaying torches, Felix Régnault asked almost enviously: 'What medical hypnotist could produce a stage like this?' Henri Berteaux even mocked rabid unbelievers. 'Why do they demand miracles that go against nature? Lourdes unites broken bones, heals suppurating cavities, makes the lame walk and, above all, relieves pain; but it does not claim to induce amputated limbs to grow again or wizened ancients to turn into

15. Before Pasteur discovered his ubiquitous microbes, which cause death and disease (as well as many beneficient chemical changes), the hygienic aspects of the pilgrimage would not have exercised French doctors overmuch.

16. See Emile Zola, *Lourdes* in *Mes Voyages* (Paris, 1958) p. 28; and Jean de Bonnefon, 'Faut-il fermer Lourdes?' *Les Paroles françaises et romaines* (1 July 1906), 3.

17. H. Bernheim, *Hypnotism, Suggestion and Hypnotherapy* (Paris, 1891), p. 210. A remarkably sane book written at a time when sanity was at a premium.

babies. Why should miracles occur *in defiance* of science? Is their faith in science so weak?'[18] It was the most telling shot, then as (perhaps) now.

But there was no doubt that Lourdes represented a revolt against nineteenth-century scientific progress. To men like Emile Zola it was an insult to reason, an affront to the dignity of evolved man. Those who exalted it were the same people who sent an innocent man to Devil's Island as a spy to preserve the sham 'mystique' of 'la Patrie' or, worse, to cover up their own corruption. They were the exploiters of the weak, the upholders of superstition, the living obstacles to progress. Many listened. His scurrilous attack on the shrine sold more than a million copies in France alone. Financially it was by far his most successful publication. It was translated into many languages. But as the century drew to a close it was Zola's world which seemed to be in retreat. In 1905 his former ally and devil's disciple, Joris-Karl Huysmans of the harrowing novel *À Rebours*, wrote almost awestruck in *Les Foules de Lourdes:*

> This hospital is at once a bodily hell and a spiritual paradise. Nowhere have I seen such appalling illnesses, so much charity, so much pain and so much grace. From the point of view of human mercy Lourdes is a wonder: there, more than anywhere else you see the Gospels in action.[19]

In a historical context Lourdes and what Lourdes represented undoubtedly drew sustenance from questionable sources – the resentment of the extreme political Right, the vicious anti-Semitism of the anti-Dreyfusards, legitimist myth-making. But it was also sustained by the piety of the poor, the charity of ordinary people and, above all, the yearning of the sick for hope. Strangely, perhaps miraculously, Bernadette Soubirous' figure seemed proof against religious bigotry (and later against Pétainist exploitation or Hollywood's glamour machine) as surely as it resisted the onslaught of unbelievers. Beyond all attempts to romanticise, glamorise, politicise or medicalise the shrine, there always remained a fixed point: a poor, cold, ignorant peasant girl in almost constant pain, kneeling in adoration before a beautiful apparition who smiled and spoke to her kindly.

### *III*

Of course in the forward march of the new Science of Suffering – *die Wissenschaft des Schmerzes* as some medical writers in Germany were beginning to refer to it – Lourdes was a sideshow. It showed the continuity of a thread that seemed to have been broken in the middle of the seventeenth

18. H.E. Berteaux, 'Lourdes et la science', *Revue de l'hypnotisme*, 1895 (9), 214.
19. J.-K. Huysmans, *Les Foules de Lourdes* (1905; Grenoble, 1993), p. 234.

century.[20] But it did not impede scientific progress. Hysteria was only one of Charcot's many interests. He and his colleagues continued to probe the neuropathology of motor and sensory symptoms, visual abnormalities, tics, migraine, seizures, abnormalities of speech, somnambulism, hallucinations, word blindness, mutism, contractures and above all pain in its infinite variety. 'Individual cases come to us like so many sphinxes, defying our knowledge of anatomy and pathology but also challenging us. No two pains are quite the same,' Charcot wrote. His style (like Claude Bernard's) was at times baroque but more readable than the robotic jargon of much scientific prose today. He was intensely interested in the work of past investigators and nearly always clarified it. Following up Parkinson's studies – it was he who called the illness Parkinson's disease – he described the characteristic tremor 'as when a pencil is rolled between the thumb and fingers or when crumbling a piece of bread'. Often with uncanny precision he traced unusual pains to lesions in the brain and spinal cord. Or pointed to the absence of such lesions. [21]

He was not alone. Multiple sclerosis, reflecting seemingly random areas of degeneration in the cord, cerebellum and even the brain, came together as a distinct clinical and pathological entity in Jean Cruvelhier's magisterial *Anatomie pathologique du corps humaine*, still the greatest hand-drawn atlas of human disease. Cruvelhier was also an authority on pain, classifying and reclassifying different kinds and trying to locate their origin.[22] Aphasia and dysphasia, disturbances and abnormalities of speech, some almost unbelievably bizarre, were studied by Peter Paul Broca.[23] Searching for the part of the brain which appeared to be involved, he found that the third frontal convolution on the left side, now known as Broca's area, was more developed than the corresponding area on the right and identified it as the centre of articulate speech.

John Hughlings Jackson, a dignified Yorkshireman with a beard bushy even by Victorian standards, eventually migrated to London and became one of the city's most sought-after doctors. His name is linked today to the kind of epilepsy which is triggered by a focal – that is localised – abnormality in the brain. Careful observation of the progress and especially of the onset of a fit sometimes provided a clue to the exact site of the lesion. But he frowned on those who boasted that they could locate a minute scar on the cortex from the first muscle to go into spasm. 'It shows a fundamental misconception. The

20. With the Treaty of Westphalia which concluded the Thirty Years War in 1648 – the most cynical compact of the 'haves' to impose their private beliefs on the 'have-nots' – the fires of faith which had been burning for 150 years of religious conflict in Europe seemed suddenly to became wisps of smoke.
21. J.M. Charcot, *Leçons sur les maladies du système nerveux* (Paris, 1873), Vol. V, p. 42.
22. Cruvelhier died in 1878, aged eighty-three.
23. Broca died in 1880, aged sixty-six.

brain knows nothing about muscles: it knows only about movements.' Like Charcot, he worked in many fields. He analysed for the first time the seemingly simple but immensely complex function called speech. He soon distinguished 'motor aphasia', the inability to speak because of the malfunctioning of the muscles involved in speech, from 'sensory aphasia'. In the latter patients were not deaf in the ordinary sense, for they heard noises and sounds, but spoken words sounded to them like a foreign and unknown language. Their 'speech activation' mechanism could also be disordered. Words would become jumbled and mutilated, a disability of which, because of their aphasia, they themselves remained unaware. Such disabilities sounded bizarre; but Jackson showed that mild cases were not uncommon and were often missed. This could make the diagnosis of painful illnesses difficult.[24] Hector Bastian in France and Carl Wernicke in Germany explored the condition in depth.[25] The extraordinary growth of neuropathology was reflected in the increasing bulk of standard textbooks. Sir William Gowers' *Manual of Diseases of the Nervous System* grew from a slim and elegant volume to a tome requiring the strength of weightlifters to shift.

. . .

Pain remained a cardinal theme. An important aspect of it was investigated by Emil Kraepelin whose neuropsychiatric clinic in Munich became a place of pilgrimage for budding neurologists and psychiatrists. Instead of trying to fit mental abnormalities into preconceived 'compartments' (some dating back to Graeco-Roman medicine) he approached every new patient as a 'syndrome carrier', classifying each case by its 'core abnormality'. He also broke new ground by following up the evolution of these syndromes over many years, making 'longitudinal diagnoses'. Such follow-ups revealed the link between mania and depression and distinguished them from developing dementia praecox, the forerunner of schizophrenia. His work on the interpretation of pain, though less well known, was also significant. He showed how mental illnesses could both exaggerate yet also virtually demolish painful experiences. Depression in particular emerged as a powerful modulator of symptoms.[26]

Among Kraepelin's disciples and later colleagues, Alois Alzheimer, the son of a small-town Bavarian notary, showed that the disease now named after him was not a feature of 'old age' but an illness in its own right, associated with degenerative changes in the brain. The first patient in his series was a

24. Hughlings Jackson eventually became a member of the staff of the new National Hospital for Nervous Diseases in Queen Square, London, and died in 1911 aged seventy-six.
25. Bastian died in 1915, aged seventy-eight; and Wernicke died in 1905, aged fifty-seven.
26. Kraepelin died in 1926, aged eighty.

fifty-year-old man, which led Alzheimer to call the condition '*pre*senile dementia'. He warned that when the 'melancholy and depression of old age' are accompanied by a 'disinclination to remember facts and retain impressions', the onset of a specific disease should be suspected.[27] Both Alzheimer and his friend Wernicke also gave classical descriptions of general paralysis of the insane and of a number of genetically determined diseases like Huntington's chorea.

Hypnosis experienced a revival, this time bedecked with academic credentials. It was no longer needed as a form of surgical anaesthesia; but to Charcot it was a useful diagnostic tool. It also provided him with the mainstay of his Sunday morning clinical demonstrations. The theme of hypnotic revelations would be taken up by young Sigmund Freud, a visitor from Vienna. They would lead him step by step into the murky world of dreams and then to gaze into the dark but profitable abyss of the Unconscious.

. . .

Sadly, the treatment of most of the newly identified diseases lagged behind diagnosis. By the end of the century the optimism which fired early studies of neuropathology – 'an essential preliminary to rational management' – had run into the sand. A German psychiatrist summed up the state of the art in 1905: 'We now know a lot but can still do nothing.' But if this was true – or partly true – of neurological and mental disorders, it was not true of pain in other branches of medicine and surgery. One of them at least deserves a short chapter.

27. Alzheimer died in 1915, aged sixty-one.

# Chapter 33

# THE ACUTE ABDOMEN

Until the 1860s the risk – indeed, the near certainty – of infection made the abdominal cavity virtually inaccessible to surgery.[1] To cut through its delicate inner lining, the peritoneum (which also envelops the organs inside the cavity), would have been courting disaster. It was occasionally done, in desperation, but almost always with the predictable result of general peritonitis and death. Wounds too which ripped open the peritoneal cavity were uniformly fatal. General peritonitis was a terrible death. Patients usually remained conscious, in pain and terrified to the end. Not unnaturally, few such cases were reported in the literature. There was little to report. Surgeons might not be blamed for operating, often at the urging of distraught parents; but nor would another death enhance their reputation.

'Acute abdomens' (as later known in medical jargon) could have many causes. Two of the commonest may serve to illustrate the preanaesthetic pre-Listerian disease, one as described by a lay onlooker and one by a surgeon. 'Perityphlitis', the condition known today as appendicitis, was described in 1759 by Louis Mestivier, who both operated on a patient and gave an account of the post-mortem findings.[2] By the 1820s the condition was probably becoming common.[3] Then as now, it could occur at any age but it most often affected children and young people. The clinical history was almost certainly

1. See Chapter 19.
2. Little is known about Mestivier except that he described what is still recognisably a case of appendicitis. He died in 1768, aged sixty-five.

   The classic and excellent paper on the illness was by Jean Louyer-Villermay of Paris, published in 1824 in the *Archives générales de médecine*. The name 'appendicitis' was introduced in 1886 by Reginald Heber Fitz of Boston (died in 1913, aged seventy).
3. The ups and downs of the incidence of appendicitis are among the minor mysteries of surgical pathology. It seems to have started to increase around 1820; and during the first half of the twentieth century the disease was by far the commonest 'acute abdomen'. It then began to decline and became quite rare in 1970s. Since about 1980 it has been growing more common again.

similar to what it is today. The illness would begin with sickness, vomiting, constipation and pain in the epigastrium, the area just above the umbilicus.[4] At this stage the pain was often 'colicky', coming in waves, rising to a crescendo and then subsiding until the next wave.[5] Over the next twenty-four hours it would get worse and gradually migrate to the 'right iliac fossa', the lower right quadrant of the abdomen. It was usually treated with hot poultices and opium (or later morphine), occasionally and disastrously with a dose of purgatives. There often followed a lull.

The lull sometimes raised false hopes. In one particular case described by Countess Caroline Eszterházy the imperial physicians attending the fourteen-year-old Archduke Leopold Salvator in Vienna on 5 September 1823 pronounced the emergency to be over.[6] A thanksgiving *Te Deum* was sung in the Church of the Capuchins, the Imperial House's family chapel. But next day the pain returned. It was worse than before, no longer colicky but continuous. The slightest movement was now excruciating, the lightest pressure on the abdomen intolerable. A little reluctantly, a surgeon was called in. The profession was still not quite acceptable at Court and the countess did not record the name. He had no doubt about the diagnosis. From the start the case had been one of perityphlitis.[7] The condition was not yet famous but was already known to experienced practitioners. The lull had been a well recognised stage in the evolution of the disease.

What happens inside the abdomen is a complicated sequence. The appendix is a narrow blind sac, 5–10 centimetres long, opening into the caecum, the first part of the large bowel. Neither its purpose nor its origin has ever been established. The inflammation is usually initiated by some kind of obstruction at the mouth of the sac. There is progressive engorgement as the pressure in the sac increases. This is the stage when pain is referred to the epigastrium. Soon the peritoneal covering of the organ becomes involved and the pain now moves to the right iliac fossa. The arteries to the sac become blocked. Beyond the block the tissues becomes gangrenous. An abscess forms. In an obvious attempt by the body to limit the inflammation, the abscess is

4. A good example of 'referred' pain, that is pain experienced at some distance from the lesion because of a shared nerve supply.
5. The word is derived from the Greek κωλικός *kolikos*, a pain in the colon – i.e., large intestine. The Romans tended to divide abdominal pains into 'ileus' and 'colic' which later changed into 'ileal passion' and 'colic passion'. The meaning of both has changed; but 'colic' has been widely used in England since the fourteenth century to describe severe abdominal pain 'coming on in waves'. The wave sensation accurately reflects peristaltic waves moving down the intestine (or any hollow muscular organ), trying to propel its contents and encountering and trying to overcome some obstruction.
6. A. Torday (ed.), *Eszterházy Karolina Naploja és levelei* (Budapest, 1909), p. 342.
7. *Typhlos* τυφλός in Greek is the Latin caecum, the first part of the large bowel: hence perityphlitis.

walled off from the main peritoneal cavity by layers of omentum, a mobile apron formed by two layers of peritoneum. The abscess bursts, relieving the pressure inside the appendix. With the relief of pressure the pain subsides. The burst abscess is at first contained by the omentum wrapped round it. This is the interval of the lull. At this stage the temperature may fall. But nothing has happened to stop the infection. The pressure inside the walled-off abscess once again begins to mount until eventually the abscess bursts through its protective layers. Pus and blood teeming with pathogenic organisms spill into the general peritoneal cavity. The clinical picture becomes one of general peritonitis. The peritoneal membrane, unlike many internal organs, is highly sensitive to pain. Any touch or rubbing against its inflamed surface is agonising.

Today even at this stage the situation would be far from lost. The condition would be diagnosed as an 'acute abdomen' and the patient would be taken to hospital and the operating theatre. If at operation the appendix was still recognisable, it would be removed. If not, the abscess and the peritoneal cavity would be drained. A drain of some kind would be left behind to allow any further pus to escape. Long before the advent of chemotherapy and antibiotics, the operation was usually successful. The patient would start to recover almost at once. With good nursing and supporting treatment he or she would be back to normal in a week or two. The mortality today is lower than one in 10,000.

No such course was open to doctors in 1823. To open the peritoneal cavity deliberately would have been unthinkable or at least unwise: it would have converted a 'natural' into a 'surgical' death. Nothing could be done in the way of a cure. Before the discovery of the syringe and needle, even opiates could be given only by mouth. They were usually immediately vomited. Family and doctors could only look on, apply hot or cold compresses, hold the patient's hand, whisper encouragement into his ear and pray. In the case of Imperial Highnesses a mass was usually sung in the sick-room. The last rites were administered. Then another waiting period. The pinched face, the anxious eyes staring out of deeply hollowed orbits, the ears turning cold and blue, would haunt those who had not seen it before. The image was well known as the 'Hippocratic facies'. Delirium and loss of consciousness, often late, came as a blessed relief. The young archduke died a day after lapsing into unconsciousness, five days after the diagnosis was made.

...

Cases of perityphlitis/appendicitis occasionally recovered. Women with burst ectopic pregnancies almost invariably died. The emergency was not uncommon. The ovum is fertilised in one of the Fallopian tubes and may get embedded there instead of migrating to the uterus. For a time the pregnancy may seem to proceed normally; but sooner or later it will erode a vessel. This also presents as an 'acute abdomen' requiring an emergency operation. On opening the peritoneum there is blood instead of pus. Fortunately, stopping the bleeding and removing the tube is usually simple. If the other tube is healthy, many young women go on to subsequent normal pregnancies. But, short of a near-miracle, in the early 1850s the condition was uniformly fatal.

Robert Lawson Tait was born and trained in Edinburgh but spent most of his surgical career in Birmingham. A faun-like little man with a lion's head and mane, he was known as a brilliant operator.[8] Soon after taking up his appointment in Birmingham he saw within a week two young mothers die in his ward from ectopic pregnancies. Blood outside vessels is a merciless irritant and bleeding into the peritoneal cavity is agonising. Lawson Tait's generation of surgeons had to be inured to such suffering; but two women in quick succession bleeding to death in agony prompted a resolution: he would operate on the next case even if the operation itself killed the patient. Fortunately, Tait was something of an eccentric. He thought little of the newfangled idea of 'germs' (emanating from Glasgow), but he had recently become converted to the 'cold-water school' of surgery.[9] This consisted of washing and scrubbing hands thoroughly with ice-cold water and soap and subjecting the operating theatre and operating instruments to the same treatment. It was, unknown to its advocates, an aseptic regime of sorts. Since the Board of Governors of the Infirmary refused to subsidise expensive eccentricities, Lawson Tait usually operated at his private nursing home.

By coincidence, a third and private patient with an ectopic pregnancy was rushed to the nursing home a week later. Lawson Tait hesitated. But the patient was visibly sinking – 'she was already as pale as the bed sheet' – and barely breathing. The husband implored him to 'do something'. Could he not see? His wife was bleeding to death in front of their eyes. Lawson Tait overcame his dislike of chloroform anaesthesia, another newfangled device but at least emanating from Edinburgh, not Glasgow. He opened the woman's abdomen and found it full of still liquid blood.[10] The bleeding had temporarily stopped, presumably because of the fall in blood pressure, but

8. His physical resemblance to James Young Simpson made many suspect that he was Simpson's son. Tait himself was called 'Father of Pelvic Surgery' by William Mayo. He was something of a womaniser and died from a cause not perhaps unrelated in 1899, aged forty-four.
9. The regime's most fervent advocate was Sir Thomas Spencer Wells (died in 1897, aged seventy-nine), a distinguished London surgeon who became president of the Royal College of Surgeons in 1883. The 'cold-water regime' was a passing fad which eventually gave way to antisepsis and then to asepsis. In 1880 Spencer Wells published a series of 1,000 cases of ovariotomy (removal of the ovary or a cyst of the ovary).
10. For a combination of reasons blood shed into the peritoneal cavity often does not clot.

would undoubtedly start again if the patient ever recovered. This seemed unlikely. Her respiration had become reduced to laboured infrequent gasps and the pulse was unrecordable. Lawson Tait evacuated the spilt blood with a spoon. He located the bleeding point. He removed the tube. He closed the abdomen. The patient survived.

In reporting the case Lawson Tait drew three conclusions.[11] First, he did not suggest that ectopic pregnancies should be treated by operation as a *general rule.* This, he suggested, would be foolish. The operation was rightly 'known to lead to peritonitis and death'. Second, he felt – perhaps rightly – that his 'cold-water regime' contributed significantly to the happy outcome in this particular case. Third, the procedure had taken over half an hour and would not have been possible without a general anaesthetic.

. . .

But Lawson Tait was not to be the last surgeon to operate successfully on a ruptured ectopic pregnancy. Whatever he himself thought about the far-fetched idea of 'germs', by the time of his successful 'one-off' operation, the new concept was changing the face of surgery. Its effect on pain relief would be as dramatic as that of ether and chloroform.

Antisepsis had its origin in a walk home by Joseph Lister, Regius professor of surgery in Glasgow, in the company of the professor of chemistry, Thomas Anderson. Anderson, a well-read man, was intrigued by recent reports by a French chemist, Louis Pasteur. Pasteur had shown that the 'visible universe' was teeming with billions of 'invisible organisms'. These belonged to a wide variety of species and, under appropriate conditions, could bring about hitherto mysterious chemical transformations. Lister wanted to know more, borrowed the relevant journals and battled his way through Pasteur's chemical French.[12] At some stage during or after his reading he had his moment of illumination. The invisible germs described by Pasteur were the obvious causes of 'hospital disease'. The main trouble with previous explanations – toxic miasmas, lunar radiation, even ordinary dirt – was that they could not account for the apparent growth and spread of the infection. Yet this was the killing feature of wound sepsis. But the objection would not apply if the causative agents were alive. Living creatures could grow. They could multiply. They could spread. They could account for the galloping progress of septicaemia and pyaemia. Indeed, Pasteur had shown similar growth and spread in

11. Lawson Tait, 'A Case of Tubal Pregnancy treated by Emergency Operation', *Surgical Journal*, 1860 (10), 325.
12. This was not as arduous as it would be today. Pasteur's papers were written in simple, elegant French, 'comprehensible even to the president of the Republic' (as the President said at Pasteur's 70th birthday celebration) and to anybody with school French.

milk and alcoholic brews. But Lister went further. If something lived, it could also be killed.[13]

It took Lister only a year or two to develop a strategy for killing the living germs which, he rightly suspected, caused the various forms of wound infection. It is fashionable today to belittle or at least to patronise his achievement. It has been described as a 'cul-de-sac' because in time it was superseded by other and better strategies. But being superseded has been the hallmark of every scientific advance, from rolling logs to relativity. Lister's strategy was based on the killing of organisms by an antiseptic spray and by the frequent drenching of the operation site with the same liquid.[14] The unpleasantness of the procedure was not denied.[15] Carbolic acid is an evil-smelling, irritating liquid. The procedure was tedious. Few surgeons working in the soothing ambience of an operating theatre today with Vivaldi's *Four Seasons* wafting from sterile loudspeakers would tolerate operating in a haze of a cough-inducing, throat-burning, eye-watering and vision-obstructing vapour. Few anaesthetists today would sit calmly through long hours of tedium in such a poisonous atmosphere. Many surgeons and anaesthetists resisted in Lister's time too. But the strategy worked as nothing had done before. And it worked because anaesthesia had abolished time as a limiting factor.

As often happens, Lister's innovation was for a time more enthusiastically received abroad than in his native Britain. On the Continent, especially in Germany, antisepsis inaugurated the most exciting decades in the history of surgery. Lister's friend, Karl Thiersch of Leipzig, developed the modern skin graft and revolutionised the treatment of burns.[16] Richard von Volkmann in whose clinic in Halle carbolic acid was sloshed about from watering cans – 'if dirt there has to be, at least it should be antiseptic dirt' – transformed orthopaedic surgery.[17] Friedrich von Esmarch of Kiel did as much as any general to win the Franco-Prussian War for Germany by introducing Listerian principles into field surgery.[18] Theodor Billroth, Prussian born but later professor of surgery in Vienna, became a commanding figure. His main field

13. Until Lister wrote to him, Pasteur had no particular interest in medical or surgical matters. His initial work was on bacterial fermentation and on the spoiling of wine and beer.
14. See Dormandy, *Moments of Truth*, p. 241.
15. Lister himself eventually abandoned the spray, realising that the airborne contamination of wounds was comparatively unimportant.
16. Thiersch was professor in Leipzig for twenty-eight years, and died in 1895, aged seventy-three.
17. Richard von Volkman was the third generation of Herr Professor Volkmans in Halle and the most famous. He died in 1889, aged fifty-nine.
18. Friedrich von Esmarch, the inventor of the 'exsanguinating' tourniquet still used in orthopaedic surgery, was largely responsible for introducing Lister's principles into the Prussian army. The care of the wounded was far more advanced on the Prussian than

was abdominal surgery, where he developed the first operations for gastric and colonic cancer and for non-malignant ulcers.[19] Pirogoff introduced the antiseptic technique in backward Russia.[20] Eventually even Britain and the United States caught up. Paul, Moynihan, Lane, McBurney and the Mayos – father and sons – are still household names in surgical practice.[21]

In no period were so many new operations devised and performed on parts of the body which until only a few years earlier had been 'no-go' areas. A few were curative: many afforded temporary relief. Antisepsis was succeeded by asepsis. Such dramatic – even intoxicating – advances are rarely made without occasionally overstepping the mark. The first half-century of the new surgery was also the age of heroic but useless operations. Many new procedures became feasible without pain and without death, but the feasibility of an operation is not an indication of its performance. This was often forgotten. Thousands of 'floating' kidneys, allegedly the cause of pain in the lower back, were 'hitched up' and 'fixed in place' by an operation called nephropexy. Floating kidneys probably do not exist. If they do, they are almost certainly harmless. Yet medical students and prospective fellows of surgical royal colleges failed in their examination if they did not palpate the lower poles of 'dangerously descended' organs. Sir Arbuthnot Lane, Bart, pride of Guy's Hospital and indeed of British surgery, invented a whole series of abnormal 'kinks' in the bowel which, he claimed, were the cause of digestive symptoms. Even when they were causing no *overt* pain, their prophylactic surgical repair might save the patient (or future patient) endless trouble. When the unkinking of non-existent kinks failed to produce the expected result (which happened occasionally but not often), other operations were indicated. By 1908 Sir Arbuthnot claimed to have performed a staggering thousand total

on the French side in the 1870–71 war, one reason why wounded French soldiers, among them Pasteur's son, were not averse to be taken prisoner. (Even eminent military historians seem to be blind to the importance for morale of the standard of surgical care in the field behind the front line.) Esmarch died in 1908, aged eighty-five.

19. Billroth was a friend of Brahms, an accomplished pianist and flautist. He died in 1894, aged sixty-five.
20. See Chapter 24.
21. The list is almost random. Frank Thomas Paul developed the standard operation for cancer of the colon and died in 1941, aged ninety. Lord Moynihan of Leeds became the second surgical peer (after Lister) and died in 1936, aged seventy-one. He was most acclaimed for operations on the gall bladder. Sir Arbuthnot Lane, Bart, was surgeon to Guy's Hospital and died in 1943, aged eighty-seven. Charles McBurney of New York was a pioneer of abdominal surgery and died in 1913, aged sixty-eight. Charles H. and William J. Mayo, sons of Dr William Mayo, were founders of the Mayo Clinic in Rochester and died in 1939, aged seventy and seventy-seven respectively.

colectomies – the complete removal of the large bowel – for what a physician colleague acidly described as 'chronic remunerative constipation'.[22]

But the field where the combination of anaesthesia and antisepsis achieved its first and greatest triumph was and remained the acute abdomen. After the 1860s young women with ectopic pregnancies no longer bled to death. Anxious young men with perforated duodenal ulcers did not develop general peritonitis. Children with perityphlitis/appendicitis began miraculously to survive. Even some 'bad-risk' old people did.

. . .

Royalty has occasionally set the seal on medical advances.[23] Two weeks before his planned coronation in 1901 Edward VII, King and Emperor, developed severe abdominal pain. Even in an age of culinary excess unparalleled since the reign of Nero, his must have been one of the most sorely tried appendices in the Empire. The diagnosis was straightforward. But for some days, especially during the characteristic lull, the king refused to have an operation. He developed an abscess. A surgical consultation, presided over by Lister (recently ennobled) decided that only surgery could save the royal life. Bertie was by then developing the severe pain of incipient peritonitis. His loss of appetite made the gravity of his condition clear both to Queen Alexandra and to his principal mistress, Mrs Keppel. It even alarmed him. He relented. Under chloroform anaesthesia Sir Frederick Treves, surgeon to the London Hospital, drained the abscess. The king recovered, was duly crowned and happily over-indulged for another eight years.[24]

22. He may have been the original of Bernard Shaw's Mr Cutler Walpole of the 'nuciform sac' in *The Doctor's Dilemma* (1906).
23. Queen Elizabeth the Queen Mother was probably the oldest patient on whom a hip replacement was performed (no waiting) at the age of a hundred. Queen Victoria popularised chloroform. But being attended by a gaggle of royal physicians has not always been a blessing. Sir James Clark's misdiagnosis of Prince Albert's typhoid may have contributed to the prince's death. Lord Dawson of Penn recorded that he timed and administered the last dose of heroin to the dying George V so that his death should first be reported in a 'dignified manner by The Times newspaper rather than in the evening sheets'.
24. Lister was rewarded with the newly instituted Order of Merit. 'Without you I would not be here today,' the king joked, trying to put the ribbon round Lister's head. But ribbons are expensive (and provided by the donor) and the head was too big. Eventually Lister had to carry the gong home in his hand.

# Chapter 34

# OLD DRUGS, NEW DRUGS

By the time of Edward VII's fraught accession in 1901, opium and its derivatives were no longer unchallenged in the treatment of pain. A few doctors were voicing concern about them. In 1898 John Davey, an obscure backbencher from Wales, suggested in the House of Commons that their sale should be made a 'criminal offence'. The idea provoked hilarity.[1] 'In Wales or in China?' was one of the witty interjections. Yet new indications for 'poppy juice' were still being discovered. Around the turn of the century their value in two diseases was much discussed. One was diabetes, the other insanity.

. . .

Diabetes mellitus was still an uncontrollable disease – and would remain so for another sixty years – when in 1869 F.G. Pavy of Guy's Hospital in London showed that opium reduced the loss of sugar in the urine.[2] Glycosuria was regarded at the time as the underlying cause of all the other symptoms and signs and ultimately as the cause of death. If the 'waste of sugar' could be reduced, a general improvement might follow. But almost more important to patients was that opium sometimes eased the neuralgic pains that could be such a distressing feature of the illness. It was not always effective and rarely completely so; but nothing else was on offer. Others pointed to the dangers of

1. Nevertheless, the Society for Promoting Legislation for the Control and Cure of Habitual Drunkards, established by Dr Norman Kerr in 1876, was renamed nine years later the Society for the Study of Inebriety. Its primary aim was to reduce alcoholism, especially alcoholism caused by gin, which was 'sadly rampant among the poorer classes'; but in its new guise it included 'other narcotising agents'. Kerr believed that addiction to opium was less dangerous than addiction to alcohol but more difficult to cure. He was probably right.
2. Thomas Willis in the seventeenth century tasted the urine of a patient who was drinking water all the time and passing large quantities of urine. Because her urine tasted sweet, he called the condition diabetes (passing through) mellitus (honeyed). In his footsteps the disease remained centred on glycosuria until the 1920s, when it was shown to be caused by insulin deficiency and treatable in most cases by insulin supplementation.

an addictive drug in a condition that, even untreated, could last for years. Nobody claimed that it was a cure; but William Osler, soon to become a knight, Regius professor and an oracle on matters therapeutic, became an advocate. In his *Principles and Practice of Medicine*, the bible of generations of doctors first published in 1892, he stated that 'opium alone stands the test of experience as a remedy capable of limiting the progress of the disease'.[3] It was still being used when Banting and Best embarked on their historic experiments with insulin in the early 1920s.

...

In mental illness opium seemed a possible alternative to cages, shackles, chains and other forms of physical restraint, some unbelievably horrible.[4] Few wanted a return to the old regime of 'beating and terror' – or at least few voiced such opinions – but, as John Ferriar, a Manchester doctor, had predicted, the reforms had created a new need. If not force, then what? For 'violent and greatly disturbed patients' there was an urgent need for effective medication.[5] John Conolly, physician to the lunatic asylum in Hanwell and one of the leading advocates of a more humane approach to mental disease, recognised this but still warned against a too ready recourse to opium.

> With some patients laudanum acts like a charm; others derive comfort for long periods . . . to some the *liquor opii sedativus* is the only drug tolerated. Whenever such a sedative is used, the dose should be large or very large. In mental illness less than a grain of morphia has no effect whatever.[6]

Others, led by Henry Maudsley, 'a fresh and commanding voice', disagreed, arguing that opium and morphine should be used with caution. Maudsley was not against them. 'The sleeplessness, depression and strange feelings of alarm which often precede regular insanity' often benefited from 'stupefying drugs'; and they were 'valuable in melancholia and depression,' he wrote; but he felt that a 'precise diagnosis' should always come first. Sadly, a precise diagnosis

3. W. Osler, *The Principles and Practice of Medicine* (London, 1892), p. 332.
4. In 1815 the Parliamentary Committee on Madhouses visiting Bethlem (or Bedlam), London's main lunatic asylum, found that one patient, James Norris, had been restrained with 'a stout iron ring riveted round his neck, from which a short chain passed through a ring made it slide upwards and downwards on an upright massive iron bar, more than six feet high, inserted into the wall. Round his body a strong iron bar about two inches wide was riveted; on each side of the bar was a circular projection which, being fashioned to and enclosing each of his arms, pinioned them close to his side'. This was not unusual, Bethlem physician, Thomas Munro, told the Committee; but he also reassured them that 'chains and fetters were only for pauper lunatics: a gentleman would not like them'.
5. J. Ferriar, *Medical Histories and Reflections* (London, 1910), p. 136.
6. J. Conolly, 'The Principal Forms of Insanity', *Lancet*, 1845 (2), 526.

was not always possible (then as now), but Maudsley was right in principle. The mode of action of opium and its derivatives was ill understood; and they were as likely to cause mental illness, especially mania, as to cure it.[7] Thomas Clouston of the Cumberland Asylum in Carlisle won the gold medal of the Medical Society of London for a paper in which he argued that *Cannabis indica* and bromide of potassium were more effective as well as safer.[8]

Like so much else in Victorian medicine, it was partly a matter of class. Sir Ronald Armstrong-Jones of the Claybury Asylum noted that morphine was more useful among the 'private class of patients' than the 'rate-aided' ones. The observation rested on the 'rock solid scientific evidence of comparative anatomy'.

> There is a marked physical difference between the brains of those in the private and in the rate-aided class . . . not only is the brain in the former considerably heavier but there is also an added complexity of the convolutional pattern . . . and those differences carry with them a higher sensitiveness to pain and greater responsiveness to opiates.[9]

Ultimate rehabilitation also had to be considered. The poor could ill afford to maintain an expensive habit; and after their release from an institution, if it ever came to pass, their addiction might plunge them into a life of debt and crime. The very fabric of civilised society might be imperilled. Alternatively, continued handing out of the drug might become an 'intolerable financial drain' on charitable foundations.

. . .

One of the two drugs recommended by Clouston in his medal-winning paper was becoming popular outside asylums too. Cannabis was of course an ancient remedy. It had been used in China for thousands of years and had been fashionable at various times in classical Greece and Rome as well as in Persia, India and Egypt.[10] The word 'assassin' was said to derive from one of its long list of synonyms – hashish. But, though medieval names like

7. H. Maudsley, 'Opium in the Treatment of Insanity', *Practitioner*, 1869 (2), 1. Even in his lifetime Maudsley's name was synonymous with new departures in mental disease and his opinion was influential. He died in 1918, aged eighty-three.
8. T.S. Clouston, 'Observations and Experiments on the Use of Opium, Bromide of Potassium and Cannabis Indica in Insanity', *British and Foreign Medico-Chirurgical Review*, 1870 (46), 493 and 1871 (47), 203.
9. R. Armstrong-Jones, 'Drugs of Addiction', *Morning Post*, 10 June 1914.
10. The bewildering number of its names and synonyms shows how widespread was its cultivation and use. Hashish is called *esrat* in Turkey (meaning secret) and *kif* in Morocco or *madjun* when it is made into sweetmeat. The material derived by chopping the leaves and stalks is marijuana. Cannabis is the generic name for hemp, its full botanic name is *Cannabis sativa*, given to it by Linné in 1753.

'Hempfields' suggest that the plant had never entirely disappeared, its medical use as a painkiller in European medicine dates only from the mid-nineteenth century. In Britain its launch was largely the achievement of one man.

William O'Shaughnessy of Limerick came from an old Catholic family. His father was a doctor: he himself was one of ten children. He graduated from Edinburgh in 1830 and a year after graduation recognised the potential value of intravenous fluids in cholera.[11] John Snow thought well of the idea; but neither of the two men had the chance to put it into practice. O'Shaughnessy then joined the medical service of the East India Company and, three years after qualifying, was appointed 'Surgeon to Bengal' and professor of chemistry at the Medical College in Calcutta. To be the surgeon to roughly 30 million Indians and professor at India's largest medical school were plainly regarded by the Company as not requiring the full-time attention of a capable doctor; and O'Shaughnessy duly diversified. He would eventually be knighted for laying the foundations of the Indian telegraph system. Before then, as a hobby, he developed an interest in cannabis, widely used by ordinary people in parts of the subcontinent. He learnt several languages and travelled to Tehran, Kabul and Kandahar to talk to local sages and to consult ancient texts. He then studied the effect of extracts of the plant on stray dogs of which 'there is an over-abundant superfluity in the streets of this city' (Calcutta). A dog given ten grains of hemp 'became stupid and sleepy, dozing at intervals, wagging his tail as if extremely happy, consuming his food greedily and staggering about in a positively blissful way'. This was promising. He then tried the extract on 'numerous patients with rheumatism, tetanus, cholera and epilepsy'. It proved a valuable painkiller. Given to a man who had been bitten by a rabid dog it did not save the man's life but at least 'this awful malady was stripped of its worse horrors'. Above all, it was safe. In 1842 O'Shaughnessy published a paper in the *Transactions of the Medical and Physical Society of Calcutta.*[12] In it he described Indian hemp as an 'anticonvulsant, sedative and analgesic remedy of the greatest value'. While noting that the drug could induce 'a singular form of delirium . . . a strange balancing gait . . . a constant rubbing of the hand . . ., perpetual giggling, and a propensity to caress the feet of all bystanders *of whatever rank*', he nevertheless warmly commended the extract. Peter Squire of Oxford Street, London, Liston's friend and apothecary, converted the resin into an alcoholic tincture which he distributed among his customers. It never began to rival opium in popularity; but when rumour got about that Sir John Russel Reynolds had prescribed it to Queen Victoria for 'woman's cramps', it

11. See Chapter 27.
12. W.B. O'Shaughnessy, 'On the Preparation of Indian Hemp, etc.,' *Transactions of the Medical and Physical Society of Calcutta*, 1842 (8), 421. He also published a paper in the *Lancet* in 1839 (2), 239.

acquired a following in the treatment of dysmenorrhoea and 'female neuralgias'. In the field of mental disease it was recognised as both a possible cure and a potential cause. By the 1850s it was listed in a number of British and European pharmacopoeias as well as in the *United States Pharmacopeia* and it retained its place in most of them until the 1920s.

. . .

Beside new roles for old remedies, three new painkillers slipped on to the medical scene. The first and by far the most important provided relief from cardiac pain for the first time in history. The discoverer, Thomas Lauder Brunton, came from Roxburghshire in Scotland, the only child of elderly parents. He was taught at home by his scholarly father, a retired teacher, and graduated from Edinburgh in 1860. A scholarship enabled him to spend two formative years travelling around the medical centres in Europe. In Leipzig he worked in Ludwig's famous Department of Physiology and absorbed the idea that laboratory science and especially pharmacology could provide the basis of much improved clinical practice. On returning to Edinburgh he spent a year as an unpaid clerk in the wards of the Royal Infirmary and then migrated south. At St Bartholomew's Hospital, London, he was given a disused scullery as a 'pharmacological laboratory'. He remained on the staff of Bart's for fifty years, achieving every distinction open to a medical man in England, including a baronetcy, fellowship of the Royal Society, a clutch of prestigious lectureships of the royal colleges and an invitation to the royal enclosure at Ascot. At the turn of the century he was the most sought-after consultant in the capital. His textbook, *Pharmacology, Therapeutics and Materia Medica*, laid the foundation of modern applied laboratory science and passed through twenty-two editions.[13] But, though he continued to make useful contributions to clinical practice, he made his one great discovery in 1864 as an unpaid clinical clerk in Edinburgh.

While on duty at night he noticed that in one patient attacks of anginal pain in the chest always coincided with a rise in blood pressure.[14] Brunton had learnt in Leipzig that a rise in blood pressure was often a reflex response of the heart to widespread constriction of the peripheral arteries. The reason had been explained by Ludwig himself. The constriction of the peripheral arteries raises 'peripheral resistance', that is the resistance which the pumping action

13. Obituary in the *Lancet*, 16 September 1916, p. 659. He appears to have been a kindly and deeply religious man, happily married to the daughter of a Scottish clergyman for fifty years. One of their three sons was killed in the First World War.
14. Angina (cf. Greek ἄγχειν *anchein*) is from the Latin meaning choking (*angere* is to strangle). The term is Hippocratic in origin and originally referred to constriction of the throat. It was used for quinsy in English till comparatively recently. *Pectus* is Latin for chest.

of the heart has to overcome to provide tissues with arterial blood. By a reflex action the rise in 'peripheral resistance' provokes the heart to pump blood with greater vigour. The heart pumping with greater vigour is shown by the rise in blood pressure. Brunton thought that his patient's angina might also result from the constriction of arteries, in particular of the arteries of the heart. He also reasoned that this might be capable of being reversed. This seemingly simple train of thought has to be seen against a background of centuries of futile guesswork.

Known by a variety of other names, angina pectoris was already an old disease when in 1632 the Earl of Clarendon described an attack in his own father.

> He was seized on by so sharp a pain in the left arm that the torment made him as pale as if he were already dead. He later said that he had passed the pangs of death and that he would die from the next attack . . . which he did.[15]

In a classic description in 1768 William Heberden gave the attacks their name; but he noted that *angor animi* or 'fear of the soul' was as important a feature as angina pectoris or 'strangling of the chest'.[16] It is not known who coined the phrase which describes the sensation as one of 'impending dissolution'. The cliché hints at the feeling of terror which often accompanies the pain.[17] In 1685, almost a century before Heberden, Giovanni Battista Morgagni had suggested – an astonishingly inspired guess – that the condition was associated with obstruction of the vessels of the heart.[18] (At that time the term 'coronary', that is 'garland-like', was still applied more commonly to vessels of the

15. Quoted by H. Skinner in *Origin of Medical Terms* (Baltimore, 1949), p. 25.
16. In the *Transactions of the Royal College of Physicians*, often reprinted in collections of 'classical' papers. William Heberden (died in 1801, aged ninety-one), was much admired in his day. He was Samuel Johnson's doctor. Dr Johnson called him the *ultimus Romanorum*; his son, another William, became an almost equally distinguished physician.
17. In Chapter 11 two works of art were mentioned which powerfully portray physical pain. There have been others, though few quite so striking. Music, by contrast, is unsuited to representing physical as distinct from mental suffering. The deaths of the consumptive heroines of Verdi and Puccini are affecting but do not even hint at physical discomfort. Cavaradossi's screams in Act II of *Tosca* are non-musical. Madame Butterfly expires in silence. Isolde's *Liebestod* is a joyful rather than a painful occasion. In his *Suite tragicomique* Telemann brilliantly depicts the woes of gout, perhaps his own, but cannot quite convey the sickening pain of an acute attack. None of the great Passsions, not even Bach's, and none of The Last Words on the Cross, not even Vittoria's, begins to convey the physical, as distinct from the spiritual, torment. One exception comes to mind. Sung by a great bass, the death of Boris Godunov in Mussorgsky's original score strikingly illustrates both the agony and the anguish caused by a massive coronary occlusion.
18. See Chapter 13.

stomach.) The severity of the pain explains the long list of herbal and other remedies that had been tried in an attempt to mitigate it. Claims of success had been numerous. Repeatable successes had been non-existent.

Brunton reasoned that anything that would dilate peripheral blood vessels might also dilate the arteries supplying the heart. Of course such a dilatation would also diminish the force with which the heart was pumping the blood. That might be bad for the perfusion of the heart too. But the danger could be outweighed by two gains. Vasodilatation of the coronary arteries might improve the blood supply to the heart; and the diminished force of the pumping action might lessen the workload of the organ. Was the balance good or bad? The outcome was by no means a foregone conclusion.

There were practical difficulties too. Brunton had absorbed the physiological principles of Ludwig, but at the time not many substances were known which dilated blood vessels. (Many more were known to constrict them.) One of the somewhat doubtful ones was 'the nitrate of amyl', a yellow liquid that had been used in animal experiments in Leipzig but never in humans. It had been discovered in 1844 when the French chemist Antoine-Jérôme Balard passed nitrogen fumes through amyl alcohol and recovered a curious liquid. Balard was one of the notable eccentrics of nineteenth-century French academic chemistry, a field and period not short of notable eccentrics.[19] The liquid he had recovered had a pungent smell and inhaling its vapour made him blush. 'Nothing else has ever done that to me,' he confided to his colleague, Dumas. 'I am a shameless character. I don't blush easily.'[20] More vigorous inhalation of the vapour made them both faint. Balard concluded that the compound dilated the blood vessels of the skin and precipitated a drop in blood pressure. He was right as usual though he had no idea what to do with his discovery. That was for the doctors to find out.

Brunton speculated that anything that dilated the blood vessels of the skin might have a similar effect on the blood vessels of the heart. So he tried to give a spoonful to his anginal patient: it worked, as he put it 'like magic'.[21] It worked in other patients too. He then tried inhalation. That too was a success. It was only the first step. Over the next twenty years numerous related

19. Even as a professor at the Sorbonne Balard lived like a bohemian in an unheated garret over his laboratory where he cooked for himself and slept on a bench (when not entertaining his aristocratic lady friends). Starting, like Jean-Baptiste Dumas, as a pharmacist, he was twenty-five when he discovered amyl nitrite. He was a devoted patron of the young Pasteur who, in return, placed him second only to Dumas in his personal pantheon.
20. He was twenty-one at the time.
21. T.L. Brunton, 'On the Use of Nitrite of Amyl in Angina Pectoris', *Lancet*, 1867 (2), 97.

compounds were tried by Brunton and others, some more and others less effective than amyl nitrite.[22] It was soon realised that neither the tablets nor the inhalations were curative: nor did they significantly prolong life.[23] But to have some means of counteracting or, even better, preventing anginal pain in critical moments was a blessing. Victorian ladies would carry drops of the liquid in thin glass 'pearls' housed in a pretty golden box. The glass pearls could be crushed in a handkerchief by the daintiest hand. Held to the nose, a whiff of the vapour would relieve the pain triggered by some unforeseen event, a social solecism perhaps, or an inconveniently sudden death. Gentlemen by and large preferred to place a tablet under the tongue. Mr Gladstone was seen doing it before embarking on the third hour of one of his speeches in the House of Commons. His admirers hastened to follow his example. Lord Rosebery nearly became an addict. (Perhaps it accounted for the vacuous – or statesmanlike – silences which became the distinguishing mark of his premiership.) Neither the liquid nor the tablets were expensive. In late Victorian and Edwardian days angina was still a middle- and upper-class disease. It would not stay so for ever. Increasing material affluence meant increasing coronary pain. Brunton's stand-by is still used by millions.

. . .

The second pain-killing novelty recommended by Clouston in mental disease was not really a drug but an element. Bromine is chemically close to iodine, chlorine and fluorine. The four are often grouped together under the name of 'halogens'.[24] By coincidence – or not – bromine too was discovered by Antoine-Jérôme Balard. It came in 1847, three years after amyl nitrite. Iodine had then recently been recommended by Coindet, a Geneva doctor, for the treatment of goitre, endemic in the Alpine regions of Switzerland. It was also advocated by

22. W. Murrell, 'Nitroglycerin as a Remedy for Angina Pectoris', *Lancet*, 1879 (1), 80. It has long been thought that nitroglycerin and amyl nitrite (and several related compounds) have a common mode of action, being converted to nitric oxide in the vascular lining. More recent work (Bauer *et al.*, *Journal of Pharmacology and Therapeutics*, 1997 (280), 326) suggests otherwise. Contrary to what often creeps into popular papers, nitric oxide is not nitrous oxide or 'laughing gas'. There are two 'acids of nitre' (as recognised by Lavoisier, Cavendish and Priestley), a stronger one containing three atoms of oxygen ($HNO_3$, nitric acid) and a weaker one containing two atoms of oxygen ($HNO_2$, or nitrous acid). There are correspondingly two oxides, nitric and nitrous.
23. Anginal pain can be caused either by a reversible spasm of a coronary artery or by the actual occlusion of the vessel, usually by a clot or a combination of the two. Amyl nitrite relieves the first but not the second. Coronary thrombosis and the anginal pain associated with it remained untreatable till the development of coronary surgery in the 1950s.
24. Literally the term means 'salt former' from the Greek ἅλος (*halos*) for salt and γεννάω (*gennao*) for I generate. The name was proposed by Berzelius since he believed that *halos* referred primarily to sea salt and the halogens were present most abundantly in the sea.

Jean Guillaume Lugol, physician to the hôpital Saint-Louis in Paris, for treating the rash of secondary syphilis. But pure iodine was expensive. Balard tried to find a cheap way of extracting it from sea sponges. The extraction proved difficult and eventually failed, but it led to the identification of a 'chemical sibling'. He first called the sibling 'muride' but later changed the name to 'bromine', from *bromos* (βρώμος), stench in Greek. (This was a little unfair: by chemical standards the smell of the fumes was not particularly obnoxious.) Because of the chemical similarity, bromine too was tried in thyroid disease and in syphilis. In both conditions it proved ineffective. The difference provided a first glimpse of the extraordinary biological specificity of the 'essential trace elements'.[25]

But like many negative results in medical research, the studies revealed an unexpected property of bromine not shared by any of the other halogens. It calmed anxiety, diminished fatigue and produced what Aubrey Beardsley would describe as 'a pleasing sense of aloofness and imperturbability'. In other words, it made pain bearable rather than making it disappear. One was almost as good as the other. Those kept awake by anxiety allied to comparatively mild physical discomfort found it an excellent inducer of sleep (which could nevertheless be resisted), waking up fully rested. In this respect it was – and remains – superior to many other narcotics which enforce rather than enable loss of consciousness and often lead to a sense of heaviness on waking. The Hungarian writer Mór Jókai, kept awake by the suspense of his own novels, found it a 'sovereign remedy against mental agitation'. It was 'infinitely better than the cursed poppy' (which made him constipated) or the 'useless and foul-tasting chloral' (which impaired his creative imagination). It was too cheap and unpatentable to be taken up by industry but 'bromine' became part of the language. Tom Laycock of Edinburgh found it the best prophylactic in epilepsy, controlling fits without impairing mental performance; and it remained the first-line drug in the everyday management of the disease until the introduction of the more effective but also more toxic barbiturates.[26] It later acquired an undeserved reputation as a calmer of uncalled for sexual urges and a deserved one as an occasional cause of irritating rashes.

...

25. Just 11 out of the 92 natural elements make up 99.9 per cent of all human and animal material. (The 11 are: carbon, oxygen, hydrogen, nitrogen, sulphur, phosphorus, sodium, chlorine, potassium, magnesium and calcium.) Part of the remaining 0.1 per cent are the 'essential trace elements'. Some of these are required in microgram amounts for only a single molecule (e.g. iodine for thyroxine, the hormone of the thyroid gland, and cobalt for Vitamin B12); but for that particular slot no other element will do.
26. See Chapter 40.

The third newcomer, chloral hydrate, dismissed by Jókai as 'useless and foul-tasting', came into use in 1862, a few years after the first trials with bromine (or its salts, known as bromides). Johann Liebreich, a German physician, recommended chloral hydrate as an alternative. This relatively simple and cheap compound had been discovered thirty years earlier by the great Justus von Liebig. It belongs to a chemical series known as the 'aliphatic hydrocarbons' which differ from each other mainly by the number of carbon atoms in their carbon–carbon (–C–C–) chain. Many but not all have a hypnotic action. Their effectiveness is not directly related to the number of carbons or to any other easily recognisable structural difference. Their action is often but not always enhanced by the introduction of one or more atoms of chlorine into the molecule. The introduction of bromine has a similar but less powerful effect: the introduction of iodine none at all. The largely unpredictable link between chemical structure and pharmacological potency began to puzzle chemists and pharmacologists of the nineteenth century. It puzzles them still.

Somewhat like bromides, chloral hydrate promotes deep sleep which refreshes rather than induces drowsiness. It was, until the Second World War, widely regarded as the safest of sedatives, especially suitable for children. Its foul smell and taste added moral uplift to the relief of pain. Children were expected to earn relief from their sore throats by a few bad moments bravely borne; and by and large they accepted this. Old-fashioned ward sisters too liked to see evidence of grit in their charges. Chloral hydrate may have tasted horrible but was all the better for it; medicines which sound in advertisements like a gourmet's dream are a modern invention. But there is no such thing as a perfectly safe sedative. Chloral hydrate occasionally caused excitement and even delirium; and despite its unpleasant taste, it had its notable addicts. It was prescribed for Elizabeth Siddal, beautiful consumptive mistress, model and eventually wife of the Pre-Raphaelite painter and poet, Dante Gabriel Rossetti. It helped her sleep in the last desperately painful stages of her illness.[27] Much in love with her, Rossetti may have started to take a few drops to encourage Elizabeth and keep her company. This was a common Victorian practice, perhaps a useful one. Mothers often shared spoonfuls of cod-liver oil to encourage their offspring. But Rossetti became addicted; and after Elizabeth's death needed increasingly large doses. He was weaned with difficulty and only with the help of laudanum.[28]

...

27. Contrary to operatic lore, tuberculosis in its final stages, especially when it involved the larynx, was a terribly painful condition.
28. The painter Amedeo Modigliani, suffering from the same illness, also became addicted to chloral, as did Claude Debussy in his last years, 1915–18, when laudanum became difficult to obtain in Paris

Chloral hydrate was followed on the pharmacy shelves by a series of more or less related compounds – chloralose, chlorobutanol, chorembal, sulphonal – all similar in action and complications. But the next analgesic with a select following, introduced towards the end of the 1880s, struck a new note. It went by the name of phenacetin and came in the form of small blue tablets. Imported from Germany, it was not cheap but, to fastidious users like Mr Henry James, 'worth every penny'. It neither smelt nor tasted foul. It was not messy. It offered relief not only from headaches, toothaches and such-like but also from fever and joint pains. There was nothing to touch it in incipient delirium tremens. It was quick in action. In its occasional unpleasant side-effects, mainly nausea and indigestion, it was not unlike its rivals. Yet it represented a new departure.

For centuries medical remedies were invented by herbalists, doctors, quacks, alchemists, chemists, witches, witch-doctors, shamans, monks, high priests, oracles or kindly gods. Their purpose was to relieve symptoms, even to cure patients. With phenacetin, this was no longer true. Other things being equal, its makers hoped that it would make sick people feel better. They also hoped that it would not cause too many troublesome complications such as death. But both hopes were incidental. The drug was not even intended to bring fame and glory to its discoverers. That too might be an acceptable bonus. But the makers of the new drug were businessmen. What was essential was that the drug should sell. Phenacetin was the first product of a creature few people had heard of in the 1880s. It would soon be known as the International Pharmaceutical Industry. Though the drug would acquire many functions and have numerous side-effects, some of them beneficial, its primary purpose was to make money.

## Chapter 35

# THE BARK OF THE WILLOW

Phenacetin was only a harbinger. The drug which would soon become the emblem of the new industry – more of it would be sold than of any other drug in history – had its roots in the prehistoric past. The bark of the willow had been among the thousands of herbal remedies recommended – never with particular fervour – by Sumerian, Egyptian, Greek and Roman physicians. In one of the Hippocratic writings an extract is praised for combating the pains of childbirth and reducing fever. Celsus mentions it as useful in mitigating his famous signs of inflammation. Nicholas Culpeper in his popular though in its day much reviled tome, *The English Physician*, wrote about it at length.[1] The bark, he maintained, was effective in stanching bleeding. It was a useful diuretic and an improver of sight. It was of 'proven effectiveness' in the treatment of warts. It also reduced unseemly lust in both men and women. Culpeper was an admirable character; but to the modern reader none of this makes sense. Credit for having launched the willow in modern medicine must go to the Reverend Edward Stone of Chipping Norton in Oxfordshire.

1. Culpeper who died in 1654, aged thirty-eight, had no university degree but set up as a doctor to the London poor. He supported Cromwell during the Civil War and in 1643 boldly produced a *Physical Directory, or a Translation of the London Dispensatory*, an unauthorised translation of the College of Physicians' closely guarded Latin *Pharmacopoeia.* He regarded (rightly) the college's monopoly as unchristian because it made medical care unnecessarily costly. The college was outraged. 'By two yeeres drunken labour he hath Gallimafred the apothecaries book into nonsense . . . and to supply his drunkenness and lechery with a thirty shilling reward endeavoured to bring into obloquy the famous societies of apothecaries and chirurgions.' Indignantly Culpeper published in 1653 an even more controversial version of the *Dispensatory. The English Physician Enlarged.* After many editions this became *Culpeper's Herbal*, still in print. It contained 500 prescriptions to deal with all human afflictions from adder bites to wind. Fashionable physicians were berated as toadies of tyranny. Princes, priests and physicians equally infringed the freedom of the people. In Culpeper's day Latin was still the tongue of monopoly, greed and obscurantism rather than that of sublime scholarship. Three centuries later Culpeper found a worthy champion in the late Roy Porter.

Stone, the only son of a modestly prosperous yeoman farmer of Princes Risborough in Buckinghamshire, was born in 1702. Nothing is known about his childhood; but at some stage he must have been offered the chance of advancement by studying for holy orders. This required a university degree, and he duly enrolled at Wadham College, Oxford. He graduated in 1727. After briefly serving as vicar of Charlton-on-Otmoor, Oxford, he returned to Wadham and rose to become fellow, librarian, bursar, dean and subwarden.[2] He married the daughter of a wealthy Buckinghamshire landowner, thus acquiring the pleasant living of Horsenden, Drayton and Bruern. He eventually settled in a handsome house in Chipping Norton in 1745. It was thirteen years later, at the age of fifty-six, that he made his discovery. On a summer's day, after 'a contemplative walk' along the stream known locally as the Common Brook, he rested under a willow tree planted by the town council some years earlier. What exactly prompted him to break off a piece of bark and pop it into his mouth is not known; but as his tongue registered the bitter, astringent taste, he was reminded of another and more famous tree.

. . .

The ague or intermittent fever accompanied by sweats and shivering had been a resident illness in England, as in much of Western Europe, for centuries. It had probably included a number of different fevers, but malaria was both the commonest and the most lethal.[3] The disease was known to flourish close to dank and stagnant waters and was widely believed to be caused by foul air and miasmas rising from the depth. 'This is evident,' William Buchan wrote in his famous *Domestic Medicine* in 1769,

> from [the illness] abounding in rainy seasons and most frequent where the soil is marshy, as in Holland, the Fens of Cambridgeshire and the Hundreds of Essex . . . The intermitting fever begins with pain in the head and loins, weariness of the limbs, coldness of the extremities . . . sometimes with great sickness and vomiting to which succeed shivering and violent shaking. Afterwards the skin becomes moist with profuse sweat . . . Sometimes the

2. But never made it to warden. As sub-warden he initiated proceedings against his superior, Warden Robert Thistlethwayte who had been accused of sexually assaulting an undergraduate (i.e. 'commoner'). Thistlethwayte fled the country, having inspired the verse: 'There was once a warden of Wadham / Who approved of the folkways of Sodom / For a man might, he said / Have a very poor head / But be a fine fellow at bottom'. But when he was being replaced, the sub-warden was passed over.
3. Shakespeare refers to it frequently, distinguishing between quotidian, tertian and quaternary forms which reflect the duration of the life-cycles of different species of the *Plasmodium* parasite. In *Henry V* (II, i, 118) he even uses it to illustrate the silliness of the Hostess of the Tavern who gets them mixed up: 'As ever you come of women, come quickly / to Sir John. Ah, the poor heart! He is so shaked / of a burning quotidian tertian that is / most lamentable to behold.'

> disease comes on suddenly when the person thinks himself in perfect health but more often it is preceded by listlessness and loss of appetite.[4]

The disease was no respecter of rank, struck at the virtuous as well as the dissolute and sometimes reached epidemic proportions.

The first and only effective treatment against the illness was the bark of the cinchona tree of South America, described in 1633 by the monk, Father Antonio de la Calaucha in his *Chronicle of St Augustine.*

> This tree, called the fever tree, grows in the country of Loxa . . . Its cinnamon-coloured bark made into a powder the size of two small silver coins and given as a beverage . . . produces miraculous results.[5]

Whether or not the bark had been used by the indigenous population before the arrival of the conquistadores is uncertain. In 1647 Pope Innocent X commanded the Jesuit priest, Juan de Lugo, to test its effectiveness; and, when the tests proved encouraging, approved its use in Catholic lands.

The Jesuit connection counted against the bark in Protestant England. Then an English apothecary, Robert Talbor, a rogue but of unimpeachably Protestant descent and utterances, claimed to have discovered an alternative. With his concoction he cured Charles II from a fever and was knighted and appointed royal physician as a reward. Leaving behind a supply, he then set out on a European tour. Grateful patients on his travels included the dauphin of France and the queen of Spain. After his death the magic brew was revealed as Jesuit bark which he seems to have acquired from Dutch privateers and whose bitter taste he had successfully disguised with sweet wine. By then religious fervour had sufficiently abated to make the drug acceptable to all faiths – provided that it was available. This remained the catch. Despite countless attempts to establish the tree in Europe, the cinchona would grow only in its native South America. This meant that, apart from the occasional loot, supply depended on imports to and re-exports from Spain.

Much of this may have passed through the mind of the Reverend Stone who had in the past used Peruvian bark (still the preferred synonym of 'Jesuit') for his own rheumatic pains and aches and occasional fevers. As a former librarian to Wadham College he probably knew about Paracelsus' doctrine, that Nature always provides a remedy for its afflictions in the region where the

4. Buchan's book became a fixture in well-ordered American households. On his death in 1805, aged sixty, he was buried in Westminster Abbey.
5. Quoted in D. Jeffreys's *Aspirin* (London, 2004), p. 25. The tree was named after the Countess of Chinchon, wife of the Spanish Viceroy who was given the concoction by a native servant and recovered from her feverish illness.

affliction is most prevalent.[6] This made the willow tree, at home on river banks and swamps, the natural source of any cure for the ague. The bitter taste of the bark, reminiscent of that of the chinchona tree, was no accident. Stone conducted his own experiments among his family and parishioners as well as on the household of his scientifically minded patron, Jonathan Cope of Bruern. Feverish pains were common enough. Children died from them. Old men and women were crippled with rheumatism.[7] All of these pains were, in Stone's opinion, part of the same affliction. And there was no doubt: the willow bark, whether crunched between the teeth or made more palatable as an alcoholic tincture, would often ease the pain and bring the fever down. It was not, perhaps, as effective in high swinging fevers and sweats as the Peruvian bark; but it was readily available. The Reverend Stone was a public-spirited man; and, after conducting his trials for three years, he penned a letter to George Parker, second Earl Macclesfield, president of the Royal Society.

> There is a bark of an English tree, which I have found by experience to be a powerful astringent and astonishingly efficacious in curing agues and other intermitting feverish disorders . . . On accidentally tasting it, its bitter taste immediately reminded me of the Peruvian bark . . . As this tree delights in moist soil where agues chiefly abound I could not but help applying the general rule of Providence providing special remedies not far from the causes of diseases.[8]

The letter was read at the meeting of the Society on 2 June 1763 and deemed worthy of publication in the *Transactions.* The Reverend Stone was congratulated on his ingenuity and told to report any further observations. There were none, at least none related to the willow tree. Stone was a man of many interests, and now he turned his fertile mind to the doctrine of parallaxes, the path of comets and problems of algebra. For the rest of his life pain did not figure among his pursuits.

. . .

6. Paracelsus also maintained that Nature always gave a hint of the specific curative powers of plants. The orchid could be used in venereal disease because of its resemblance to the testicle and the blue eyebright plant was useful in eye diseases.
7. From the Greek ῥεῦμα meaning a flowing or a flux. The term is Hippocratic; but Hippocrates made no distinction between rheumatism and catarrh, both referring to 'flowing down' which in turn referred to one of the humours of the body. It is in this sense that Shylock in *The Merchant of Venice* says 'and void his rheum in my beard'. Guillaume de Baillou (died in 1616, aged seventy-eight) published a book on 'rheumatism', a general disease. In the modern sense of primarily joint disease the term may first have been used by Sydenham in 1666.
8. E. Stone, 'An Account of the Success of the Bark of the Willow in the Cure of Agues', *Philosophical Transactions* [of the Royal Society] (London, 1763), p. 67.

The *Transactions of the Royal Society* were widely read by the enlightened; and over the next few decades the bark of the willow was occasionally tested. A Hertfordshire doctor, Samuel Jones, reported to the Royal Society that it was 'remarkable [sic] effective in childhood fevers'. William White, pharmacist to the City Infirmary in Bath, testified that the introduction of the willow to replace the chinchona had saved the Guardians of the Infirmary £20 a year. 'The willow, moreover, had proved in no way inferior.'

None of this amounted to much until historical developments began to shape events. The Napoleonic Wars and the Continental Blockade made the importation of cinchona to Europe more uncertain than ever. Spain was at first Napoleon's ally; but its naval power was crushed at Trafalgar together with France's. For some years Europe was at the mercy of England for its supply. Whether or not as a direct response to this, French, German and Italian scientists in the Napoleonic decade turned their attention to the active principles of plants. Once their interest was aroused, it continued after the political urgency. The restored Bourbons in France presided over a torpid regime, and the Holy Alliance in the rest of Europe was hostile to any kind of original thought; but the chemical achievements of the age were dazzling. In France alone, within a single lifetime, Louis-Nicolas Vauquelin, Pierre Joseph Pelletier, Joseph Caventou and a handful of others isolated strychnine, brusine, veratrine, caffeine, nicotine and quinine.[9] Stone's discovery was by then well known on the Continent; and the bark of the willow inevitably moved into their sight. The first important advance came from Germany, where Johann Buchner, professor of pharmacy in Munich, isolated a minute amount of bitter-sweet tasting yellow crystals. From the Latin name of the willow, *salix*, he named it salicin. Next a French chemist, Henri Leroux, using a simpler procedure, obtained about 25 grams from around 2 kilos of bark. He was in turn trumped by the Italian Raffaele Piria, who simplified the method further and named the crystals salicylic acid.

The chemical frenzy was not confined to the universities. For a few astonishing decades there seemed to be only a small number of village pharmacies in Europe where chemical experiments of some sort were *not* being conducted. Many proved successful. But perhaps it was not altogether surprising. Both in towns and in villages pharmacies attracted young men with inquisitive minds but too poor to go to a university. A few, like Claude Bernard, Dumas and Balard, eventually had brilliant academic careers. Many maintained that their experience in pharmacies stood them in good stead. 'One met real people there with real ailments,' Balard wrote. 'And they were grateful for any help we could give.' To accommodate their discoveries new

9. Isolating them did not mean that they could be synthesised.

chemical journals sprang up in most European countries. Germany soon had four, France three, Switzerland and Italy two each. Even Russia began to publish the proceedings of its Imperial Academy of the Sciences, though the language of publication remained French until the Crimean War in 1856.

Among those who remained in their village shops, Johann Pagenstecher, a pharmacist in the Bernese Oberland in Switzerland, took an interest in the meadowsweet flower, *Spiraea ulmaria.* A simple extract of the fresh plant, he found, relieved both toothache and rheumatic pains. But the unprocessed plant had to be fresh. A preparation which would be stable on the shelf might save a back-breaking expedition every spring. (Mountaineering for pleasure was still a pastime for mad Englishmen.) The plant had helped his own rheumatism and nothing stimulates a man's interest more. After labouring for some years in his shop, he prepared a stable essence. He identified it as an aldehyde and wrote a report for the Swiss journal, *Die Pharmazie.*[10] The article caught the eye of one of the new breed of 'alkaloid' chemists. Karl Jacob Löwig of Berlin travelled to Switzerland and obtained some of Pagenstecher's tincture.[11] After a good deal of experimentation he converted the aldehyde to a fairly pure acid which he called *Spirsaure* from *spiraea.*

It was a few years later that Löwig realised that what he had extracted from meadowsweet was in fact salicylic acid, the substance which Piria had already isolated from the willow bark. But Löwig's experimental protocol was simpler; and, whatever the source, the compound obviously had potential.

. . .

Scientific endeavours now begin to intersect with the industrial revolution. One of the earliest and most splendid discoveries of that productive upheaval was made by a young Scotsman with no scientific training. Relaxing by his fireside one evening twenty-year-old William Murdoch added coal dust to his pipe and laid it among the embers. A beautiful white flame shot out. He guessed that it was due to coal gas burning in a vacuum. Within a few years that flame would transform domestic interiors and the urban landscape in Britain and then in Europe and America. Murdoch also noticed that a dark amorphous mass was left behind by the burning gas.[12] He described it as 'uninteresting'. But there he was wrong. No dark amorphous mass was uninteresting to children of the age of chemistry. Despite its unpromising look, a young German chemist, Friedlieb Ferdinand Runge, soon showed that it

10. Usually highly reactive substances possessing the chemical group –C–H = O.
11. Pagenstecher's dates are unknown. Löwig died in 1890, aged eighty-three.
12. He later greatly improved Watt's steam engine and invented the high-pressure locomotive. He wrote comparatively little but accomplished much and died a wealthy man in 1839, aged eighty-three.

contained several intriguing substances. Runge was another prodigy. By the age of twenty-five he already had several discoveries to his credit, including the isolation of caffeine.[13] He then isolated a black substance from the coal-gas residue, and called it aniline. From aniline he prepared a derivative he called phenol. This, under its pseudonym, carbolic acid, would blossom into Lister's great antiseptic. Runge also prepared what would later be recognised as the first industrially useful organic dye: aniline black. He earned the admiration of his fellow chemists but aroused little interest outside academic circles. He himself had neither the ambition nor the means to exploit his discoveries commercially.[14] This would have to wait.

But not for long. Another development which would have an impact on Löwig's *Spirsaure* is linked to the career of William Henry Perkin. Born in 1838, Perkin was the son of a London builder and a precociously clever youth. His chemistry teacher at the City of London School recommended him for a bursary at the Royal College of Chemistry. The college was a new foundation, sponsored by Gladstone, Peel, Michael Faraday, the Prince Consort and other public-spirited citizens for a specific purpose. England, once the cradle of revolutionary industrial advances, was felt to be increasingly lagging behind Germany and even France. The college, promoting young talent unable to go to university, would remedy that. Here Perkin shone. He became the personal assistant and protégé of one of the imported German professors, August Wilhelm von Hofmann; but he also conducted private and secret experiments in the kitchen of his digs in London's dockland. They had to be secret and private because Hofmann did not approve of 'applied' chemistry. Once again the search was on for a substitute for that elusive and wonderful substance, quinine. There is no colour to correspond to a gardener's 'green' fingers in chemistry: perhaps in honour of Perkin it should be mauve. As generally happens, he had a succession of failures and survived the usual contretemps of experimentation in the kitchen – fires, breakages and explosions. But he also made intriguing chance observations. In this last category he noticed that when the black sludge left behind by aniline was extracted with spirit and then washed with water, it turned the test-tubes a brilliant purple. He made the inspirational leap from drugs to dyes (eventually to be reversed) and went on to isolate and later to patent the most successful colour of the century. Not far removed from purple, once the prerogative of emperors, harlots and cardinals, it was comparatively cheap to produce and still looked grand. Most importantly, it set off the luminous dark eyes of the Empress Eugénie of France and would flatter even her dumpy sister of Great Britain. Originally christened mauvein, it was soon abbreviated to mauve. Unplanned by Perkin

13. He worked independently from Pelletier, who beat him to the post by a few months.
14. Runge died a professor of chemistry in Berlin in 1867, aged seventy-one.

or anyone else, the discovery proved a significant advance in therapeutics as well as in ladies' fashion.[15]

...

During the decades which followed the launch of mauve, trials with salicin and salicylic acid ('*Spirsaure*') continued. They were not immediately productive. The difficulty was the irritating nature of both substances. They could be swallowed in small quantities, accompanied by copious libations of milk; but the larger and effective doses which were both analgesic and fever-reducing produced a burning sensation in the gullet and stomach and sometimes led to the vomiting of blood. One researcher got tantalisingly near the solution. Charles Gerhard was professor of chemistry in Montpellier and salicylic acid was one of his passing obsessions.[16] To clarify its chemical structure he allowed it to interact with acetyl chloride, a reagent known to substitute hydroxy (-OH) with acetyl (-$CH_3$–COOH) groups. He produced acetylsalicylic acid which, to his surprise, he found non-irritating. He was in fact not particularly well read: the substitution of hydroxy by the acetyl group had already been observed to have an emollient effect on phenolic compounds. Unfortunately it also, as a rule, destroyed their chemical usefulness. But the procedure using acetyl chloride was tedious and Gerhard, a noted archaeologist, was easily bored by chemistry. Others were less so. Hermann Kolbe of Marburg University simplified Gerhard's procedure;[17] and one of Kolbe's pupils, Friedrich von Heyden even established a small factory to exploit it. The factory floundered. As would happen again with the sulphonomides and penicillin, all pieces necessary for a coherent picture were now available. What was lacking was medical credibility.

It was to be provided by Dr Thomas Maclagan. The son of a prosperous general practitioner in Scone in Perthshire and his wife, the daughter of a rich plantation owner in Jamaica, Maclagan was a graduate of Glasgow University in the humanities and of Edinburgh University in medicine. After qualifying he travelled at leisure in Europe, putting in stints of postgraduate study in Vienna and Paris. On his return in 1864 he was appointed superintendent of the Dundee Royal Infirmary. This was an impressive new edifice in the Gothic style, occupying a commanding position over the city. The jute, whaling and marmalade industries, traditional mainstays of Dundee's prosperity, were booming; and no expense had been spared over the statuary and the stucco

15. Perkin died a rich man, a knight (for his philanthropic work) but not a fellow of the Royal Society, in 1904, aged sixty-nine.
16. Gerhard died in 1872, aged sixty-three. He is remembered more for his Roman finds around the charming town of Saint Rémy-de-Provence than for his chemistry.
17. Kolbe died in 1876, aged fifty-three.

embellishments. The stained glass in the chapel was much admired. Inside Maclagan found typhus and typhoid rampant, as were tuberculosis and fevers nobody had troubled to identify. During his four years in the post he did good work. He established separate fever wards for the first time, one of them later to be named after him. He also learnt much that was new to him.

Among the new and unexplained diseases of childhood, 'acute rheumatism' was becoming common.[18] It was an agonisingly painful acute inflammation of the medium-sized joints, most often the knees and elbows. The pains were accompanied by high fever and 'galloping anaemia': only diphtheria among children's diseases could cause such a rapid general decline. But the illness was more than an unusually severe form of arthritis. It was soon realised that, whatever was the underlying cause, it also involved the heart, sometimes the kidneys and occasionally the nervous system.[19] It was the first which was most feared. Somebody soon coined the tag: rheumatic fever licks the joints but bites the heart. Even after apparent recovery from the acute joint pain and swelling the smouldering inflammation of the inner lining of the heart continued. Valves are the keys to the proper functioning of any pump, and these seemed to be the main targets. As the inflammation progressed, they became thickened, narrowed, ineffective.[20] Months, sometimes years, after the initial attack, the heart would begin to fail. The patient would become progressively short of breath. The lungs would become congested, the liver enlarged and tender, the ankles swollen. It was a miserable progression in young people in their prime – and inexorable. Most of the patients would be dead by their mid-twenties.[21] Controlling the joint disease by any means

18. 'Acute rheumatism' by this time was rheumatic fever. Confusingly, this disease is unrelated to rheumatoid arthritis. It was later recognised as one of the immune responses to the haemolytic streptococcus (the cause of scarlet fever, certain forms of acute nephritis and severe sore throats). For reasons which remain almost entirely guesswork, it became extremely common in the second half of the nineteenth century and remained common for about a hundred years. It is rarely seen today.
19. Kidney involvement manifested itself in haematuria, loin pain and sometimes oedema. It occasionally progressed to chronic renal disease and death. The neurological complication was known as Sydenham's chorea (from the Greek χορεἰν *chorein*, dancing). It consisted of convulsive, uncoordinated movements and an inability to perform tasks requiring fine muscular control like threading a needle. It added to the patient's profound exhaustion.
20. The valve most commonly affected was the mitral valve separating the right auricle from the right ventricle. Its progressive narrowing (stenosis) led to right-sided heart failure, most commonly manifested in difficulty in breathing and swelling of the ankles and of the abdomen.
21. Mitral stenosis remained incurable until the advent of cardiac surgery in the 1950s. Splitting the valve with a special knife fitted on to a ring was at first performed blind, the surgeon inserting his 'ringed' finger through a small incision in the heart muscle. The special knife, with luck, recreated the original mitral opening. The initial mortality was high.

available was therefore a matter of life and death. Whether such control would also prevent later heart and kidney complications was anybody's guess; but one could hope.

But control of the acutely inflamed, painful joints was difficult. Poultices, splinting and the application of such well-tried remedies as oil of wintergreen or liniment of menthol were ineffective. Opium dulled the pain but not the joint swelling. Nor did it prevent the appearance of an ominous heart murmur – lubb-*sssh*-dubb, lubb-*sssh*-dubb – indicating cardiac involvement. Maclagan had come across salicin on his continental travels. He now purchased some at his own expense. He started to experiment with self-designed concoctions, trying to make it more palatable. Children seemed to tolerate it better than adults: or perhaps they were less inclined to complain. But discouraged by the immensity of his task at the Infirmary – he was not by nature a go-getter – he did not renew his contract after four years. He set up in practice on his own and prospered. He then moved to London and opened consulting rooms in fashionable Cadogan Square. With Buckingham Palace and other royal caravanserais nearby, he soon had minor royalty on his books. A well-read and cultivated man, he also attracted the literati of Chelsea, including the terrible Carlyles.[22] But his interest in childhood fevers and their possible management with salicin continued. Eventually he embarked on what even today would pass as a well-planned controlled trial. He divided his patients into more or less matched treated and control groups and carefully charted their progress. This left no doubt in his mind about the merits of the drug. On 4 March 1876 he published his findings in the *Lancet*.

> Quite apart from the antipyretic action of salicin [he concluded], it is in my experience the most effective treatment of pain in acute articular rheumatism. It may even show itself specific for the disease and prevent the dread complication of mitral valve disease.[23]

. . .

If in modern times the Reverend Edward Stone was the 'discoverer'of salicin as a potential drug, it was Maclagan who gave it medical standing. He was a good communicator both in writing and in person. As a successful practitioner and much liked personally, his article was well received. For once, an objective measure of this intangible commodity exists. During the months following publication the price of salicin in London pharmacies doubled and then trebled.

22. Both Thomas and Jane were deeply neurotic and prey to countless minor and major ailments.
23. *Lancet*, 1876, i, 283.

Rheumatic fever seemed to be getting almost as common in the South as it had been for some time in the North. It was as heart-rending to watch as its deadly twin, diphtheria. To still the pain it sometimes required salicin in larger doses than anyone would have contemplated giving to children in the past. The drug was not free from side effects, mainly gastric irritation and bleeding, but the risk was dwarfed by the need to deal with the acutely inflamed joints. And salicin was undoubtedly effective: the only treatment that was.

Medicine was becoming cosmopolitan. There was a good international response to Maclagan's article. German doctors, Solomon Stricker and Ludwig Reiss, and a Paris practitioner, Germain See, wrote letters to the editor of the *Lancet* about the good results they had obtained not only with salicin but also with salicylic acid. To clinch matters, a Dr Ensor wrote from Cape Town, telling readers that the 'Hottentots of Southern Africa' had been eating willow bark for aches and pains for many generations. They were wholly insensible to its foul taste and held it in great esteem. The ground was prepared for the industrial offensive.

...

In Germany the opportunities unveiled by Perkin's discovery had been eagerly grasped. The textile industry there had long resented British dominance in the manufacture of dyes; the Ruhr was providing ample coal; and the country's universities were churning out keen and capable chemists. Among the smaller dye factories that sprang up was one established by Friedrich Bayer and Friedrich Westcott in Barmen near Cologne. Both founders came from families of weavers; and in the 1860s both realised the potential of the new synthetic dyes. When Bayer died in 1880, followed a year later by Westcott, the reins were picked up by Bayer's son-in-law, Carl Rumpff.

Rumpff, a much-travelled visionary, began to recruit young chemists. The laboratory facilities of the factory – in reality, no more than a workshop – were modest but they still offered new graduates a chance to prove their worth. In return they strengthened the scientific base of Bayer's. Among the recruits was Carl Duisberg, a local lad and chemistry graduate of Göttingen University. He was hungry, clever and ambitious, then as now the ideal combination for an aspiring industrial chemist.[24] After some initial success with dyes – Duisberg synthethised Congo Red within a year of starting with the firm – Rumpff put him in charge of the company's grandiosely entitled but still virtually non-existent 'pharmaceutical division'. There Duisberg observed with interest a new drug being launched by a rival company in the small town of Hoechst, only a few miles up the Rhine. The drug was in fact new*ish* rather than new:

24. Carl Duisberg died in 1935, aged seventy-three, just before the factory and firm largely created by him became a tool of the Nazis.

the novelty was mainly in the green colour. Though not without unpleasant side-effects, under its revamped title, 'Antipyrine', it did unexpectedly well. To Duisberg the event revealed the power of a good brand name. The lesson was soon reinforced by another new-old drug marketed by Kalle and Company under the name of 'Antifebrin'. 'Antifebrin' was in fact a confidence trick.

A year earlier two Strasburg doctors, Arnold Cahn and Paul Hepp, had ordered naphthalene from wholesale suppliers. Naphthalene was the standard treatment for intestinal worms and a consignment duly arrived. Unfortunately – or fortunately – confusion was endemic in the mushrooming and therefore chronically understaffed warehouses, and what had been dispatched was not naphthalene but acetylated aniline, known as acetaniline. This was a by-product of coal-tar distillation and another discovery of Perkin's. It was cheap and beginning to be used in the dye industry. More interestingly, though it proved powerless against tapeworms, it unexpectedly lowered the temperature of several patients suffering from *la grippe*. It also seemed to ease their aches and pains. Nobody died from it. Paul Hepp's brother was an industrial chemist with Kalle and Company and saw his chance. Sadly, acetaniline was both generally known and widely available: there was nothing to stop Kalle's rivals from manufacturing it too. The solution was to disguise this by giving it the attractive brand name 'Antifebrin'. It was simple and self-explanatory. That 'Antifebrine' was acetaniline was not kept a secret but nor was it 'trumpeted about'. There was no need for the word acetaniline to appear anywhere on the label or in the promotional literature. It would, Kalle and Company maintained, merely confuse the public. For a time at least the drug sold well.

'Providentially', as the inspiration was later described by Duisberg, it was at that time that he remembered some 30,000 bottles of another dye waste product called para-nitrophenol skulking around in the yard of the Bayer plant. He gave two doctoral fellows the task to find out whether that too might have anti-pyretic properties. It did. After a preliminary trial on six volunteers, an innocuous blue dye was added to the liquid and it was renamed 'Phenacetin'. The name was memorable as well as possessing a pleasing scientific ring and the tablets appealed to the fastidious. Although demand at first dangerously outstripped capacity – the first batches had to be brewed in discarded beer bottles – the company made a great deal of money. Rumpff was less than totally happy – though a visionary, he also possessed a puritanical streak – but Duisberg was soon on the look-out for other promising waste products. Once again 'providentially', Rumpff died. The success of phenacetin persuaded the Bayer Board to give Duisberg the go-ahead. Duisberg realised that to tempt the best of the output of Germany's universities, an imposing building programme was necessary. Doubts were raised by some of the more staid members; but the *Zeitgeist* was against them. To the strains of Haydn's hijacked and corrupted imperial anthem Germany was building a glorious future. The vision was shared by factory hands and the Kaiser, Prussian Junkers and Rhineland poets. And of course by far-seeing industrialists. The navy came first; but next to the navy, industry. A state loan was negotiated. By 1890 the palatial new laboratories of Bayer's in nearby Leverküsen were ready to receive the cream of the Reich's science graduates.

As Duisberg had predicted, young men were tempted by the bricks and mortar (not yet glass, steel and potted palms) as well as by the relatively generous pay. Among the newcomers were two young doctors of chemistry, Arthur Eichengrün and Felix Hoffman, as well as a young professor from Göttingen University, Heinrich Dreser.[25] Which of the three had the crucial role in developing the first new drug on the agenda is still a matter of debate. Eichengrün was the brightest, Hoffman probably the most industrious. (The problem is sifting through mountains of fake and falsified documentation.) In truth, Maclagan's clinical success with salicin and Stricker's with salicylic acid provided an obvious target. The difficulty was to overcome in a novel and, above all, patentable way the still troublesome side-effects. To what extent the Bayer chemists were influenced by previous work by Gerhard is uncertain. But according to in-house Bayer histories and histories commissioned by Bayer's for external consumption, on 10 August 1897 Felix Hoffman made the historic jotting in his laboratory book:

> When salicylic acid, 100 parts, is heated with acetic anhydride, 150 parts, for three hours under reflux, the salicylic acid is quantitatively acetylated. After distilling off the acetic acid one obtains the above product in the form of needles . . . My acetylated product no longer gives a reaction with ferric chloride which distinguishes it from salicylic acid. Its lack of corrosiveness makes it superior to the parent substance.[26]

All seemed set fair when a last obstacle loomed. It would be unfair to describe it as a red herring. The obstacle itself would prove a notable development in pain control. But it requires a brief digression.

. . .

25. Under the Nazis Eichengrün, a Jew though married to an 'Arian' woman, was sent to the concentration camp of Theresienstadt which he survived – just. He died in 1949, aged eighty-two. Even after the Second World War he continued to be written out of Bayer-sponsored histories, including the company's current website. According to these, Hoffman and Dreser were the only inventors of aspirin. The key role of Eichengrün was established only through the brilliant researches of Walter Sneader, medical historian at Strathclyde University, Glasgow.
26. Quoted in Jeffreys, *Aspirin*, p. 85.

In 1874 C.R. Alder Wright, a chemist working in the laboratories of St Mary's Hospital in London, was experimenting with morphine derivatives, hoping for a compound equally effective but less addictive. Like Hoffman two decades later, he tried acetylation and obtained a white powder which he rightly assumed to be diacetylmorphine. He gave it to his dog, which was violently sick, became hyper-active and nearly perished. Alder Wright liked dogs. In disgust he threw the powder away but reported his findings in a short paper in the *Chemical and Pharmaceutical Journal.*

There the matter might have ended if Dreser at Bayer's had not been a compulsive browser through old journals. This was in fact his forte. He came across Alder Wright's article. Acetylation struck a chord in his mind. Quickly he convinced himself that he had discovered another winner. He instructed Hoffman to try the acetylation process on morphine. A little reluctantly – he had his hands full with salicylic acid – Hoffman complied. Within a few days he reproduced Alder Wright's white crystals. Instead of trying it on dogs, Dreser fed it to frogs and rabbits. They liked it. Dreser then persuaded four workmen in the factory to try taking it. Nothing at Bayer's was to be done by halves. The human guinea pigs consumed a huge dose. The results were spectacular. All four stated that the product made them happy, resolute and divinely confident. Even the meekest among them felt positively *heroisch* (heroic). The last experience immediately suggested a name. Diacetylmorphine would be known as 'heroin'.

By 1898 'in-depth clinical trials' of heroin – another two workmen and the uncounted patients of Bayer's two tame general practitioners – had proved 'uniformly successful', and Dreser reported his discovery to the Congress of German Naturalists and Physicians. The new product was ten times more effective against cough than codeine and ten times less toxic. Compared to morphine, it was both five times more potent and 'completely non-habit-forming'. (Nobody seemed to wonder how the last property had been tested in a trial lasting three and a half weeks.) Even Duisberg was caught up in the excitement. Still, getting a new wonder drug tested for general consumption and putting it on the market was becoming complicated, expensive and time-consuming. Legislation in Germany was late and half-hearted; but, though the distraction was resented, bypassing it was unthinkable. Bayer's resources were still limited. Duisberg and Dreser had to decide between acetylsalicylic acid and diacetylmorphine. Both would be developed; but which would get priority? They decided to put acetylsalicylic acid on the back burner.

Hoffman and Eichengrün were furious. Without authorisation, they got in touch with Bayer's Berlin representative, Felix Goldman. Could Goldman, another young Turk, arrange a 'discreet' clinical trial for acetylsalicylic acid? The results surpassed expectations. The new compound not only reduced fever but made toothache and headache vanish almost before the powder had been swallowed. No side-effects were reported. Countess von Roon had administered a dose to Maxi, her ailing poodle, and the effect had been miraculous. Maxi was back to his normal gregarious self. The drug would clearly revolutionise medical practice. It would certainly banish hangovers.

It was now Duisberg and Dreser's turn to feel aggrieved. Their juniors had gone behind their backs. Dreser thought Goldman's report 'typical Berlin boasting'. But the Countess von Roon, a second cousin of the Chancellor, could not be ignored. Independent clinical reports were no less enthusiastic: the new drug was more effective than any previous salicin or salicylic acid preparation. Grudgingly, Duisberg and Dreser accepted the case for giving it priority. Perhaps they were swayed by the first case of 'severe craving' which developed in the wife of a local pastor who had been given large doses of heroin for eczema of her hands. Perhaps diacetylmorphine was not quite so *heroisch* after all. Or too much so. Better to pursue the less spectacular acetylsalicylate.

The final task was to find a brand name. A round robin was circulated among Bayer's management, asking for suggestions. After some wrangling – 'Euspirin' was the chief rival – Eichengrün's suggestion was accepted. With a nod to the genus *Spiraea* (meadowsweet) rather than to the English willow, the winner was 'Aspirin'.[27]

27. The later commercial and political developments of Bayer and later I.G. Farben are briefly described in Chapter 45.

# Chapter 36

# COCAINE

Of the two momentous events in pain control which ushered in the twentieth century the first, aspirin and the rise of the pharmaceutical industry, tiptoed in quietly. The second, cocaine and local anaesthesia, was recognised at once as an advance rivalling ether and chloroform. Today, of course, cocaine means a stimulant, a substance of addiction, a social problem, a criminal conspiracy and the *raison d'être* (if the word reason can be used in this context) of one of the world's most expensive and ineffectual international police operations. None of these were part of the original picture. The drug was initially acclaimed as the means of controlling pain locally, an action Claude Bernard had thought was impossible. Today's dental anaesthetics, nerve blocks, epidural anaesthesia in obstetrics and countless 'minor uses' (like the painless insertion of large needles in blood transfusion and bone marrow sampling), all followed from the first drop of a 2 per cent cocaine solution into a frog's eye.

Like so much else in the history of pain-killing drugs, the coming of cocaine to the West is bound up with the search for quinine. In 1852 the Foreign Office sent letters to its consular staff in South America, instructing them to try and identify cinchona trees whose seeds could be planted in British colonies to create 'home-grown' sources of the precious anti-malarial. At about the same time the Dutch government dispatched a single botanist, Justus Charles Hasskarl, from their scientific station in Java with the same objective. The British government followed up their enquiry by appointing a high-powered team of explorers under the leadership of Clemens Markham. It included the as yet unknown but later famous Amazon explorer, Richard Spruce. No equipment was available to measure the quinine content of barks. There are at least 38 different species of cinchona tree and bark, only four of which contain sufficient quinine to be worth growing. The Markham expedition knew nothing about this. It included only one member who spoke Spanish. He was sidelined. Markham himself, an arrogant and obtuse but manipulative and ambitious former naval officer, lost no time antagonising local experts. The expedition eventually returned home with a collection of seeds which they had decided would grow the right tree. The seeds collected by Hasskarl were equally arbitrary. Markham was fêted in London. Hasskarl was welcomed in the Hague. Both were knighted. The Dutch seeds were planted in Java. Hasskarl became head of the quinine programme there. A year later, when it became obvious that he had come back with the wrong seeds, he was demoted. Markham's haul took longer to unravel. By 1900 more than a million of the wrong cinchona trees were growing in Ceylon (today's Sri Lanka). Their yeild of quinine was so low and their upkeep so labour-intensive that they would soon have to be uprooted at great cost. By then their champion was slithering toward the presidency of the Royal Geographical Society. With all the authority at his command Sir Clement declared that the right cinchona plant would grow only in South America.

Ten years after the dispatch of Hasskarl and the Markham expedition a British trader living in Peru, Charles Ledger, acquired 7 kilos of the right seeds from his native contacts.[1] He sent them to London. Both the Botanical Garden at Kew and the British government expressed themselves deeply uninterested. They already had a cinchona forest growing in Ceylon. The Dutch, who had seen the light sooner, bought the seeds and planted them in Java. The trees had a bark with the highest quinine content of any species known. The island eventually became the world's largest supplier.

. . .

During their exploration both Hasskarl and Markham were impressed with another plant that was being grown in Peru and neighbouring countries and whose leaves were revered by the natives. The European scientists learnt that it had been the sacred plant of the Incas for at least five hundred years before the arrival of the conquistadores. It had been regarded by the people as a gift of the gods and had initially been a royal monopoly, handled exclusively by a priestly caste. It was too desirable and too ubiquitous to remain so. By the time Pizarro penetrated Peru in 1532 the coca leaf had permeated every walk of life. It was chewed, made into cakes, consumed as a refreshing beverage, smoked, rubbed into wounds as a healing salve and incorporated (as bread was in Europe) in all solemn religious ceremonies. Unlike tobacco, it was at first ignored by the Europeans. Because it was the means local priests and witch doctors used to induce a trance-like state, the Catholic Church declared it an abominable vice. Only when they discovered that the natives would not work in the gold mines without it did it become less abominable. The conquerors

1. The plant is now known as *Cinchona ledgeriana* and its bark contains 15 per cent quinine, more than any other known variety.

appropriated all existing plantations and established new ones. Instead of food or money, coca was now to provide the wages of native labour. The drug had many actions and was credited with many more; but most importantly it enabled inadequately nourished slaves to sustain great physical effort and even to endure starvation. There was of course both a limit and a backlash to this. The practice of substituting cheap coca for expensive food resulted in the partial extinction of the Indian population. As the workforce began to decline, a succession of laws was enacted to regulate the trade. Eventually coca became a monopoly of the Spanish crown. This did not last long. Throughout the political upheavals of the nineteenth century the leaves sustained native forces fighting both for independence and each other. European interest in it remained sporadic until the quest for quinine drew attention to the plant.[2]

Both Hasskarl and Markham tried to interest their respective governments in the new stimulant, 'more effective than coffee'. Official responses remained lukewarm – there were too many stimulants already about – but the curiosity of medical and chemical researchers was aroused. An Italian doctor, Paolo Montegazza, who had practised in Peru and had studied coca, published a book on his return to his native country. He vividly described the hallucinogenic and stimulating effect of the plant but also enthusiastically recommended it for a wide range of painful afflictions, from toothache to neuralgia. About the same time the active principle, cocaine, was isolated by Albert Niemann of the University of Göttingen. By the 1870s interest in the effect of the leaves was widespread. Alexander Hughes Bennett of Edinburgh obtained coca leaf through a local pharmacy 'with some difficulty'. He demonstrated 'the similarity of the physiological effect of coffee, tea and cocoa' but expressed the view that cocaine was definitely the most potent.[3] A stream of articles followed in both the *Lancet* and the *British Medical Journal*, extolling the

2. The Jesuit Father, Don Antonio Julian, strongly encouraged the introduction of this 'perla de America' as a substitute for tea and coffee in 1695; and five years later Don Pedro Nolasco Crispi, a doctor, urged the crown of Spain to introduce it for the use of its sailors. There was no response: J. Nevitny, *Das Cocablatt* (Vienna, 1886), reprinted by S.B. Karch in *A History of Cocaine* (London, 2003).

   In England Abraham Cowley, physician and poet, famously celebrated the virtues of coca in his *Book of Plants*, published in 1662: 'Endowed with leaves of wondrous nourishment, whose juice sucked in, and lo the stomach taken long hunger and long labour can sustain: from which our faint and weary bodies find more succour, more they cher the dropping minds, they can your Bachus and Ceres joined'. Cowley's too remained a lonely voice.

   In 1735 the French botanist Joseph de Jussieu who accompanied the explorer and mathematician La Condamine to Quito sent specimens home; and the plant received the botanical name *Erythroxilum coca*, the 'erythro-' referring to the red colour of the bark.
3. A.H. Bennett, 'An Experimental Inquiry into the Physiological Actions of Theine, Guarinine, Cocaine and Theobromine', *Edinburgh Medical Journal*, 1874 (19), 323.

remarkable action of the leaves on fatigue, depression, hunger and even 'bashfulness of a severe nature'. Sir Robert Christison of Edinburgh who had investigated the action of opium during the first decades of the century and was now past eighty, became an enthusiast. He wrote that after climbing and reaching the top of Ben Vorlich:

> my companions were provided with an excellent luncheon but I contented myself with chewing two-thirds of one drachm of cuca-leaves and I went down the long descent with an ease like that which I used to enjoy in my mountainous rambles as a youth. At the bottom I was neither weary, nor thirsty, nor hungry, and felt if I could easily walk home for miles.[4]

Sir Clifford Allbutt took a supply with him on a walking tour of the Alps and succeeded in 'impressing his friends' with his endurance. An anonymous 'Doctor Sportsman' reported in the *Lancet* that his prowess in 'downing birds right and left' after filling his hip flask with a coca tincture instead of brandy and taking 'refreshing swigs' from time to time became 'prodigious'. The scene was set for the coca craze of the 1880s.

. . .

It is usually impossible to say why a drug which has been known for many years suddenly becomes the rage; but the late 1880s saw such a frenzy for cocaine. Dr Sigmund Freud of Vienna was only one of many (and at the time more eminent) personages who was swept off his feet.

The Freuds in the 1880s were a poor but upwardly mobile Jewish family. Sigmund's parents were just beginning to look down on more recently arrived *Ostjuden* or Eastern Jews, the obligatory first step to total assimilation as good Austrians. Sigmund entered Vienna University in 1875, moving from the Natural Sciences to Medicine 'out of a kind of curiosity', and finishing the course in a leisurely manner with the help of charitable grants. He travelled to Manchester, Trieste and Germany visiting both relatives and medical centres and for a time toyed with the idea of pursuing full-time research. His informal and then formal engagement to Martha Bernays, the 'fresh and sweet' daughter of an Orthodox Jewish family, put an end to such plans. He needed money to marry, especially against the prejudices of Martha's parents, and reasonable career prospects. He was also puritanical in sexual matters and the possibility of living with Martha (or anyone else) outside marriage never arose. He gave up his plans for research, never very compelling, and became a *Nervenarzt* instead.

4. R. Christison, 'Observations on the Effect of the Leaves of Erythroxylum Coca', *British Medical Journal*, 1876 (1), 527.

Though as a capable clinician in a sought-after specialty – neurosyphilis in many guises was rampant, affecting the highest and richest in the land – his distant prospects were good, they were definitely distant. In April 1884, aged twenty-seven, he thought up a scheme which, he hoped, might speed up his professional advancement and solve both his financial and his sexual problems. He began to experiment with cocaine. His interest in the drug was aroused after a Bavarian army doctor, Theodor Aschenbrandt, had reported that, during the autumn manoeuvres, he had issued cocaine to five recruits, with 'striking results'. After being blots on the regiment's escutcheon with their 'laziness and sloppy turnout', they began to display exemplary martial qualities. Indeed, they became, while the effect of the drug lasted, 'almost too ready to pick a fight'. Freud obtained some cocaine from the pharmaceutical firm of Merck & Co. of Darmstadt and started to take doses of 50 milligrams dissolved in water. They left him confident and cheerful. He then read American reports suggesting that cocaine might be a cure for morphine addiction. He decided to try it on his friend and colleague, Ernst von Fleischl-Marxow.

Von Fleischl-Marxow had had his thumb amputated following an infection two years earlier and developed severe phantom pains. The pains drove him to morphine and he became an addict. He had told Freud a year earlier that when his parents died he would commit suicide. He now became one of Freud's guinea pigs. He was not the only one. Freud handed out small doses to his friends and continued to take the drug himself. First reports were enthusiastic. Freud declared that he was ready to try the drug in nervous exhaustion, neuralgia, heart disease, rabies and diabetes as well as in seasickness (to be induced experimentally on swing-boats in the Prater). By correspondence Martha, removed by her parents to Hamburg, was drawn into this world of enchantment. Sigmund wrote to her about the brilliant success of the drug with 'all kinds of stomach trouble' and hinted at its fantastic aphrodisiac qualities. She herself would soon experience them, though only in the form of chaste but passionate kisses. In July 1884 he published 'On Coca', a hymn of praise, dedicated to his fiancée.[5]

The infatuation did not last more than a year and a half. Some of Freud's colleagues disapproved from the start. Albrecht Erlenmeyer, a prominent psychiatrist, accused him of 'unleashing a scourge of humanity'. Von Fleischl-Marxow became an addict to cocaine instead of morphine, having to take enormous and increasing doses. There was no need for him to shoot himself. Cocaine killed him before the death of his parents. In 1885 in a paper read to the Vienna Psychiatric Association Freud had recommended cocaine 'without

5. R. Byck (ed.), *S. Freud: The Cocaine Papers* (New York, 1974) contains reprints of all Freud's papers relating to cocaine.

reservation'. A year and a half later he cautiously suggested that the drug 'given by injection' was certainly dangerous. He then quietly dropped it altogether, though he took with him a small supply when a year later he set out for Charcot's clinic in Paris. But before his departure, cocaine dealt him another blow.

. . .

Carl Koller came from a background similar to Freud's, slightly more genteel perhaps but also poorer. Born in Schüttenhofen in Bohemia, then part of the Austro-Hungarian Empire, he was educated by the Jesuits and qualified as a doctor at the University of Vienna in 1882, aged twenty-five. While still a medical student he completed a research project on the development of the chick embryo in the department of the famous anatomist Solomon Stricker. For the rest of his life he regarded his exploration of the chick blastomere as his most useful contribution to medical science. Entering the ophthalmic clinic of the university as an unpaid 'Aspirant', he was impressed when Professor von Arldt deplored the lack of a local anaesthetic in eye surgery. Koller realised that any advance in that direction might help him to secure a paid assistant job, the first step on the career ladder. He experimented on frogs and rabbits privately with all analgesics known at the time, including morphine, chloral and bromine. None worked.

By then a shared background had led to a friendship with Freud. Freud, eighteen months older, had just completed his paper 'On Coca' and was about to depart to visit Martha in Hamburg. In his absence, would Koller test the effect of cocaine on grip strength? When Koller gave the drink to a friend, the friend commented on the numbness of his tongue as he swallowed the liquid. Yes, Koller replied, many people had noticed that. Then, in one of those moments of inspiration for which Hollywood would later invent the celestial choir, he realised that it was the anaesthetic he was looking for. He went to Stricker's laboratory and, with the help of Stricker's assistant, Gustav Gärtner, dissolved cocaine powder in distilled water and trickled it into the eye of a frog. After one minute, the frog allowed its cornea to be touched and manipulated without any reflex action. Similar experiments followed with a rabbit and a dog. Eventually, Koller and Gartner trickled the solution into each other's eyes. After a minute or two they could indent each other's corneas with a pin without any sensation. The anaesthesia lasted for about half an hour.

Koller was a modest and methodical man and he searched the literature for similar reports. The numbing effect on the tongue had been described some years earlier by a Peruvian doctor, Thomas Moreno y Maiz. He suggested that the drug might be useful as a local anaesthetic; but there was no record of him actually trying it. Alexander Hughes Bennett of Edinburgh had also described the numbing effect but without realising its potential significance. Vassily

Konstantinovich von Anrep of the Pharmacological Institute of Würzburg had described the effect of cocaine on the pupillary reflexes but not on corneal sensation.

Koller could not afford the fare to the next meeting of the German Ophthalmological Society scheduled for September 1884 in Heidelberg. An older and more affluent friend, Josef Brettauer from Trieste, read the historic paper for him. He then demonstrated the effect of Koller's 2 per cent solution at the Ophthalmic Hospital. The response was sensational. Henry Noyes, a New York ophthalmologist who attended the meeting, published an account in the *New York Medical Record* on 11 October. A week later Koller read his paper to the Vienna Medical Society. He credited Freud with introducing him to the drug; but there was no question of who discovered its analgesic action.[6]

Koller's days in Vienna were numbered – perhaps providentially. A senior assistant whom he had corrected about some detail during an operation had called him 'ein frecher Jude' (an impudent Jew). Belying his meek manner, Koller punched the assistant on the nose. A sabre duel followed in which Koller inflicted a wound (much sought after by students in German universities at the time but not in Vienna) on his opponent's face. There were no legal consequences but Koller's chances of promotion in Vienna dissolved. He set out first for Holland and then for the United States. In New York he was elected to the staff of several hospitals, including Mount Sinai. Over the next decades he gained a high reputation as an ophthalmologist but pursued no further research into local anaesthesia. He died full of honours in 1944, aged eighty-seven.

...

Among those who read about Koller's discovery was a young New York surgeon visiting the medical centres in Europe. William Stewart Halsted in his thirties was an all-American hero, a former captain of football at Yale and, according to Osler, the most brilliant operator of his day. He began his trials of blocking nerves in Paris in the late 1880s; but did not publish his results till after his return to New York. The details have always been a little vague. What he undoubtedly discovered was that while cocaine had no effect on the intact skin, it could be injected under the skin and provide local anaesthesia. But there was more to his discovery. If cocaine was injected along the course of a sensory nerve, it could block the whole area supplied by that nerve. It was a revolutionary step forward from Koller's direct application to mucosal surfaces and the cornea, even though the full potential of the method

6. See B.P. Block, 'Cocaine and Koller', *Pharmaceutical Journal* 180 (1954), 69; and J. Sharp 'Coca and Cocaine Studied Historically', *Pharmaceutical Journal* 1909 (1), 28. See also, V. Berridge and G. Edwards, *Opium and the People* (London, 1981).

would not become apparent until the replacement of cocaine by a non-habit-forming analogue. Halsted was unaware of this and experimented on himself and younger colleagues. All became addicted. Three never recovered and died within the next few years. Halstead himself was rescued by his friend, William Henry Welch, professor of bacteriology and dean of Johns Hopkins University, Baltimore. Welch hired two goons, had his friend abducted and put him on a cruiser. The cruiser, complete with crew including a chef, took him on an enforced journey to the Windward Islands. As far as the world was concerned, Halstead had disappeared. When he re-emerged, Welch persuaded him to move to Baltimore as professor of surgery. First Halsted checked into Butler's Hospital in Rhode Island to complete the cure. He was weaned of cocaine; but he arrived in Baltimore a 'strangely altered' character. The gay extrovert had become withdrawn and deeply antisocial. He was still a brilliant surgeon, though more a theoretician than a practical man. Some of his most important operative innovations (as well as his marriage to his theatre sister) were still in the future. He was recognised as a formidable thinker. He introduced the rubber glove into operative surgery.[7] But he remained a man of mystery until he died. What in fact had happened was not revealed until the contents of Sir William Osler's *Secret History* were first leaked and then published in the *Journal of the American Medical Association*.[8] Halsted was able to shake off the cocaine habit only by becoming addicted to morphine. His story was in fact the reverse of what had happened to Freud's friend Von Fleischl-Marxow. Only a few colleagues closest to him, including Welch and Osler, knew the truth. It is not mentioned in the two more or less official biographies – or hagiographies – published after his death. Morphine became an addiction which he could never entirely shake off.

Perhaps because Halsted's experiments with local anaesthesia had led to his addiction and illness, his pioneering work with local anaesthesia went almost unnoticed for a time. It was a New Orleans surgeon, Rudolph Matas, who drew attention to it in 1902. By then others were taking an interest; and Matas warned against the risks. He stressed that, even more than with general anaesthesia, local anaesthesia with cocaine required psychological preparation of the patient. There was, he thought, room for combining different forms of anaesthetic, a 'balanced programme'. It was a small but important contribution.

7. He did so to protect the sensitive hands of his theatre sister and later wife, not to promote antisepsis; but the innovation caught on.
8. Osler died in 1919 and bequeathed his entire library to McGill University where he had started his career. Among his books was a locked 'Secret History' which, he instructed the librarian, Dr William Francis, was not to be opened till 1989. Handwritten by Osler, it contained a detailed account of Halsted's addiction. Francis impiously decided to publish the manuscript in 1959 'since all concerned are dead'; but he himself died a year later, before publication. The contents were not eventually made public till 1969.

In the meantime Koller's invention was extended in another direction. James Leonard Corning, a New York surgeon, was probably the first to inject cocaine, into the human spine. It must have been daunting. The indications are still not entirely clear but Corning's patient's main symptoms were 'spinal weakness, seminal incontinence and masturbation addiction'. Corning injected a 3 per cent cocaine solution, inserting the needle between the 10th and 11th thoracic vertebrae, probably into the epidural space. When the first injection had no effect, he injected another 3 millilitres. The patient's legs then 'went sleepy and remained that way for an hour'.[9] Corning wondered whether the technique might be suitable for anaesthetising the lower limb for operations; but, like, Humphry Davy with laughing gas a hundred years earlier, he did not pursue the idea.

A German surgeon did. August Bier was born in Helsen in north Germany in 1861 and graduated from Kiel University in 1886. He then became assistant to Friedrich von Esmarch and was probably the main mover in changing from antiseptic to aseptic surgery in north German university hospitals. He learnt from Heinrich Quincke, a physician friend, that with a fine hollow needle filled with a sharp stilette for insertion it was now possible to withdraw a few millilitres of cerebrospinal fluid for diagnostic purposes. He conceived the idea of anaesthetising the lower part of the body by replacing 3 millilitres of cerebrospinal fluid with a solution of cocaine. A patient deemed suitable, a thirty-four-year-old man with disseminated tuberculosis who needed an amputation above the ankle, presented himself. On 16 August 1898 Bier laid the patient on his side and inserted a fine, hollow needle into the lumbar subarachnoid space. (He later claimed to have practised the manoeuvre 'extensively' on dogs.) He withdrew 3 millilitres of fluid and injected the same amount of a 5 per cent solution of cocaine. Sensation in the lower limbs was lost after twenty minutes and Bier performed a painless amputation.

Six patients later – all apparently successfully anaesthetised below the hip – Bier persuaded his younger colleague, Otto Hildebrand, to inject 2 millilitres of a 1 per cent fluid into his (Bier's) spinal fluid. When the syringe did not fit the needle and cerebrospinal fluid started to leak out, the experiment was terminated; but now it was Hildebrand's turn to submit. He was duly injected with the cocaine solution and, for about forty minutes, he lost all sensation in his legs. Next day both suffered from severe headaches and were covered in bruises. The experiments were temporarily suspended.[10]

9. R.M. Schorr, 'Needles: Some Points to Think About', Parts I and II, *Anaesthesia and Analgesia*, 1966 (45), 509 and 514.
10. A.K. Bier, 'Uber einen neuen Weg local Anasthesia in den Gliedmassen zu erzeugen!', *Archiv für klinische Chirurgie*, 1908 (86), 1007.

Six months later a French surgeon, Theodore Tuffier, created a stir at the International Medical Congress in Paris when he described his first experiences with spinal anaesthesia. His technique on three patients was similar to Bier's. Numerous American surgeons attended the meeting and returned enthusiastic. There was for a time a danger that the procedure might be tried wholesale without mastering the far from easy technique. Both Tuffier and Rudolph Matas warned against this. Fortunately perhaps, the growing reputation of cocaine as an addictive and toxic drug remained a deterrent. At least one patient was known to have died, though the case was never published. The delay would not be long.

In 1901 a French chemist, Ernest Fourneau, synthesised the first, virtually non-toxic substitute. He also inaugurated the convention of ending the names of new local anaesthetics with '-aine'. He called his own discovery 'Stovaine', freely and modestly translating his own name into English. Two years later Alfred Einhorn in Germany synthesised procaine, better known by its trade name 'novocaine'; and his surgeon friend, Heinrich Braun, introduced it into clinical use. Its merits were quickly recognised: it was effective, fast-acting and non-toxic though its action only lasted for about twenty minutes. For half a century it became the most widely used local anaesthetic both for spinal anaesthesia and for nerve blocks and dental work. Longer-lasting formulations, like procaine suspended in peanut oil, were soon developed. In 1940 William Thomas Lemmon described a new method of performing spinal anaesthesia, using a soft needle which could be left in the subarachnoid space during operations, allowing more anaesthetic to be injected when required.[11]

...

At this point the development of local anaesthesia and the use of cocaine as a stimulant, painkiller and calmer of nerves diverge. Outside the field of local anaesthesia cocaine enjoyed an unprecedented vogue for a few years. During his love affair with the drug Freud declared (in a pamphlet circulated by Merck, perhaps without authorisation) that cocaine 'will in a few years do away with all lunatic asylums and alcoholics'. But the craze was even more virulent in France, England and America.[12] Angelo Mariani, an Italian-born Parisian chemist, entrepreneur and distant cousin of Count Cavour of Piedmont, made a fortune selling coca extracts not only as Vin Mariani but also as Elixir Mariani, Pastilles Mariani, Pâté Mariani and Thé Mariani (non-alcoholic). He was among the first to have his products endorsed by a string of celebrities, including Sarah Bernhardt, Lucien Guitry, Rudyard Kipling, Edward Elgar and the beautiful and notorious Marquise de Torroba. The

11. In J.R. Maltby, *Notable Names in Anaesthesia* (London, 1998), p. 216.
12. In Berridge and Edwards, *Opium and the People*, p. 221.

divine Mistinguette was presented with a lifetime's supply of Vin Mariani after singing the praises of 'une goutte de Mariani' at the Alhambra. In London Messrs Ambrecht, Nelson and Co. of Duke Street, St James, offered five varieties of coca wine for sale, including a sweet Malaga for ladies and children and a Burgundy coca for gouty and dyspeptic gentlemen. Coca sherry and coca port were also to be had and 'were stocked in the best London clubs'. In the United States, Coca Tonic, a frothy alcoholic drink containing cocaine, was marketed in 1885 by a Georgia chemist, J.S. Pemberton. A year later, to comply with the new 'dry' laws of the state, he eliminated alcohol but added cola nuts (which contain caffeine) and citrus essence. He introduced the new concoction, Coca-Cola, as 'the supreme remedy for all ailments' and 'the greatest calming drink'. It would certainly be the greatest liquid money-spinner in history. Pemberton eventually sold his share to another druggist, A. Griggs Candler, who added the barrel container and the refrigerated spout. The American drug store was born.

But doctors were not backward in their praises. Sixty-seven articles in the first volume of the *British Medical Journal* (half a year) of 1885 extolled the utility of cocaine both as a local anaesthetic in operations on the vagina, urethra, mouth, ear, nose and larynx and in a range of diseases like abscess of the breast, cancer, scalds, circumcision, neuralgia, hay fever, senile gangrene, nymphomania, seasickness and the removal of needles from the foot. Never was there such a wonder drug.[13] Not to be outdone, the *Lancet* recommended coca 'confidently in all cases where a nervous restorative was needed'.[14] William Martindale of the highly respected *Extra Pharmacopoeia* of 1884 devoted an entire supplement to commending it for the 'temporary but effective relief of hunger, thirst, fatigue, exhaustion, distaste for food, nervous depression and weakened digestion'. In 1890 Burroughs Wellcome introduced their celebrated 'voice tabloids' to impart a 'clear and silvery timbre' to singers, actors and public speakers. Used by 'leading statesmen and the greatest stars of the operatic and tragic stage throughout the civilised World', they were composed largely of cocaine with a dash of chlorate of potassium and bromides. The 'pastils or tabloids' could be kept under the tongue while actors delivered taxing Shakespearian soliloquies and singers sang coloratura arias by Bellini.[15] W. Golden Mortimer dedicated his *History of Coca: the 'Divine Plant of the Incas'* (1901) to Angelo Mariani.[16]

. . .

13. *British Medical Journal*, 1885 (2), 1344.
14. *Lancet*, 'Coca Pastils', 1884 (2), 1078.
15. Quoted in Berridge and Edwards, *Opium and the People*, p. 346.
16. W. Golden Mortimer, *History of Coca, the 'Divine Plant of the Incas'* (New York, 1901).

By the turn of the century the craze was abating. The gradual disillusionment can be traced in the changing attitude of the great detective, Sherlock Holmes. His creator, Alfred Conan Doyle, a doctor, was almost certainly himself a user and initially an enthusiast. In 'A Scandal in Bohemia' (1886) and 'The Man with the Twisted Lip' (1887) Holmes dismisses Watson's disapproval with light-hearted nonchalance. 'It's splendid stuff, Watson, splendid.' But Watson's concern – and probably Conan Doyle's – steadily increases. In 'The Five Orange Pips', 'The Yellow Face' and *The Sign of Four* Holmes himself has 'no more use for the drug'. In 'The Missing Three-Quarter' (1896) he suddenly refers to the hypodermic syringe as 'an instrument of evil'. Watson now mutters darkly about the 'drug craze' and Holmes does not contradict him. By the turn of the century cocaine was no longer chic.

Outside fiction too, doctors began to warn against the 'too ready use of this inebriating substance'. The *British Medical Journal* now pontificated against the 'extravagant pretensions advanced on behalf of cocaine'.[17] To Sir Clifford Allbutt, a former enthusiast, the addiction had suddenly become a form of 'slavery worse than that of morphine or indeed death'. What worried many was that the drug habit was beginning to percolate down the social scale, especially in America. The main anxiety there revolved around Southern Blacks 'forgetting their places'. The 1900s saw the first full segregation laws and the first high tide of lynchings. The 'coke-crazed nigger', his mind set on raping white women, became the image to justify every form of repression. Stories abounded of the superhuman strength of 'the black coke-fiend'. 'Ordinary shootin' don't kill him,' an Atlanta police officer declared in *Everybody's Magazine* in 1902. The standard issue .32 revolvers were duly upgraded to .38s. In Britain and France the use of cocaine by the poorer classes caused little concern until the poorer classes began to provide the machine-gun fodder of Flanders and Gallipoli. Cocaine-fuelled confidence could inspire heroism and win the Victoria Cross but it could also lead to desertion, absence without leave and wholesale killing by friendly fire. Strict regulation of the sale and use of the drug then became a matter of national urgency. Legislation was introduced to 'stamp out the use of cocaine world-wide'. The programme turned out to be more expensive and less successful than envisaged. In 1918 the total production of cocaine world-wide was about 10 tons. In 2005 the figure is an estimated 1,100 tons and rising.

17. 'Cucaine Habit and Cucain Addiction', *British Medical Journal*, 1885 (1), 1183.

# Chapter 37

# HIGH VICTORIAN PAIN

It is time to turn from drugs to people and the interface between the two. The most common interface is illness. Illness is sometimes represented as something timeless and unchanging. In fact, illness in the modern sense is a nineteenth-century middle-class invention. Since pain is one of its chief manifestations, its origins deserve a glance.

Before the end of the eighteenth century a European middle class hardly existed. Squashed between the rulers above and the masses below, its influence was negligible. What since time immemorial mattered to the rulers was the dynasty, the continuation of the lineage. Provided this was assured – or not – the ailments and indispositions of individuals got little attention.[1] What mattered to the masses was survival. It was a struggle in which temporary hovering between life and death was barely noticed. Of course the great epidemics could not be ignored. But the great epidemics were not illnesses in the modern sense. They were natural disasters like famine, drought or the depredations of war.

For reasons which are outside the scope of this book, the class between the rulers and the masses began to emerge as a significant force as the eighteenth century advanced. Astonishingly and unexpectedly, by the end of the nineteenth it virtually ruled Europe. That, for practical purposes, meant the world.[2] The things they valued suddenly carried weight.

What above all else distinguished the middle class – or middle classes, since in-class stratification started almost at once – from the rulers above and the masses below was reverence for the individual. For the first time private

1. This is not to say that private anguish and grief did not exist in all classes and in all ages: see next chapter. It is indeed a reaction that man shares with many other species. But private anguish and grief are not necessarily matters of general interest.
2. By 1900 only the United States and Japan were independent great powers outside Europe; and in Europe only Russia was a declared absolute autocracy. In the rest of Europe, often under an aristocratic veneer, the middle classes ruled.

destinies mattered. Illnesses are among the formative influences on the unfolding of these destinies. They shape and mould character. They determine achievement. Though intensely private, their effect often reaches beyond the sufferer. In a middle-class family the illness of one member might touch the fate of every other. For all those reasons they were suddenly of profound concern

The portraits of two young women dying of the same disease illustrate the separation of classes. The ice-cold splendour of Ingres' *Princesse de Broglie* is hieratic in its remoteness. A victim of tuberculosis, she is already doomed. She will be dead within two years. She will no doubt be mourned by her husband and missed by her friends. But having given birth to an heir to the dukedom and a male second-in-line, her destiny, indeed the main purpose of her earthly existence, has been fulfilled. Her illness as such is of no great moment. Her portrait on the wall of the ducal gallery will be a memorial to her sojourn in this world.[3]

Only a generation later the Munchs of Oslo – or Christiania as it still was – were a typical middle-class family. Their lineage was respectable, no more. It meant little to them.[4] But they were not anonymous members of an underclass. One of the children, Sophie, died after a lingering illness at the age of eighteen. The memory of her suffering – her hopes, her sadness, her beauty and her courage – remained with her brother Edvard all his life.[5] The picture he painted of her, *The Sick Child*, has become emblematic of the illness which killed her. But it is wholly personal, the image of a much-loved sister, child, young woman. To Edvard, as to the patient, the illness was a unique experience. Her individual destiny and therefore her individual illness was deeply significant.[6] A few months before he himself died, fifty years later, he was still painting her remembered likeness.

Of course the outcome of an illness was not always death. It was often not even particularly tragic. But to members of the middle classes its manifestations and ramifications always deserved and commanded attention.

...

3. She was born Joséphine-Eléonore-Marie-Pauline de Galard de Brassac de Béarn, a lineage even more ancient and noble than that of the De Broglies. The painting is now in the Metropolitan Museum in New York.
4. The artist Edvard Munch's father was a doctor. Other members of the family were lawyers, academics and civil servants. Norway had no indigenous hereditary aristocracy.
5. He painted several versions. The best is in the Munch Museum in Oslo, the second-best in Tate Modern, London.
6. It was tuberculosis in its heyday. The current resurgence of the disease is clinically very different (Dormandy, *The White Death*).

Medical historians tend to bemoan the patchiness of meaningful data about diseases of the distant past.[7] With the emergence of the middle classes the grounds for that particular moan disappear. What was a trickle of often deliberate obfuscation, permitting no more than more or less inspired guesses, all at once becomes a flood of detailed – if sometimes euphemistically expressed – medical data. This is of course true only of members of the middle classes. For royalty and the rabble the ground rules never changed. But the clinical histories of the real-life Forsytes, Thibaults, Pasquiers and Buddenbrookses can often be charted in extraordinary detail. Extraordinary detail hints at extraordinary tedium; and indeed, the piling up of medical records of personages no longer of general interest can be wearing. For the purposes of the present book one will suffice, chosen for the excellence of the available biographies and the less than total unreadability of the relevant literary and personal memoirs.

. . .

Mrs Humphry Ward, born Mary Arnold, was the granddaughter of Rugby's famous headmaster, Dr Thomas Arnold, and niece of the poet Matthew Arnold.[8] She was, in other words, born into the purple of the Victorian upper middle class, a privileged destiny. Privileged, but not necessarily easy. The great doctor himself was a hard act to follow. After Oxford, his second son, another Thomas, emigrated to Tasmania. There he married the daughter of a local merchant family, Julia Sorrell. It was in Hobart that Mary Augusta and most of her eight siblings were born. Tasmania, one of the world's most beautiful islands, might have been a child's paradise in which to grow up. It was not to be. Most of Mary's girlhood was spent at English boarding-schools of varying degrees of ghastliness. Yet her time in these institutions was not entirely wasted. It was largely in the school sick-rooms that she learnt 'under the dire stress of boredom' to amuse herself a good deal by developing a natural capacity for dreaming awake. Hour by hour she followed an endless story of which she was always the heroine. Fiction was for her (as it was for the young Trollope and many other future story-tellers) an act of self-preservation. In her case it was a means of escape from the 'sensory deprivation and punitive immobility of Miss Davies's "sanatorium"'. At twenty-one she married Humphry Ward, a socially acceptable Oxford don

7. The present writer is no exception.
8. The most useful source for this section has been John Sutherland's excellent biography, *Mrs Humphry Ward, Eminent Victorian, Pre-eminent Edwardian* (Oxford, 1990). For a different slant, J.P. Trevelyan, *The Life of Mrs Humphry Ward* (London, 1923) is revealing. Mrs Ward published her own reminiscences, *A Writer's Recollection* (London, 1918). Added to this there has been a certain amount of family lore originating from Mrs Ward's friend, Miss Maud Sepsworth who was a much-loved friend and governess in the present writer's mother's family.

whom Evelyn Waugh might have described as 'dim'. He lost his fellowship at Brasenose College on marriage and became leader writer and later art critic of *The Times.* Had he been highly valued by the dons, his fellowship would undoubtedly have been renewed. He was in fact offered a tutorship without being re-elected a fellow, but declined. His image as paterfamilias was always carefully nurtured; but his role was largely ornamental and his speculative business ventures as an art dealer remained a drain on the family finances. One of Mary's sisters, Julia, married Leonard Huxley and became the mother of Julian and Aldous. The boys were often entertained by the Wards.

Mary Ward was thirty-seven when she published her first novel, *Robert Elsmere* (1888) and forty-three when her most famous work, *Marcella* (1894), saw the light of day in three volumes.[9] None of her books are much read today; but after the first she was recognised as an intellectual force and after *Marcella* she became the most successful woman author in England. She contributed articles and essays to *The Times* and numerous magazines; she wrote an introduction to the Haworth edition of the Brontës' work; and she was, for most of her middle years, what today would be described as an 'A List' celebrity. Her 'Thursdays' in their house in Bloomsbury attracted the literary, artistic, social and political lions of the day: Henry James, the Frederick Harrisons, Mr Gladstone, Arthur Balfour, Lord Curzon, the Tennants, the Burne-Joneses, Robert Browning, the Du Mauriers, the Alma-Tademas. George Goschen, the Toynbees, the Edward Greys, Dr Elizabeth Garrett Anderson and the Oscar Wildes were among those who regularly graced her drawing room. The family eventually acquired an ugly but imposing country seat, the Stocks (purchased from the Edward Greys), as well as a metropolitan base in Grosvenor Place, and, during the first sunlit years of the new century, commanded an army of servants. To maintain such a standard and intermittently to bail out delinquent members of the family Mary continued to write voluminously and – apart from her failure on the London stage – successfully. She was fêted in the United States where she was entertained by the Theodore Roosevelts, and travelled on the Continent. She lectured to universities (on 'The Peasant in Literature' to Glasgow) as well as to literary societies. She was restlessly active in politics and church matters, mainly as a committed anti-suffragette but also as a tireless supporter of good causes. Today her name is best known as the founder of the Mary Ward Settlement, Queen Square, one of the late-Victorian charitable foundations where members of the affluent middle classes were expected to help by personal involvement with the lives of the poor. The Mary Ward Settlement soon included the first children's day centre in England.

9. *Robert Elsmere* is controversial and still interesting, *Marcella* is a hard and long read.

'The wonderful Mrs Ward', as *The Times* more than once referred to her, promoted schools for the handicapped; and her clever lobbying was largely responsible for the introduction of special provision for disabled children in the first Education Bill in 1917. During the First World War she dug for victory, twice visited the Western Front and once the fleet in Scotland. Her nephew Tom was killed in Flanders. Her war novel, *Missing*, was a huge success. In 1920 she became one of the first female magistrates. (She never actually sat on the Bench.) She had three children. Dorothy remained a maiden lady and became her mother's constant companion. Theodore became an unsuccessful politician and briefly a Member of Parliament, but eventually a morphine addict and an alcoholic. Janet married into the Trevelyan dynasty, distinguished academics, writers and civil servants all.

It might be thought that in an age when nothing was done in a hurry or in a perfunctory way, when dinner parties, like parliamentary speeches, often lasted for four or five hours, when house parties went on for days, and when a lady in society changed her dress and hat two or three times a day, a life like Mrs Ward's would require the constitution of an ox. In fact she was almost constantly ill or ailing and when she had a moment to spare from her own afflictions, the illnesses of her family claimed her attention.

Her problems started as a child when her gums and teeth required several operations. Thereafter her teeth constantly troubled her until she had the last two removed at the age of fifty-five. She had several sets of false teeth, one manufactured for her in Gronau in Germany, an internationally known workshop-laboratory specialising in difficult dentures for a cosmopolitan aristocratic clientele. None were entirely satisfactory. Her father, not a robust man himself, noted with irritation 'her perpetual colds and hacking coughs . . . she is not a healthy child'.[10] Fortifying syrups, some made locally in Hobart, some sent by concerned grandparents from England, all tasting vile, had no notably beneficent effect. Bizarrely, her 'chestiness' was one reason for sending her from Tasmania to boarding-school in England. At school she would desperately plunge her head into cold water to try and stop her toothache. A medicine for young girls prepared by Miss Emily Davies herself, probably containing lashings of laudanum, made her feel 'dopey but happy'. But a dose of the elixir had to be earned by several hours of brave and silent suffering; and its effect lasted only two or three hours. Later her infected teeth were blamed for her 'general seediness' as well as for excruciating attacks of facial neuralgia.

10. He was also, as he confided in a letter to his mother, somewhat concerned that 'Mary [seven years old at the time] showed a slight tendency to story telling . . . A timely whipping checked this.'

Headaches were another regular complaint, especially bad in the mornings, requiring in later life various unnamed tablets and strychnine drops usually taken in thimblefuls of cordial. The latter were highly valued as general tonics as well. Mary continued to be prone to inflammations of the nose and throat: according to her daughter, Dorothy, she suffered from 'septic' and 'inflamed tonsils' five or more times every winter. Dorothy herself was 'forever sniffling and snuffling with colds in the head, a great trial for those around her'.

Mrs Ward's 'terrible rheumatic ailments' started in her twenties when she first recorded a 'tingling and stiffening pain in her fingers' and 'a vexing pain in the back'. By the 1890s she was chronically ill 'with arthritis'. Her 'writer's cramp' and painful 'neuralgia' necessitated the use of a special writing stand. She was incapable of lifting anything heavy: even for picking up a bulky reference book or for arranging flowers in a vase a maidservant had to be summoned. Her writing sometimes had to be done with the left hand. She experimented with dictation to secretaries, but 'all but the last were useless'. It was a comfort that her sister, Julia Huxley, was similarly afflicted. They experimented together with a range of herbal remedies and proprietory preparations and eventually found a 'Chinese draught based on nuts' the most effective. A constantly changing tribe of doctors argued about the aetiology. For a time they blamed an enlarged gland in the armpit – but the gland disappeared and the pains continued. Later in life partial relief came from 'gentle stroking' – i.e. massage of some sort – at first administered by a Swedish professional but later mainly by Dorothy. She also tried 'gymnastics', prescribed by a German doctor, a Dr Wolf, and 'electric vibrations', administered by another German doctor, a Dr Maschlik. The early results with the latter were promising but they then started to 'depress her spirits' and were discontinued on the advice of a third German doctor, a Dr Fürth.

In her forties Mary developed 'piles which were yet not quite piles'. For once a detailed description is lacking; but the condition was probably a prolapsed rectum.[11] She had an operation under anaesthesia but, according to Janet Trevelyan, it was 'clumsily and horribly performed . . . After it she lay for days in agonising pain which the doctor had neither foreseen nor prophesied . . . The ordeal was aggravated by the ministrations of a drunken nurse.' She was later given a succession of more or less satisfactory appliances to wear. Her attacks of abdominal pain were for many years borne with 'stoic fortitude', eased (or not) by linseed oil, laudanum and 'arsenical tonic' as well by a 'nutritious diet' specially designed by M. Hyppolite Châtelet, a French chef specialising in invalid regimes. At first thought to be a gastric ulcer, her condition was eventually diagnosed as gallstones by her brother-in-law 'that clever Harry

11. John Sutherland opts for an anal fissure.

Huxley'. Several consultations were held to decide whether the condition justified 'surgical interference'. It was felt that she was too frail to benefit or even to survive an operation.[12]

The most intractable, persistent and mysterious of Mrs Ward's illnesses was what she termed her 'side' and often referred to as 'the old enemy'. Giving a name to and creating a personality for persistent and untreatable pains was a common Victorian practice. Nietzsche called his tabetic crises 'mein Hund'. 'The pain,' he wrote, 'is just as faithful, just as obtrusive, just as shameless and just as clever as any other dog . . . and I can scold it and vent my bad mood on it as others do on their servants and wives.' A 'black dog' of depression sometimes snarled at Winston Churchill and Scott of the Antarctic. Strindberg referred to his paralysing rages as 'the Hun'. Queen-Empress Elizabeth of Austria-Hungary called her attacks of migraine 'visits from Torquemada'. Did the demonising help? Perhaps. Mrs Ward's 'sides' started in 1892 as an 'obscure but vicious neuralgia in the right breast and side'. Her principal London doctor, then a Dr Hames, diagnosed it as 'rheumatic gout', another doctor as 'internal catarrh', whatever either of those nonsensical terms meant. The diagnosis was later refined to a 'chronically floating kidney', an ailment then recently invented. The kidney floated up (or down) with great regularity, often on holidays. On such occasions the slightest motion of a train or boat became 'horrific torture'. In Lucerne her Swiss doctor diagnosed 'upper sciatica . . . a very rare condition' and prescribed large doses of morphine and brandy. The kidneys, he had no doubt, were stationary and blameless.

Later treatments included another assortment of diets, mainly of raw beef and 'all-milk', none much good and often 'tasting horrible'. The attacks usually persisted for a few days to a week at a time and they would be accompanied by moderate temperature and often by diarrhoea. During the paroxysms the intermittent 'jabs of pain' were 'quite unbearable' and required a cocktail of morphine and other painkillers. For a time sulphonal and trional were favoured; but after 1898 Phenacetin, recommended by 'dear Willie James', was the chief stand-by. In 1900 she also started to take 'a couple of Aspirins' regularly, and these were appreciated for the quickness of their action. But the attacks or the drugs or both made her so 'stupid' that sometimes she could hardly recognise people who came to visit her. The underlying cause may have been chronic cholecystitis. Today she would have had her gall bladder removed, perhaps with good results. But diaphragmatic hernias must also have been exceedingly common and would fit the clinical history. The condition was not recognised at the time; but one need only glance at Manet's

12. The diagnosis was probably correct but X-ray confirmation was not yet available. Cholecystectomy was, however, an operation recently perfected in England by Berkeley (later Lord) Moynihan of Leeds.

brilliant *Nana* or a page of a ladies' fashion journal displaying an assortment of hourglass corsets to visualise numberless stomachs being squeezed through the diaphragm into the chest.[13]

Another long-standing and recurrent ailment was her 'eczema'. During attacks, as Dorothy recorded in her diary, 'her arms and face swell up and she develops most painful blisters and sores'. Sun, wind, cold or anxiety could all bring on an attack. They were sometimes followed by boils. In later life her face became scarred. Face creams were shunned before the First World War, being regarded as 'common'; but nightly packs with Carlsbad mud, imported from what is today the Czech Republic, helped.

In around 1900 she began to take large quantities of quinine, prescribed for her by Harry Huxley, who seems to have assumed that she had developed 'swamp fever' during a visit to Venice. She consumed and injected increasing doses of morphine – which worried her. She had seen the effects on her brother Willie. (It was the only drug that partially relieved his 'locomotor ataxia', a synonym for tabes dorsalis.) But for some time she had no such compunction about cocaine. During the 1880s, while the drug was 'widely fashionable', she took it regularly in 'moments of stress', maintaining that she could not get through a 'nerve-wrecking lecture' without it. It was usually taken with a glass of port. Travelling to Paris in 1889 to attend an Anglo-French literary gala, the cocaine lozenges 'had the most wonderful effect'. But by 1900 she was off the drug (as were most of her circle) and later became a vocal advocate of its 'scheduling'. She 'strongly urged' Dorothy to banish it from her travelling pharmacy. In Paris, as elsewhere where she travelled, she had at least one but usually two English or English-speaking doctors attending her. There was never any shortage of those. Practices in fashionable continental spas like Ems, Ischl, Aix or Baden-Baden were among the most expensive to purchase but rarely went begging. Capital cities like Paris, Brussels or Munich had English nursing homes.

This is by no means an exhaustive list of Mrs Ward's disabilities. Like all Arnolds, she had a 'weak heart' – her famous grandfather had died at forty-eight from a coronary attack – and she eventually died from 'pulmonary disease'. But, despite those, and thanks presumably to the ministrations of a long list of devoted relations, nurses, secretaries, doctors and friends, she remained active in the literary and charitable fields until her final illness at the age of seventy-one.

...

13. The painting, an illustration to Zola's famous novel, is now in the Kunsthalle in Hamburg.

No comparable account of illnesses and remedies could be given about a member of the masses – in late Victorian days beginning to be called the 'working class' – except under one condition. However obscure and unimportant the individual, one road to fame (or at least to a permanent record) was open to them. They could become interesting cases. This happened to the twelve-year-old girl whose history became, on 27 August 1887, the subject of a Letter to the Editor of the *Lancet*. 'The case of hysterical spasms simulating tetany' by Edward Kershaw, general practitioner in Barnsley and medical officer to the workhouse hospital, is not too long to be reproduced with only minor cuts.

> M.P., a female aged twelve and a provisional pauper, came under my notice about the end of June with symptoms of gastrointestinal disturbance produced – as is so commonly the case, especially in this season of the year – by taking food of bad quality and insufficient quantity but voraciously. She lives in one of the lowest parts of the town amidst bad sanitary surroundings . . . Besides her family the three-room cottage accommodates several Irish lodgers and is in a very dirty state. My patient herself is intelligent with fair hair and blue eyes, but small and slightly built with a pinched hungry face . . . There is no evidence of syphilis or worms, nor has she had any catamenial [vaginal] discharges. During the previous months she had nursed a younger brother, one of six or seven siblings, through an attack of scarlet fever; and, when she is in her usual health, she does most of the housework . . . She has attended my surgery from time to time, complaining of occasional vomiting, pain behind the sternum and in the abdomen as well as constipation. I have treated her with mild saline purgatives and opium . . .
>
> On 9 August about 7 p.m. I received a message to go to see her at once as she was 'in fits'. I attended almost immediately and when I opened the door the following scene met my gaze. On a bed under the window, close to a great fire where culinary operations were in progress, lay my patient, surrounded by six or seven exceedingly dirty women and two or three dirty men, each of them holding a different part of her body. She was quite still when I entered and watched me approach her; but, as soon as I touched her, the following convulsions occurred. The depressors of the lower jaw contracted, opening her mouth to its fullest extent, the facial muscles round the mouth contracted exposing her teeth in both jaws. As soon as this was completed the post-cervical muscles [muscles at the back of the neck] commenced to contract and drew back the head until the tip of the nose rested on the pillow. The wave of contractions then passed slowly down the spinal [back] muscles; the thorax bulged forward, then the abdomen and then the legs became rigid. Although the child is not more than four feet six

inches in height, I shall be well within the mark in saying that the lumbar region rose some twelve inches from the bed and she rested solely on her forehead and heels. The legs were now drawn up and then violently extended . . . forcing the face up against the iron bed-head, and, had not her shoulders been in a semi-erect position, her face would have been crushed (as happened once or twice but only to a slight extent) . . .

The attacks which were accompanied by high-pitched screams lasted two or three minutes and were repeated several times during my visit . . . My suggestion that she would be better treated in the workhouse hospital not meeting with approval, I ordered strict quiet in the room from which sightseers were to be rigorously excluded and left without ordering any other treatment.

The next day, Wednesday, I found that the convulsions had continued throughout the night and were, if anything, more severe than when I last saw her. I injected a quarter of a grain of morphine hypodermically but found later in the day that it had had no effect . . . On Thursday the girl was still in convulsions which, I was informed, had continued at intervals of half an hour or so since my last visit. She seemed much exhausted. I ordered her a mixture of 5 minims of tincture of Calabar beans and 10 grains of bromide of potassium every two hours. On Friday I was told that the patient took my medicine with great reluctance, complaining that it was too salty. At that point I started to suspect hysteria; but her semi-somnolent manner also made me fear commencing meningitis; and there being some constipation, I left with her two powders each containing a little calomel with half a minim of croton oil, to be taken at three-hourly intervals. When I saw her in the evening she was not convulsing spontaneously but pressure on her ovarian region induced a pretty severe convulsion.

On Saturday I heard her screaming as I came up the street: 'Oh mother! Cut my legs off! Oh my legs and my back! Mother, be quick! They are killing me!' As soon as she ceased to cry out she went into a convulsion similar to the one I have described. She appeared quite unconscious of what went on around her. On Sunday I found her singing, shouting and beating time with her arms and legs heedless of my presence. I was told that she had slept well during the night and seemingly awakened all right, but that on an attempt being made to give her the mixture I had prescribed, the present state of things commenced. The attack passed off during my visit, giving me the opportunity to take her pulse and temperature . . . Both were normal.

I now became satisfied that I was dealing with hysteria, and thereupon decided to stop all treatment. Later in the day, at the urgent request of her friends I sent her a mixture of burnt sugar with water and was assured on my next visit that it had done her a deal of good. I then turned to her mother and told her, making sure that the girl could hear me, that I decline having anything more to do with the girl and that if she has any more pains or convulsions, I shall insist on her going to the workhouse. Since that time I have heard nothing of her.[14]

One's first comment must be that Dr Kershaw's diagnosis was almost certainly correct, though what exactly he meant by 'tetany' is uncertain. Tetany is still sometimes confused with tetanus though the two conditions are unrelated.[15] Tetanus is the result of infection with the micro-organism *Clostridium tetani* and was, in the 1890s, still a common complication of contaminated wounds.[16] Though the micro-organism does not spread, it elaborates a powerful toxin which ascends along nerves and acts on the motor cells of the spinal cord. These trigger off painful spasms, culminating in the whole body being forced into an arched 'rainbow' position. Death follows paralysis of the respiratory muscles. Tetany is also characterised by painful muscle spasms, most commonly in the hands and feet – 'carpopedal', in medical jargon – but the cause is a low circulating blood calcium. Armand Trousseau, the great French clinician who described the condition in 1867, rightly attributed it (without knowing anything about blood calcium) to neuromuscular hyperexcitability. In Dr Kershaw's time the condition was common in undernourished children with incipent or established rickets – that is, Vitamin D deficiency – in the slums of the industrial North.[17] (The illness was called the 'English Disease' in many continental countries, but it was more common in Scotland than in England.) There can be little doubt that M.P. had witnessed spasms and convulsions caused by tetany and perhaps by tetanus as well. It was, as Charcot had shown, a common and highly dramatic hysterical manifestation.[18]

One's second comment must be that Dr Kershaw was clearly pleased not only with his own perspicacity but also with his management of the patient.

14. E. Kershaw, *Lancet*, 1887 (2), 256.
15. Both terms derive from the Greek *teinein* (τεἰνειν), to stretch.
16. Today two kinds of immunisation have made the condition extremely rare. Active immunisation, almost universal, depends on giving children an attenuated dose of the organism to which they themselves will develop protective antibodies. Passive immunisation is the injection of an antitoxin preparation, administered prophylactically when contamination of a wound is feared. Neither was available till the 1890s when the Japanese microbiologist Kitasato prepared a pure culture of the organism and developed a method of active and later of passive immunisation.
17. Two sources of Vitamin D are available to the body. The vitamin is contained in various foods like animal and fish meat and liver; and it is converted by sunshine in the skin from an inactive precursor. Lack of both was the common cause of rickets till the Second World War. Vitamin D facilitates the absorption of calcium necessary for bone formation. Calcium in the body fluids is also essential for normal neuromuscular transmission.
18. See Chapter 32.

The crafty use of a placebo unmasked the hysteric. His threatening her with the workhouse hospital was probably effective. Dickensian horrors were in the past; but the place was still close enough to a prison to frighten a highly strung girl. And last, Dr Kershaw's frequent attendance must be commended: only the very rich could afford such regular visits today.

But a chasm separates Charcot's insistence on the 'reality' of hysterical pain in the 1870s and Dr Kershaw's dismissal of it as a form of blackmail twenty years later. To ascribe the difference to the Channel would be misleading. Professional codes in France and England differed – as they still do – but in both countries, as the Victorian age was reaching its zenith, medical and lay attitudes were hardening. Hysterical convulsions, however painful, were becoming forms of mischief in children and of criminal malingering in adults. In Britain, by the 1890s, a sinister alliance between eugenicists, imperialists and aesthetes began to make complaining even of 'real' pain vaguely reprehensible.

Writing in *Blackwood's Edinburgh Magazine* in 1894 war correspondent G.W. Steevens expressed his horror at modern man's inability to face physical pain. His heroes were Rider Haggard's Henry Curtis in *King Solomon's Mines* and the real-life Colonel Frank Rhodes, correspondent of *The Times*, 'who bore a bullet wound in his shoulder with his usual humorous fortitude'.[19] The 'gospel of painlessness' was 'throttling patriotism and virility of character'. For Steevens and for many like him, builders all of the greatest and most beneficent empire the world had ever seen, civilisation and especially medical progress was having a debilitating effect. 'We break bicycling records but we are no longer prepared to converse coolly while having our legs cut off as was the way of our great-grandfathers.'[20] These sentiments claimed no kinship with the religious exaltation of pain, still preached by many churchmen, nor with the alleged medical benefits of pain, still a matter of controversy among doctors.[21] Pain had to be faced, not avoided, simply 'to put backbone back into the Victorian Briton'. Equally, it was necessary to inflict pain. 'We became and are an Imperial Race by dealing out necessary pain to other men, just as man became a powerful species by dealing out necessary pain to lower species.' Steevens' sentiments were shared by the schoolmaster in George Meredith's 1879 novel, *The Egoist*. In response to the suggestion that the cane should

19. G.W. Steevens, 'The New Humanitarianism', *Blackwood's Edinburgh Magazine* 1898 (98), 163.
20. Quoted and informatively discussed in L. Bending, *The Representation of Bodily Pain in Late Nineteenth-Century English Culture* (Oxford, 2000).
21. A unsigned leader in the *Lancet* in the issue in which Dr Kershaw published his report posed the question 'What is Pain'? After some rather woolly philosophising it came to the conclusion that pain was indefinable. A slightly livelier correspondence ensued, arguing over the occasional value of pain in making the diagnosis. Very little that was new transpired.

teach the work-shy boy Crosby a lesson, he exclaims: 'No sir, no; the birch! the birch! We English beat the world because we take a licking well.'[22]

At the seemingly opposite end of attitudinising, 'effeminate' aesthetes too were beginning to deplore undue concern with physical suffering. In *The Picture of Dorian Gray* Lord Henry expresses the view that 'there is something terribly morbid in the modern sympathy with pain ... The Nineteenth Century has gone bankrupt through an overexpenditure of sympathy.' Did Oscar really mean this? Probably not. But whether he did or did not, his creation struck a malignant chord in the minds of many.[23]

22. G. Meredith, *The Egoist: A Comedy in Narrative*, 3 vols (London, 1879), Vol. I, p. 147.
23. O. Wilde, *The Picture of Dorian Gray* (London, 1891), p. 59.

# Chapter 38

## THE POWER OF PAIN CONTROL

Late Victorian Britain saw yet another development in pain control. For the first time the power it could give to unscrupulous individuals was widely discussed. George du Maurier's *Trilby*, published in 1894, became a bestseller not only because it was readable trash but also because it gave voice to previously unexpressed fears. In the book the young woman Trilby suffers incapacitating neuralgic pains in her eyes, pains which only the evil Svengali can relieve. In a parody of the Christian faith – not surprising perhaps in an anti-Semitic novel – Svengali claims that he has taken over Trilby's pain and has transferred it to his own elbow. 'But I love it,' he adds, 'because it comes from you.' He achieves complete ascendancy over her despite her flashes of insight and struggle to resist. The price of being free from pain is that she 'shall see nothing, hear nothing and think of nothing but Svengali, Svengali, Svengali!'[1]

Such relationships undoubtedly existed (as Freud would soon describe at even greater length than Du Maurier); and they sometimes involved doctors. The risk was certainly perceived as real. Advances in the fields of drugs and anaesthesia were awe inspiring. Public debate about how to control the dangers raised by these advances became urgent. All pointed in one direction. Step by step the medical profession would gain control over the provision of effective painkillers, sharing that power only with organised crime. Such a state had never existed before. Even after the Pharmacy Act of 1868 opium and morphine remained available to all, provided that the customer could pay. Commercial interests ensured that opium-based patent medicines could be bought over the counter until the 1920s. But the writing was on the wall.[2] The

1. G. du Maurier, *Trilby*, 3 vols (London, 1894).
2. To guard against misuse of power by doctors and protect the public the Medical Act of 1858 established (among other safeguards) the General Medical Council. Similar legislation followed in most European countries, including Russia, and in the United States within the next thirty years. In Britain the Council was at first entirely the creature of

development was not entirely welcome even to some leaders of the profession. Expressing an Enlightenment attitude surprising in an eminent Victorian (but of course he was more than a Victorian prototype), Lister felt that drugs should be either banned or be available to everybody. 'They should not be doled out at the whim of doctors.' But he was by then a 'fossil'; and he and like-minded survivors of the Age of Reason were fighting a losing battle. That most disastrous and stupid of public perceptions, that 'something must be done' (without the faintest idea what that something should be or indeed if such a something existed) seemed irresistible. And history would soon provide an example of power through pain control, more frightening than either Du Maurier or Freud could dream up.

. . .

On the Continent the death of Queen Victoria in 1901 marked no symbolic break. There the Long Peace lasted for another thirteen years. During its final years the suffering of a little boy changed the course of history.

The only son of Nicholas II and Alexandra, the long-prayed-for heir to the imperial throne of Russia, was a haemophiliac.[3] Like most haemophiliacs before the 1950s, he was subject from an early age to bleeds into his joints, especially the knees. The slightest injury could set them off. They were excruciatingly painful. Nobody could stop them. Nothing could relieve the pain. The boy's parents were distraught. The condition was kept a secret; but rumours circulated. An unschooled and unwashed 'wandering monk' or *starets* was introduced to the Tsar and Tsaritsa by the Grand Duchess Milena, one of two silly Montenegrin sisters married to two Russian grand dukes.[4] His home was in a remote village in Siberia where he had a wife and four children

the unreformed royal colleges, self-perpetuating oligarchies based on the London teaching hospitals. Agitation by the British Medical Association and other bodies eventually led to reforms; and reforms have continued ever since. But self-regulation has been distrusted from the start by many.

3. The source of the mutant haemophilic gene was almost certainly Queen Victoria. The disease is transmitted by symptomless female carriers to their male offspring. The Tsaritsa, a princess of Hesse, was Queen Victoria's granddaughter on her mother's side. The queen had two other carrier daughters (as well as a haemophilic son); and the daughters transmitted the gene to members of the English and Spanish royal families.
4. Her father was a Montenegrin brigand who eventually made it to the throne. All his daughters married into European royalty, earning him the sobriquet 'Europe's Father-in-law'.

   Most *staretses* were genuine and poor holy men. Whatever else he was or was not, Rasputin was a fake *starets*. There is a vast literature on him. The best from the haemophilia and pain-control point of view is R.K. Massie's *Nicholas and Alexandra* (London, 1968), a beautifully written and moving double biography.

   Haemophilia did not become effectively treatable with regular transfusions, first of blood plasma and then of various concentrates of the antihaemophilic globulin, till the 1950s. It is still not curable. Transmission of the HIV virus then turned the advance into a tragedy.

and where the villagers and his own family regarded him as a wastrel. But he had hypnotic powers; and he used them to good effect. On more than one occasion when the boy, Alexis, was in pain, his ministrations seemed to stop the bleeding. At the least, they reassured the boy and his mother. Prince Felix Yusupov who hated the *starets* from the start and would later lead the conspiracy to murder him wanted to find out the man's secret for himself. Pretending to be suffering from intractable backache, he asked for Rasputin's help. By then Rasputin was known for his greed and the prince offered him a princely sum if he could cure him. Yusupov's description is useful because it is certainly not that of an admirer.

> The *starets* made me lie down on the sofa; then, staring intently at me, he gently ran his hand over my chest, neck and head; after which he knelt down, laid both hands on my forehead and murmured a prayer. His face was so close to mine that I could see only his eyes. He remained in this position for some time, then rising brusquely made some 'passes' over my body . . . I suddenly felt as if a warm current was flowing into my whole being. I fell into a torpor, and my body grew numb; I tried to speak but my tongue no longer obeyed me; and I gradually slipped into a drowsy state as though a powerful narcotic poison had been administered to me. All I could see was Rasputin's two glittering eyes; two phosphorescent beams of light melting into a great luminous ring which at times drew nearer and then moved further away. I heard his voice but could not understand what he said . . . I remained in this state without being able to cry or move . . . My mind alone was free and I fully realised that I was falling into the power of an evil man. I fought him in my mind . . . yet to move I had to wait until he ordered me to get up. He then closed the interview, saying with a laugh: 'Well, my dear, that'll be enough for the first time'.[5]

But direct hypnotic power was not Rasputin's only weapon. One of many incidents in the Tsarevitch's suffering (and one of the few which are well documented) occurred while, in response to pressure from ministers, Rasputin had been temporarily banished to his native village. In 1912 the country was celebrating the centenary of the 'Great Patriotic War' against Napoleon. One of the commemorations took place near Borodino, the river crossing where, during the retreat from Moscow, Napoleon's army suffered terrible casualties.[6] The imperial family were staying at the hunting lodge in Spala, once the retreat of the kings of Poland. It was a gloomy old building and Alexandra

5. F. Yusupov, Prince, *Rasputin* (New York, 1927), p. 82.
6. It was technically a victory for Napoleon since the crossing was achieved; but the end of the *Grande Armée* was in sight. Russian historians always regarded it as a victory for Russia.

became worried about her son, convalescing from a previous fall, being cooped up in a dark dank room. She arranged a drive. Not long after setting out, Alexis winced and began to complain about his lower leg and abdomen. Frightened, the Tsaritsa ordered the driver to return to the lodge immediately. Alexandra's friend and lady-in-waiting, Anna Vyrubova, later remembered the ride as an experience in horror.

> Every movement of the carriage, every rough place, caused the child to cry out in agony. By the time we reached the house he was pale and contorted, almost unconscious with pain.[7]

Botkin, the chief court doctor, diagnosed a haemorrhage in the thigh and groin. A stream of telegrams was dispatched; and, within the next hours, eight other doctors arrived, some by special train, from St Petersburg and Moscow. Their presence added worried faces, whispered consultations, frequent examinations and an air of despair to the scene. They could do nothing and could not disguise their impotence. The pain got steadily worse.

> Over the next days the little boy's screams pierced the walls and filled the corridors. Many in the household stuffed their ears. Unable to sleep or relax, the poor darling suffered intensely. When he stopped screaming and whimpering he just repeated: 'Oh God, oh God, have mercy on me. Let me die, oh God.'[8]

For eleven days Alexandra hardly left the boy's side. When she had to sleep, she lay back on the sofa next to his bed and dozed. Awake, her eyes seemed constantly on the contorted, pale little body.

> After a while his groans and shrieks dwindled to a constant whimper. Through the pain he called to his mother, 'Mama, help me, oh help me! . . . Mama, after I am dead it won't hurt me any more, will it?' . . . The Empress's golden hair seemed visibly to turn gray. Yet she stood the ordeal better than the Tsar. When Nicholas came into the room his courage often gave way and he rushed out weeping.[9]

News of the illness of the heir to the throne plunged Russia into grief. Shops and offices in St Petersburg and Moscow closed; the churches filled and then the squares in front of the churches. At Spala more than once it seemed that the end had come. At lunch on the twelfth day the Tsar was handed a note from the Empress. Alexis was suffering so terribly that she knew the end had come. The Tsar hurried to the bedside, as did the doctors. The Last Sacrament

7. A. Vyrubova, *Memoirs of the Russian Court* (New York, 1923), p. 68.
8. Ibid., p. 100.
9. Ibid.

was administered. A bulletin was ready for release announcing the death of the Tsarevitch. It was then that Alexandra asked Anna Vyrubova to send a telegram to Rasputin in Pokrovskoe, begging him to pray. Everybody disapproved but nobody dared to voice their disapproval. Rasputin immediately cabled back: 'God has seen your tears and heard your prayers. Do not grieve. The Little One will not die. Do not allow the doctors near him!' Against horrified protestations, the doctors were dismissed. Alexandra went to sleep on the sofa 'with a smile on her face. Before then she whispered in Alexis' ear: "Father Gregory has reassured us. All will be well, my darling".' The haemorrhage stopped. Next morning the boy was wasted, pale, utterly spent and still in some discomfort – but alive.

. . .

The episode has been much discussed both in the lay and in the medical literature. Bleeds can stop even in haemophiliacs. Old textbooks mention that fainting from anaemia may reduce the blood pressure and this may arrest the haemorrhage. Calm and reassurance can have an astonishing effect. By contrast, 'haemophiliac patients bleed more profusely under emotional stress'.[10] Rasputin's advice not to let the doctors hover around in gloom was astute. Justified or not on medical grounds, his reassurance of Alexandra, the only person with whom Alexis had a close emotional tie, was probably lifesaving. Confronted with Rasputin's telegram, doctors, courtiers and the grand dukes and grand duchesses muttered about an 'eerie coincidence'. But the Empress, the Tsar and Alexis himself were in no doubt. A month later Alexis had recovered sufficiently to be moved back to Tsarskoe Selo. Every yard of the roads to the station at Spala and from the station to Tsarskoe Selo was smoothed out, every stone was removed. The imperial train crawled at 15 miles per hour. Almost a year would pass before Alexis could walk again; and his right knee was permanently bent. (In official photographs he was always shown resting his foot on something as if the bend was natural.) But Rasputin would never again be allowed to leave the patient for more than twenty-four hours. His family were installed in St Petersburg. No royal mistress or fancy-boy in the history of the Romanovs ever wielded such power.

. . .

It was Rasputin's success in keeping Alexis fit and, above all, free from pain which gave him a hold over the boy's parents, especially over the Empress. They came to regard him not only as a blessed healer but also as a saint, 'the Soul of Russia'. He shamelessly exploited his sway. He flaunted it. His amorous

10. D. Lukin, *A Vérzékenység* (Budapest, 1938), p. 34.

and financial exploits were the scandal of St Petersburg and soon of Europe. But more important, even if perhaps partly imaginary, was his political influence.[11] The Tsar was an autocrat, not a constitutional monarch. He was also commander-in-chief. Yet even during the first years of the war everybody knew – or thought they knew – who wielded the real power. Eventually, in 1916, the *starets* was murdered by a conspiracy of courtiers, aristocrats and a grand-duke cousin of the Tsar.[12] By then he had done irreparable harm to the prestige of the monarchy. When the crunch came a year later, few among the aristocracy or even the ever-loyal middle classes would risk their lives for the imperial family. In this they were not alone. In Britain George V, scared out of his wits that any 'foolish act or even show of assistance might lead to a bolshevik revolution in our own empire',[13] vetoed all bids to rescue his 'beloved cousin Nicky'. In 1918 the whole Russian imperial family, including the sick boy, were murdered by the Bolsheviks. The bodies of the imperial couple, their four daughters and Alexis were shovelled into a well. The Empress' body was still trying to shield her son. By a few seconds he probably survived the rest.

11. Apart from opposing any yielding of absolute imperial power to an elected body, Rasputin was not interested in politics. He was opposed to the war; but so was the Tsar. It was the Kaiser who declared war on Russia.
12. It was probably the worst-executed assassination in history. Yossoupov (in Italian opera he would have been described as the First Assassin) lured Rasputin to his villa. There Rasputin was given wine and bread liberally laced with cyanide. He did not turn a hair. He was then shot three times; but he staggered on. Eventually, still alive, he was dragged to the frozen Neva and drowned through a hole cut in the ice.
13. Lloyd George who was prepared to mount a rescue operation was moved to describe Europe's royalty as the most squalid of trade unions.

# PART V

# YESTERYEARS

# Chapter 39

# SEMINAL YEARS

Few of today's trends in the arts, music, literature, architecture, philosophy, political thought, mathematics, physics and chemistry cannot be traced back to the astonishing last decade and a half of the Long Peace, the years before the outbreak of the First World War. Musicians thrilled to the sound of *The Rite of Spring;* visual artists were dazzled (or scandalised) by Kandinski's first abstracts and the invention of cubism by Braque and Picasso; in the theatre Chekhov and Stanislavsky lifted the curtain on the modern stage; in literature Proust created a new way of writing; the modern decorative arts were born in Vienna and Glasgow; in the sciences atomic physics and psychoanalysis were launched about the same time. Man started to fly and the air was beginning to be filled with radio waves. The list could be continued. Medical practice did not change significantly; but the scientific basis of medicine did. In particular, the seminal years saw the transformation of the neurosciences, including the nuts and bolts of pain.

The founders, Camillo Golgi and Santiago Ramón y Cajal, loathed each other with an abandon which in the more inhibited (if no less venomous) twnety-first century would be hard to match. With a fine sense of occasion the Nobel Academy of Stockholm awarded them the Nobel Prize jointly in 1906 but had to be circumspect in seating them at the official banquet at a safe distance from each other, separated by three elderly members of the Swedish royal family and a former laureate of the Peace Prize. On the official photograph they stand (rather grim-faced) at either end of the festive line-up.

Camillo Golgi, the elder of the two, was born in Corteno, near Brescia in Italy, the son of a doctor. After qualifying from the University of Pavia in 1872, at the age of twenty-nine, he accepted the post of chief (and only) medical officer to the Hospital of the Chronically Sick in Abbiategrasso. Happily, the clinical care of 2,500 chronically sick left him plenty of time to install a laboratory in his kitchen and to devote himself to what had already become his consuming passion, the microscopy of the nervous system.[1] His great discovery – among several notable ones – was a method of staining nerve cells and their ramifications using silver nitrate, which he described as 'the black reaction'.[2] For the first time under his microscope the constituent and basic units of the brain, soon to be known as neurons, took shape. They had never been seen before, at least not with anything approaching such clarity.[3] His adaptable stains showed up not only the cells themselves and the structures within the cells (one of them known today as the Golgi apparatus) but also the branching processes which establish the all-important links between them. So complex and fine was their appearance that Golgi conceived the notion that the whole of the brain was an interconnected network of fibres forming a fantastically elaborate lattice or 'reticulum'. The only flaw with this brilliant notion was that it was wrong.[4] The man who demonstrated Golgi's error was Santiago Ramón y Cajal. For this he was never forgiven.

Cajal, the youngest son of a professor of anatomy in Saragossa in Aragon, was apprenticed to a barber and then to a cobbler and in his heart of hearts always wanted to become a painter. (How could he not, with Goya's birthplace only a few miles up the road?) He remained all his life a superb draughtsman, his illustrations to his own papers among the finest in the great tradition of anatomical drawings. His father, whose anatomical specimens he illustrated, eventually agreed to the son becoming a doctor. In 1874, at the age of twenty-two, Santiago graduated from his home university and became an army surgeon. During the disastrous Spanish expedition to Cuba of 1874–75 – there were several, all disastrous – he acquired both malaria and tuberculosis which precluded him from pursuing a clinical career. Indeed he was regarded as moribund, the more so since, then and later, he consistently ignored medical advice. To keep him out of mischief, he was appointed director of the Saragossa Anatomical Museum, an academic sinecure. In that capacity he began to take an interest in the brain and its architecture. Recovering from his ailments he took the MD degree of Madrid University and eventually came to occupy the chair of pathological anatomy there.

1. His interest had been kindled by his teacher in Padua, Bizzozero, another pioneer whom Golgi eventually succeeded and whose daughter he married.
2. He also used his staining techniques in making useful observations about malaria.
3. The first to describe cells in the brain was the Czech anatomist, Jan Purkinje who died in 1869, aged eighty-two.
4. Golgi eventually became rector of the University of Pavia (where a hall is dedicated to his diplomas and innumerable medals today) and a senator of the Kingdom of Italy. He was much liked by his juniors and an excellent teacher. During the First World War he created a pioneering centre in Pavia for the treatment of nerve injuries. He died in 1926.

From the moment of his arrival in Madrid Cajal's life was dedicated to the study of the cellular structure of the brain.[5] He made excellent use of Golgi's stains as well as divising some of his own; but, as regards the overall design of the central nervous system, he came to a conclusion different from Golgi's. As shown in a beautiful series of microscopical slides, nerve cells do not form a continuous interconnected reticulum but are individual and distinct. They do, however, possess two kinds of 'processes'. One kind sprouts out and extends like a branching tree. Tree is *dendron*, δένδρον in Greek and these processes are now called dendrites.[6] This is the route by which incoming signals or impulses reach the parent cell. The other kind is a single process, called axon, the route by which the nerve cell transmits its own signals. Some axons become the nerve fibres of peripheral nerves. Some of those – for instance the nerves supplying the toes – are more than a metre long; but they are still part of the parent cell. The nerves bunched together and descending in the spinal cord as 'tracts' are the axons of cells situated in the brain. Nerves ascending are axons of cells situated in the grey matter of the spinal cord. The axons and small branches of the dendrites can attach themselves to the membranes of other nerve cells; but there is no anatomical – that is physical – continuity between two neurons.[7] There are a staggering 1,000 billion neurons in the grey areas of the brain, that is in the cortex and a number of subcortical 'basal nuclei'.[8] The white matter consists of fibres – that is, the axons of the neurons – their colour derived from the insulating white myelin sheath. This myelin is itself contained in living cells.[9] The whole is supported by an elaborate scaffolding of so-called glial cells, even more numerous than the neurons.[10] Though these are not themselves part of the nervous system, they are more important than

5. His interest was obsessional. While he was attending a conference in London and stayed with the Sherringtons for the duration – only a week – his hostess was mortified to find her guest room converted into a travelling laboratory, their Sheraton side-table transformed into a histological bench, complete with microscope and bottles of indelible stains.
6. A neuron may have as many as 100,000 dendrites which is of course far more than any tree has branches, or even leaves.
7. This discovery was the more remarkable since the physical gap between a nerve cell and an 'attached' dendrite or axon from another cell is of the order of a 200,000$^{th}$ of a millimetre and could not be clearly seen until the advent of the electron microscope.
8. The name nuclei can be confusing. Subcortical nuclei (like the thalamus) are a collection of neurons appearing as grey masses. They have nothing to do with cell nuclei.
9. See Chapter 32.
10. From *glia* γλία Greek for glue. Some are mobile scavenger cells, others line the duct system and the cavities inside the brain and spinal cord.

had been thought until comparatively recently.[11] Much of this picture was virtually complete by the time Cajal died in 1934, aged eighty-two.[12]

. . .

The work of Golgi, Cajal and a few other neuro-anatomists had an effect on the neurosciences comparable to the impact of the anatomical discoveries of Vesalius and his contemporaries on the medicine and surgery of the Renaissance. Both Golgi and Cajal were essentially descriptive anatomists, neither venturing far into the possible functions of the structures which they described. But once the structures were established, the way was open for neurophysiologists and neuropathologists and for new interpretations of sensations like pain.

During his long and full life – he died in 1952 aged ninety-seven, a little frail but as bright as ever – Charles Scott Sherrington became something of a *beau idéal* of British medical scientists. He rowed for his college, Gonville and Caius, in Cambridge, became a pioneer skier in Grindelwald, was an accomplished (but not indecently so) flautist and wrote poetry.[13] He himself described the work which remains his main claim to fame, the tracing of nervous pathways from muscles to the spinal cord, from spinal cord to the brain and back again as 'labours of the most tedious and pedestrian kind'. Tedium is an unavoidable part of all medical research; but there was more to Sherrington's labours. As was also true of William Harvey, his capacity for painstaking work was allied to considerable powers of imagination; and, no less important, to a gift for communication. Most of Sherrington's seminal studies were completed as a young demonstrator in anatomy in Cambridge and then as professor of physiology in Liverpool (where the budding American neurosurgeon, Harvey Cushing, stayed with him for a formative six months); and he published his classic, *The Integrative Action of the Nervous System*, in 1906. During the First World War, as a member of the Industrial

11. Susan Greenfield has called them 'Cinderella' cells. See *The Brain Storm* (London, 2000).
12. Among numerous other honours Cajal delivered the Croonian lectures and became a fellow of the Royal Society in 1923.
13. On publication of *The Assaying of Brabantius and Other Verse* in 1925 'Miss Sherrington' was congratulated on her imagination and sense of rhythm by the *Times* poetry critic. His Rede Lectures, delivered in Cambridge in 1933, *The Brain and its Mechanism*, are still a great read. Apart from every conceivable honour in his own country, he was the recipient of seventeen honorary doctorates abroad and shared the Nobel Prize with E.D. (later Lord) Adrian in 1932.

Such a paragon should be beyond criticism; but the present writer would be untrue to himself if he failed to mention that Sherrington's 'decerebrate cat' and other classical 'preparations' (reproduced for decades in every textbook of physiology), showing cats in different attitudes of spastic and flaccid paralysis after having different parts of their central nervous system sectioned or extirpated, has always filled him with horror.

Fatigue Board, he studied the stresses of labour by working in a shell factory, a daily shift of 13 hours and a Sunday shift of 9 hours, for a year. This involved taking study leave from the Waynfleet chair of physiology in Oxford to which he had been invited in 1913. (He had applied for it unsuccessfully eight years earlier and was in two minds about accepting it when it was offered to him.)

In retrospect Sherrington's main contribution was to demonstrate the extraordinary complexity of the nervous mechanisms which govern even the simplest of voluntary and reflex actions. In particular, every movement depends on the perfect synchronicity of two opposing sets of nervous commands. One set commands a group of muscles to contract. The other set commands the opposing group of muscles to relax. If this did not happen, a reflex like the knee jerk would break the thigh bone. Cooperative and synchronised nervous commands govern not only obvious voluntary movements and 'simple' reflexes like the knee jerk but also the continuous maintenance of muscle 'tone' and 'posture'. They also operate in involuntary muscles, ensuring such astonishingly precise 'waves' of muscle contractions as the peristaltic movements which propel food and secretions through the gastrointestinal tract or ensure the passage of urine down the ureter, through the bladder and then the urethra. A blockage to these peristaltic waves and the increasing effort of the proximal muscles to overcome the obstruction gives rise to the pain known as colic. Perhaps the most remarkable nervous coordination is shown by the nerve supply of sphincters, the circular muscle rings whose contraction and relaxation must synchronise with the opposite action of the hollow organ whose contents they control. These complementary and coordinated actions largely depend on spinal reflex arcs. They obey a number of laws of nerve transmission which were put into precise, almost legal terms by Sherrington. In contrast to most contemporary legislation, there has been little need to modify them since. He showed that spinal reflexes are normally under the continuous influence of higher centres situated both in the subcortical basal nuclei of the brain and ultimately in the cerebral cortex. These too he did much to localise anatomically.

Quite early in his career Sherrington was asked by his chief, Sir Michael Foster, to help him with the new edition of Fulton's famous textbook, *Physiology of the Nervous System.*

> I had begun [the editing] and had not gone far before I felt the need of some name by which to call the junction between nerve cells or their processes (because the place of junction had now entered physiology and carried considerable functional importance). I wrote to Foster of my difficulty and my wish to introduce a specific name. I suggested 'syndesm' from συνδέσμος [*syndesmos* a band]. He consulted his Trinity friend Verrall, the Euripidean scholar. Verrall suggested 'synapse' from συγάπτω ἄψω

[*sygapto-axo*] meaning clasp; and since that yields a better adjectival form – i.e., synaptic – it was adopted for the book.[14]

Euripidean scholars too have their uses.

...

More or less overlapping with Sherrington's work on 'ordinary' reflexes, a different kind of reflex action, no less important for the understanding of pain mechanisms, was studied in Russia by Ivan Petrovich Pavlov. In the last years of his long life and for some decades after his death he too became an idol, though in an entirely different setting from Sherrington's.

Pavlov was born in 1849 in Ryazan (Khazan) in central Russia, the son of a village priest. He was educated at the local church school and seminary with a view to becoming a priest himself; but under the combined influence of the reformist literary critic, D.I. Pisarerv and the charismatic founder of Russian physiology, I.M. Sechenov, he changed course and enrolled in the Faculty of the Natural Sciences. From the start his career sparkled, a combination of obsessional hard work and intelligence winning him gold medals and scholarships at every level. By the 1880s he was head of laboratories in the university department of the leading Russian clinician of the day, S.P. Botkin, in St Petersburg; and in 1890 he was invited to direct the Department of Physiology in the Imperial Institute of Experimental Medicine. Except for the dropping of the 'Imperial' from the title in 1917, he remained in this post for forty-five years.

It was during the decade 1892 to 1902 that Pavlov developed the experimental surgery necessary for creating the salivary, gastric and intestinal fistulas in dogs which enabled him to study the function of these organs under relatively 'normal' conditions.[15] The studies led him step by step to evolving the concept of 'conditioned nervous reflexes' to explain the 'psychic' secretions recognised in everyday life but not in science.[16] At the 14th International Medical Congress in Madrid in 1903 he presented a historic paper entitled 'The Experimental Psychology and Psychopathology of Animals'. He argued that conditioned reflexes should be regarded as basic psychological phenomena and that they were the mechanism of most responses in animals and almost certainly in humans to their environment. The thesis became the guiding light of materialistic physiological teaching in explaining virtually all

14. Quoted in H.A. Skinner, *Origin of Medical Terms* (London, 1949), p. 311.
15. Such a dog was one with a gastric fistula who could be shown to start secreting gastric juice at the sight or smell of food. Such preparations were not dissimilar to the patient studied by Beaumont in the United States in the 1840s (see Chapter 22).
16. A cardinal point in Pavlov's message was that conditioned reflexes were the basis of 'normal' life.

activity of living organisms. Any external agent, by coinciding in time with an ordinary reflex could become the conditioned signal for the formation of a new conditioned reflex. Over the next few years Pavlov and his pupils showed that these conditioned reflexes, far from being 'primitive' animal responses, originate in the cerebral cortex which 'functions as the prime distributor and organiser of all activity of the organism'. It is, in particular, 'responsible for the delicate balance between a living organism and its environment'. Pain and pleasure are, in the Pavlovian universe, almost entirely conditioned responses, which start to be acquired from the moment of birth or even in the womb, to sensations associated with inborn reflexes. So are faith and denial.

The concept was by implication uncompromisingly subversive of any orthodox faith and doctrine and might have led Pavlov and his followers to the stake three hundred years earlier. During the first iconoclastic decade of the twentieth century, it brought him little but international acclaim.[17] He was awarded the Nobel Prize in 1904, elected member of the Russian Academy of Sciences in 1907, became the recipient of an honorary doctorate of Cambridge University in 1911 and was made officer of the Legion of Honour in 1913.[18] But it was after the Revolution that in Russia he became an oracle. A special government decree signed by Lenin in 1921 noted the 'outstanding services of Academician I.P. Pavlov which are of enormous significance to the Working Class of the whole world'; and Russia's greatest living painter, Repin, was cajoled to paint in his Finnish exile an official portrait of the great man.[19] After his death and in the final demented years of Stalin's reign of terror no director of a physiological institute was safe unless he had at least one 'Pavlov dog' in a cage prominently displayed to all visitors, whether such a creature was necessary for the work of his unit or not, dribbling saliva or gastric juice at the sound of the people's democratic whistle. It was unintentionally but deeply symbolic.

...

One other scientific development of the Seminal Years was to have a decisive influence on the pharmacological control of pain towards the end of the

17. But George Bernard Shaw eloquently railed against 'Nobel-Prize-winning animal torturers' who had to mutilate cats, dogs and other dumb creatures in order to demonstrate what every schoolboy knew – that the sight of candy induced salivation. He also memorably wrote to a scientist friend: 'How do you know that it was not Pavlov's dog which trained Pavlov to ring a bell every time it felt like salivating?'

    There was much truth in this. On the other hand, most great discoveries in science have been perfectly obvious since time immemorial.
18. Even in tsarist Russia a few decades before Pavlov, his brilliant teacher, I.M. Sechenov, was persecuted by the authorities for his materialistic teaching and his publications were confiscated.
19. It is one of Repin's few failures.

century. When Henry Dale and Otto Loewi embarked on studying the chemical transmission of nerve impulses, the idea of such mediation was not entirely unheard of. The Cambridge physiologist, T.R. Elliott, had suggested in a short note in 1904 that some nerve fibres might exert their action by releasing the hormone adrenaline (epinephrine);[20] and the American physiologist W.H. Howell had speculated in 1908 that the action of the vagus nerve on the heart was mediated by the release of potassium. But it was Dale and Loewi who in a series of experiments stretching over two decades demonstrated the crucial role of what today are called 'neurotransmitters', norepinephrine in the case of Loewi and acetylcholine as shown by Dale.

Otto Loewi was born into a professional Jewish family in Frankfurt am Main in 1873 and Henry Dale into an academic family in Cambridge two years later. Both had successful undergraduate careers; and they met and became lifelong friends as holders of scholarships in Ernest Starling's famous Department of Physiology at University College, London. (Their host there had recently coined a new word to describe a family of compounds nobody had heard of before and few believed existed. The word was 'hormones'.) From University College Dale went on to become director of the Wellcome Physiological Research Laboratories in 1906; and it was there, over a period of six years, that he made his most important discoveries. He showed that autonomic nerve fibres belonging to the parasympathetic system but also some belonging to the sympathetic system exerted their effect by liberating minute amounts of acetylcholine at the nerve endings.[21] Later Dale became director of the National Institute of Medical Research in London, the laureate of countless academic honours, the recipient of prestigious medals as well as the Nobel Prize, a knighthood and the Order of Merit.[22] He died in 1968, aged ninety-three.

Loewi became professor of physiology in Graz in Austria where his department acquired world renown. He showed that the effect of most sympathetic autonomic nerve fibres was mediated by the liberation of minute amounts of norepinephrine (noradrenaline), a substance which, in larger amounts, is also secreted by the medulla of the suprarenal glands. After Austrians voted overwhelmingly to join Nazi Germany he was allowed to leave, though not until he had his Nobel Prize money transferred from a Swedish to a Nazi-controlled bank. After spending some time as visiting professor in Brussels and then in

20. Epinephrine is the proper chemical name for adrenaline and norepinephrine for noradrenaline.
21. They are called 'cholinergic'. It was eventually shown that acetylcholine is liberated at all ganglionic synapses. Anatomically some nerves – such as those supplying sweat glands – belong to the sympathetic nervous system but are functionally 'cholinergic'.
22. He married a cousin, Ellen Hallett. Their eldest daughter married Lord Todd, Nobel Laureate in Chemistry in 1957.

Oxford he became professor of physiology in New York University and in 1946 a United States citizen. He too received numerous honorary doctorates including a clutch from reborn post-war Germany and Austria. He died in 1961, aged eighty-eight.

Though the first two neurotransmitters, noradrenaline and acetylcholine, were recognised as important from the start, the significance of their discovery can be appreciated only by an unhistorical forward glance. The expansion of the number and range of chemical neurotransmitters has probably been the most important development in the neurosciences, and their practical application in the last fifty years. Their appearance put an end to the complicated but still naïvely simplistic picture of the brain as a giant computer. Neurotransmitters do not simply operate on–off switches but can turn quantitative dials of almost infinite gradations. Their action, moreover, depends on specialist receptor sites which are themselves of widely varying sensitivity. The complexity of the unfolding picture might have seemed daunting had it been envisaged from the start – and that picture is still far from complete – but the advances have already transformed pharmacological perspectives.

Acetylcholine, the neurotransmitter discovered by Dale, is known to be deficient in Alzheimer's disease. It has been shown to act as a mediator not only between nerves and muscle cells (as demonstrated by Dale) but also between neurons within the brain. There appears to be a 'fountainhead' of the mediator near the centre of the brain from where axons fan out to connect both with the cortex and with subcortical nuclei. As well as being involved in a negative way in Alzheimer's disease, the transmitter probably plays a key role in sleep, wakefulness, arousal and the emotional response to pain. The drug hemicholinum which inhibits acetylcholine reduces dreaming sleep.[23] The enzyme acetylcholinesterase, which ensures the rapid removal of acetylcholine once it has transmitted its message, can be dramatically influenced by drugs and poisons.[24] Norepinephrine, the substance shown by Loewi to be released at sympathetic nerve endings, has also been shown to be released by a cluster of about 20,000 neurons inside the brain. This release too seems to be organised like a fountain, the cell axons reaching into anatomically distant parts. A burst of norepinephrine release makes the subject feel wide awake and often oblivious of pain. The transmitter also stimulates sexual activity. Drugs like the amphetamines and cocaine probably stimulate or simulate its action.

23. Dreamless sleep produces slow regular electrical brain waves on the electroencephalic record, whereas dreaming generates faster irregular activity. Dreaming also produces rapid eye movements (REM) and is therefore sometimes referred to as REM sleep.
24. Nerve gases like organophosphates probably act by blocking the action of acetylcholinesterase. It has been suggested that the Gulf War Syndrome is linked to inhibition of this enzyme.

Norepinephrine is chemically related to dopamine, one of the second generation of neurotransmitters. Dopamine deficiency may be an underlying cause of Parkinson's disease and may play a part in the aetiology of schizophrenia. Pain perception can be powerfully altered by it as it often is in mental disease.

Another of the second generation of neurotransmitters, serotonin, is produced by small clumps of cells arranged along the central axis of the brain. By long axons these cells too reach down to the far end of the spinal cord as well as up towards the subcortical nuclei and the cortex. Their secretion is important in governing two basic mental processes: sleep and pain. Serotonin has spawned a still proliferating number of new drugs acting on mood, anxiety and response to chronic pain.

. . .

In the 1960s a further group of neurotransmitters were discovered, belonging to the chemical family of aminoacids. One among those, gamma-aminobutyric acid or GABA, seems to be particularly involved in transmitting nerve impulses governing movement. The excessive, uncoordinated and often painful movements of the inherited disease, Huntington's chorea, may be related to a drop in the production of this comparatively simple substance.

If the 1960s was the decade of aminoacids, the 1970s saw the advent of neurotransmitting peptides, chains formed from the link-up of several aminoacids.[25] It seems that peptide neurotransmitters may be released by 'hyperexcited' cells not only at synapses but also along the length of axons, an apparently new mode of action.

. . .

There were other discoveries during the Seminal Years which seemed to be ground-breaking for a time but which, in the long run (or until the present), did not fulfil their promise. Theodor Meynert in Vienna and Alfred Campbell in England made an attempt to correlate the appearance of cortical cells with different kinds of cerebral function. It has so far proved a blind alley. Much ingenious but so far inconclusive research was also devoted to finding an area in the brain specially concerned with the registration of pain, an anatomical pain centre.

At one time studies of the 'thalamic syndrome', first described by two French neurologists, Jules Déjérine and Gustave Roussy, seemed promising. The thalamus is one of the subcortical nuclei at the base of each hemisphere; and clinical lesions as well as experimental work in animals suggested that it

25. Beyond a somewhat ill-defined arbitrary number a chain of amino acids becomes a protein.

might function as a kind of crude or primitive brain. Déjérine and Roussy found that lesions in this area could be associated with severe hyperalgesia, that is the perception of comparatively mild peripheral stimuli as severe pain.[26] Their work was carried further in England by Henry (later Sir Henry) Head and Gordon Holmes who concluded that the 'thalamic response' was important but more complicated than the French workers had originally thought.

> The characteristic thalamic response does not consist of an excessive response to painful stimuli only. In suitable cases we have shown that the response to pleasurable stimuli, such as warmth, is also greater on the affected side. Moreover, the manifestations of general mental states both of pleasure and discomfort may be more pronounced in thalamic lesions . . .
>
> The thalamus may be the centre of consciousness for certain basic and violent sensations. It responds to all stimuli capable of evoking either pleasure or pain . . . The feeling tone [or emotional component] of somatic or visceral sensations is the product of thalamic activity; and the fact that a sensation may be devoid of it shows that the impulses which underlie the sensation bypass the thalamus . . . All stimuli which appeal to the thalamic centre have a high threshold. They must reach a high intensity before they reach consciousness, but, once they have risen above that threshold, they tend to produce a response of excessive duration and intensity which it is the business of the cortical centres to control.[27]

For a time the idea that the thalamus acted as a kind of uncensored 'primitive' brain, the seat of simple and violent emotions and responses, acquired a literary as well as a scientific vogue. In the 1930s André Gide described an emotional over-reaction as a *réponse thalamique;* and Aldous Huxley liked to refer to people who disagreed with him as 'thalamic in their denial of the obvious'. The concept may be useful rather than rigorously scientific.

. . .

Head, together with Gaskell and Langley in Cambridge, also pursued fundamental studies in classifying sensations; and for the best part of the twentieth century their division of sensations into 'visceral' (arising from internal organs), 'protopathic' (dominant and crude) and 'epicritical' (fine and discriminatory) became part of standard medical-school teaching.[28] To verify

26. J. Déjérine and G. Roussy, 'Le Syndrome thalamique', *Revue neurologique*, 1906 (4), 521.
27. H. Head, 'Sensory Disturbances from Cerebral Lesions', *Brain*, 1911 (74), 102.
28. H. Head, 'The Afferent Nervous System from a New Aspect' *Brain*, 1905 (68), 99.

his ideas Head cut and then 'resutured' his own radial nerve and then observed the stepwise reacquisition of various kinds of sensation.[29] He found that painful ('protopathic') sensations were regained first, and the finest discriminatory ('epicritical') sensations last or never.

Following Head's lead, W. Trotter and H.M. Davies also espoused self-mutilation (though on a less drastic scale) and did much to discredit the seductive ideas of Max von Frey. Around the turn of the century, this Austrian neurologist, who had already acquired fame for his physiological research in the University of Würzburg in Germany, described in considerable detail special microscopic 'pain receptors' in the skin. He even drew elaborate maps of their distribution.[30] The receptors looked excitingly like little insects and were widely reproduced in medical textbooks until the 1950s. Unfortunately, as Trotter and Davies showed, they almost certainly do not exist. It is not clear what Von Frey saw. Perhaps they *were* insects. Another German neurophysiologist, Alfred Goldscheider, who also studied skin sensibility, eventually opted for an alternative pain mechanism. He suggested that as yet unidentified receptors registered a 'caress', a 'touch' or a 'pain'.[31] The different sensations depended on the intensity of the stimulus and were 'translated' into different rates of impulse. The controversy between the two concepts led to a number of 'compromise' hypotheses. It is in fact surprisingly easy to demonstrate evidence supporting most of these mechanisms if one puts one's mind to it.

...

More important for the basic understanding of pain were the investigations of Walter Cannon and his group in Harvard. They showed how pain, together with hunger, fear and other 'basic emotions and sensations', triggered off a set of 'drastic alterations in body economy, adapted to the individual's welfare or even survival'.[32] Some classical responses to 'pain' could even 'anticipate' the sustaining of painful injuries. A scream and the sensation of a blow, together with adaptive and measurable reflexes, could precede a stimulus that might never actually come. Fear and pain were 'in some kind of rapport'. Cannon, a master of understatement, thought it was all 'very odd'. And the more he and his colleagues studied the relationship, the odder it got. 'Alarm' was not a purely neurological mechanism. It also involved coordinated and apparently synchronous changes in the vascular and

29. The nerve is a mixed sensory and motor nerve which supplies about half of the hand and two and a half digits on the thumb side.
30. M. von Frey, 'Untersuchungen über die Sinnesfunctionen der menschlichen Haut', reprinted in H.O. Handwerken (ed.), *Deutschsprächige Klassiker der Schmerzforschung* (Heidelberg, 1987), p. 91.
31. A. Goldscheider, *Das Schmerz Problem* (Berlin, 1920), p. 77.
32. W.B. Cannon, *Bodily Changes in Pain, Hunger, Fear and Rage* (New York, 1915).

endocrine systems. Changes in blood flow could occur even in parts of the body deprived of their nerve supply. Sphincters too could contract without nervous command. Necks prickled. Hairs stood erect. A cliché inevitably emerged. The concerted response was to be known as the 'fight and flight reaction'.[33] Such verbal aids are balm to the scientist's soul but tend to leave basic questions unanswered.

33. W. MacDougall, *Introduction to Social Psychology* (London, 1908), p. 204.

# Chapter 40

# THE GIFT OF SAINT BARBARA

Advances in clinical pain management were less spectacular but in some areas they were important. It was during the Seminal Years that the first modern pain-killing and mood-setting drugs emerged. Their potential was not immediately appreciated; but during the second half of the century they would transform the medical landscape, even the moral climate, of the Western world. At a more theoretical level, they not so much bridged as abolished the gap between 'physical' and 'mental' pain.[1] Essentially the same drug which induced near-instant 'physical' anaesthesia could also cope with paralysing fears. Tablets which controlled epileptic fits also relieved depression. The story began with a pious conceit.

On 4 December 1863 Johann Adolf von Baeyer, a twenty-eight-year-old German chemist working in the department of the famous Friedrich Kekulé von Stradonitz in Ghent in Belgium, discovered a new compound.[2] It had an interesting ring structure and was the product of two 'historical' organic molecules, urea and malonic acid.[3] It was also the first new compound he had synthethised himself; and the obvious term of malonylurea struck him as far

1. Much ink and a certain amount of venom has been spilt between 'monists' and 'dualists' of pain research. So far as the present writer can make out the difference, dualists (like Sir Karl Popper) tend to distinguish between an apparatus concerned with registering and transmitting pain and a 'higher' system concerned with perceiving and interpreting it. 'Monists' (like the late Patrick Wall) believe that the two are inseparable and one. This may be an oversimplification and to the present writer the argument appears rather semantic. More importantly perhaps, 'monists' have always insisted that there is no fundamental difference between mental and physical pain, a view with which the present writer agrees but many disagree.
2. Kekulé's idea of a ring structure to explain the properties of benzene was a landmark in organic chemistry. He was also among the first to propound the doctrine of valency. From Ghent he moved to Bonn as professor of chemistry. He died in 1896, aged sixty-seven.
3. Urea was the first organic molecule to be synthethised by Friedrich Wöhler, in 1828; malonic acid among the earliest.

too prosaic to mark such an achievement. Still brooding over a suitable name, that evening he celebrated his discovery at his local tavern. There artillerymen garrisoned in the city were toasting their patron saint. They willingly told him her story.

St Barbara was the beautiful Christian daughter of the pagan and evil Dioscurus in third-century Nicodemia in Asia Minor.[4] When she resisted her father's blandishments to marry a rich but repellent merchant, she was immured in a tower. In answer to her prayers, a tunnel miraculously opened in the fortifications. Protected by an ostrich feather given to her by an angel, she escaped to a neighbouring province. There she performed numerous miracles, including saving a newborn infant from drowning after it was thrown into the river by its distressed child-mother. After escaping many perils, she was eventually betrayed by a jealous woman. She was delivered to her father, who personally beheaded her. He was instantly struck dead.

In centuries to come the reputation of the saint blossomed.[5] In the Middle Ages and the Renaissance she became the patron saint of artillerymen, construction workers, miners, prisoners, casters, bell-ringers, hatters, chefs, bricklayers, butchers, stonemasons, grave diggers, builders of towers and fortifications but above all of tunnellers. Since a charming statue was erected in her honour in Sangatte by French workers on the Channel Tunnel in 1987, one should surely add illegal immigrants to her list of protégés.[6] Baeyer joined the revellers and learnt that they were celebrating the saint's feast day. This 'wonderful coincidence' inspired him to name his new compound 'barbituric acid'.

Contrary to Baeyer's initial hopes, neither the name nor the compound aroused much interest for the next forty years. He nevertheless advanced rapidly on the academic ladder and was, by 1900, *von* Baeyer and an important professor in Strasburg (then part of the German Reich). Under his wing two of his former pupils, Emil Fischer and Josef von Mering, were testing a

4. Or Heliopolis in today's Syria.
5. There are many representations of her in medieval and Renaissance art, including a great painting by an anonymous fourteenth-century German artist now in the National Bavarian Museum in Munich and a gorgeous ikon in the Uspenski Cathedral in Helsinki.
6. In the early 1990s the French Red Cross helpfully established one of its largest 'transit camps' for migrants, asylum seekers and refugees in Sangatte, within walking distance of the French end of the Channel Tunnel. The majority of those in the camp hoped to make their way to the United Kingdom and tried their luck night after night on trains entering the tunnel. Many made it, much to the distress of Mr Blair's government on the receiving side of the Channel. A soupçon of malice on the part of the French was suspected. A bargain was eventually struck between the two governments but never disclosed and the camp was moved a few miles further south.

wide range of compounds for their possible pain-relieving action.[7] Out of respect for their teacher, they included his youthful discovery in their experimental protocol. A close derivative of barbituric acid, now called 'barbital', showed promise in rabbits. The two young chemists took out a patent and offered it to the pharmaceutical firm E. Merck & Co. in Darmstadt. 'Barbital' struck the directors as 'ein bischen unheimlich' (a little spooky). Could the clever Dr Fischer think of a cosier brand name? Fischer had just returned from a romantic holiday in Italy, and the name of Verona still evoked happy memories in his mind. What about 'Veronal'? The suggestion was accepted and the drug proved a success. The century of the barbiturates had begun.[8]

Veronal was followed by a succession of proprietary variations on the chemical theme, the products differing from each other mainly in speed and duration of action. Phenobarbital or Luminal was introduced in 1919 and amytal in 1923. The addition of a sodium atom to one of the carbons in the ring made the compound soluble in water; and six years after amytal, sodium amytal became the first barbiturate to be used as an intravenous anaesthetic. At the same time as amytal, another derivative, 'nembutal', was offered for trial by the makers, Abbotts, to John Lundy, anaesthetist at the Mayo Clinic. Lundy tried it on his wife, on Charles Mayo and on Jack Dempsey, the world heavyweight boxing champion. All three found it to be the best combined analgesic and soporific yet. With such endorsements, the drug became an instant success. But what in fact did it do? It undoubtedly promoted sleep: so it was a narcotic. It also calmed nerves: so it was a tranquilliser. But it also relieved ill-defined pains and aches, more the result of 'tension' perhaps than of physical hurt. This made it an analgesic as well. The demarcation between physical and mental pain was beginning to get blurred. With each new barbiturate – and then with some of their rivals – the blurring would become more dense. Today most neuroscientists find the distinction unhelpful, though ordinary people still tend to draw a line between the two.

In 1934, by replacing an oxygen atom with a sulphur atom, Volvier and Tabern in France created a new 'ultra-short-acting' barbiturate, thiopentone or Pentothal. Once again it was John Lundy who first tried the new drug as an

7. Emil Fischer came from a wealthy business family but chose chemistry as his vocation. Partly for his work on barbiturates he won the Nobel Prize in Chemistry in 1906, three years before his mentor, Von Baeyer, did. Five of Fischer's pupils became Nobel laureates. He died in 1919, aged sixty-seven.

Josef von Mering is best remembered for producing diabetes in dogs by injecting phloridzin which selectively destroys the pancreatic islet cells. This paved the way for the work of Banting and Best on the insulin treatment of diabetes. Von Mering died in 1908, aged fifty-nine.

8. D.A. Cozanitis, 'One Hundred Years of Barbiturates and their Saint', *Journal of the Royal Society of Medicine*, 2004 (97), 594. A brief and excellent review.

intravenous anaesthetic.[9] It proved a hit and remained so, despite the setback of being used inappropriately after Pearl Harbor and adding to rather than reducing the number of casualties.[10] It is still the drug of choice for the induction of general anaesthesia; and to many who have experienced it (and few today reach old age in the developed world who have not) it embodies the magic of surgical anaesthesia. As the liquid enters the vein, patients lose consciousness without any discomfort. They do not normally regain it until they are in the recovery room after the operation.

But quick surgical anaesthesia was only part of the barbiturate saga. By the 1950s no fewer than 2,000 different derivatives of barbituric acid had been patented world-wide and over sixty were being manufactured and prescribed. There was of course never any need for more than six or eight to cover the entire range of speed, duration and intensity of action; but there is no need for more than six or eight brands of toothpaste either. The multiplicity of coloured pills gave a foretaste of bigger changes to come. Barbiturates were by then no longer the only analgesics and mood-setters. Starting back in the 1930s, they had set off a pharmacological fusillade.

...

The amphetamines, the first on the scene, were initially sold over the counter in the 1930s. Their official selling point was their beneficial action in relieving nasal congestion, the discomfort of a blocked nose. But they did more. They were soon recognised as stimulants, more powerful than cocaine and much cheaper. They reduced hunger and especially the comfort eating of the anxious and the worried. This made them weight-reducers. Most strikingly, they built up confidence and endurance. They were capable of enhancing performance even in such allegedly 'unalterable' markers as the standard IQ. (The snag was that the change could be in either direction, accounting for

9. J.S. Lundy, 'Intravenous Anaesthesia. Preliminary Report of Two New Thiobarbiturates', *Proceedings of the Mayo Clinic*, 1935 (10), 536.

10. F.J. Halford, 'A Critique of Intravenous Anaesthesia in War Surgery', *Anaesthesiology*, 1943 (4), 67. Some factual medical papers can conjure up wartime experience as vividly as any old newsreel. The author, a civilian Honolulu general practitioner, was called in to help with the wounded the day after Pearl Harbor. 'The injuries we were attending were all severe. The casualties had not only been blasted by demolition bombs but machine-gunned: most injuries included traumatic amputations of one limb or more, compound comminuted [fragmented] fractures, penetrating chest and abdominal wounds and head and jaw injuries. In several cases both buttocks had been blasted away . . . Nevertheless these were mostly healthy young men with initially good vascular and respiratory capacities.' Equipment for administering oxygen was inadequate and most of the wounded suffered from anoxia. Several hundred were given intravenous barbiturates – evipal or pentothal – and more than thirty died before surgery could begin. The author attributed this to lack of oxygen and circulating blood volume and warned against the use of intravenous barbiturates in this class of patient.

unexpected examination disasters as well as fantastic successes.) They would also reduce the impact of pain on morale. They tended to still 'conventional' doubts, scruples and misgivings. The last made them a boon to military commanders. As with aerial bombs and nerve gases, the Spanish Civil War provided the ideal testing ground. Amphetamines were lavishly doled out by both sides to overcome flagging morale and battle fatigue. They were politically impartial but may have prolonged the conflict by a few months.

But it was the Second World War which saw their potential fully used. Millions of metamphetamine tablets were distributed with the food rations to British, German, Italian and Japanese troops. From Britain most of the initially limited supply went to hard-pressed troops in Africa. It made the heat and the sand more bearable. Did it help to stop Rommel's advance? Perhaps. But the drug also sustained morale on the Home Front, easing the suffering both of the bereaved and of those trapped under the rubble of collapsing buildings. A tablet or two helped to pass the night in crowded improvised air-raid shelters. In 1941 the *Evening News* carried the headline: 'Methedrine wins the Battle of London'.[11]

By 1943 Japanese production reached such a level that at the time of the country's surrender in 1945 an estimated hundred billion tablets were stashed in military warehouses. The stocks disappeared within weeks, flooding the streets first of Tokyo and then of cities in the United States with cheap capsules. They reached millions of delirious users, engendering an estimated 100,000 suicides. For some years they were thought to be partly at least responsible for half the murders in California and the same proportion of self-inflicted mental illnesses. Even by enthusiasts the last was recognised as a distinct drawback. Even small doses in those susceptible could cause delusions, megalomania and violent symptoms of persecution indistinguishable from schizophrenia. In some the mental illness persisted even after the drug was discontinued.

The post-war period shifted the pattern of users to lonely old people, tired housewives and despairing university students. 'Two pills are better than one month's vacation,' one advertisement ran. Dextroamphetamine was still advertised mainly as a painkiller. 'Shrug off your discomforts! Shrug off your exhaustion! Shrug off your pain!' It was probably in the mid-1950s that competitive sportsmen and women began to embrace it. In 1958 a champion cyclist died during a practice ascent in the French Alps fuelled by methedrine. A week later ten participants of the Tour de France fell sick with 'acute amphetamine intoxication'. Two had to be admitted to a mental hospital to

11. Should it have been 'Never in the field of human conflict was so much owed by so many to so small a tablet'?

control their schizoid symptoms. The regulating bodies of international sport began to take note. Routine urine testing was beginning to be canvassed.

...

Following the amphetamines in the late 1930s came the first synthetic opiates, most of them cheap and easy-to-manufacture petroleum by-products. Meperidine, better known as Pethidine or Dolantin, was introduced as a non-addictive analgesic and sold in the United States without prescription for five years. It was a superb painkiller as well as a calmer of nerves, especially effective against intestinal and renal colic. Panic regulation followed reports of the first thousand addicts and a growing number of fatalities. Even after reclassification 19 tonnes of it were being sold every year in the United States alone under eighteen different names.

Methadone was discovered by a German army chemist in 1942 and said to have been named Dolofin by Hitler personally. After initial trials on inmates of the Dachau concentration camp it was deemed too 'unpredictable' to be issued to the *Wehrmacht*; but it became widely available on the black market in most of occupied Europe. Smuggled supplies offered a last escape route to the starving troops trapped in Stalingrad. (The last surrealist order by Field Marshal Paulus made suicide a capital offence.) After the war came ketobemidone, allegedly seven times more active than methadone and classified as 'extra dangerous'. Dextromoramide or Palfium, a euphoriant three times more potent than heroin, was at first advertised as 'manageable, powerful and non-addictive'. Its chemical cousin, meprobamate, was known (at least to its makers) as 'the happy pill granting moral tranquillity and total freedom from chronic pain without addiction'. By 1960 seventy similar preparations – including normethadone, morphinone and dihydromorphinone – were listed in pharmacopoeias. All had been marketed for some years as 'unrelated to opium and therefore non-addictive'. It was against such a background that thalidomide was launched in 1957, a wonderful and wonderfully profitable sedative to relieve pregnant women of morning sickness 'without any of the usual side-effects'.[12]

...

A new breed of 'pain-killing anxiolytics' made a discreet entry in the 1950s and 1960s. A 'beatific hallucinogen' had been synthesised by Albert Hofmann

12. The drug, as is well known, was advertised as particularly suitable for pregnant women. Women taking it during the first trimester gave birth to babies without fingers, hands, arms, legs or all four limbs. The United States was saved from the disaster by a middle-ranking employee in the Food and Drug Administration whose suspicions were aroused by stray reports and who 'sat' on the licence to sell it. She was sacked before being acclaimed a year later as 'Woman of the Year'. Nobody was ever prosecuted.

in 1943 from a fungus of the ergot plant. Chemically it was the diethylamide of lysergic acid, soon to be known as LSD 25 or simply LSD. This extraordinary substance acted in doses of thousandths, not hundredths, of a gram and, since the mould was ubiquitous in grain cultivation, its cost was negligible. After the war Sandoz began to market it under the name of Delysid and distributed samples as a free gift to psychiatrists and psychotherapists 'throughout the civilised world'. As the promotional literature put it, the drug was 'to provoke liberation of repressed psychic material in the patient, to provide mental relaxation . . . and to induce short-term experimental psychosis in normal subjects'. In 1956 the Chemicals Division of the CIA embarked on a massively funded research project to test its potential in case of a 'surprise attack on the United States by anti-American elements'. The project continued until 1960, by which time most of those involved had become users or addicts. Since the report's main conclusion was that 'the effect is essentially a joyous disturbance of the ego function', its military potential was finally judged to be negligible if not counterproductive.

Though in the early 1960s the number of papers published about LSD every year was said to exceed the number devoted to any other drug, the 'hallucinogenic painkillers' always remained a minority cult.[13] The most famous propagator of the 'psychedelic' experience was Aldous Huxley whose book *The Doors of Perception* became a bestseller. He wrote in 1959:

> I suspect that these drugs [mescaline and similar preparations] are destined to play a role in human affairs at least as important as that of alcohol up to now and an incomparably more beneficent one . . . It enables one to know by experience that 'God is love' feeling, that, in spite of death and suffering – everything is, in some way and in the final instance, all right.[14]

For some years the drug became an emblem of the 'counterculture' in universities on both sides of the Atlantic. In his last article, published in *Playboy* at its peak circulation, Huxley wrote:

> With the help of these drugs – as well as rigorous intellectual discipline – the individual could selectively adapt to his culture, reject the undesirable, the stupid and the senseless, while accepting the treasures of accumulated knowledge, rationality, compassion and practical wisdom. The way might be open to universal and perpetual peace.[15]

13. A historical precedent was the 'Club des hashishiens' founded by Baudelaire which enjoyed a vogue among the 'happy decadents' of the Paris Left Bank in the 1890s. Yeats and O'Neill among others claimed to have been inspired by 'drug-induced dreams'.
14. A. Huxley, *The Doors of Perception* (London, 1956).
15. Quoted by A. Escohotado in *A Brief History of Drugs*, translated by K.A. Symington (Rochester, NY, 1996), p. 112.

Huxley himself knew that he was dying of cancer of the throat and asked for the drug as terminal therapy. A prophet of peace, he passed away peacefully under its influence, believing that the world was at last heading for sanity. The news of his death was eclipsed by the assassination of President Kennedy a few hours later.

...

Despite such rival attractions, until the 1960s barbiturates remained market leaders. Indications for the drug steadily expanded. Barbiturate coma as a therapy in mental illness had first been introduced in the late 1920s and paved the way for what became known as the 'physical treatment of psychological disorders'. Schizophrenic patients were put to sleep with a cocktail of barbiturates for several days with apparently 'brilliant results'. Barbiturate coma was soon followed by insulin coma, epileptogenic drugs, electroconvulsive therapy and finally lobotomy. All were received with enthusiasm and not put out of business until the advent of the antihistamine sedatives in the 1950s.

The more conventional use of barbiturates as mild analgesics and pleasant soporifics continued its upward trend. In 1939, 100 tonnes were manufactured in the United States. By 1960 the figure had doubled. There seemed to be no plateauing. The sales department at Lilly's reported that 'some jobbers were ordering barbital in lots of $1000 each'. One specific use of the drug became 'visible'. Phenobarbitone became the most widely used and most effective antiepileptic drug, controlling fits as no other treatment had done before. Epileptic convulsions in schools, theatres, on beaches and on public transport became a rarity. Before the 1950s few schoolchildren growing up in big cities like London, Paris or New York had not seen a fit. After the 1950s few had.[16]

But this was not the whole picture. Over the ten-year period from 1928 to 1937 an analysis of one million hospital admissions to ten centres in the United States showed that 143,000 were for barbiturate intoxication or overdose. The corresponding figure for the six-year period 1940–46 was an ever more alarming 200,000. The statistics did not include 'successful' cases of suicide where the victims were not admitted since they were DOA – dead on arrival. In New York City alone one death every thirty-six hours was due to barbiturate 'misuse' or abuse (the misuse covering unsuccessful suicides). In

16. With the decline of epileptic fits in public places came the decline of fits as a manifestation of hysteria (conversion disorder, in today's politically correct parlance). Phenobarbitone was in short supply in many Communist countries and the frequency of epileptiform seizures in Hungarian refugees streaming to the West after the 1956 uprising and often under severe mental stress astonished Western doctors.

Hollywood the multitude of victims included comedienne and sex symbol Marilyn Monroe. Nembutal was her (or her murderer's) favoured drug. The figures reflected not the frequency of epilepsy, pain or neurological disease but a revolutionary new phenomenon. Millions of apparently normal individuals today are permanently on analgesic, sedative, tranquillising, 'mood-setting' or similar drugs, often several. The barbiturates blazed the trail.

Attempts at legal restriction began during the last years of the Second World War but ran (or rather crawled) an uphill course. Any form of legislation was opposed by the pharmaceutical industry; and in the United States uniform state laws, initially believed to be easier to administer than federal legislation, proved too expensive to enforce. The 'illegitimate' use of barbiturates then came to austerity Britain and war-ravaged Europe. Everywhere ordinary people – ordinary that is except for being in pain, desperate, frustrated, lonely, sad, foolish or all of those in varying degrees – were gulping 'red dogs', 'nebbies', 'goof pills', 'yellow jackets', 'angels', 'reds' and 'downers' at a rate never seen before with any other drug. Lilly's quinalbarbitone or Seconal was especially popular. Tens of millions of 'seccies' were imported illegally to the United States from Mexico and to Western Europe from North Africa. Eventually, the 1971 US Controlled Substances Act divided barbiturates and other narcotics and analgesics on the basis of their potential for abuse and habit-formation and attempted to regulate their prescription accordingly. (Ultrashort-acting barbiturates like thiopentone are useful in anaesthesia but have virtually no habit-forming effect, while long-acting ones like amylobarbitone have the highest addictive potential.)

By the late 1960s Britons, increasingly prosperous and cool, had caught up with their transatlantic cousins. About 24,000 prescriptions for barbiturates were being issued by doctors in England and Wales every year. Phenobarbitone combined with an amphetamine, the first combination known as the 'purple heart', was a star performer. It caught the upbeat if also at times desperate mood of the swinging decade. The Campaign for the Use and Restriction of Barbiturates (CURB) was launched and perhaps did some useful educational work. Yet in Britain the number of hospital admissions for barbiturate overdose did not begin to fall till the early 1970s. The decline was then rapid, whether as a public response to government advertising – the first venture of its kind – or, more probably, because of the introduction of new and more effective successor drugs.

...

A new and valuable use for phenobarbitone was discovered in 1980s for newborn babies with jaundice. Acting on the still immature liver the drug promotes the induction of glucoronyl transferase (and perhaps of other enzymes) necessary for the excretion of the yellow pigment, bilirubin.[17] Slight jaundice due to the destruction of foetal red cells is normal after birth; but when the circulating bilirubin level is too high or the elevation is too prolonged, deposition of the pigment becomes a possible cause of brain damage. Here at last was a development which might have pleased St Barbara.

17. J.H. Hutchinson, *Practical Paediatric Problems* (London, 1980).

# Chapter 41

# TIC DOULOUREUX

Sometimes a particular disease, not very common itself but 'representative' and dramatic, comes to dominate basic research and clinical developments in a wider field. This was true for more than a century of the condition known as tic douloureux. The syndrome is one of many in which pain originates in the nerves themselves. Comparisons in terms of severity have often been attempted between different kinds of physical suffering. The results have been unconvincing. There is nevertheless one generalisation on which most observers agree: pain generated in the pain-transmitting organs can be more devastating than any other. The term 'neuralgia' has been applied to this group of illnesses since the early ninteenth century, but the inventor of the name is unknown. Many and varied causes are now recognised.[1] One reason for the evil reputation of these afflictions is their resistance to treatment. Not surprisingly, they have been the syndromes in which the most drastic form of symptomatic therapy, the surgeon's knife, has been tried earliest and most often. For long neurosurgery has in fact been the surgery of pain relief, and pain relief most commonly meant relief from tic douloureux or trigeminal

1. The 'causes' are more correctly associated diseases: how they cause the neuralgic pain is in most cases unknown. Post-herpetic neuralgia follows shingles, a virus infection, sometimes by many months or even years. It can be tormenting. Causalgia follows trauma to a nerve and can be devastating and resistant to almost any form of treatment. 'Phantom limb' pain, a sensation felt in a part of the body which has been removed by injury or by surgery, has been recognised since Ambroise Paré and remains distressingly common both after civilian and war injuries. It can even occur in a 'phantom breast' after mastectomy. An intractable form of neuralgia sometimes accompanies metabolic diseases which are otherwise treatable or at least controllable. President Nasser of Egypt spent much of his last years travelling to medical centres in Europe and the United States trying and hoping in vain to be cured of his painful diabetic neuropathy.

neuralgia.[2] It is another form of treatment which dates back to the astonishingly productive decades immediately before the First World War. But the history of the disease, the most famous – or notorious – of the neuralgias, reaches further back.

To understand the symptomatology and the main forms of treatments attempted it is necessary briefly to outline the underlying anatomy. The two trigeminal nerves convey sensation from almost the whole of the face and the head, including the corneas of the eyes, much of the mouth and the teeth. The name derives from the fact that each nerve has three branches. The first or supraorbital branch 'emerges' from inside the skull through an opening above the orbit. It supplies with sensory fibres one side of the scalp and forehead and one cornea.[3] (Vision is conveyed from the retina separately by the optic nerve.) The second and third, the maxillary and mandibular branches, emerge through separate small openings in the base of the skull. They supply the rest of the face, most of the earlobe, the lips and the gums. The dividing line between the maxillary and mandibular territories runs roughly from the angle of the mouth to the external opening of the ear. Inside the skull the three branches converge on a structure known as the Gasserian ganglion.[4] About the size of a pea, it is situated in a small hollow at the base of the skull, covered from view when the skull is opened by the temporal lobe of the brain. Because of their position near the midline, the ganglia are difficult to access. The sensory nerves are the axons of cell bodies situated in these ganglia. From the ganglia new sets of fibres depart backward as single nerve *trunks* and convey impulses to two collections of neurons in the midbrain. These are know as 'trigeminal nuclei'.[5] From there a third set of axons radiates to the sensory cortex. The mandibular branch of the trigeminal nerve also carries motor fibres to the muscles of mastication but these play no part in trigeminal neuralgia. The ophthalmic and maxillary branches have no motor component. (The non-masticatory muscles of the face are supplied by the facial – or

2. The head, including the sense organs like the eyes, the ears, the buccal mucosa (for taste) and the nose (for smell) are supplied with both sensory and motor nerve fibres by twelve pairs of cranial nerves. The trigeminal nerve is the 5th. Neuralgia very similar to the trigeminal variety can occasionally occur in the other cranial nerves (like the 9th or glossopharyngeal nerve).
3. Since the supraorbital nerve is a purely sensory nerve carrying impulses from the periphery to the brain, it would be more correct but a little pedantic to describe it as 'entering' the skull.
4. 'Gasserian' is one of the most widely used eponyms in medicine, yet all that is known about Johann Laurentius Gasser is that he was a professor of anatomy in Vienna who died still in his thirties in 1765. The trigeminal ganglion was named after him by his admiring pupil, Aloysius Hirsch.
5. To repeat what has been said in an earlier chapter, these 'nuclei' at the base of the brain are collections of neurons and have nothing to do with cell nuclei.

7th cranial – nerve.) The neuralgia can affect any one (or, very rarely, two or all three) of the branches of the nerve on one side, the supraorbital branch being most commonly involved.

Trigeminal neuralgia has probably been known for centuries but the term 'tic douloureux' was coined in 1757 by the French surgeon Nicolas André. He described it vividly as

> un douleur atroce avec des grimaces hidieuses qui mettent un obstacle invincible a la réception des aliments, qui éloigne le sommeil et qui lient l'usage de la parole . . . Dès qu'on considère une personne qui ressent actuellement une attaque on la voit tenir les sourcils foncés, les deux paupières fortment comprimées, et la commissure des lèvres retirée vers l'oreille comme dans un rictus Sardonique . . . Sa respiration est lente et comme suspendue, souvent elle n'ose faire le moindre cri, ni proférer une seule parole. Elle semble même redoubter le plus léger movement du tronc . . . Ses attitudes forcées, et son état presque extatique expriment mieux que tous les discourse la force surhumaine et incomparable de la douleur. Aussi dès qu'elle a libre, elle ne manque guerre de se plaindre du ton le plus lugubre, les termes les plus énergiques. . .' [an atrocious pain accompanied by hideous grimaces which create an irremovable obstacle to the intake of food, which prevent sleep and which inhibit speech . . . And when one sees a person during an actual attack, one sees the victim with the forehead creased, the eyes tightly shut and the mouth drawn back toward the ears as in a sardonic rictus. Respiration is slow, nearly suspended. Often she dares not utter the least cry nor say a word. Indeed, she may not dare to make the slightest movement with her body. Such a state, such a tormented attitude expresses better than any description the terrible, indescribable severity of the suffering. Moreover as soon as the attack is over, she inevitably starts to complain in the most dire tones.][6]

Twenty years later Samuel Fothergill, a Quaker physician practising in Yorkshire, gave a masterly and more detailed description. He stated that the condition afflicts mainly people over fifty, women slightly more commonly than men, and starts for no obvious reason.[7]

> The pain comes on suddenly and is excruciating. Each pain lasts for a short time, perhaps a quarter or half a minute, but then returns at irregular intervals, the whole attack lasting for quarter or even half an hour. There may be two, three or more repetitions during a day . . . The attacks are more

6. N. André, *Observations sur les maladies de l'urèthre et sur plusieurs traits convulsifs* (Paris, 1756), p. 23.
7. Fothergill was right about the age distribution but the condition can occur in adolescents and the present writer has seen it first develop in a woman of ninety-two.

> frequent during the day than in the night, probably from there being fewer triggers of irritation; and they are more frequent during conversation than silence; and still more frequent at times of mastication . . . Talking or the least motion of the muscles of the face affects others. The gentlest touch of the hand or a handkerchief will sometimes bring it on while strong pressure may have no effect . . . The eye may be red, inflamed and watery as in extremely severe odontalgia: in other cases dry. Copious flow of saliva may succeed a paroxysm . . . There is no known way of terminating the attack.[8]

Later workers have added details like the occasional twitching and the involuntary running of tears. The muffled, terrified scream. During the attacks the victims cannot talk. Afterwards they may describe the experience as 'like fireworks exploding inside my face'. Most have experienced other pains. This is more agonising than any other. The paroxysm can be triggered by a whiff of cold air, by washing the face, by eating, by putting in or taking out dentures, by any sudden jarring motion. There may be certain 'trigger zones' like the margin of the lip or the edge of the nostrils; but the pain is always one-sided and strictly confined to the area supplied by the affected division of the nerve.

> Such patients come into the doctor's consulting room walking rigidly, speak in a soft frightened monotone without moving the jaw, indicate the site of the pain with the forefinger without touching it. Often the cheek on the affected side shows accumulated desquamated skin cells and dirt from lack of washing and when the third [mandibular] division is affected one side of the tongue may be furred. After an attack has lasted one or two weeks the patients will be wasted from suffering and from lack of food and sleep and may beg to be put out of their misery with a quick-acting poison. Some take their own lives.[9]

It is necessary to grasp the suffering to understand the risks of the attempted countermeasures. It is also necessary to appreciate that, despite much and varied research, the underlying cause is still not known and therefore cannot be treated. By the time of Fothergill's description, a variety of remedies, including electrical shocks, had been tried and found wanting. René Leriche stated without mentioning the source that Jean-Baptiste Maréschal, Louis XIV's surgeon, tried to cut the nerve which appeared to lead to the affected area, probably the supraorbital.[10] Several others have reported

8. S. Fothergill, *A Concise and Systematic Account of a Painful Affliction of the Nerves of the Face Commonly Called the Tic Douloureux* (1777; reprinted in London, 1804).
9. E.C. Warner (ed.), *Savill's System of Clinical Medicine*, 13th edn (London, 1950), p. 1223.
10. R. Leriche, *La Chirurgie de la douleur* (1937; translated by A. Young as *The Surgery of Pain*, London, 1939), p. 65.

sectioning the 7th or facial instead of the 5th or trigeminal nerve based on a disastrous misinterpretation of the anatomy. This would have caused almost total paralysis of one side of the face without affecting the pain or sensation in any way.[11] Not surprisingly, the reports tend to refer rather vaguely to experiments by 'other practitioners'.

Not till the work of François Magendie in France and Charles Bell in England was the distinction between sensory and motor nerves understood and the sensory supply of the head more or less accurately mapped out. Some of the most eminent surgeons during the second half of the nineteenth century – including Malgaigne in France and Langenbeck in Germany – tried various kinds of nerve resection usually as the nerve emerged from the skull but sometimes removing part of the skull to access the nerve in its intracranial course. In some cases the pain was temporarily eased but almost invariably recurred a few weeks later even in areas which had been deprived of all normal sensation. The condition became known as *anaesthesia dolorosa*, or pain without sensation. In 1858 James Syme of Edinburgh wrote:

> The profession now being fully convinced of the inefficacy of dividing nerves for tic douloureux have abandoned all such attempts and trust entirely to the influence of a regulated diet together with tonic medicines affecting the general state of health of patients.[12]

...

Not all surgeons were so pessimistic (or wise). Many, especially in Germany, continued to try and remove segments of the nerves both inside and outside of the skull, getting as close as possible to the Gasserian ganglion. The results were almost uniformly dismal. Even when the attacks of pain temporarily subsided, they almost always returned, sometimes with greater frequency than before. The rate and range of complications resulting from loss of normal sensation were formidable. Insensibility of the cornea in particular often led to infection, scarring and blindness. Most standard textbooks of the mid-nineteenth century refer a little vaguely to 'scores of operations' which have been tried unsuccessfully and have never been reported in detail in the surgical literature.

...

Neurosurgeons in most countries regard Victor (later Sir Victor) Horsley as the founder of modern neurosurgery. To the average medical student or

11. The 7th or facial nerve is a purely motor nerve which supplies most muscles of the face. 'Bell's palsy', still a common and usually transient condition, is a partial paralysis of this nerve.
12. J. Syme, *Principles of Surgery*, 3rd edn (London, 1842), p. 436.

doctor his name is associated – if it is known at all – with Horsley's wax. Though not his main claim to fame, it does illustrate his ingenuity. One of the obstacles to successful operations inside the skull was the continued oozing of blood from the cut edges of the skull bones. The bleeding points could neither be tied nor cauterised. In 1894 Horsley carried out a series of experiments on rabbits which convinced him that ordinary modelling wax smeared on the edges of the bones would control the haemorrhage. With the help of P.W. Squire, the Oxford Street pharmacist, he evolved a simple formula: 7 parts beeswax and 1 part almond oil. It is still used in cranial surgery.

Horsley was then twenty-eight, already, surgeon to University College Hospital and the National Hospital, Queen Square, London, and widely recognised as one of the rising stars of surgery in England.[13] He was also known as one of the rudest and most demanding. Assistants who failed to read his thoughts, anticipate his wishes or correctly interpret his grunts at operation were reduced to tears within five minutes of the first incision. Yet he could be extraordinarily tender and compassionate to frightened patients, especially those who, he thought, had been ill-treated by the profession. Three years later, in 1897, he removed a tumour of the spinal cord accurately diagnosed and located by his medical colleague, Sir William Gowers. The patient survived. It was a landmark operation, the first carried out successfully inside the central nervous system. By then for many years he had been conducting animal experiments to see how tic doloureux might be improved if not cured. By the 1880s it was clear that the best if not the only hope for victims lay in an attack not on individual nerves as they emerged from the skull but on the ganglion itself or the nerve root leading from the ganglion to the brain. Operations had been tried for some years, trying to pull out the ganglion through burr-holes in the skull; but success had been variable and the complications and death rate formidable. Horsley embarked on a new and even more audacious procedure. He made a large hole in the side of the skull, lifted up the temporal lobe of the brain and located the posterior root of the nerve emerging from the ganglion. He cut it, a procedure later known as a retrogasserian neurotomy. His first patient died of shock but two others survived.[14] Other surgeons in other countries, notably Fedor Krause in Germany and, above all, Harvey

13. He carried out valuable work on the surgery of the thyroid gland and on the control of rabies. Politically active, he was an ardent supporter of women's suffrage and other liberal causes. His support was not always appreciated. He did not suffer fools gladly and often antagonised even his fellow campaigners. Lady Horsley recalled after his death that he could silence with a glance not only ill-informed students but 'his long-suffering wife' as well.
14. V. Horsley, R.H. Clarke and V.S. Colman, 'Remarks on the Various Surgical Procedures Devised for the Relief of Trigeminal Neuralgia (Tic Douloureux)', *British Medical Journal*, 1891 (2), 1139.

Cushing in the United States, took up and developed the operation. By the early years of the twentieth century it was even possible to cut the trigeminal root only partially, trying to preserve the nerve supply to the cornea.

Horsley died in 1916 while serving with Allenby's army in Mesopotamia;[15] but numerous other surgical procedures were developed in the inter-war years, some aiming at the nucleus of the fifth nerve inside the mid-brain or even at connections with the sensory cortex. An alternative approach was pioneered by Pitres in 1902 who injected alcohol into the Gasserian ganglion.[16] This was followed by temporary improvement. Alcohol was later replaced by glycerol, less corrosive of surrounding structures. Under X-ray control it later became possible to approach the ganglion with a needle or electrode without actually exposing it. So-called stereotactic surgery, requiring high-technology equipment, can now destroy the ganglion partly or completely by focusing intersecting irradiation on the site from several sources. Some normal sensory loss is inevitable whatever surgical method is used.

Hope for a medical cure has emerged over the past twenty-five years. There is a certain similarity between idiopathic epilepsy in which a group of neurons fire off involuntary and purposeless motor impulses giving rise to a fit and sensory neuralgias in which apparently normal sensory cells generate unprovoked painful sensations. The similarity has led to trials of antiepileptic drugs in neuralgias. They now provide at least partial or temporary relief in about half to two-thirds of patients and are generally the first line of treatment.[17]

. . .

Trigeminal neuralgia is fortunately not common but operations for it led the way to other forms of surgical pain relief. These in turn opened the door to 'psychosurgery', the surgical treatment of mental disease. It will be considered in a later chapter.[18]

15. He died of exhaustion and heatstroke in Amarra in 1916, aged fifty-nine, and was given a funeral with full military honours. Despite his sharp tongue, he was (as many later testified) much admired by ordinary soldiers, who trusted his professional judgement. By contrast, he was 'deplorably careless of uniform and military etiquette': Field Marshal Lord Allenby described him as 'the brainiest but worst-turned-out medic in the British army'. H. Bailey and W.J. Bishop, *Notable Names in Medicine and Surgery* (London, 1946), p. 98.
16. Described in A.E. Walker (ed.), *History of Neurosurgery* (London, 1951), p. 343.
17. J.D. Loeser (ed.), *Bonica's Management of Pain*, 3rd edn (Philadelphia, 2001).
18. See Chapter 44.

## Chapter 42

# TWILIGHT SLEEP

One episode of pain relief during the Seminal Years was, on the face of it, less significant than the launching of modern synthetic analgesics or the beginning of neurosurgery; but it was a portent. Few descriptive names in anaesthesia had such a beguiling ring as *Dämmerschlaf* or Twilight Sleep. In 1906 two obstetricians from the University Frauenklinik in Freiburg in Germany, Carl Gauss and Bernhardt Krönig, reported a new method of inducing a state of 'clouded consciousness' in childbirth which they described by this evocative term. They had followed the procedure in 500 carefully observed cases and, by using it, they were able to 'limit the suffering of women in labour to the lowest minimum imaginable'.

> The objective was attained without disagreeable secondary effects, without substantial interference with the labour itself, without danger to the mother and without injury to the child.[1]

The method was simple, at least on paper: it consisted of an injection of a small dose of morphine and the alkaloid scopolamine. Six years later the news hit America through articles in *McClure's Magazine*, the *Ladies' Home Journal* and the *Woman's Home Companion.* The last two had sent special correspondents to Freiburg. The emissaries talked to mothers who had been delivered by the Freiburg method and one of the two correspondents, Mary Boyd, seized her chance and had her own baby there. Their experience confirmed the unanimous testimonies of 'dozens of other mothers'. One woman described the pleasant drowsiness merging into an overpowering sleepiness. 'There is never a disagreeable sensation associated with it.' Another gave a more detailed description:

1 Quoted in Robinson, *Victory of Pain* , p. 268.

> I did not even feel the injection of the scopolamine, for they first used cocaine on the spot before using the hypodermic needle. Very soon after I found myself getting drowsy and, in about half an hour, I went to sleep as naturally as I do any night when going to bed. The next thing I knew was hearing the sympathetic voice of Dr Krönig telling me that all is well. I thought to myself 'I wonder how long before I shall begin to have the baby' and, while I was still wondering a nurse came in with a pillow and on the pillow was my baby . . .[2]

The issues of the magazines which carried the articles sold in record numbers; and the editors' mail-bags were mountainous. All the letters voiced outrage or at least puzzlement. Not only was it clear to readers that the method should be adopted at once in the United States. Many also wanted to know why womanhood outside Germany had been kept in the dark by the medical profession for so long. Why had the public had to learn about so momentous an advance through women's magazines? The news was taken up by the British press and caused an uproar almost as shrill.

There were grounds for indignation; but the issues were not quite so simple. The combination of morphine and scopolamine as an anaesthetic had been tried before Gauss and Krönig by Robert von Steinbüchel in Graz in Austria and, less thoroughly, in some other obstetric units.[3] None were impressed. Scopolamine on its own had been known since Antiquity as a poison, causing in sub-lethal doses excitement, confusion, hallucinations and occasionally coma. In combination with small doses of morphine its most striking effect was the erasure of memory. Although the patient might be aware of the pains and indeed scream in agony, she would have no recollection of what had taken place under the drug's influence. Gauss and Krönig would not let any member of the patient's family remain within earshot since, they readily admitted, the patient's cries might be as distressing as if she had received no painkiller. The story was told of one mother who had violently abused her doctor all through her Twilight Sleep, wildly declaring that 'if this is your Twilight Sleep I don't want any of it', until the doctor resolved never to try the method again. Yet some hours after the baby's birth the same woman would hardly believe that labour was over and fulsomely expressed her surprise, delight and gratitude. She would never have a baby by any other method.[4]

2. M. Tracy and C. Leupp, 'Painless Childbirth', *McClure's Magazine*, 1914 (413), 31.
3. R. von Steinbüchel, 'Vorläufige Mitteilungen über die Anwendung von Skopolamin-Morphium-Injectionen in der Geburtshilfe', *Zentralblatt für Gynaekologie*, 1902 (30), 1304.
4. D. Caton, *What a Blessing She Had Chloroform* (London, 1999), p. 159.

Some doctors argued that it was wholly misleading to call Twilight Sleep 'painless' since all it abolished was the memory of the pain; but Gauss and Krönig maintained that it was the memory which mattered most. This was an important general problem (as it still is) and Twilight Sleep provided the first full-dress occasion for its airing. Unfortunately, doctors arguing general philosophical propositions are rarely at their most illuminating. Gauss and Krönig admitted, indeed they insisted, that their method had to be followed to the letter if it was not to disappoint. Experiments trying to standardise the dose of scopolamine were unsuccessful, so that no routine could be prescribed. This meant that Twilight Sleep required the continuous attendance of an obstetrician or a nurse trained in the technique. Whether the patient had reached the 'twilight stage' or not could only be established by 'memory tests' started half an hour after the first injection of scopolamine and carried out every twenty minutes thereafter. Even though the patient might remain rational in every other respect, she had reached the twilight stage only when she could not remember an object that was shown to her seven or eight minutes earlier. Gauss and Krönig would not compromise.

> Where it is impracticable for the obstetrician or . . . for a skilful and trained obstetric nurse to maintain close observation for the whole of the labour, nothing but repeated failures can be looked for.[5]

Even in those leisurely days this was a tall order.

. . .

But the passions aroused by Twilight Sleep transcended the medical argument. In different ways, some local and some national, American womanhood had been on the march since the end of the Civil War. Associations like the New England Women's Medical Society, the Visiting Nursing Service of the City of New York, the New York Nursery and Child Hospital, the New England Hospital for Women and Children, the Philadelphia Women's Health Association, the Washington Women's Health Society and others set up on such models but, known only to their local community, had been supporting women's hospitals, promoting nurses' training, agitating for the right of women to enter the caring professions. But their influence was in fact wider. They publicised the suffering of abused mothers and the grinding misery of overworked housewives. In the past, religious differences and diverging political aspirations had kept them apart. Now, for the first time, their sense of mission coalesced around the most universal of women's experience, the pain of childbirth. The National Twilight Sleep Association cut across social

5. W.H.W. Knipe, 'The Freiburg Method of Dämmerschlaf or Twilight Sleep', *American Journal of Obstetrics and Gynaecology*, 1914 (70), 884.

barriers. Eight of the twenty board members were listed in *Who's Who among American Women;* but the list also included an elementary school teacher from Spokane, a dental nurse from Dallas and a 'mother of twelve and wife of a miner' from New Hampshire. Of course this was tokenism before the word was invented; but it also had real significance. Shared suffering can be a powerful unifier. The torment of labour was an issue on which women in all walks of life could agree. Nor were the declared aims of the Association narrowly limited to childbirth: 'Women are more sensitive both by nature and by experience to mental and physical suffering. This makes them better than men at relieving pain.'[6] As for the alleged 'benefits' of labour pains, the old shibboleth with its faintly biblical overtones no longer cut much ice. Times had changed. Men were no longer obliged to go out with clubs and face the mammoth. Nor would the average congressman in Washington or stock-broker in Wall Street shine at such a task. Nor, happily, was it necessary for them to do so. But why should not women too benefit from thousands of years of advances in science and technology?

To many the answer to the last question was all too clear. Articles and editorials in some of the country's most staid newspapers accused American doctors of rejecting Twilight Sleep because it had been promoted by patients rather than by the medical establishment, because they were too prone to procrastinate to learn about new developments and because, deep down, they were callously indifferent to the pain of women in labour. The normally restrained and dignified *New York Times* declared:

> Had doctors thought less of their own comfort, convenience and bank balance and more of sparing women pain that was always agonising and peril that was often fatal, they could long since have freed childbirth of its terrors and primal curse.[7]

Searing words. Doctors, it was also claimed, were rejecting Twilight Sleep as they had rejected other major medical advances, such as vaccination against smallpox and the contagious nature of puerperal fever. Twilight Sleep was safe though it needed more care than doctors were currently prepared to devote to their patients. It was time to change all this.

The outcry in Europe was more muted. Perhaps Freiburg was nearer; perhaps European womanhood was more conditioned to respect professional wisdom (or more cowed, as their American sisters would have it). There was nevertheless a groundswell of opinion demanding action; and in 1906 the Berlin Charité sent Franz Hocheisen, an experienced obstetrician, to Freiburg to study the method. Returning to Berlin after several weeks at the shrine,

6. Quoted in Caton, *What a Blessing*, p. 169.
7. *New York Times*, 17 September 1914, p. 18.

Hocheisen carried out a series of experiments of his own. He eventually reported to a meeting of the German Obstetrical and Gynaecological Society, the most prestigious of its kind on the Continent. He had found that in a hundred cases of *Dämmerschlaf* the 'expulsion period' averaged 6 hours 15 minutes against the normal 1 hour 45 minutes and that the protracted birth 'must be expected to cause an increased risk of asphyxia'. The Charité had a high reputation and Hocheisen's report was widely quoted. Gauss and his partisans, on the other hand, justly pointed out that the Berliners had 'as usual' not followed the method in every detail and could therefore not be relied on to evaluate it. This was not surprising 'since Herr Doktor Hocheisen had spent most of his time in Freiburg on the tennis court'. Eventually Hocheisen declared that

> if I am right, scopolamine and Gauss's cleverly advertised *Dämmerschlaf* will disappear from the scene as many other loudly advertised methods have done. If he is right, then within a few years all women will share painless birth without risk and I shall have to admit that I have been wrong.[8]

...

But once again, the matter was not quite so simple. Twilight Sleep in a less than dedicated large hospital was undoubtedly beset with difficulties. It required from those conducting the birth the acquisition of an exacting technique and exceptional devotion to the ideal. 'Twilight Sleep,' Gauss said, 'is a narcotic condition like a narrow mountain crest. To the left lie the dangers of too great a depth with the absence of birth pains, to the right too shallow an effect with sensibility to pain.' Even in Freiburg the method was not foolproof. In Gauss' series of 3,000 cases reported in 1911, 82 per cent of those accommodated in well-furnished sound-proofed dimly lit rooms had a successful painless delivery but only 52 per cent of births in open wards were entirely satisfactory. The procedure was wholly inapplicable to home delivery. Gauss himself wrote that the environment had to be carefully adjusted. The dose of scopolamine had to be judged as each labour progressed. Even in the first decade of the century one person rarely stayed on duty in a hospital for more than twelve hours. This meant that not just one obstetrician or midwife but an entire team had to be conversant with the method and in tune with each other's practices.

...

8. F. Hocheisen, quoted in Robinson, *Victory over Pain*, p. 273.

But the innovation had created a new situation and started to change fundamental attitudes. As the *Literary Digest* commented:

> Probably the lay public has never interested itself so much in what, on the face of it, was a purely surgical matter. We have not in the past seen articles in newspapers and magazines calling on the medical profession to adopt this or that procedure in operations for appendicitis or cataract. But many lay writers clamour loudly for this particular method and assert that doctors who do not agree with them are malicious, negligent, incompetent or lazy.[9]

Nor were the 'many writers' people of no consequence. From the United States Mrs C. Temple Emmet, a granddaughter of the Astors, travelled to Freiburg for the 'wholly painless' delivery of each of her three children. She was not alone. 'Freiburg babies and their happy mothers', many of them distinguished, wealthy and socially concerned, continued to feature in magazines, their advocacy finding an echo in other women's movements. The testimony of the mothers was unequivocal. Krönig was an aggressive and charismatic man – apart from Twilight Sleep he was the first to use X-rays in obstetrics, against stiff opposition – and he was particularly scathing about the argument that what was good enough for countless generations of women should surely be good enough today. Nor did he accept the logic that labour pains never deterred any woman from having a second baby.

> The modern women on whose nervous system far greater demands are made than was the case in former times responds to pain more rapidly with nervous exhaustion and paralysis of the will.[10]

Greater susceptibility to pain made such women more needful of relief; and Twilight Sleep offered just that in a safe way.

In response, safety concerns were overemphasised by the profession. This always happens when doctors feel sheepish. But some undoubtedly feared that opioids would 'paralyse the nerves' responsible for smooth labour. On this topic leaders of the profession could quote up-to-date physiological research as well as, more ominously, 'centuries of experience'. In academic circles nervous regulation was the flavour of the decade. Nerves and reflexes had recently been shown to control the heart rate, peristalsis, respiration, perhaps even thought. Sir Michael Foster of Cambridge, head of a famous school of physiology, had stated categorically: 'The whole process of parturition may be broadly considered a reflex act.' W.H. Howell at Johns Hopkins, agreed. So did

9. *Literary Digest*, 1912 (23), 68.
10. B. Krönig, quoted in Robinson, *Victory over Pain*, p. 273.

his clinical colleague, J. Whitridge Williams, professor of obstetrics.[11] In his influential textbook Williams quoted 'extensive experimental evidence', in itself a highly innovative, almost revolutionary departure in a clinical textbook. It 'implicated brain nuclei', especially the thalamus, in the close control of every stage of labour. This was awesome. Most middle-aged obstetricians had never heard of the thalamus. Williams described the nervous pathways which ensured that dilatation of the vagina led to uterine contractions and speculated that anaemia and stress, among other 'nervous influences,' might be responsible for premature labour.[12]

They were all wrong. Within ten years parturition would be revealed as being almost entirely under hormonal, not nervous control. Patients with complete paraplegia could have a normal labour and epidural and related methods of anaesthesia blocking local reflexes had virtually no effect on uterine contractions. But this was in the future: indeed, the complex relationship between hormonal and nervous control is only now beginning to be unravelled.

A stronger ground for concern was the possible effect of opioids on the foetus. The painstaking work of Paul Zweifel, a Swiss physician in the 1880s, had left little doubt that many drugs, including such habit-forming ones as heroin, could cross the placental barrier from mother to foetus. Babies could be born not only syphilitic but alcoholic too. Cases of 'neonatal narcosis and asphyxia' were reported after Twilight Sleep by keen young obstetricians not averse to pleasing their sceptical chiefs. This was probably nonsense. Even by the most cautious standards the amount of morphine used by Gauss and Krönig in the mother in the single initial injection posed no threat to any infant. But that could not be known with certainty at the time.

. . .

The campaign for Twilight Sleep collapsed almost as suddenly as it had started. The reasons had little to do with reflexes, foetal narcosis, opioids or even memory. One of the most effective canvassers of the method, Mrs Francis X. Carmody, died during her second labour from haemorrhage. Both her husband and her obstetrician declared that this was wholly unrelated to Twilight Sleep; but it cast a pall. The death also provided an occasion to rehearse all the other objections. Was it safe? Was it effective? Probably more important than Mrs Carmody's death was the outbreak of the First World War. The Kaiser's troops streaming into neutral Belgium aroused widespread animosity in the United States. For long Germany had stood for scientific

11. J.W. Williams, *Obstetrics*, 2nd edn (New York, 1908), p. 324; W.H. Howells, *Textbook of Physiology*, 7th edn (Philadelphia, 1919), p. 998.
12. Williams, *Obstetrics*, p. 345.

excellence and professional rectitude. However unreasonably, that faith was now shaken. How could the Hun be trusted with reforming childbirth when it was murdering babies in Flemish villages? Twilight Sleep did not immediately or entirely disappear. But it lost its dynamic. This did not mean that the clock was put back.

. . .

Thirteen years after the demise of the Twilight Sleep Association in the United States the National Birthday Trust was launched in Britain. Its objectives were broader; and from the start its leadership worked hand in glove with the College of Obstetricians and Gynaecologists. It entered the stage in a flurry of pamphlets and articles in the popular press. They revealed the shocking state of obstetric care generally. A leader in the *Daily Express* reported that, whatever choices were available to the rich,

> an ordinary woman in Great Britain still has to have her baby without any relief, whatever her wishes. She cannot afford a nursing home and there are too few beds in hospitals. The midwife who attends her is usually unqualified to handle an anaesthetic apparatus. Even when qualified she has no access to one.[13]

This was painfully true. In 1936 only 14 out of 186 local councils in England and Wales provided a single anaesthetic machine for their home midwifery teams. And that was often obsolete or out of action. The *Daily Mirror* deplored the 'red tape, parsimony and diehard attitudes of local authorities responsible for the suffering of hundreds of British mothers each day'.[14] This gave doctors a chance to practise their favourite incantation: 'Patients' lives are being put at risk because nobody listens to us.'

Conditions were slightly better in some European countries, especially in Holland, Belgium and Scandinavia, but worse in those still crippled by the war, like Austria and Hungary. The deficiencies were slowly being rectified when an incendiary book by a young Woking obstetrician raised a new storm.[15]

13. A. Edwards, 'Mothers Should Make a Fuss', *Daily Express*, 16 February 1946, p. 7.
14. 'We've Changed Our Minds about Having Babies', *Daily Mirror*, 15 May 1945, p. 15.
15. See Chapter 47.

# Chapter 43

# DOLORISM

The stench of the trenches of Flanders may have pushed the pains of motherhood to the back of people's minds. It may even have contributed to the last great wave of the exaltation of pain on moral, philosophical, religious and even on medical grounds. As had often happened in the past, the centre of the stirrings was France; but during the inter-war years ideas hatched in cafés on the Left Bank found resonance in many countries.

The term 'dolorism' was coined in 1919 in a review of Georges Duhamel's moving war novel, *La Possession du monde.*[1] It gained wider currency and a following after Julien Teppe, a previously obscure writer and journalist, published his *Apologie pour l'anormal* or *Manifeste du dolorisme* in 1937. It remains one of the most powerful and silliest affirmations of a creed which has intermittently flourished since biblical times and perhaps before.[2] A declaration of war against those 'sluggish and torpid individuals' whose 'empty lives seem to be dedicated to the cultivation of physical well-being', it rants against the 'tyranny' of those who wish to do away with pain. Far from being an experience to be avoided, it is only through pain that an individual truly learns about him- or herself: it is the liberating encounter which alone can reveal the divine and heroic in the human soul, mind and spirit. In short: 'I am in pain, therefore I am.'

Quoting supporting authorities ranging from Ecclesiastes to Rimbaud's sickly sublime poetry, the *Manifeste* is a hymn to physical suffering, to its

1. Georges Duhamel (died in 1966, aged seventy-five) qualified as a doctor and served in the First World War as a surgeon, but he never intended to practise and so became a full-time writer. The review in *Le Temps* applied the term 'doloriste' to his book in a pejorative sense. Duhamel's fame rests today largely on his multi-tome novels published in the 1930s; but his greatest chronicle of mental and physical suffering is his *Vie des martyrs*, published in Paris in 1917, followed by the slightly less compelling *Civilisation* two years later.
2. Dolorism is described in more detail in Rosalyne Rey's *Histoire de la douleur*, p. 374.

creative power, to its self-revealing capacity and to its ability to advance both the individual and the species. Its publication was followed by the launching of the *Revue doloriste*, an irregular periodical which, over the next few years, numbered among its contributors Gide, Valéry, Benda, Colette, Léautaud and Daniel-Rops. In France the influence of intellectuals has always been powerful (just as in England their vacuous denigration still earns royalties); and the *Doloriste* or at least its arguments reached a wide and influential audience. The Archbishop of Rheims, Monseigneur Duloupis, rejoiced in some though not all of its message.

. . .

The present book is a history, not a polemical tract; but since, in one form or another, the 'dolorist' creed has been a recurrent historical theme, its most recent re-emergence offers an opportunity for the writer to declare his personal creed. There are three, and only three, arguments which can be advanced in favour of physical pain. Books have been written about each; but all three can be summarised in a few sentences.

First, within strict limits pain is undoubtedly a precondition of life. This is most clearly shown by the few tragic cases of a rare inborn abnormality, congenital analgesia. Individuals who suffer from the defective gene feel no pain of any kind. They are crippled almost from birth, soon develop multiple disabilities from unperceived injuries and usually die young from conditions like painless but fatal appendicitis.[3] On a more circumscribed scale in Charcot's arthropathy mentioned earlier,[4] loss of pain sensation leads to a painless but disorganised and eventually useless joint. Such examples could be multiplied. They show that pain has a basic protective function which man shares with animals. This is the background to the widely parroted 'warning bell' argument.

Second, pain relieved, like a much-loved lost object found, provides one of the keenest pleasures of life. It is a powerful but irrational response since the absence of pain, like the continued possession of the much-loved object, gives no pleasure whatever. Despite this irrationality, the response is displayed by infants only a few hours old and continues till the last breath of life. In this restricted sense pain becomes the most fertile source, almost a precondition of

3. Two such cases are well summarised in Wall and Jones' *Defeating Pain* p. 95. Miss C's father was a doctor and her story is therefore particularly well documented. As a child she frequently burnt herself without realising it (climbing on hot radiators and touching hot kettles) and her body was extensively scarred by the age of eight. She also bit off her tongue quite painlessly. She survived to go to university but died of painless and therefore untreated osteomyelitis at the age of twenty-nine. See also R.A. Sternbach, 'Congenital Insensitivity to Pain', *Psychological Bulletin*, 1963 (60), 252.
4. See Chapter 33.

pleasure, just as prolonged pleasure often ends in pain. Let this be called the 'lost-penny-found' argument.

Third, pain can and does sometimes ennoble. Of the three arguments in favour of pain this is the most difficult to prove; yet this too rests on common, perhaps universal experience. Men and women whose life seems to be one of unclouded happiness – if such freaks exist – tend to be shunned by their fellow beings, and understandably so. At best they are crushing bores. At worst they become willing dupes of hair-raising ideologies. It is suffering which makes people sufferable; and it is shared suffering which can establish the closest bond between two unrelated individuals not actually in love. One may describe this as the 'ennobling' argument.

None of the three arguments should lead to the conclusion that pain itself is 'good' or that it should be cultivated rather than fought against. The opposite is true. Milton called it the 'worst of evils' and he was right. He was, one may be sure, aware of the fact that in this world there is no such thing as pure evil any more than there is such a thing as pure good. There is, in other words, some 'merit' in pain just as there is often much evil in pleasure. The limiting factors are nevertheless worth reviewing.

As regards the 'warning bell' argument, pain is an essential but also a highly imperfect signal. An animal with a broken bone will rest – that is, voluntarily immobilise – an injured limb. This will give the bone fragments a chance to unite. Man has more sophisticated ways of immobilisation; but the bandage, the sling and the plaster cast serve the same purpose. Both in animals and in man the guide to effective immobilisation is pain, or rather the cessation of pain. Today, when an inflamed appendix or an ectopic pregnancy can be removed, the pain which draws attention to such conditions can also be described as useful. Most usefully, 'pains' arising all the time, like the discomfort of a dangerously hyperextended joint or the strain of an over-exercised muscle, are hardly registered as pain. It is these barely noticed advance warnings of how far one can go without inflicting self-damage which have the greatest survival value. But, even allowing for such benefits, there are strict limits to the validity of the concept.

Most obviously, vast areas of human pathology are painless, at least to start with. Some cancers can be symptomless until they reach an incurable stage. The pain which may accompany advanced cancer and many other illnesses serves no useful warning purpose whatever. No patient needs to be reminded that he or she is suffering from a fatal disease. Nor is there any merit in being reminded that one is a victim of post-herpetic neuralgia, migraine or backache. Even a useful warning bell ceases to be useful once it has been sounded.

The limitation of the 'lost-penny-found' logic is obvious. Most normal people would forgo the pleasure of finding a mislaid object for an assurance that essential objects will never be mislaid. Much as one may relish the end of a toothache, no toothache would be preferable. In any case, the conditions leading into and out of this kind of pain cannot be created by the individual. The 'dolorist' scenario implies a choice. Real life offers none.

A similar fundamental limitation applies to the 'ennobling' argument. Even a saint like Marguerite Marie Alacoque who professed to 'love pain' did not advocate inflicting pain on herself to get a thrill from it. In the Cornaro Chapel in Rome Bernini created a shattering image of the ecstasy in pain; but the pain is inflicted by an angel, not by St Teresa herself. Of course there have been and there still are creeds that preach self-inflicted pain, and there are individuals who crave it. But pain inflicted as part of such a creed or craving is not ennobling. For pain or suffering to have such an effect it must be an Act of God (or chance, as atheists will have it). This is at least implied even by modern Christian apologists. C.S. Lewis goes a long way to 'justify' or 'explain' pain on moral grounds and quotes, obviously approvingly, a doctor saying:

> Long-continued pain is often accepted with little or no complaint and great strength and resignation are developed in response to it. Pride is humbled or, at times, results in determination to conceal suffering . . . Pain provides an opportunity for heroism. The opportunity is seized with surprising frequency.[5]

This is sailing perilously close to the dolorist wind; but even the anonymous doctor does not advocate the deliberate acquisition of pain for the kindling of heroism.

. . .

But what about physical suffering kindling genius? In Hemingway's 'The Snows of Kilimanjaro' (1938) the dying author recalls how his creativity has declined since his material comforts have improved. And in a letter to Scott Fitzgerald he makes the same point in more personal terms:

> You have to hurt like hell before you can write seriously. But when you get the damned hurt, use it – don't cheat with it. Be as faithful to it as a scientist.[6]

Hemingway was a 'fantasist' and something of a fraud as well as a great writer; but similar statements by otherwise sane and sober if less gifted artists and writers have been made from time to time and are still being made. The present writer finds them totally unbelievable. Of course suffering has been the theme of some of the greatest creations in the arts, literature and music. Of course Milton's testimony – and that of hundreds of lesser poets – must have been

5. C.S. Lewis, *The Problem of Pain* (New York, 1944), p. 82.
6. Quoted by H. Neal, *The Politics of Pain* (New York, 1978), p. 92.

based on personal experience. Since the personal experience has been instrumental in creating a masterpiece, it has been fruitful. Indeed, it is at least arguable that without having experienced pain no artistic or literary communication of any value is possible. Some illnesses associated with intermittent pain (like a painful cough and a low-grade fever in tuberculosis) have undoubtedly been sources of high poetic inspiration. But the *simultaneous* perception of severe physical pain and any kind of creative work is impossible. To put it crudely, no great poem has been written by a poet in the grip of renal colic.

. . .

There are nevertheless reasons for the delusion. Most commonly it stems from failure to distinguish between 'real pain' and 'professed pain'. The profession of pain, like pain itself, can serve and has always served a variety of purposes. In the Western world, until comparatively recently, a 'headache' has been the most popular method of contraception. During the Second World War a credible history of duodenal ulcer pain was the surest way to exemption from front-line duty. But these are trivial examples of a deeper, more complex and more universal function.

Pain can arouse sympathy, compassion, sometimes even love. These are among the most precious possessions in life. In some lives they are all too rare. In the absence of 'real' pain the profession of pain may have to serve as a substitute in trying to acquire them. The substitution may work, at least for a time. An unspoken compromise may be reached. A show of concern for the profession of pain may become a replacement for real concern for real pain. After a time the distinctions may become blurred. A not unhappy state of mutually agreed and only half-conscious understanding of make-believe may ensure a measure of contentment.

There are other links between 'real' and 'professed' pain. Pain can be deeply disabling. Or frighteningly enabling. Pain outranks even money in the power which it can give to some individuals over others. This is no less true of the profession of pain. Both can provide the means and the licence to exercise tyranny of the most merciless kind. World literature would be bereft without the portrayal of such depressing relationships. But they are not always depressing. There will always be some who welcome – even relish – being tyrannised. The relationship can then become mutually fulfilling.

At its most debased, the profession of pain can be as financially rewarding as pain itself. And not just to patients. Lawyers would be queuing at job centres if a clear distinction could always be made between the two.[7] One of

7. In 2004 a worker on the London Underground went sick, unable to walk with a painful ankle. Next day he was spotted entering a squash club. He was sacked. His union claimed unfair dismissal. The case almost paralysed London Transport.

the difficulties is that some people seem genuinely pain prone as others seem to be accident prone. They are repeatedly disabled by pains without any detectable lesions or evidence of injury, just as the accident prone trip over non-existent rugs and burn their hands on lukewarm kettles. The pain prone sometimes blame their suffering on someone close to them. Sometimes their suspicion is justified, though not in the sense that the pains are physically caused by that other person. But physical battering is not the only way of inflicting pain.

Professed pain can also become part of an elaborate pain-game of the my-pain-is-worse-than-your-pain variety. He gets a pain in the chest, she gets a cramp in the stomach. She gets a chill, he gets pneumonia. It can be harmless fun. Or it can be deadly.

. . .

Last, there are the true neuroses of which pain is usually a cardinal symptom. Two have always been common and still are. Thornton Wilder unkindly but accurately described hypochondriacs as 'people who listen to their body as if it were a Stradivarius'.[8] They listen and they inevitably hear a seductive tune. It can be an itch, a twinge, a numbness, an ache or a simple sensation of some marginal malfunction. Every human being is a gallery of potential symptoms offering themselves for inspection. The contemplation of the daily faeces – their shape, their consistency, their colour, their smell, their frequency – can become a wholly absorbing pastime. (There have been great and famous people as well as celebrities who have preserved their excreta in jars for months.) From symptom to disease and from disease to serious disease are but small steps. What makes these obsessive phobias depressing, disruptive and dangerous is their resistance to treatment. True hypochondriacs will always reject even the most convincing negative finding: medical reassurance merely shows up the incompetence of the doctor. To 'catch him (or her) out' in some small factual error will then become the patient's sport and delight. It does not make for a happy relationship. Yet hypochondriacs continue to haunt surgeries, outpatient departments and consulting rooms, often ending up as the hard-core problem cases of pain clinics.

The second pain-centred neurosis used to be called hysteria. During the second half of the twentieth century the name fell into disrepute. There have been two reasons for this. First, the word is derived from ὑστέρςα *hystera*, the womb, implying a disease of women. Nobody believes that any more. If a

8. Quoted in Neal, *Politics of Pain*, p. 14, note 6. 'Hypochondria derives from ὑπό *hypo* (below) and χόνδρος *chondros* (rib). The word commemorates the Hippocratic belief that the liver is the organ from which most complaints arise. Hence also 'liverish' and 'melancholia', meaning black bile.

gender difference has ever existed, equality in the workplace, in the professions and in public life has abolished it. Second, the term has come to mean a condition in which the sufferer is out of control, sometimes screaming, often behaving outrageously, as if 'putting on' a fit. This does happen. But outrageous behaviour has never equalled hysteria and hysteria has rarely been overtly outrageous. Indeed, the typical sufferer has always gazed into the world with calm resignation, almost serenity. Charcot called it *la belle indifférence.* This is still true. The victims may suffer, they must certainly be seen to suffer, but they suffer 'in silence'.

The politically correct though cumbersome replacement term for hysteria is 'conversion disorder'. It implies the unconscious conversion of a seemingly insoluble problem into a physical disease – blindness, paralysis, backache. It is something on which patients can focus their own and other people's attention. Pain is often though not always a prominent feature. It can conceal loneliness, rebellion, frustration, an almost infinite variety of dissatisfactions – that is, real and terrible pains but pains of a different kind from the one the individual complains of, and undoubtedly experiences.

Conversion disorders are different from hypochondria. The diagnosis of hypochondria is not usually difficult. Conversion disorders, by contrast, can be among the hardest abnormalities to diagnose.[9] Treatment and prognosis also differ. Hypochondria may be easy to diagnose but is often untreatable. Conversion disorders can respond to compassion, care and competent psychotherapy.

9. Much work on the psychoneuroses (by whatever name) was done in wartime on service personnel. Contradicting the trend of making the diagnosis largely by exclusion, long an expensive procedure, I. Douglas-Wilson reviewed the positive features which should point to the correct diagnosis: *British Medical Journal,* 1944 (1), 4342.

# Chapter 44
# RENOIR

It was suggested in the previous chapter that the simultaneous perception of severe *acute* pain and creative work is impossible. This raises a more interesting question. Can creative work *suppress* a comparatively low-grade pain? The answer must surely be yes. Medical case histories of the distant past are not always credible; but, having reached the beginning of the twentieth century, one has a choice of illustrative examples. They do not explain how being in the throes of a musical composition, of painting a picture, or even of writing a book on pain, can suppress a backache or a painful toe; but they do establish the fact beyond reasonable doubt. One example must suffice. It requires a brief digression into art history.

. . .

When the Renoirs arrived in Les Collettes above Cagnes-sur-Mer in the South of France, Auguste was already sixty and crippled by a painful and still advancing illness. Visitors often described his struggle against the disability as 'heroic'; but he punctured the compliment at once.[1] Heroism implied a choice, and he had none. Anyway, he did not think much of heroism if that meant pitting oneself against one's destiny. That was, he thought, always a mistake. Nor did he condemn those who might be labelled cowards. Everybody and everything had a function in life. Some fulfilled their function less easily than others. But most tried. It was fortunate for him that his

1. Many great artists painted lovely portraits of their children. Few of the children had the talent or the opportunity to repay their parent in kind. The film director Jean Renoir had and did in his affectionate memoir, *Renoir, My Father* (translated by R. and D. Weaver, London, 1962).

There is of course a vast literature on Renoir and an even vaster one on the Impressionists. In the former category the best ones (in the writer's opinion) are W. Gaunt, *Renoir* (London, 1952) and B.E. White, *Renoir: His Art, Life and Letters* (New York, 1954). In the latter category J. Rewald, *The History of Impressionism* (New York, 1946) remains a classic.

function was to paint, for it also happened to be his joy. Even more blessed, his function and joy was also his defence and unfailing remedy against the physical pain that might otherwise pitch him into despair.

His had been an odd and in some ways improbable career. He was four when his family moved from Limoges to Paris in 1845. His father was a tailor. Parents and five children lived in two rooms above the shop in the row of medieval houses which at that time separated the Louvre from the Tuileries. Queen Amélie would sometimes lean out of the first-floor window and ask the urchins rampaging in the gutter to be a little less boisterous. She and her husband were entertaining foreign royalty whose French was often hard to understand.[2] At fifteen Auguste finished school; and soon he was decorating porcelain plates in Monsieur Lévy's small pottery shop. He loved it, quickly progressing from Chantilly sprigs on the rims to portraits of Queen Marie Antoinette as a shepherdess after Mme Vigée-Lebrun in the centre well (or, for those loyal to the Bonaparte dynasty, Napoleon I in profile surmounted by an eagle).[3] From there he moved on to Monsieur Junot's ecclesiastical emporium, where he painted devotional awnings on transparent paper, in great demand as a substitute for stained glass by the missionary fathers in Indo-china. Eventually he saved enough money to enrol in the fashionable painting academy of Monsieur Gleyre, packed with plaster casts of decorously draped Greek gods and goddesses, stuffed horses and reproductions of the sublime mythological conceits of Monsieur Ingres. It is in this Temple of the Muses that he met Monet, Bazille, Sisley and Pissarro and that Impressionism was born.

More precisely, Impressionism was born on outings from the Temple to the forest of Fontainebleau, the banks of the Seine and later to busy pavements of the Paris boulevards. It was the beginning of a long struggle, joyous at times but in material terms almost unbelievably tough. Of the first exhibition of the group Albert Wolff, the critic of *Figaro* wrote:

> An exhibition of what has been described as paintings has opened at Duran-Ruel's. Attracted by the banners outside it may draw a few innocent pedestrians. What cruel spectacle will meet his eyes! Five or six lunatics – one of them a woman [Berthe Morisot] make up a group of poor wretches who have succumbed to madness and ambition and dared to put on an

2. Opinions about King Louis-Philippe differed (as they still do) but everybody loved his queen. She would reward the urchins with a coin wrapped in scented tissue paper.
3. All his life he would pour scorn on the idea that 'commercial art' was something intrinsically different from the 'fine arts', an expression he hated. The idiot accusation of painting chocolate boxes sometimes levelled against him today would not have dismayed him. What was wrong with chocolate boxes?

> exhibition of their work . . . It is a horrible and pathetic spectacle, reminding me of the inmates of the Ville-Evrard Asylum.[4]

Such vituperation was not an isolated example. The abuse continued for years, one flop after another.[5] France was prosperous. Old friends were settled, earning respectable wages and salaries, many of them married with children, living in bourgeois comfort. Commercial and academic art beckoned. 'But I had tasted the forbidden fruit; and I was not going to give up.'[6] Renoir was already fifty, well past middle age in those days, when he met Aline Charigot, aged nineteen. He was living in cheap unfurnished lodgings in Montmartre, a rough but inexpensive neighbourhood; and he was having his one proper meal a day in a small crèmerie in the rue Saint-Georges. The shop was run by the Widow Clément from Dijon who was a dear friend of Mme Charigot, Aline's mother, who also came from Burgundy and who had been deserted by her husband many years earlier. 'We must find a rich protector for your beautiful daughter,' Mme Clément had said to Mme Charigot, 'who will maintain her and perhaps even you in proper style. A respectable married man with a family . . .' A meeting with an elderly official from the Ministère de la Marine was arranged; but plans went awry. Aline had eyes only for the customer at the next table, a penniless, threadbare artist, 'polite but so thin that it wrung your heart'. As for Renoir, 'I recognised at once not only the freshness, beauty and youth of your mother but also the poise and dignity of my own mother', he would tell their son, Jean, twenty-two years later.[7]

It was to be a wonderfully happy marriage; and slowly, very slowly, not only students from the Beaux Arts but a few of their elders too began to see the light. Paul Duran-Ruel remained the Impressionists' main prop, showing their pictures in London, Boston, New York, Berlin and Antwerp as well as in Paris. To Renoir's surprise, in New York (though never in London) some of his paintings sold well. In Paris a few rich and eccentric collectors – the Bernheims, Louis Petit, fellow artist Caillebotte and the Prince Bibesco – became patrons. By solid bourgeois standards the family's finances remained precarious, but worries about next Sunday's joint were beginning to recede. They acquired a wide circle of friends and spent happy summers in Argenteuil, with the Monets, Aix-en-Provence with the Cézannes and in

4. *Figaro*, 1 April 1876. Wolff later changed his mind and became a patron. Manet painted his portrait.
5. Of course incomprehension often greeted new and unorthodox art; but with Impressionism comprehension flounders. Even allowing for the novelty of the technique, how could anybody fail to be ravished by the sheer sensuous beauty of Renoir's *La Loge* or *La Grenouillère*? But unravished the critics remained, regarding such works as unworthy even of execration.
6. Quoted in J. Renoir, *Renoir, My Father*, p. 86.
7. Ibid., p. 136.

Essoye in Burgundy with Aline's family. It was in Essoye that on a summer evening in 1890, cycling home from a day's painting in the vineyards, Renoir fell off his bicycle and injured his knee.

. . .

There is no evidence that progressive rheumatoid arthritis is caused or even initiated by physical injury; but patients often date the beginning of their illness to an accident.[8] It was so with Renoir. After 1890 the disease progressed by fits and starts for almost thirty years, with periods of quiescence and sometimes even with apparent though never long-lasting improvements. The reasons for these fluctuations are still obscure. Sometimes following a minor illness like a cold or some gastrointestinal upset but at other times out of the blue, he would be struck down with agonising joint pain and swelling, often accompanied by fever and sickness. The acute attacks would usually subside after five or six days but leave behind stiffness and deformities. Bones, nerves and skin all gradually become involved; and for many years, after several bouts of painful conjunctivitis, his eyesight too seemed to be threatened. Even during remission and at rest, pain was rarely wholly absent and the simplest movements often became a struggle. Fortunately, he did not develop any of the systemic manifestations most feared today: he did not become severely anaemic and his kidney function, so far as one can tell, remained good. Perhaps the absence of such life-threatening complications was due to the 'primitive' but harmless treatment available at the time. There were no steroids, no gold, no immunosuppressants and no powerful synthetic anti-inflammatory agents which might ease symptoms but also introduce new risks. Only salicylates in various early forms were on offer, unless patients in despair started to take opium or morphine. Renoir never did: even with salicylates he was sparing. He had always been a light eater and drank only the occasional glass of wine; but he lost weight over the years until he weighed less than 40 kilos. This made nursing easier in some ways, but loss of the cushioning effect of subcutaneous fat also made his skin over bony prominences vulnerable. Despite the devoted care of his family and nurses, he was to suffer in his last years from intensely painful pressure sores.

The illness was a calamity but for some years it did not entirely restrict the family's activities. Renoir consistently made light of his pains. Jean Renoir was born in 1891 – Pierre was already five – and Claude, known to all as Coco,

8. Rheumatoid arthritis may have existed for centuries but in its clinically recognisable form it did not emerge till the middle of the nineteenth century. It then replaced gout, with which it was for some years often confused. Why gout became rare and rheumatoid arthritis common is a mystery. Osteoarthritis, the degenerative joint disease, was not separated from the other two till Virchow's researches in the 1880s.

seven years later. From Rembrandt to Picasso artists have painted their loveliest paintings of their children and Renoir was no exception. Coco in particular was a perennial source of delight. Just before Jean's birth Aline imported a second cousin, Gabrielle, from her village. She became the artist's favourite model.[9]

But from the mid-1890s he was increasingly confined to a wheelchair; and, as his condition deteriorated, his patrons scoured Europe for anybody who might offer hope of reversing or at least arresting the illness. Eventually the Bernheims hit on Professor Rudolf Schade of the grand imperial city of Vienna. Schade had acquired the reputation of a miracle worker. He presided over his own sanatorium in the Vienna Woods patronised by more than one imperial highness as well as by Brahms and the Bernheims' cousins, the Rothschilds.[10] But nothing could persuade Renoir to travel to another spa. He had already spent a few weeks in Vichy and in Ems and vowed never to go near a bandstand or *Kursaal* again. But the Bernheims were undeterred as well as immensely rich and took their cue from the Prophet and the Mountain. Schade was summoned to Paris and, having examined Renoir from head to toe, expressed the opinion that he could make the artist walk in a fortnight. He charged a fee that was more than generations of Renoirs had earned in their combined lifetimes; but he was no charlatan. Installed in the Bernheims' palatial mansion in the Faubourg, he visited the Renoirs every day. He administered mysterious herbal remedies. He took daily measurements of joint movements. He radiated not optimism but certainty. He designed and supervised every meal himself but professed to be won over by Aline's homely remedies and dishes. He even made extensive notes of her recipes. Would Mme Renoir permit him to pass those on to the chef at his sanatorium in Baden bei Wien? She would. Contrary to the Bernheims' fears, the Renoirs loved him. He never arrived without flowers for Aline and *bonbons* for the boys. He was charmed by Gabrielle's new outfits and asked permission to present her with an antique Viennese brooch. He admired on the walls not the Renoirs, the Monets, the Sisleys and the Corots but the bird embroideries of Mme Charigot.[11]

Jean Renoir described the day when his father rose from his wheelchair. For rise he did.

> The professor lifted him out of the chair. 'Now walk', he ordered. All of us watched transfixed as my father mustered all his energy and took a step.

9. She remained part of the Renoir household till 1914 when she married the American painter, Conrad Slade, and went to live in California. She died in 1944.
10. G. Frankl, *Rudolf Schade, Wunderdoktor* (Vienna, 1924). Schade died in a boating accident in 1897, aged sixty.
11. They still exist and are in fact charming.

> Then another. And another – even though lifting his foot off the ground was obviously agonising. He walked around his easel and back to his invalid chair. 'Yes, I can do it', he sighed, 'but it would take all my energy. I would have none left for my painting. And, if I have to choose, I would rather paint'.[12]

He sat down in his wheelchair and never got up again. But from then on, for the next twenty years, it was 'fireworks all the way; as if his love of life and beauty would now be emanating from his brush'.[13]

...

The kindly and infinitely wise Dr Gachet warned them against the polluted air of Paris; and they took his advice.[14] In Cagnes-sur-Mer their first home was the old post office, a real village shop which precariously clung to the medieval ramparts above the road to Vence.[15] Three years later they moved to Les Colettes, an isolated farm a few miles outside the town. Aline was worried. The place was beautiful but remote, and remoteness was not Renoir's scene. But she had a small guest-house built in the orchard; and there was a constant stream of visitors. Of course they were mainly the younger generation.[16] Young Matisse came, then Bonnard, Vuillard, Rippl-Ronai, Augustus John and Cézanne's son Paul who married a local girl, Renée Rivière. Maillol, a little gauche, sculpted in one sitting the best portrait bust of the painter. In response to the family's compliments he enquired: 'But why does the Master himself not try his hand at modelling? He is a much greater artist than I will ever be.' One can picture the look Renoir gave him. By then his hands were so painful and deformed that the brushes had to be 'implanted' in his palms on small cushions of gauze. He could manage to paint – just. But sculpture! The young Catalan was a charming fellow but he should not mock a cripple. But Maillol would not take no for an answer. And a few weeks later a pupil of his, Guino, made his entry into the Renoir household, bringing a present of a barrelful of the best clay from Maillol's home in Banyuls just north of the Pyrenees. Renoir should dictate his sculpture: Guino would be his fingers. It

12. J. Renoir, *Renoir, My Father*, p. 242.
13. Ibid., p. 295.
14. The kindly Dr Gachet, doctor, herbalist and amateur painter, seems to have looked after every Impressionist and Expressionist genius of his day without charge. Van Gogh painted him most famously; and, less well known, he painted Van Gogh on his deathbed. He awaits a worthy biographer.
15. Not the concrete monstrosity which serves as a post office today and sells stamps with Renoir's portrait on them.
16. By the time the Renoirs moved to Cagnes, most of the friends of his youth were either dead or disinclined to travel. Sisley had died in poverty in 1899, Pissarro, also poor and blind, died in 1903, Cézanne in 1906. Degas was totally blind. Monet came once in 1908 but did not leave Giverny after that.

was a unique collaboration and it is still impossible to envisage exactly how it worked. But work it undoubtedly did. Not surprisingly, the enchanting little terracottas of nudes and washerwomen became Jean Renoir's favourites among his father's *oeuvre.*

One day the young Hungarian painter, Jozsef Rippl-Ronai surprised the old man adding a few 'touches' to one of the figurines. Rippl-Ronai recorded the conversation in a letter home. 'How are you Master?' 'In terrible pain as always.' 'But does it not hurt to press into the clay?' 'Pressing into clay never hurt anyone. It makes the pain better, did you not know? Now let me get on with it, young man.'[17]

. . .

Both Jean and Pierre volunteered in 1914 and both were wounded on the Marne; but the unexpected casualty of the First World War was Aline. She had developed diabetes which she neglected. She went to nurse Jean, perhaps saved his life. But she died in her sleep the day after her return to Cagnes. She was fifty-six. When in 1918 Jean came back to look after his father, he found the house desolate, even sinister. The tangerine trees which his mother had planted and the vineyard had gone wild. His father's suffering seemed intolerable. Yet the more he suffered, the more he painted. 'It is the only treatment that helps.' And slowly the place flickered into life again. Friends in Nice found a new model. Sixteen, red-haired, plump, Andrée's skin 'took the light wonderfully'. Her joyous youth erupted on canvases among the olive trees with their silvery reflection.[18]

But Renoir's nights were becoming frightful. The slightest rubbing of the sheets would start a sore. In the morning it would have to be cleaned, powdered and dressed. He could hardly hold anything any more. 'It's like fire, like touching blazing coals.' He would rage when his nose ran and had to be wiped for him. 'Oh, I am a disgusting object!' He was not. He was clean as he had always been and as he wished to remain. Valerian drops at night helped. But the new barbiturates he shunned. A herbal medicine prescribed by a Nice doctor may have contained some opium. He was washed and dressed while still half asleep, but he insisted on sitting up at a table for breakfast. Then he would be transferred to a specially designed sedan chair and, according to the light and the weather and depending on what he was working on, taken to a glazed shed or seated in the open in front of an ingenious construction which enabled him to roll and unroll a canvas. The transfer invariably cost him a good deal of pain but he brushed aside any suggestion of giving up work to rest for a day or two. Then, according to Jean:

17. J. Rippl-Ronai, *Levelek* (Budapest, 1938), p. 56.
18. She later married Jean Renoir.

> While my father is put in position the model takes her place on the flower-spangled grass. Somebody prepares his palette while he adjusts his stricken body to the hard seat. It is painful but it allows him to keep upright and a certain amount of movement. The piece of protective lint is folded into his palm and he points to a brush. 'That one there . . . no, the other one.' It is given to him. Flies circle in a shaft of sunlight. 'Ah, those flies,' he exclaims in rage, 'they can smell a corpse . . . or not a corpse but a body in pain . . .' But then they stop bothering him; and for a moment or two he seems somnolent, hypnotised by a butterfly or the distant sound of cicadas. 'It is intoxicating.' He slowly stretches out his arm and dips the brush into turpentine. But the movement is terribly painful. He waits for a few seconds as if asking 'why not give up?' 'Is it worth it?' But then a glance at the scene in front of him restores his courage. He traces on the canvas a mark of madder red that only he understands. 'Jean please push that curtain aside a little more.' Then, in a stronger voice: 'Ah yes, it is divine.' He smiles as he calls everybody to witness the conspiracy which has just been arranged between the grass, the olive trees, the sunlight, the model and himself. After a minute or two he starts humming. He is no longer in pain. Another day of happiness has begun for him, a day as wonderful as the one which had preceded it and the one which will surely follow.[19]

. . .

During the summer of 1919 for a few weeks the disease became quiescent and Jean and Pierre took their father to Paris to see some of his own paintings exhibited at the Luxembourg. After their wartime exile the Museum's treasures were being reinstalled, and Renoir, accompanied by the director and specially invited guests, was wheeled along the galleries which were covered with scaffolding. News of his arrival spread. 'I feel like a pope,' he said, but it was more like the progress of a tribal chief. His approach was signalled from afar by workers whistling, clapping and hitting their step-ladders with their hammers. Visitors cheered as the procession passed. That should have been enough. But after being lifted back into their car he expressed the wish to be driven to Chatou, to the inn known as La Grenouillère.[20] Jean and Pierre expostulated – it had already been a long and tiring day – but it was, their father insisted, the only reason why he had agreed to the journey in the first

19. J. Renoir, *Renoir, My Father*, p. 367.
20. The name *La Grenouillère* derived not from the thousands of frogs which populated (and still populate) that particular stretch of the river bank, nor even from one of the delicacies on offer, but from a class of young women, not exactly prostitutes but 'good sorts', often witty and amusing as well as pretty, ready to satisfy every whim of poor struggling artists but moving with ease from garrets in the Batignolles to mansions in the Champs-Elysées. They were the true inspiration and belles of the *belle époque*.

place. He had never felt better. So they were driven to the spot on the river bank where, many years earlier, he had painted the *Déjeuner des canotiers* (better known as *La Grenouillère*) in which Aline, one of the gay young company, had made her first appearance on canvas (on the left holding her little dog). He was too weak to move; but 'as tears began to roll down his cheek, we all got out and for a few minutes he was left alone with his memories'.[21]

21. J. Renoir, *Renoir, My Father*, p. 367.
In November he developed pneumonia. On 2 December he asked for his paint box and whispered a few words about having just discovered something. But then, brush in hand, he dozed off and a few hours later he stopped breathing.

# Chapter 45

# PILLS AND POISONS

Though old stick-in-the-muds like Renoir kept to their valerian drops and Mme Charigot's herbal medications, by the time of his last illness both traditional painkillers and surgical anaesthesia were moving on. Aspirin, arriving with the new century, created little stir at first.[1] Unlike opium, morphine or cocaine, it did not promise a never-never land of delights. It did not even induce sleep. But within a few years the chaste little pills had become as much part of everyday life as breakfast cereals. They stopped the agony of toothaches. They calmed headaches. They eased sore throats. They relieved acutely inflamed joints. They soothed hangovers. Slowly at first but then dramatically, sales began to rise. By the end of the 1900s, Bayer found itself making profits in five continents on a scale never dreamt of in the history of pharmaceuticals. Since the formula was simple and could not be kept a secret, they also found themselves entangled in labyrinthine legal battles over patent rights, brand names and unlicensed imitators. These would continue throughout the century.[2] The salesmen's wars lacked the gore of the battlefields but they were almost as lethal.

The peace treaties of 1918–19 abolished the dominance of the German parent company but not of aspirin or the name of Bayer. When the Spanish flu struck in the summer of 1918 the drug was the only one to promise a measure of relief.[3] It lowered the temperature and eased the agonising cramp-like pains in the limbs that preceded death. The epidemic itself was the most

1. See Chapter 35.
2. They are well described in Jeffreys *Aspirin. The Remarkable Story of a Wonder Drug*, and in even more graphic detail in C.C. Mann and M.L. Plummer's *The Aspirin Wars. Money, Medicine and 100 Years of Rampant Corruption* (New York, 1991).
3. The epidemic was called 'Spanish' because it was first openly mentioned and discussed in the Spanish press. Spain was neutral and had no press censorship. In belligerent countries on both sides of the trenches the news was censored and suppressed. During Armistice week in November 1918 in London alone 14,000 people died from the epidemic; but the first mention of it in the English press was on 1 December.

devastating the world had ever seen.[4] Because it came in the wake of the greatest man-made carnage and was accompanied by the crash of toppling thrones, it is also the least memorialised in history books. The disease progressed so quickly that aspirin made little difference to the outcome. But occasionally the slight symptomatic improvement rekindled the patient's 'will to live'. How big a difference this made to the death rate nobody could tell then or can tell now. What is not in doubt is that millions reached for the little pills at the slightest sign of the illness. It was their best, indeed their only hope.[5]

At the height of the pandemic thoughtful observers in the United States – worse hit than Europe – had apocalyptic visions. In October 1918, when the number of deaths exceeded 190,000 a week, Victor Vaughan, eminent microbiologist and surgeon-general, wrote: 'If the epidemic continues its rate of acceleration, civilisation as we know it will disappear from the face of the earth in a matter of weeks.'[6] It did not happen, though perhaps only just. After waxing and waning for almost a year, the virus subsided as suddenly and as mysteriously as it had struck. The supportive role aspirin had played lingered in people's minds. It had not been a life-saver but it had eased the suffering. The memory cemented the drug's reputation as no advertising campaign could have done.

But the tablets were no longer exclusively Bayer products. The German patents were thrown open to famously rigged auctions.[7] In 1919 in a long and complicated judgment (which in many countries did not end litigation), Judge Hand of New York ruled that 'aspirin' was now a generic name. In the United States it became plain aspirin. The ruling opened a no-holds-barred

The virus disappointed the Kaiser. On 1 October 1918, during the two-week lull between the disease striking the Allies and then Germany, he expressed the hope that it might win him the war his generals were about to lose. Starving Germany and Austria-Hungary eventually became two of the worst-hit countries in Europe.

4. In terms of the number of deaths it greatly exceeded the Black Death of the Middle Ages, though the Black Death killed a higher proportion of the population of Europe.

   Among the celebrity victims of the epidemic were Sir Hubert Parry, composer of 'Jerusalem', the King of Tonga, Edmond Rostand, author of *Cyrano de Bergerac*, Louis Botha, first premier of the Union of South Africa, the painter Egon Schiele and his wife in Vienna, both in their twenties, 'Admiral Dot', one of Phineas Barnum's midgets and Harold Lockwood, star of the silent screen. Among those who caught the virus but apparently recovered was President Woodrow Wilson. At least one biographer attributes his mishandling of the Peace Conference to his lingering mental and physical ill health.

5. The Maori of New Zealand were devastated by the epidemic. F.G. Wayne, health superintendent, secured a large consignment of aspirin for one of the villages. He was later invited to a christening where the newborn was baptised Aspirin Wayne.

6. Altogether about 100 million people may have died during the pandemic, East Africa, India and Indonesia being the most severely affected.

7. Most of the American assets of Bayer, including several factories, went to two small-time dye manufacturers, William Weiss and Arthur Diebold, for just over $5 million. They had spent another $5 million on bribes.

publicity war between rival manufacturers. One strand in the hostilities was the appearance of the tablets in a slightly revamped form known as 'Aspro'. First marketed in Australia, it marked the beginning of modern mass advertising in pain relief. The success of the campaign vindicated the genius of a New Zealand businessman, George Davies, recently forced to leave his native country to escape prison for fraud. 'An Aspro tablet now and then / Builds up and soothes the greatest men.' The jingle would appear alongside pictures of Lloyd George, President Wilson, the Pope, Charlie Chaplin or whichever benefactor of mankind happened to be in the news, effectively putting the words into their mouths. It worried Davies not at all that few of the celebrities portrayed could have distinguished an Aspro tablet from chewing gum. The dead were enlisted as well as the living. 'Hats off to the great Abe Lincoln because he was a stickler for the truth. He would have recognised at once the wonderful discovery that is Aspro.' In whatever form, aspirin or Aspro sold by the trillion. As an ingredient of Jeeves' incomparable pick-me-ups, Bertie Wooster maintained, 'it regularly brought back the bloom of roses to anyone's cheek short of an Egyptian m.' Cocaine often set the mood of twenties' partying but it was aspirin that stopped heads bursting the next morning. The Spanish philosopher, José Ortega y Gasset called the decade not the 'Jazz Age' or the 'roaring twenties' but the 'Aspirin Years'.[8]

. . .

In the late 1920s Carl Duisberg and the revived German company became the leading players in an amalgamation of German chemical manufacturers, to be known as I.G. Farben. Bayer remained the dominant partner and aspirin retained the iconic Bayer cross. The firm's close association with Hitler and the Nazis began on 2 February 1933, a few hours after Duisberg switched on the famous logo in front of the new factory in Leverkusen, the world's biggest illuminated display at the time. The Board committed itself to bank-rolling the Party's election campaign in exchange for its unique position being

8. It became part of the literature and the artistic scene of the inter-war years, mentioned among others by Franz Kafka (in *Der Prozess* [*The Trial*]), George Orwell (in *The Road to Wigan Pier*), Graham Greene (in *Stamboul Train*), Edgar Wallace (in *The Door with Seven Locks*), Hemingway (in *For Whom the Bell Tolls*), and Agatha Christie (in virtually all her books). Caruso would not go on stage until he swallowed two aspirins. Noël Coward rhymed 'aspirin' with 'naughty twin'.

The possible toxicity of the drug was frequently debated in the medical and lay press. In a sniffy editorial the *Lancet* concluded in 1931: 'Not the least contribution of chemistry to human well being is the gift of aspirin, now manufactured and swallowed by the ton. Its analgesic effect and its safety are so widely known that its use has long passed into the control of the public, with much extravagant encouragement from the manufacturers and lamentably little guidance from the medical profession.' The first vending machine installed in the House of Commons (after heated debate) dispensed containers with four of the magic pills.

recognised by the Führer. Duisberg died in 1935. I.G. Farben grew into one of the props of the Nazi economy. Jewish employees were dismissed. Eichengrün was written out of the firm's history. The Bayer cross appeared on the wings but the Hakenkreuz on the tail of the company's fleet of private aircraft; or, as the intellectual giant of the regime, Dr Goebbels, put it, I.G. Farben and the Third Reich marched 'im gleichen Schritt und Tritt' (to the same tune). The firm was duly rewarded. During the Second World War it became the largest employer of slave labour in Germany, proud providers of the Zyklon B gas used in Auschwitz. This led to the directors facing the War Crimes Tribunal in Nuremberg in 1944. Eleven out of twenty-three were sentenced to terms of imprisonment, despite the memorable claim of production chief Fritz ter Meer that 'concentration camp victims of scientific experiments were not subjected to unacceptable suffering since they were going to die anyway'. It was no more than a contretemps. I.G. Farben was broken up, being replaced by new/old companies like Hoechst, BASF and a 'reborn Bayer'. The last once again became a leading manufacturer of pharmaceuticals. Once again aspirin became a top seller. In 1956, a few weeks after his release from prison, Fritz ter Meer became the company's new chairman. He was a man of charm and as dependable in his faith in liberal capitalism as ever. Pain was still big business.

. . .

But there were niggles. One of the earliest revelations of the gastroscope, still a cumbersome and occasionally dangerous instrument,[9] was the damage aspirin could do to the lining of the stomach. In some patients Arthur Douthwaite of Guy's Hospital found fragments of the tablets taken by the patient an hour or two earlier still clinging to the wall of the organ in small pools of blood. This explained why in habitual users the drug could cause severe anaemia. It also inspired the search for soluble and non-irritant formulations. Working for Reckitt and Colman, a comparatively small pharmaceutical company in Hull whose plant had been almost totally destroyed by the Luftwaffe, Harold Scruton, an industrial chemist, hit on the idea of combining acetylsalicylate with chalk. He found that, provided proportions and timing were just right, the complex could be made stable enough to pass through the stomach and not fall apart until it entered the bloodstream. Even after the chemistry was mastered, manufacture in the shell of a factory, using gimcrack equipment, remained daunting. But on 20 November 1948 the new soluble aspirin, named 'Disprin', was launched. Oddly but not in retrospect surprisingly, its uniform success with doctors was not at first shared by the public.

9. Inflamed and ulcerated areas in the lower oesophagus could be perforated. The mishap was rare but catastrophic.

The fact that Disprin was 'aspirin without the deadly side-effects' was the first most people had heard of the deadly side-effects.

But the days of the virtual monopoly of acetylsalicylate in whatever form were numbered. Ironically – an overused word but unavoidable when chronicling the tortuosities of the international drug industry – it was Bayer in reborn Germany which produced the first serious competition. Scientists at Yale and in New York had been studying an old Bayer product, Phenacetin, and related compounds, and found that most of them metabolised in the body to another aniline derivative, acetaminophen. When acetaminophen itself was tested by Bayer researchers in Germany, it proved a remarkably effective analgesic without many of the side-effects of the parent drug. Amidst the obligatory publicity 'Panadol' or 'Paracetamol' was introduced to Britain in 1956 and quickly eclipsed both aspirin and Disprin. In the United States Bayer (now in the hands of a company called Sterling) refused to compete against their own star product. Their decision came to haunt them in the shape of cute matchbox-size red fire engines. A Pennsylvania firm, McNeil Laboratories, had spotted a gap in the market, a convenient children's liquid analgesic, and launched the easily soluble acetaminophen dispensed in the unwanted stock of a bankrupt toy manufacturer. The little fire engines did not extinguish fire but did almost everything else, including make millions for McNeil Laboratories. When the firm were bought by the giant manufacturers of Band-Aid, Johnson and Johnson, the liquid inside the container was resolidified into tablets and reintroduced as Tylenol. Nothing thereafter could shake it from its perch. When in 1982 it had to be withdrawn because 'an ill-wisher' had injected cyanide into some packages which killed eight people, makers of aspirin expressed shock-horror while launching an invigorated advertising campaign. (The older drug was hailed as 'safely uncontaminated'.) After a dip, their hated rival easily regained its old primacy. Throughout these metamorphoses the question of how acetaminophen alias Panadol/Paracetamol/Tylenol worked in the body remained a mystery and remains so still.

The next player on the analgesic market – or battlefield – was the fruit of long and costly experiments conducted by Boots, the chemist.[10] Their final product, Brufen, launched in 1962, belongs to a family of chemical compounds unrelated to aspirin or phenacetin, all derived from the comparatively simple substance, propionic acid. The drug quickly proved effective,

10. Boots had been founded by Jesse Boot, an enterprising Nottingham chemist, in 1883. By the time of his death in 1931 the chain had 600 retail outlets in Britain; but the firm did not start its own research and development programme until the end of the Second World War.

especially in rheumatoid arthritis and other forms of musculoskeletal aches and pains. It was also gratifyingly free of lethal complications.

. . .

One of the side-effects of the discovery of cortisone and other powerful steroids was the creation of an artificial class of analgesics collectively known as 'non-steroidal anti-inflammatory drugs' (NSAIDs). Besides aspirin, Paracetamol, Brufen and their numerous legitimate and illegitimate offspring, they included at various times such bizarre medications as injections of gold and the metal-binding agent penicillamine.[11] Some were effective in painful conditions like rheumatoid arthritis but too toxic for comfort. Today standard textbooks list more than a hundred NSAIDs as both available and in use. Some are identical but marketed under different names. (Nature is no less prodigal in minute variations on a theme but does not produce the same substance under a dozen different names.) What all NSAIDs have in common is that they kill pain not by acting directly on the transmission or perception of pain but on the inflammatory response of which pain (as Celsus had observed almost two thousand years ago) is a cardinal feature. Their so far inescapable limitation is the direct result of this.

Inflammation is the basic defence response of all tissues to any kind of injury. Even plants respond to physical or chemical damage with local and even distant changes that in some respects resemble inflammation in animals and man. While Celsus admirably described the outward manifestations of the response, only in the last thirty years have scientists started to explore its chemical complexities. These involve non-enzymic as well as enzymic reactions not only at the site of the injury but also in remote parts of the body;[12] In addition to limiting the damage, they also prepare the ground for subsequent reconstitution and healing. Many of the anti-inflammatory agents are now known to suppress a particular enzyme – cycloxigenase 2 (often abbreviated as COX-2) – which promotes the conversion of arachidonic acid to a local hormone, prostaglandin. The prostaglandins in turn are responsible for some of the vascular changes in inflammation and also for initiating impulses perceived as pain. Ideally analgesics should suppress the pain without affecting any of the other responses. The number of currently competing

11. In many cases the analgesic and antifebrile actions of these drugs were first noticed as unexpected side-effects in the treatment of some other condition (the treatment of tuberculosis in the case of gold and mustard gas poisoning in the case of penicillamine).
12. The means of communication whereby a localised inflammation in the big toe can activate, often within minutes, the temperature-regulating centre in the mid-brain (causing fever), the mechanism regulating the production of white blood cells in the bone marrow (causing a sharp rise in the numbers produced) and several other remote functions is one of the mysteries of inflammation. There are many others.

NSAIDs is evidence that the ideal has remained elusive. Nor does a COX-2 probably exist which has no untoward side-effects in some people. The question is: how much risk is acceptable?

In August 2005 a court of law in Texas awarded punitive damages against the giant drug company, Merck, amounting to an unprecedented $253 million. The plaintiff was the widow of a man who had died in 2001 after taking the company's flagship COX-2 painkiller, Vioxx. It was alleged that the drug had been rushed on to the market in 1999 to beat rival firms about to produce their own brand of cycloxygenase inhibitor, that the results of some preliminary trials were distorted or suppressed, that doctors and medical charities were bribed with freebies and lavish donations, and that uncomfortable facts were airbrushed out of submissions to statutory bodies like the Food and Drug Administration in the United States and the Medicines and Healthcare Products Regulatory Agency in the United Kingdom. The fine opened a floodgate of claims from other alleged victims of the drug or their relatives. The sum of $253 million would dent but not seriously deplete the reserves of a company the size of Merck. Hundreds or thousands of such awards would. Bad news for some is of course good news for others. As the shares in Merck took a dive, the shares in rival companies making similar drugs rose. But how similar are they?

Astronomical figures were soon being bandied about. What is not in doubt is that Vioxx has already been taken by millions. Many (apart from Merck) have hailed it as a wonder drug against severe arthritic pain. It has transformed their lives. The majority might happily run an added risk rather than abandon what may be the only successful treatment available to them. But how big an added risk? Anyway, they are given no choice. A drug on which they have come to depend is suddenly taken away from them.

Some have estimated the number of Vioxx-related deaths to be around 60,000. (Increasingly the great modern fudge word '-related' crops up in reports. Any statement containing it should be treated with the utmost scepticism.) In a torrent of sensational news, as well as some tragic individual case histories, some disturbing facts tend to be overlooked. There is no defence against deliberately falsifying statistical results; but, contrary to what is often implied in the media, large foggy patches still hover over drug trials. This is especially so when the complications looked for are, like strokes and heart attacks, common in the general population. Glib comparisons with thalidomide are misleading. The disastrous effects of that drug were virtually unknown in non-thalidomide infants. Complications and side-effects may take years to declare themselves. There is no reliable way of telescoping such time lags. In matters of life and death statistics are easy to generate but hard to assess. Are thirteen avoidable strokes more important than twelve avoidable strokes? Not to the victims. Or to the unaffected beneficiaries. But the statistical difference can determine the fate of a drug. Yet statistical differences do not exist in a vacuum. Academic and professional reputations may hang in the balance. Pharmaceutical companies are not charitable institutions: they must make a profit or go under. Tens of thousands of jobs may be at risk. No statistical formula can insure against wishful thinking. No mathematical law is proof against self-delusion. In the prevailing moral, economic and political climate (which its crusading champions clearly regard as justifying any number of holy wars in remote parts of the world) no government agency will ever ensure that the rules which govern the manufacture and sale of pain-killing drugs are significantly different from the rules which govern the manufacture and sale of landmines.

...

Not all unexpected side-effects of drugs are harmful. Aspirin was already seventy years old when two researchers at the Institute of Basic Medical Sciences of the Royal College of Surgeons in London, John Vane and his assistant, Priscilla Piper, showed that the drug inhibited the formation of the prostaglandin thromboxane A2. Thromboxane A2 is a promoter of the aggregation of blood platelets.[13] This makes aspirin in low concentrations an inhibitor of coronary and cerebral thrombosis. The resulting renaissance of the drug is only distantly related to its original role as a painkiller in feverish joint pains. Will the drug's second innings also reveal new complications?

...

General anaesthesia too advanced after John Snow created the specialty in Britain. Frederick William Hewitt stepped into Snow's shows when he administered the anaesthetic for the drainage of Edward VII's appendix abscess in 1901. He also stepped *out* of Snow's shoes when he insisted on examining the patient before putting him to sleep. Lord Lister tut-tutted at such presumption; but Hewitt got his way and made sure that the king had his appendix abscess drained in the safer semi-recumbent rather than in the flat lying-down position.[14] By then the 'rag and bottle' used by Snow and his contemporaries had begun to evolve into trolleys with cylinders, valves, gauges and coils. Colour-codes ruled (and sometimes misruled). It became possible to mix gases in more or less known proportions. Most usefully, Edmund Andrews of Chicago showed that mixing nitrous oxide with pure oxygen rather than with

13. J.R.Vane, 'Inhibition of Prostaglandin Synthesis as a Mechanism of Action for Aspirin-like Drugs', *Nature*, 1971 (231), 232.
14. Hewitt set out to be a surgeon but his poor eyesight made him take up anaesthesia instead. After his services to Edward VII he became the first 'Anaesthetist to the King' and was duly knighted. He died in 1916, aged fifty-nine.

air made it possible to use the gas for relatively lengthy procedures without inducing dangerous anoxia.

. . .

Two other advances enhanced the safety and enlarged the scope of general anaesthesia. The first was a stepwise development which began in 1869. A patient admitted to the University Surgical Clinic in Leipzig needed a complicated operation on the face. Such operations had always made the conventional administration of chloroform through a mask difficult. Friedrich Trendelenburg was one of the commanding figures of surgery in Germany during the last third of the nineteenth century.[15] Already known for his ingenuity, he solved the problem by performing a tracheostomy – that is making an opening into the trachea below the larynx.[16] Through the opening he introduced a catheter. An inflatable ring attached to the catheter ensured that blood, saliva and mucus from the mouth and larynx did not trickle down into lungs. The catheter was connected to the anaesthetic gas container through a specially constructed funnel. The operation was successful and the procedure was repeated by others. It was not ideal. It involved additional surgery, and the tracheostomy often took a long time to heal. The seconds that elapsed between making the opening in the trachea and inflating the rubber ring were fraught. To overcome these drawbacks William Macewen of Glasgow devised a curved metal tube which a finger inserted into the patient's mouth could guide through the larynx into the trachea. The manoeuvre required skill and luck and could be performed only after the patient had been deeply anaesthetised.

The next advance was the result of a tragedy. During the summer of 1896 Joseph O'Dwyer, children's specialist to the New York Foundling Asylum, watched in helpless despair his four small children suffocating during attacks of diphtheria.[17] Over the next two years he devised a set of fine elastic tubes which could be passed through the inflamed throat of children. His invention saved many lives. It was taken up in Europe. It occurred to Karel Maydl,

15. His name is commemorated in several eponyms, most notably Trendelenburg's position in which the patient is placed on the operating table head down at an angle of about 45 degrees to facilitate operations in the pelvis. There is also a Trendelenburg's sign which is positive in congenital dislocation of the hip and a Trendelenburg's operation for varicose veins. He died in 1924, aged eighty.
16. Trachea literally means 'a hard artery' (the 'hardness' due to the cartilaginous rings) and was believed by Galen and his followers to be part of the circulatory system. Fabricius described its true function in 1584. The surgeon Brasevolus (died in 1570, aged seventy) claimed to have performed tracheotomy in two cases of 'internal strangulation', perhaps diphtheria. Trousseau advocated the operation as an emergency and may have performed it himself.
17. Similar deaths were the inspiration of Gustav Mahler's *Kindertotenlieder*. Diphtheria did not become treatable until Emil von Behring and others developed the diphtheria antitoxin in the early 1890s.

professor of surgery in Prague, that it could also be used for administering anaesthetic gases. His colleague, Victor Eisenmenger added an inflatable cuff. A third German surgeon, Franz Kuhn, devised a modified set of tubes which could be introduced through the nose and then passed through the larynx.

The possibility of blowing gases directly into the lungs led to the idea of controlling respiration by inflating and deflating the chest by means of a bellows connected to an airtight tube in the trachea. On 30 October 1898 such controlled respiration enabled F.W. Parham of New York to open a patient's chest and remove a benign tumour from a bronchus. Without 'positive-pressure' respiration the operation would have led to the collapse of the patient's lung and probably death.[18] The development of the laryngoscope by Chevalier Jackson during the first years of the twentieth century made it possible to pass a tube through the larynx into the trachea under direct vision.[19] Of course it required skill. In Britain the method came into its own during the First World War. Used by two young and otherwise still inexperienced anaesthetists, I.W. Magill and E.S. Rowbotham, it allowed the plastic surgical team of Sir Harold Gillies to try and reconstruct some of the horribly shattered faces of wounded soldiers.[20] In America the method was popularised by the New Orleans surgeon Rudolph Matas. Combined with new anaesthetic gases like cyclopropane, endotracheal intubation became the most important safety device in anaesthesia. Some of his 'chesty' patients, Lord Moynihan remarked, were safer during the hours they spent under the anaesthetic than they had been for years before or would ever be again.

. . .

The second momentous advance in anaesthesia, perhaps the most important since ether, had originally been described as Nature's 'cruellest conceit'. The quick-acting poison curare paralyses muscles by its action on the junction of

18. What keeps the lung expanded is the less than atmospheric pressure between the two layers of the pleural sac. The outer layer is attached to the chest wall, the inner to the lungs. Allowing air into the pleural space causes the lung to collapse to a mass smaller than a tennis ball. 'Positive-pressure' respiration through an endotracheal tube can overcome this, keeping the lung inflated.

19. Laryngoscopy, the indirect visualisation of the larynx by means of a mirror pressed against the uvula, was discovered by Manuel Garcia, a Spanish singing teacher living in London, in 1854.

Chevalier Jackson (died in 1958 aged ninety-three) published the first monograph on endoscopy in 1907. It was a brilliant text which popularised both laryngoscopy and bronchoscopy. He himself became the first professor of endoscopy in the University of Philadelphia and is still widely regarded as the 'Father of the Endoscope'. He was an accomplished painter.

20. Ivan Magill (knighted in 1960) died in 1986 aged ninety-eight. He gave his last anaesthetic aged eighty-four to a visiting head of state in a bathroom of the Ritz Hotel in London. Stanley Rowbotham died in 1979, aged eighty-nine.

motor nerves and muscle without affecting sensation or the workings of the brain. The substance had been used as an arrow poison by the Orinoco Indians of South America for centuries before it was noticed by some of the earliest European travellers. In his book on the discovery of Guiana Sir Walter Raleigh wrote:

> The Ariras are valiant and desperate people and possess the most strong poison on their arrows of all nations . . . For besides the mortality of the wound, the part shot endureth the most insufferable torment in the world and abideth a most lamentable and ugly death, sometimes dying stark mad, sometimes their bowels breaking out of their bellies which are presently discoloured as black as pitch and so unsavoury that no man can endure to attend them.[21]

In 1802 the German traveller, naturalist and polymath, Alexander Humboldt, obtained a specimen of the poison and nearly killed himself when he sprinkled a few drops on his leg which was freshly scratched. He recalled that

> when we came to Esmeralda most of the Indians were returning from a trip beyond the Rio Padamo where they had gathered the creeper from which they obtain curare . . . [During the ensuing festivities] we met an old Indian, known to the people as the 'Poison Master', who was in the middle of preparing the poison from the fresh plant. We found in his possession large earthenware pans for the cooking of the plant, shallower vessels for evaporation and banana leaves rolled into neat funnels for straining the fluid. The utmost cleanliness and order prevailed in his hut which served as his laboratory. He had the same affected attitude and pedantic way of talking as apothecaries have in Europe. 'I know', he said, 'that the white man understands the art of making soap and the black powder which makes a great deal of noise and scares away the animals if they are not hit. But curare, which is transmitted among us from father to son is better than anything you make over the seas. It is the juice of a plant which kills without a sound and without anybody knowing whence the shot came.'[22]

Curarine, the active alkaloid, was extracted from the plant by two French chemists, Pierre Roussingault and Jean-Jacques Roulen in 1828; and the brothers Robert and Richard Schomburgh gave it the technical name *Strychnos toxifera.* Virchow noted that the poison retained its potency in the dried state for years; but it was Claude Bernard who explored its action and stamped his personal style on its portrayal:

21. Quoted in Robinson, *Victory over Pain*, p. 309.
22. Quoted in ibid., p. 311.

> Within the motionless body, behind the staring eye, with all the appearance of death, feeling and intelligence persist with all their force. Can we conceive of a suffering more horrible than of an intelligence present after the succumbing, one by one, of all the organs imprisoned alive within the living cadaver? Poetic fiction in the past has endeavoured to move us to pity by conjuring up the spectre of sensient beings imprisoned in a paralysed body. The torture which the poet's imagination has invented is realised by Nature by the action of this American poison. But reality surpasses fiction. When Tasso depicted Clorine incarcerated [in *Gerusalemme Liberata*] in a majestic cypress tree, he left to her at least tears and sobs to bemoan her fate and to move to pity those witnessing her suffering . . .[23]

By the time of Bernard's experiments muscular rigidity had already emerged as one of the limitations of abdominal surgery under anaesthesia. Paralysis of skeletal muscles is among the last effects of deepening unconsciousness. Indeed, muscles of the abdominal wall can go into violent spasm under light anaesthesia. This can make the task of the surgeon (and that of his hapless assistants trying to keep the incision open with retractors) hard work. Here, potentially, was the answer.

In 1932 Richard Gill, a thirty-year-old American explorer, developed painful cramps in his limbs, subsequently diagnosed as multiple sclerosis. His neurologist, Walter Freeman, advised him that small doses of curarine might improve the spasticity. The drops were effective; and, during a remission in 1938, Gill (who never accepted the diagnosis) and his wife made another trip to the jungles of Ecuador. They brought back 11 kilos of crude curare as prepared by native Indians but could find nobody to refine it.[24] Eventually, the pharmaceutical chemist E.R. Squibb accepted the haul as a gift and prepared an extract. This they standardised by the 'rabbit head-dropping test': one unit was the amount of curare that caused an average rabbit's head to drop. They called the preparation Intocostrin.

A sample of the extract was sent to A.E. Bennett, professor of neurology in Omaha, Nebraska, suggesting that it might be useful in spastic children. In spastic children the drug proved disappointing; but it was unexpectedly effective in stopping physical convulsions in psychiatric patients receiving electroconvulsive therapy. (It did not prevent the mental effects of the treatment.) A young researcher, Emmanuel Papper of Columbia University, then tried it on two cats, both of whom developed asthmatic attacks and died. Papper's chief, Emery Rovenstine, chairman of Columbia, suggested that Pepper try it in a 'more robust species'. Papper chose a patient needing his appendix removed.

23. C. Bernard, *Leçons sur les anaesthétiques et sur l'asphyxie* (Paris, 1875), p. 128.
24. R.C. Gill, *White Water and Black Magic* (New York, 1940), p. 235.

The patient duly relaxed but then developed respiratory paralysis and could be revived only after prolonged artificial respiration. Papper and Rovenstine reported that the drug was too dangerous for human use. The directors of Squibb accepted the judgement; but it was the job of Lewis O. Wright, a doctor employed 'as roving ambassador' by the firm, to sell their product. He persuaded Harold Griffith of Montreal to try it in smaller doses in combination with cyclopropane anaesthesia to ensure muscular relaxation in difficult abdominal operations. Griffith and his resident, Enid Johnson, gave it to their first patient on 23 January 1942 and published their first series of twenty-five successful cases in *Anaesthesiology* a year later.[25]

In Britain Intocostrin was introduced to a young Liverpool anaesthetist, John Holton, seconded as medical officer to the United States bomber base in Burtonwood, by an American friend during the last year of the war. With a colleague, Thomas Cecil Gray, he tried it in a few patients needing difficult abdominal surgery. It worked. Their supply soon ran out but Gray remembered that a few ampoules of a 'curare-like' drug had been prepared in 1935 for research by Harold King of the Wellcome Laboratories in London. The ampoules were still on the shelf and still potent. They passed what Halton and Gray a little coyly described as 'stringent sterility tests'. (The ampoules were never actually opened but presumably contained no visible live creatures.) The two young anaesthetists reported their first series of eighteen cyclopropane and curare anaesthesias to a meeting of the Royal Society of Medicine. Their paper was subtitled: 'A milestone in anaesthesia'.[26] So it proved.

25. J.R. Maltby (ed.), *Notable Names in Anaesthesia* (London 2002), p. 234.
26. T.C. Gray and J. Halton, 'A Milestone in Anaesthesia?' *Proceedings of the Royal Society of Medicine*, 1946 (39), 400.

## Chapter 46

# THE SURGERY OF PAIN

Few of the arguments against exalting pain set out in Chapter 43 escaped the mind of René Leriche. One of the great surgeons of the twentieth century, he started his career in Lyons, moved to Strasburg and eventually to Paris. He published his classic *La Chirurgie de la douleur* in 1937 and his influential *La Chirurgie a l'ordre de la vie* in 1945. He died in 1955, aged seventy-eight. His contribution was important on two levels. He was a bold and innovative surgeon. He was also a humanist and the champion of anti-dolorism. Allowing for certain obvious uses, the 'dolorist' dogma – that pain was something to be valued, guarded and cherished – was to him a travesty of both religion and humanism. (Like Pasteur, he was a practising Catholic.) Almost always pain was horrible, useless and degrading. It required the unstinted effort of surgeons and physicians to combat.

At a practical level several of the operations he designed aimed at abolishing pain as such rather than trying to deal ineffectively with an ill-understood or intractable underlying condition. He wrote in 1937:

> Some physicians and surgeons claim or at least hint that pain is a useful reaction against disease. How? What pain? Against what disease? Against cancer which not infrequently gives little trouble until quite late? Against heart affliction which always develops quietly? Against consumption which rarely hurts until the illness is far advanced? Let us reject this concept of 'beneficial pain' with all our might.[1]

He railed against the heroic image of the Spartan boy who stole a fox and did not utter a cry while the hidden animal chewed off his arm. Insane! If some people wanted to emulate such lunacy, let them. If others drew inspiration from superhuman figures in literature and mythology who bore pain unflinchingly, almost with pleasure, good luck to them. But they should not

1. R. Leriche, *La Chirurgie de la douleur* (Paris, 1937), p. 37.

foist their private aberrations on others. In doctors and nurses even voicing such a preference could do harm. It could induce in ordinary people – especially in children and patients with a low threshold of pain – a tormenting feeling of guilt.

> Let it be proclaimed that it is not shameful to have a painful toe. Let it be said that it is no weakness to complain of angina. Let it be right to cry out when a neuralgia or even the imaginary spectres of madness seem unbearable. It is the doctor's and nurse's Hippocratic duty to help such people, not to preach heroism and self-control.[2]

At a time when this was far from 'obvious' and the reverse had fashionable advocates, Leriche was in no doubt that, with few exceptions, pain is incapacitating and often humiliating. It prevents victims from fulfilling their God-given potential. Being reduced to a 'screaming, writhing creature' can be a debasement. True heroism lay not in exalting but in fighting such a misfortune.

. . .

The impact of Leriche's crusade was partly at least due to his reputation as a diagnostician and operator. Patients incapacitated with angina flocked to him from every country. Members of the Moscow Politbureau travelled to Lyons incognito. They shared a floor (if not a room) with Nazi thugs similarly afflicted. Few left wholly dissatisfied. Most of the operations he designed were based on the idea that the sympathetic nerves, one half of the autonomic nervous system, are directly implicated in many forms of pain.[3] This is now regarded as unlikely. Sympathetic nerves are nevertheless responsible for the constriction of arteries and therefore for the regulation of blood flow to most organs. The deathly pallor of a woman in the grip of fear or rage or the pallor

2. Ibid., p. 134.
3. The autonomic nervous system (from *autos* ἀυτός self and νόμος law) consists of a sympathetic and parasympathetic component, the two being responsible for the normal balanced activity of unconscious functions. The term 'sympathetic' applied to nerves is Galenic; but the first to suggest the existence of two antagonistic systems regulating vital but unconscious activities may have been Eustachius, Vesalius' pupil (see Chapter 11). Thomas Willis described the ganglionated chain of the sympathetic system in 1664; but the first to use the term 'sympathique' to describe in remarkably modern terms the nervous regulation of the heart and intestines was the Danish anatomist and physician, Jacob Benignus Winslow (died in 1760, aged ninety-one). Claude Bernard did much fundamental work on involuntary nervous regulation (see Chapter 31). 'Autonomic nervous system' in the modern sense was a term introduced in 1898 by Langley, who also worked out the reciprocal action of its two components.

of a wounded man in shock is due to the reflex (and in shock life-saving) constriction of the blood vessels of the skin mediated by sympathetic nerves.[4]

The anatomy of the sympathetic nervous system is complex but one of its main features are two chains of small ganglia which descend in front of and on either side of the vertebral column. These ganglia contain the nerve cells whose extensions supply the involuntary muscles of internal organs as well as those of arteries. The internal organs include the heart, the bronchial trees and the gastrointestinal and urinary tracts. In 1897 the French physiologist François Franck suggested that intractable pain emanating from these organs might be helped by interrupting their sympathetic nerve supply.[5] The operations were put into practice by Leriche. He believed that sympathetic nerve impulses might best be abolished by the removal of the ganglia of the sympathetic chains. This, he thought, could relieve pain in three ways. First, it could put out of action vasoconstrictor fibres and thereby improve the blood supply to the organ causing the pain. Second, it could interrupt fibres conveying pain from that organ to the central nervous system. Third, it might favourably influence metabolic – that is chemical – activity in the target sites.[6] The second and third effects are now generally discounted; but the improved blood supply can bring symptomatic relief. Leriche recommended removal of one or several of the sympathetic ganglia in the neck to relieve anginal pain of the heart unresponsive to drugs. Today more effective drugs and coronary bypass surgery have virtually eliminated the need for the operation; but it helped many in its day. Leriche also popularised 'lumbar sympathectomy' – or the removal of the sympathetic ganglia in the abdomen – for intermittent claudication and even more for 'rest pain' in the lower limbs.[7] In England King George VI was a beneficiary of this operation. It enabled him to continue with his favourite relaxation, walking and shooting in the grounds of his beloved Sandringham. Eight years later he died, after a happy day out of doors, from haemorrhage following a recurrence of his carcinoma of the lung. More generally the operation was not altogether successful since the pain of

4. Cutaneous vasoconstriction, the cause of the pallor, directs as much blood as possible to essential organs like the heart and brain. Well-intentioned efforts to warm patients in shock with hot-water bottles have often proved fatal.
5. F. Franck, 'Signification physiologique de la reséction du sympathique dans la maladie de Basedow [hyperthyroidism], l'épilepsie, l'idiotie et le glaucoma', *Bulletin des hôpitaux de Paris*, 1897 (32), p. 85.
6. R. Leriche and R. Fontaine, 'La Chirurgie du sympathique', *Revue neurologique*, 1929 (4), 1046.
7. Intermittent claudication means intermittent limping, the cardinal symptom of inadequate blood supply to the lower limbs. The term 'claudication' to describe pain associated with a lack of blood supply is sometimes a little nonsensically applied to the arms and even to the gut.

claudication originates in the muscles of the leg and foot; and it is largely the blood supply to the skin which is improved by sympathectomy.

Leriche was not blind to the limitations of the surgery of pain, nor did he deny that in some individuals under certain conditions pain could be suppressed even without drugs or operations. Serving as a young doctor in Russia in the First World War – one of several medical teams dispatched by the French government to help their mighty but stricken ally – he described how

> at the express request of a number of experienced Russian colleagues who told me that it was useless to give an anaesthetic to Cossacks before operating on them, it might indeed frighten them more than the operation, I one day disarticulated without any anaesthetic, though with great repugnance, three fingers and their metacarpals of one wounded Russian comrade and the whole left foot of another. Neither one man nor the other showed the least tremor; and they raised their hand or turned his leg during operation at my request without the slightest sign of weakness, just as if under the most perfect local anaesthetic . . . Clearly the provision of a nervous apparatus to record pain is not itself enough to ensure its functioning.[8]

But he had no patience with 'the stiff upper lip' advocated by some of his army colleagues as 'the best anaesthetic'.

He also questioned the frequency of 'referred pain'. The term implies pain which is 'incorrectly' localised by patients. It follows the distribution of sensory fibres which happen to travel in the nerves whose sympathetic component supplies the diseased internal organ. It explains why pain arising in the heart muscle is sometimes 'felt' in the left arm and hand, pain in the liver is occasionally 'referred to' the shoulder and pain from the intestines is usually located on the anterior abdominal wall. Leriche did not dispute that such transposed sensations could occur; but they were, he claimed, less common than doctors suggested. The usual reason for 'referred' pain was that intimidated patients felt that they would be disbelieved if they stated that 'my gut hurts' or 'my gall-bladder is painful' or 'my kidney feels uncomfortable'. But 'why should we not believe those who tell us "My heart aches"? It probably does.' Though he was probably wrong, he was wrong in the company of Aristotle, St Jerome, Descartes and Bichat. He deserves to be cheered even for his mistakes.

. . .

Leriche was a vascular rather than a neurosurgeon and he felt ambivalent about 'heroic' operations on the central nervous system. These were developed

8. R. Leriche, *La Chirurgie à l'ordre de la vie* (Zeluck, 1945), p. 37.

and became popular during his lifetime; but cutting the sensory nerve roots as they entered the spinal cord was tried as early as the 1890s.[9] The most common indications were the sickening tabetic crises of neurosyphilis. The root sections were occasionally successful; but in most cases the pain recurred even after the nerves supposedly transmitting it had been cut. (Despite a number of clever explanations, the exact mechanism of these recurrences is still a mystery.) Untoward complications like loss of bladder and bowel control were common. The operative mortality from shock was high.

Surgery of the spinal cord itself had to wait for a clearer understanding of the disposition of the nerve tracts ascending in it. Around the turn of the century disseminated tuberculosis was still common; and several case-reports described patients who had lost pain and other sensations in their lower limbs and lower part of their abdomen due to tuberculous deposits in the spinal cord. Post-mortem examination in such cases as well as animal experiments eventually led to the mapping out of 'pain pathways'. In 1910 Schuller reported the sectioning of the anterolateral quadrant of the spinal cord in monkeys with a view to performing such an operation in tabetic patients.[10] The first 'cordotomy' in man, a horizontal cut into the spinal cord, about 2 by 3 millimetres in depth, was performed on 5 January 1911, using a fine cataract knife.[11] The patient had been in severe pain from secondary carcinomatous deposits and was relieved by the operation until his death six weeks later. In 1912 Beer performed a similar operation for 'intolerable pain due to metastases in the nervous plexus in front of the lower spine'.[12] The patient was free from pain and walked after eleven days.

A number of different and ingenious variations on this procedure were designed and tried both in Europe and in America during the following decades, moving the level of the operation nearer and nearer to the brain itself. Eventually sections began to be performed inside the skull in the midbrain. Indeed, resection of part of the sensory cortex with an electric needle

9. J.V. Crawford and A.E. Walker, 'Surgery for Pain', in A.E. Walker (ed.), *The History of Neurological Surgery*, p. 308.
10. M. Schuller, 'Die Verwendung der Nervendemmung der Rückenmarksstrange (Chordotomie)', *Wiener medizienische Wochenschrift*, 1910 (60), 2292.
11. W.G. Stiller and M. Martin, 'The Treatment of Persistent Pain of Organic Origin of the Lower Part of the Body by Division of the Anterolateral Column of the Spinal Cord', *Journal of the American Medical Association*, 1912 (38), 1489.
12. E. Beer, 'The Relief of Intractable and Persistent Pain due to Metastases Pressing on Nerve Plexuses by Section of the Opposite Anterolateral Columns of the Spinal Cord above the Entrance of the Nerves Involved', *Journal of the American Medical Association*, 1913 (60), 267.

for severe phantom-limb pain was hailed as an advance in the 1940s. It did not live up to expectations.[13]

By the late 1940s painful carcinomatous secondary deposits rather than neurosyphilis was the main indication for cordotomy. Some of the early reports are slightly ambiguous, concluding that following the operation the 'patient ceased to complain of pain' or that 'their life was made more comfortable'. Both complication rate and mortality remained high despite the technical virtuosity of many of the operators. Yet there is no doubt that in some patients with a limited expectation of life, the operation did provide welcome relief.

. . .

Opinion was always more divided about frontal lobotomy. The justification for the procedure for pain (as distinct from operations performed for mental illness) was the recognition that 'unbearable pain' was sometimes as much an emotional response as an 'objective' sensation. In his Harvey lecture in 1943 Wolf stated that 'dissociation between pain perception and reaction to pain can be induced not only by drugs and by strong religious beliefs but also by cerebral damage'.[14] One fifty-six-year-old woman with 'intractable pain' associated with an amputation stump and phantom limb had a bilateral frontal lobotomy – that is, the functional amputation of both frontal lobes of the brain, approximately 20 per cent of brain substance. In the months that followed 'she was confused but looked content and complained very little'. When questioned, she usually said that the arm still 'pained' but she did not spontaneously complain. During the next few months her confusion grew less and, though 'the pain was present . . . it did not concern her'. But distressing mental complications began to surface more generally. Some surgeons maintained that such complications could be avoided or lessened by making the lobotomy unilateral. This still, they claimed, improved or relieved pain. Others tried a more limited bilateral removal known as 'topectomy'.[15]

Few if any of these operations are performed today. Over the past decades the needle, the electrode and the fine catheter have replaced the knife as the neurosurgeon's basic instruments. Much of neurosurgery is now carried out under local anaesthesia, X-ray control and sometimes (as in operations for

13. H.G. Schwartz and J.L. O'Leary, 'Section of the Spinothalamic Tract in the Neck with Observations on the Pathways of Pain', *Surgery*, 1941 (19), 183; A.E.W. Walker, 'Mesencephalic Tractotomy', *Archives of Surgery*, 1942 (44), 983.
14. H.G. Wolff, 'Some Observations on Pain', Harveyan Lecture, *British Medical Journal*, 1943 (39), 35; H.G. Wolff and J.D. Hardy, 'On the Nature of Pain', *Physiological Review*, 1947 (27), 167.
15. J. Le Beau, 'Traitement des douleurs irréductibles par la topectomie', *Semaines des hôpitaux de Paris*, 1948 (60), 1946.

trigeminal neuralgia) using operating microscopes. New ideas about pain inhibitory impulses have led to new approaches. Theoretically – and sometimes in practice – it is possible to alleviate pain by stimulating nervous pathways which normally suppress or modulate its perception.[16]

One of the consequences of the discovery of natural opioid receptors in the brain and spinal cord has been a revival of interest in introducing opium-based drugs into the cerebrospinal fluid. This had been tried but with little success during the 1890s. Such morphinotherapy, aimed either at the critical level of the spinal cord or at the ventricles of the brain, can be effective and, unlike surgery, it is reversible. The amount of drug needed is usually considerably less than when the drug is given by injection or by mouth. When indicated, a reservoir can be implanted under the skin, a catheter delivering the analgesic to the cerebrospinal fluid; or the delivery system can be connected to a pump capable of being activated by the patient. None of these techniques is free from complications; and most neurosurgeons still regard surgery as a rational way of relieving pain only when medical treatment has failed.

16. See Chapter 48.

# Chapter 47

# THE SCHISM

Neither advances in painkillers and anaesthesia, nor developments in neurosurgery were transforming events. The triumphant rise of modern medicine – sometimes dubbed scientific medicine – between roughly 1945 and 1965 was. Diseases that only a generation earlier seemed impenetrable suddenly became treatable, controllable, even curable.[1] Less spectacular but closely linked was the beginning of the great schism between curers and the carers.

Let the cures be considered first. Nothing remotely comparable to the advances of the 1940s, 1950s and early 1960s had ever happened to medicine – or perhaps to any applied science – before. The impact of these developments on life expectancy was dramatic. Their impact on physical pain and suffering scarcely less so. It would be impossible to overstate the beneficial effect of the sulphonamides and then of the antibiotics on health in general and on pain in particular. An otherwise idyllically happy childhood exceptionally familiar to the present writer was blighted by yearly episodes of otitis media, infection of the middle ear, and mastoiditis, punctuated by operations known as myringotomies and mastoidectomies. The days and nights of pain even with the most devoted nursing and the use of every painkiller known at the time left an indelible memory. The disease is virtually unknown today, as are rheumatic fever, osteomyelitis, staphylococcal arthritis, meningococcal meningitis and a host of other terribly painful conditions of childhood virtually wiped out by penicillin. The flu pandemic of 1919–20 killed more people than the First World War. The causative virus might still be untreatable – nobody knows for certain – but a high proportion of the victims died of the secondary complication of bacterial pneumonia; and those today could be saved. When Robert Koch discovered the causative organism of tuberculosis in 1895 many regarded the cure of this ancient disease as imminent. In fact,

1. The best overview of this rise is J. LeFanu's, *The Rise and Fall of Modern Medicine* (London, 1999).

effective treatment took another fifty years to materialise. But when it did, the result was near-miraculous. The terrible sanatoriums emptied from one month to the next, the mutilating operations on the chest became horrors of the past, the spectre of the consumptive child ceased to haunt every home. The other great killer of the famous as well as of ordinary people, syphilis, also disappeared. No more Schuberts dying in their twenties; no more creative writers, poets and painters collapsing into insanity in middle life.

The next miracle drug, cortisone in 1949, spelt relief from the crippling pain of rheumatoid arthritis, a transformation no less unbelievable at the time. Nobody who has witnessed the first, still flickering 'before' and 'after' films shown by Dr Philip Hench at medical meetings can forget their impact. The 'before treatment' showed one or several crippled and deformed patients barely able to take a painful step or indeed to straighten up, images all too familiar not only to many doctors but also to the lay public. Up came the 'after treatment' images, provoking gasps of surprise sometimes leading to loud cheering and applause. They showed the same cripples jauntily climbing stairs, swinging their arms, doing little jigs. Of course 'steroids' proved to be not quite the unmixed blessing they seemed at the time; but they did virtually wipe out some of the most painful illnesses of civilisation.

Then came the end of some of the most terrible if also the most hidden pains of civilised society, those suffered by mental patients in lunatic asylums. Of all the great advances this was the least mythologised though it was in some ways the most astounding. In 1949 a young French naval surgeon working in Tunis was trying to prevent surgical shock due to blood loss, using a family of drugs known as antihistamines. His working assumption, that histamine was in some way responsible for the physiological changes in shock, was wrong; but in the course of his experiments he made an extraordinarily acute observation. One of the antihistamines he used, promethazine, seemed to have an analgesic and calming action, producing '*une euphorie et quiétude*' that made the use of morphine to deaden post-operative pain unnecessary. It was as if the drug 'disconnected some brain functions, producing a state of complete calm and tranquillity without depression of mental faculties or clouding of consciousness'.[2] He stimulated the interest of a French drug company; and, after extensive experimentation on animals with a series of related compounds, they chose one which seemed the most promising. Chlorpromazine did not cure schizophrenia but it made the life of hundreds of thousands of schizophrenics (and the burden on their families) bearable, almost pain-free for some time. It also put a virtual end to the

2. H. Laborit (1949), quoted by LeFanu, ibid., p. 69.

terrible 'physical methods' of treatment like electroconvulsion, insulin coma and lobotomy.

Like chlorpromazine in schizophrenia, levodopa only provided partial symptomatic relief in Parkinson's 'shaking palsy'; but it was almost like a return from the dead for thousands of sufferers. It also promised more drugs based on an increasing understanding of neurotransmitters.

Three years after chlorpromazine therapy in mental illness came open heart surgery, then cryoprecipitate for haemophilia and in 1961 hip replacement. The last followed a dozen or so unsuccessful attempts at dealing with the pain of osteoarthritis of the hip, one of the most limiting conditions of the second half of life. Introduced by John Charnley, a North Country lad born and bred in Bury and eventually orthopaedic surgeon at Manchester, its evolution entailed more than one setback and near-disaster before it became the gold standard of reconstructive joint surgery.[3]

The first successful kidney transplantations and coronary bypass operations were the main surgical achievements of the early 1960s; and in 1967 the first heart transplant sent the popular press into a frenzy, the image of the 'healing hands' of Christiaan Barnard gracing the cover of the French edition of *Vogue*. An extraordinary succession of lucky accidents and inspired guesses spread over twenty years made some forms of childhood cancer, notably acute lymphoblastic leukaemia, curable.[4]

. . .

There was a price to pay for these breathtaking advances, though this was not immediately apparent. In the early 1960s Helen Neal, a health journalist, was working at the National Institute of Health (NIH) in Bethesda, Maryland. The NIH was (and is) by far the largest institute of medical research in the world. Not only is it the home of some of the most high-powered research teams in the United States: it also supports the majority of biomedical research programmes in American universities and many such programmes in other countries. In 1964 Helen Neal's brother John developed cancer of the tongue. 'A cracker-jack salesman, his success depended not only on the quality of his merchandise but on his ability to talk about it.'[5] His treatment got off to a bad start when excessive irradiation burned his mouth and neck. The analgesics prescribed proved wholly inadequate. In desperation he asked his sister to make enquiries. Working at the hub of advancing medical science, Helen promised to find out what could be done about the pain. In this she was utterly unsuccessful. In the vast organisation dealing with every aspect of medicine

3. W. Waugh, *John Charnley, The Man and the Hip* (Berlin, 1990).
4. M.M. Wintrope, *Haematology. The Blossoming of a Science* (New York, 1985).
5. H. Neal, *The Politics of Pain* (New York, 1979), p. 10.

and surgery, including the outside units sponsored by the organisation in five continents, *there was not one dedicated to the study of or relief from pain.*

...

The proverbial visitor from Mars might have concluded that the condition for which Helen Neal sought advice was so rare that it could not be expected to command the interest or occupy the precious minutes of people engaged in advancing the health of the multitudes. The opposite was true. Chronic pain, often labelled 'untreatable', was not only exceedingly common in the technologically advanced countries of the world; it was getting *more* common by the day. In the light of the triumphs frequently paraded and sketched out above, this might seem an overstatement; but there was a sequel to every advance. The cure of childhood leukaemias was wonderful; the anguish caused by the monitoring and treatment, the frequent and painful bone marrow examinations often left to junior doctors, less so. Of course the price was infinitely worth paying, as were the side-effects of steroids and other life-saving treatments; but it was still a price. The very fact of prolonged survival often meant more prolonged pain. Helen Neal called it 'the silent epidemic'. But why was nobody taking any notice of it?

In a historical context pain was a condition that a hundred or even fifty years earlier had taken up the bulk of the thought and time of doctors. This was true even of doctors practising during the Seminal Years of the early 1900s and taking pride in the scientific achievements of their generation. They could take comfort from the thought that such advances often led to practical improvements; but these improvements usually took decades to materialise. In the meantime what doctors could do – and did – was to try to ease the suffering of their patients, most importantly to relieve their patients' pain. A glance at Luke Fildes' *The Doctor* in Tate Britain – the artist's tribute to his own doctor who looked after his dying daughter – tells the tale more eloquently than words could. To relieve suffering was difficult, sometimes nearly impossible. It always required time, experience and intense concentration of thought. But it was often all doctors could do. And many doctors were, given their limited armoury, exceedingly good at it. By the mid-twentieth century the picture was changing. Many of the painful conditions which had occupied the minds of earlier generations could now be treated, controlled, sometimes even cured. The treatment, control and cure of others was (to adopt the political jargon of the time) in the pipeline. These splendid enterprises were often complicated, requiring great skill, judgement and, above all, alertness. New developments had to be instantly spotted. So time-consuming were these obligations, yet so gratifying the potential benefits, that they came to occupy most of the doctors' attention. 'Untreatable' chronic pain was by definition not one of the newly treatable conditions. Time and care spent on such 'chrons' was almost a waste of time And so the Great Schism began. Medicine as practised by the up-and-coming avant-garde, the flower of the high tech and high-intelligence medical schools, was transformed into the realm of curing. The realm of caring was left to the old fogeys, immured in their dingy surgeries, or to blatant confidence tricksters (often referred to as 'shunts'), paid by the second for spouting reassuring platitudes in their stuccoed establishments in Harley Street. The two almost never met.

Two areas seemed to suffer most from the split. One was chronic 'non-specific' pain, 'non-specific' in the sense that the diagnosis was purely descriptive, like 'backache'. The other was the care of the terminally ill. The way in which these areas were treated – or rather ignored – by standard textbooks of the time is nearly unbelievable today. 'Unfortunately,' in 1968 one highly acclaimed American text concluded a long chapter on the causes and management of backache, 'all the causes listed above [thirty-two in total, including not only tuberculosis of the spine, secondary carcinoma and prolapsed intervertebral discs but also a selection of rare tropical ailments] still leave a significant number without an etiological diagnosis. The drugs that are effective in controlling such non-specific pain are habit-forming and must therefore be used extremely sparingly or not at all.' So what *can* be used? So nothing. End of chapter. A popular and practical British textbook entitled *The Principles and Practice of Medicine for Student Doctors*, was a manageable 620 pages. Not quite *one line* in those 620 pages was devoted to the pain of patients with advanced carcinoma. It went: 'Analgesics should be given to control pain but without leading to undue dependence or addiction.' But how could dependence and addiction be avoided? And why should it be? None of these questions were addressed, let alone answered.

To a generation of medical students brought up on such textbooks the term 'symptomatic treatment' became an expression of contempt. Of course symptoms had to be dealt with even if the underlying disease could not be cured; but this was strictly economy-class medicine. Certainly not something to occupy the quality time of specialist consultants, scanning the latest issues of the *Lancet* and the *British Medical Journal* for yesterday's (or, even better, tomorrow's) therapeutic breakthrough. In Britain these were the early decades of the National Health Service, when the prestige of general practice was touching an all-time low. Symptomatic treatment was just the thing to 'keep GPs gainfully occupied'.

...

Predictably perhaps in the light of these developments, the dramatic advances in curative medicine coincided *not* with an upsurge of faith in medicine but with the beginning of a shift of public trust from orthodox to alternative therapies. Contrary to what is sometimes implied, until the Second World War in most Western countries, unorthodox treatments were used by only a tiny minority of the public. Rightly or wrongly, the educated classes regarded them as quackery, sometimes as a form of entertainment. By the late 1950s, naturopathy, aromatherapy, reflexology, acupuncture, hypnotherapy, hypnosis, let alone a number of quasi-religious cults, had a following that put such old-fashioned heterodoxies as homoeopathy in the shade. The reason was not a change in the alternatives but a change in the orthodoxies. Almost symbolically, in most 'with-it' hospitals the kindly 'consultants in physical medicine', expert in massage, hot and cold compresses, poultices, remedial exercises, mineral baths, mild electrical stimulation and a host of other excellent but old-fashioned remedies were being replaced by 'rheumatologists', often as toxic in their arrogance as they were in their up-to-date treatment. To a fast-growing number of sufferers the care and undivided attention of the alternatives seemed infinitely preferable to being overinvestigated, overprescribed, overtreated and overborne on brief hospital encounters with the new species of consultant. Often, when the latest drug cocktail failed, patients were more or less openly blamed for the failure. 'Your pain is all in the mind' was the ultimate put-down. In sensitive patients, especially in children, it was neatly calculated to add a new and potentially even more destructive sensation to the pain: a sense of personal failure.

. . .

The first positive reactions to this trend were often incoherent and even misguided. First reactions often are. Dr Grantly Dick-Read, a young obstetrician in Woking, had a messianic streak and sensed that all was not well with the care of women in labour. He asked four questions. Why is labour so often painful? Why is it so often interfered with? Why are so many drugs used? Why should a woman be deliberately stupefied for the arrival of her children? None of this, he suggested, was biologically necessary. None of this happened among 'less cultured races'.

> I saw a woman walk away from a harvest field, exalted and laughing, with her baby wrapped up in a cloth: her baby less than half an hour old.[6]

The Natural Childbirth Movement which he launched with his most famous book, *Childbirth without Fear* (1933), swept round the world. By 1960 the text had been translated into twenty-five languages and acclaimed by women's organisations and by a sprinkling of practising obstetricians. There was never any doubt of Dick-Read's sincerity or nobility of vision.

6. G. Dick-Read, *Childbirth without Fear* (reprinted New York, 1980), p. 82.

> No woman who remembers her child's birth ever ceases to love that child, and no child who has been born in love and learnt of its mother's love, ever ceases to love its mother ... and so more unselfish love will fill the world . . . and all the actions and ambitions of men and women will be influenced from selfishness to the path which is followed by love . . . [Men and women] would then abolish poverty, distress and misery among the masses and would change the character of their society and eventually of the world.[7]

What was perhaps most remarkable was the positive or at least questioning attitude this provoked among some level-headed practitioners. Despite the wedge of mistrust which Dick-Read was incidentally but deliberately driving between women and the obstetric establishment, many of the establishment sensed that, whatever was wrong about Dick-Read's ideas, there was something even more wrong with 'orthodox' obstetrics.[8] And much of what was wrong with orthodox obstetrics was also wrong with scientific medicine and surgery.

. . .

Perhaps a minority – or were they still the silent majority? – of doctors and members of the other healing professions had always felt unhappy about the apparent separation of caring and curing. By the late 1960s, slowly, extremely slowly (or so it seemed), attitudes began to change. Three trains of events in particular were set in motion.

First, some notably well-equipped scientific and medical minds set themselves the task of reviewing and if possible improving on prevailing notions about the nature and mechanism of pain. Some of the results are complex, their understanding not always helped by mind-boggling diagrams. They nevertheless provide the basis of present and future means of dealing with clinical suffering. Second, the outlandish idea of specialised pain clinics was conceived by a professor of anaesthesia in Seattle and slowly translated into practice world-wide. Third and perhaps of greatest practical impact, the hospice movement and palliative medicine were launched for treating advanced disease. Each of these deserves a chapter.

7. Quoted by D. Caton, *What a Blessing She Had Anaesthesia* (London, 1999), p. 183.
8. The feeling was not dispelled even when some years later objective testing showed the limits of what could be achieved following Dick-Read's practical advice. Though dedicated training in pregnancy, the central idea in his doctrine, significantly diminished the pain of labour, the 'pain score' in both trained and untrained groups remained high.

# Chapter 48

# PAIN MECHANISMS

Pain has never been a theoretical or purely experimental science; but, since new ideas have influenced the beginnings both of pain clinics and of palliative medicine, the ideas have to come first. One of the signs of resurgent interest has been a search for a new definition. From practising doctors this may elicit a groan. Scientists have always had an urge to express the inexpressible. If pain (or pleasure or hot or cold or red or green) could be better defined, there would be no need for words like pain (or pleasure or hot or cold or red or green). 'What is pain?' has nevertheless been once again tossed about; and the search even for the unattainable can be illuminating. The currently fashionable formula was suggested by a group of 'concerned scientists and doctors' convened by Harold Merskey of the University of Western Ontario in 1979. It runs as follows:

> Pain is an unpleasant sensory and emotional experience associated with actual or potential tissue damage, or described in terms of such damage.[1]

More instructive than the definition (which introduces several terms in need of further definition) are the appended comments. One addresses the problems of 'real' versus 'imaginary' pain.

> Many people report pain in the absence of tissue damage or any likely pathological cause. Usually this happens for psychological reasons. There is no way to distinguish their experience from that due to tissue damage, if we take the subjective report. If they [the patients] regard their experience as pain caused by tissue damage, it should be accepted as pain.[2]

Pain, in other words, is not a direct function of the stimulus. Nor is the reverse true.

1. H. Merskey, 'Pain Terms', *Pain*, 1979 (6), 248.
2. Quoted by Wall and Jones in *Defeating Pain*, p. 178.

> Activity induced by nociceptors and nociceptive pathways by noxious stimuli is not [in itself] pain. Pain is always a psychological state, even though we may appreciate that it often has a proximate physical cause.[3]

To understand the meaning of the last comment it is necessary to return to the physiological mechanisms which are currently thought to underlie the experience of pain.

. . .

Few scientists now believe in the specific 'pain receptors' that were envisaged and widely accepted a hundred years ago. Sensory nerves have an 'ending' – not as clearly visualised as one might wish – and these endings are sensitive to *changes* in pressure, temperature, pH (that is acidity or alkalinity) and the concentration of a variety of chemicals. In the last category probably the most important are substances leaked out or generated by sick or damaged cells. They may include so-called 'free radicals', a highly unstable and active chemical species whose existence in living material was at one time doubted but which are now regularly invoked to explain otherwise inexplicable phenomena.[4] If the change is big enough, it elicits a sudden 'explosive' event in the nerve fibre. It is an all or nothing response and is called a nerve impulse. In any one nerve fibre its character never changes. What changes is its frequency. The bigger the explosion – that is, the tissue damage – the more impulses are fired off per unit time. Like the lighting of a fuse, the impulse sweeps along the nerve fibre and reaches the parent cell.[5] It may be recalled that Helmholtz assigned a standard speed to these impulses.[6] This was an inspired oversimplification. It is now known that in the largest, so-called A fibres, the speed may reach 100 metres per second. In the smallest, so-called C fibres it is nearer 0.25 metres per second. Impulses that are set up by tissue damage are called 'nociceptive'. (Descartes is currently out of fashion among pain scientists; but he would have described 'nociception' as *dolor* or pain and he would not have been far wrong. But 'nociception' satisfies the even more ancient urge for long-winded mystification.) Impulses in peripheral nerves can be prevented from reaching the spinal cord by a local anaesthetic applied to the nerve fibre at any point in its course. When this is done, no message is received. Precisely how this happens at the molecular level is still unclear.

3. Quoted ibid., p. 179.
4. See T.L. Dormandy, 'Free Radical Reactions in Biological Systems', *Annals of the Royal College of Surgeons*, 1980 (62), 188.
5. The impulse is accompanied by a reversible migration of sodium and potassium ions across the membrane of the fibre.
6. See Chapter 31.

Sensory impulses like pain are called 'afferent' because they travel towards the central nervous system. Motor commands travel in the opposite direction in 'efferent' fibres. The parent cells of the sensory or *afferent* fibres are situated in ganglia, pea-size collections of neurons which form two vertical chains in front of and on either side of the spinal cord. Extensions from these neurons enter the grey matter in the spinal cord as the so-called 'anterior nerve roots'. The grey matter in the spinal cord has a roughly X shape in cross-section, with two anterior 'horns' (at 10 and 2 o'clock) and two posterior 'horns' (at 4 and 8 o'clock). The anterior nerve roots make for the anterior horns where they synapse with a second set of neurons. The second set of neurons sends fibres to the subcortical nuclei and to the sensory cortex. The ascending fibres are bunched together to form tracts in the white matter of the spinal cord. The whiteness, it may be recalled, comes from the myelin sheets which, like insulator tape but contained inside specialised cells, envelop each fibre.[7] Sir Charles Sherrington, who named the junctions between neurons 'synapses', also showed that these are not simple relay points. Within limits they can compute, compare, control and decide what will or will not pass. Perhaps they even initiate new impulses. Sherrington's contemporary, Pavlov, demonstrated that a sensory impulse can take on a new meaning by being repeatedly associated with another sensory impulse.[8] In many ways they anticipated all recent developments.

What P.D. Wall, professor of anatomy at University College, London, and Ronald Melzack, professor of psychology at McGill University, Montreal, showed in the 1960s was that pain messages can be drastically modified at the level of the anterior horn cells. On arrival from the periphery they interact not only with other synchronous input but also with impulses descending from higher centres conveying past experiences and even future intentions. Wall and Melzack called it the 'Gate Control of Pain'. It is (as such pictorial fancies go) a good name because it hints at the way painful sensations are let through, turned back, delayed or altered after their arrival in the spinal cord. Over the years the authors have themselves elaborated their original ideas; and it is likely that Gate Control will eventually be almost entirely superseded. In contrast to works of art and literature, this is in the nature of all useful scientific creations: in science only the mistakes never change. While it lasts, the theory provides a useful model. Among its attractive features is that it explains some of the puzzling aspects of how pain is perceived or not perceived.[9]

Pain is almost never a single event. The flash of pain of a twisted ankle is followed by a dull ache and tenderness. The pain and tenderness may spread.

7. See Chapter 31.
8. See Chapter 39.
9. Wall, *Pain, the Science of Suffering.*

New sensations may arise during healing. All these are transmitted by different and specific nerve fibres, some not by 'classical' nerve impulses as described above but by chemical neurotransmitters transported relatively slowly along nerve fibres. But most usefully, the theory accounts for the extraordinary variability with which apparently the same pain can be felt or not felt by different individuals under different circumstances.

Every Victorian schoolchild was familiar with Dr David Livingstone's encounter with the lion; but for non-Victorian schoolchildren the adventure may be briefly recalled. One day, while exploring central Africa, the doctor was attacked by a lion. He noticed the beast too late and barely had time to deliver two shots. These were not enough to stop his attacker.

> The lion caught my shoulder as he sprang and we both came to the ground below together. Growling horribly close to my ear, he shook me as a terrier dog does a rat. The shock produced a stupor similar to that which seems to be felt by a mouse after the first shake of the cat. It was a sort of dreaminess in which there was no sense of pain nor feeling of terror, though quite conscious of all that was happening. It was like what patients partially under the influence of chloroform describe, who see all the operation but see not the knife. This singular condition was not a mental process . . . The peculiar state is probably induced in all animals killed by carnivora, and, if so, is a merciful provision by our benevolent Creator for lessening the pain of death.[10]

Whatever the nature of the sensation, it was the lion who collapsed and died first, perhaps as oblivious of the shots that killed him as Livingstone was of being mauled. Yet Livingstone's injuries were serious. 'Besides crunching the bone into splinters,' he recorded, 'he left eleven teeth wounds on my upper arm.'

A less familiar case was recounted by Baron András Halassy, a dashing Hungarian hussar general who in his youth served as aide-de-camp to the Archduke Carl of Habsburg. It was in the midst of the battle of Aspern in 1808 that he and a colonel were dispatched by the archduke to deliver a message to a subordinate commander on the right (or perhaps left) wing. Galloping across the battlefield Halassy noticed that a French cannon-ball had carried off the colonel's right leg. 'Your right leg, sir, has just been shot away,' he helpfully informed his senior officer. The colonel looked at his leg (or where his leg had been until a moment ago). '*Donnerwetter*, so it has,' he acknowledged, not slowing down. Not till they had arrived at their destination and had delivered their message did he slip out of the saddle, and, deathly pale, faint on the ground. In an effort to stanch the blood still pumping from the stump Halassy

10. D. Livingstone, *Travels and Researches in South Africa, Including a Sketch of Sixteen Years' Residence in the Interior of Africa* (Philadelphia, 1859), p. 16.

applied his belt as an improvised tourniquet to the colonel's thigh. As he pulled the tourniquet tight, the colonel recovered and screamed at him: 'Stop it, you fool, you're hurting me!'[11] The 'gate' which had been kept firmly shut during the ride was now clearly open.

The gate can act in the opposite way too, so that a wound which has been quiescent can suddenly be reactivated. In Heine's wonderful poem, *Die beiden Grenadiere* (set to even more wonderful music by Robert Schumann), two grenadiers of Napoleon's defeated army who had been imprisoned in Russia reach Germany. There they learn that the Emperor had been beaten and taken prisoner. 'Wie brennt meine alte Wunde' (how my old wound starts to burn) one says to the other.[12]

Gate Control partly at least explains some of the most distressing forms of pain. Amputations are often pictured as the outcome of the violence of wartime. They are in fact far more common in peacetime, especially in an ageing population prone to degenerative vascular disease. Motorcycle accidents can be responsible for some of the most horrible lesions, including the avulsion or tearing away of the whole plexus of nerves in the armpit. After traumatic amputations the immediate sensation is the result of tissue damage, the 'classical' pain mechanism. But in a high proportion of cases sooner or later a new kind of pain appears. Known as 'phantom limb' (or sometimes 'phantom breast'), it is a malign invention of the brain. It arises in the neurons in the spinal cord or cortex which have been suddenly deprived of the continuous input of sensations. Such an input seems to be necessary for them simply to be ticking over. (One is unconsciously aware of one's limbs and other body appendages all the time without actually registering their presence. Any change in their position is instantly noticed.) At first the response of the deprived neurons is not exactly painful but more an awareness of the 'phantom'. But often this progresses to a burning or cramp-like pain. Eventually it can become terrible agony. Some of it may be due to the cut nerve fibres in the stump sprouting into unfamiliar tissue, picking up unusual chemicals. (Occasionally the stump displays points of exquisite tenderness.) But most of the pain reflects the activity of hyperexcited cells in the central nervous system deprived of their previously continuous calming input. The name applied to the phenomenon today is 'deafferentiation', meaning the stoppage of afferent messages. It is by no means confined to cases of amputation. Whenever a sensory nerve is cut, damaged or destroyed – whether by injury, by a virus like herpes, by some metabolic abnormality like diabetes or by a nerve poison – the parent cells can go into a frenzy of activity. Sometimes

11. A. Halassy, Báro, *Emlékezéseim* (Kassa, 1868), p. 45.
12. Probably the greatest exponent of this song was Fédor Chaliapine who recorded it on 11 March 1929 in London.

this is perceived as no more than discomfort. However, sometimes the pain comes on as an explosion. Anxiety and grief often act as triggers, as can physical illness. The sheer 'expectation' of a paroxysm can start a paroxysm.

Deafferentiation pains remain among the most intractable: they rarely respond even to strong conventional analgesics. Understanding the mechanism offers a glimmer of hope. Surgery is almost always futile: in Leriche's words, 'the pain always runs ahead of the knife'.[12] A phantom pain may start in the stump of an amputated finger and be referred to the lost digit. The patient may ask the surgeon to cut the nerve at the level of the wrist or forearm to stop the agony. But cutting the nerve at the wrist or in the forearm merely displaces the pain to a higher level. One of Leriche's patients had half a dozen operations trying to get rid of his phantom – and was begging for some more. Ideally the solution would be to restore in some artificial way the normal – or near-normal – afferent input to the deprived cells.[13] This is what the technique of transcutaneous electrical peripheral nerve stimulation (TEPS) attempts. It is too early to say how successful it will eventually prove to be. The concept could also account for some of the effects of acupuncture, the needles setting up 'calming' impulses acting on deafferentiated nerve cells.

...

There are many kinds of pain which the Gate or any other modern theory still cannot explain. This is even more true of absences of pain. Congenital analgesia has already been mentioned and its causation remains a mystery.[14] Such abnormalities are almost invariably caused by inherited single-gene defects; these usually manifest themselves in a defective or deficient single molecule. Most modern pain scientists firmly repudiate the idea of a 'pain centre' as envisaged by Descartes and his followers. The modern 'equivalent' of an anatomical pain centre would be a specific enzyme or neurotransmitter concerned with pain only and defective in congenital analgesia. If such a 'Cartesian' molecule exists, it remains to be discovered. If it were discovered, it might upset the 'anti-Cartesian' band wagon.

...

12. Leriche, *La Chirurgie de la douleur*, p. 154.
13. The technique was developed in response to the 'Gate' concept of pain. The underlying idea is that selective stimulation of nerve fibres might 'replace' the missing normal input and suppress the generation of pain signals. TEPS is now available in three forms: continuous stimulation, pulsed (burst) stimulation or as a form of acupuncture (high-intensity bursts). It has been reported as beneficial in pain syndromes associated with bony metastases, painful vertebral collapse and nerve-root compression. For a detailed discussion see H. McQuay and H. Filshie, 'Transcutaneous Electric Nerve Stimulation in Chronic Pain', in H. Doyle, G. Hanks and N. MacDonald (eds), *Oxford Textbook of Palliative Medicine*, 2nd edn (Oxford, 1998).
14. See Chapter 43.

If a Cartesian pain molecule is still missing, other molecules, just as unexpected, *have* emerged. In the 1970s neuroscientists in several countries reported the existence of receptor sites on the surface of certain brain cells specifically for morphine and heroin.[15] This was, as one of the discoverers remarked, like an intrepid explorer entering an unexplored cave in the middle of nowhere and finding a friendly cup of tea waiting for him inside. He would be forced to conclude that the cave was not, after all, quite as unexplored as he had supposed. The metaphor may be far-fetched; but the idea of natural morphine receptors was, at the very least, surprising. By the 1970s scientists had come to accept the existence of specific receptor sites on certain cells for hormones and other products of the body; but one could not expect such natural havens to exist for drugs derived from plants, let alone manufactured in factories or the laboratory. Their presence suggested that there may exist in the brain naturally occurring substances chemically closely related to or even identical with morphine or heroin. And so it proved.

The first 'endorphine' was discovered in the early 1970s by the American neuropharmacologist, Candace Pert while still working for her Ph.D. degree.[16] A whole kinship has now been identified in key parts of the central nervous system (and perhaps elsewhere), concerned with the attenuation of pain. They are called enkephalins. The fact that they are not only biochemically but also functionally at least siblings of morphine can be demonstrated with the drug naloxone. Naloxone is known to block the specific enkephalin receptor sites. It also heightens the severity of painful sensations.[17] Enkephalins have been shown to be released under a variety of conditions, including physical exercise. Is it possible that the exhilaration experienced by joggers – or 'jogger's high' – is due to the release and action of such 'natural morphines'? And, more important, can one can get addicted to naturally occurring enkephaline in the same way as one can get addicted to morphine and heroin?

These questions are still being addressed. At a simplistic level, jogging can certainly become a form of addiction: at least, those who jog regularly feel deprived when they cannot indulge their habit. But the mechanism of pain-

15. C. Pert and S. Snyder, 'Opiate Receptors: Demonstration in Nervous Tissue', *Science*, 1973 (179), 1011.
16. C. Pert, *Molecules of Emotion: Why You Feel the Way You Feel* (New York, 1997).
17. N.J. Woolf and P.D. Wall, 'Endogenous Opioid Peptides and Pain Mechanisms: a Complex Relationship', *Nature*, 1983 (286), 155. But this is still a controversial field. B. Olausson, E. Ericson, C. Ellmarker, C. Rydenhgag, B. Shyu and S.A. Anderson, 'Effects of Naloxone on Dental Pain Threshold following Muscle Exercise and Low-frequency Transcutaneous Nerve Stimulation: a Comparative Study in Man', *Acta Physiologica Scandinavica*, 1987 (126), 299.

Naloxone also diminishes the analgesic effect of acupuncture, suggesting the possibility that mechanical stimulation by the needles might release natural enkephalins.

relieving enkephalins differs in several respects from that of morphine and heroin.[18] It may be a question of more accurate targeting. It may be a question of dosage. Natural enkephalins are released in minute doses at specific sites where pain might be registered. They are rapidly removed by local enzymes as soon as they have acted. Drugs have no such targeted distribution. They act on the whole brain. Nor are they necessarily removed as soon as they have acted. Enkephalins (and other neurotransmitters and hormones) are sometimes likened to keys fitting into keyholes represented by receptor sites. The limitation of the metaphor is that keys and keyholes do not get tired. Transmitters and receptor sites do. Their interaction, like the serial handshakes of electioneering politicians, tends to become limp after a time.

Yet there are also striking similarities. The danger of habituation and addiction seems to be significantly greater in those who take morphine and heroin for their effect on general well-being than in patients who would be in severe pain without them.[19] The implications of such a 'balance' between pain and pain relief – whether endogenous or exogenous – for palliative care are momentous.

. . .

Beyond revolutionary enkephalins and new theories with catchy names, Helen Neal – whose book, *The Politics of Pain*, has been such an effective wake-up call – would not nowadays have to search in vain for research groups engaged in studying pain. There is hardly a university, medical school or science faculty in which somebody somewhere is not exploring a corner of the field. There is a flourishing International Association for the Study of Pain which holds regular meetings in various agreeable spots on the globe. Prizes are awarded for research on pain. A respected international scientific journal, *Pain*, is dedicated to disseminating new discoveries. Such developments are not necessarily unmitigated blessings. Anybody following the evolution of a new, exciting topic in science will sometimes find that, as the topic becomes respectable and even trendy, conferences tend to become devoted to the researchers rather than to the research. But two valuable practical developments – pain clinics and hospices – have been linked to growing academic interest in the subject.

18. I.N. Robins, D.H. Davis and D. Nurco, 'How Permanent was Vietnam Drug Addiction?' *American Journal of Public Health*, 1974 (64), 83; J. Wildeman, A. Kruger, M. Schmole, J. Niemann and H. Mathaei, 'Increase in Circulating Beta-endorphin-like Immunoreactive Correlates with the Change in Feeling of Pleasantness after Running', *Life Sciences*, 1986 (38), 997; S. Greenfield, *The Private Life of the Brain* (London, 2000).
19. R.G. Twycross 'Opioids', in P.D. Wall and R. Melzak (eds), *Textbook of Pain*, 3rd edn (Edinburgh, 1994).

## Chapter 49

# PAIN CLINICS

The inventor of pain clinics, John Bonica, was born in 1917 in Filicudi, one of the Aeolian Islands off the north-east coast of Sicily. A beautiful but barren volcanic rock, it had supported a penal colony since Roman days but never much else. In 1927 his family followed generations of forebears in seeking a better life in the United States. Five years later Bonica's father died and the boy assumed responsibility for his family, now living in New York. In the circumstances his choice of medicine as a career was almost foolhardy. He shone shoes, hawked newspapers and sold fruit and vegetables from barrows. Then he took up exhibition wrestling and paid his way through medical school, travelling with circuses during the summer. He used the name Johnny Bull or the Masked Marvel in his sporting incarnation. Among the legacies of his gladiatorial career was a multiplicity of chronic joint aches which added personal authority to his later pronouncements on pain. At twenty-seven, he was made chief of anaesthesia of a 7,000-bed army hospital. He added 'pain relief' to his official brief and taught himself the rudiments of the art. Ten thousand war wounded flown in from Asia and Europe passed through his hands. In 1960 he was appointed first professor of anaesthesia at the Washington Medical School in Seattle. It was here that, with a nursing sister, Dorothy Crowley, and a neurosurgeon, Lowell White, he opened the first pain clinic.[1]

There had been sporadic attempts to establish special units to deal with intractable pain in Japan, Scandinavia and perhaps elsewhere; but the idea of a new 'multidisciplinary discipline' dedicated to the task was Bonica's. Endearingly inarticulate but impassioned, he became one of the most successful persuaders of the twentieth century. The first edition of his 1953 book, *The Management of Pain*, was ground-breaking. 'I have declared war on pain' was one of his resonant phrases. Its message proved more enduring than

1. John Bonica died in 1994.

the wars intermittently declared by politicians on communism, capitalism, corruption, crime, poverty, drugs, 'terror', let alone non-existent weapons of mass destruction. At the practical core of his idea were two beliefs. First, pain deserved to be treated even when its cause was unknown or untreatable. Second, such treatment could be effective only through the combined effort of doctors, psychologists, nurses, physiotherapists and, when indicated, other health professionals.

It was a good moment for such a venture. The Great Schism in medicine was gathering momentum. In many and unexpected fields medicine was scoring astonishing successes. But the triumphs seemed to increase rather than decrease the number of people needing long-term care for pain. And the time available to cope with such unfortunate side-effects of success seemed to be shrinking.

...

New specialties tend to be unpopular with established professions. This is not a recent phenomenon. In Britain excessive specialisation had for centuries been condemned as a form of quackery by the medical and surgical royal colleges. Even today doctors pay lip service to the ideal of the 'general surgeon' and 'general physician'. It makes sense. Patients looked after by half a dozen super-specialists in self-proclaimed 'centres of excellence' often lose out to those treated by less specialised single doctors in hospitals which do not aspire to ministerial accolades. But occasionally a new specialty seems to meet a need. This was true of chronic pain relief in the 1970s and 1980s. The top echelons in many specialties were fully engaged in saving lives. The care of patients who seemed insufficiently grateful for being alive and still harped on a multiplicity of wearisome symptoms was pushed down the career ladder. In hospitals such 'follow-up cases' rarely saw the same doctor twice. What united the ones they did see was their often well-meaning inexperience. Many doctors also felt bewildered and confused by the manically shifting battery of new analgesics, tranquillisers and sedatives. The changes provided both a negative and a positive reason why the pain clinic idea should be acceptable.

But something more than a notice on a hospital door saying PAIN CLINIC was needed. At a time when a 'multidisciplinary approach' to everything has become one of the mantras of problem-solving, the impact of the idea on mid-twentieth-century medicine is hard to convey. In London's prestigious teaching hospitals a consultant surgeon in a career spanning fifty years had probably *never* exchanged a word with a physiotherapist to whose care he regularly referred his patients, a social worker (who often had to cope with his failures as well as his successes), a psychologist or a clinical biochemist, let alone members of the *demi-monde* like osteopaths, herbalists, reflexologists, acupuncturists or homoeopaths. Professional relations between doctors and nurses were close but rigidly formalised in the mould set for all eternity by Florence Nightingale. To bring these people together on an equal footing dedicated to a common task was not going to be easy. The very way of presenting a patient's case history to a group of professionals created divisions. Doctors traditionally eschewed personal observations. In medical case notes even such comments as a 'a nice chap' or 'a little loquacious' were frowned upon: they were deemed irrelevant to the patient's angina or migraine. Nurses, by contrast, tended to describe patients in personal terms, even as good or bad company, as well as repositories of symptoms and signs. Niceness to them was important. Inevitably, the multi-professional teams sometimes came to grief, at least initially. B. Tuckman identified five stages in their evolution: 'forming, storming, norming, performing and mourning'.[2] But a common purpose can work wonders; and in a surprising number of cases a constructive *modus operandi* did emerge.

Opposition to the establishment of such clinics was surprisingly muted. Some motives were ignoble. Pain clinics would act as sinks for difficult patients. But there was also a genuine feeling of helplessness among established specialists, of a need not met. Pain had never been part of the curriculum. The role models in hospitals seemed to be as helpless as their flock. In less then ten years the new specialty became part of most large hospitals in Britain and most of Western Europe. Some pain clinics were initially staffed by volunteers; today they are usually financed by hospital budgets. A more important question is: do they work? They probably work as well (or as badly) as any more established hospital department. Patients who fit into the organisational framework, the ethos and the style of the clinic benefit enormously. Those who do not will be disappointed. But even if the staff of pain clinics do recognise 'difficult' cases – they would have to be saints not to – it is a matter of principle not to label any as untreatable. And across the board, the clinics have significantly raised awareness of the growing and multidisciplinary problem of chronic pain of uncertain origin.

2. B. Tuckman, 'Development Sequence in Small Groups', *Psychological Bulletin*, 1965 (63), 384.

# Chapter 50
# HOSPICES

The modern hospice movement and the development of palliative medicine may one day be seen as the most important advance in pain control since general anaesthesia. It may also be recognised as one of the few wholly positive achievements of the second half of the twentieth century.[1] The term 'hospice' itself is ancient, one of the long list of words derived from the Latin *hospes*, meaning a guest or stranger in a foreign land. Institutions dedicated to the care of the dying probably existed in most early civilisations; but the term was first used in a recognisably modern sense in early Christian Europe when the task was identified by the Church as a duty ordained by Christ.[2] Many of these early hospices were part of larger religious foundations and were either dissolved or transformed into homes for the destitute or hospitals with wider medical aims during the Reformation. Paradoxically (or not), it was the age of the religious wars, one of the most hag-ridden and intolerant in European history, which saw charitable foundations blossom as never before.[3] The Age of Reason, the industrial revolution and the Age of Science seem, in retrospect, to have promoted individual and somewhat abstract rather than institutional benevolence. The word was revived in 1842 when a young French widow, Mme Jeanne Garnier, who had also lost her two children, opened the first of her refuges for the dying in Lyons and called it 'Hospice' or 'Calvaire'.

1. This may seem a wilful exaggeration in the light of the breathtaking advances of scientific medicine and science in general; but few of those have not had their debit side, even if greatly outweighed by the benefits.
2. See Chapters 7 and 9.
3. This is an aspect of baroque Europe which has been oddly neglected by historians. It cut through the religious divide. In the Protestant Netherlands charitable foundations embraced the care of cripples, the homeless, the mad, the leprous as well as the sick and the old and the *Heilige Geist* institutions, financed mainly by lotteries, were dedicated specifically to non-Calvinists. Foreign visitors like Descartes gasped at the sums collected in Sunday boxes in churches even in small towns. But Catholic France was not far behind with its 'Invalides' (not only in Paris) for disabled soldiers, and its beautiful almshouses.

Other similar refuges were established later in other parts of France by her disciples, who were known as Dames de Calvaire. Some still exist as modern hospices.

Independently of the venture in France, the Irish Sisters of Charity also chose the name 'hospice' for their home for the dying which opened in Dublin in 1879. The Order had been active in helping the poor and incurably sick since its foundation earlier in the century, but this was their first house dedicated to the task. The need for such institutions quickly became apparent. By the time the Sisters established St Joseph's Hospice in east London in 1905, three Protestant homes with similar objectives had already opened in the capital. One was the Friedensheim Home of Rest, later to be renamed St Columba's; the second was known as the Hostel of God, established in 1891; and the third was St Luke's Hospital for the Dying Poor, founded by Dr Howard Barrett and the Methodist West London Mission in 1893. The last received a volunteer part-time nurse in 1948. The tall, short-sighted, gawky young woman, then twenty-eight, professing a rather unfashionable brand of evangelical Christianity, was destined to transform these few isolated ventures into a world-wide movement.

Cicely Saunders came from a prosperous but tense north London middle-class family and was packed off to an exclusive boarding-school at the age of ten. There she was friendless and felt isolated, suffering from a bad back which excluded her from many school activities.[4] She was no happier in wartime Oxford, where she read politics, philosophy and economics with a view to becoming perhaps the secretary of a respectable but not too ambitious Tory politician. It all seemed unsatisfactory; and yet, as was also true of Florence Nightingale, the only woman reformer to whom she can be compared, her education contributed to that intangible air of 'class' which in the England of the mid-twentieth century was still essential for taking on the establishment.[5] Throughout her career she successfully persuaded, cajoled and sometimes chided medical grandees, archbishops, ministers, permanent secretaries and of course royalty with the same assurance as she showed when approaching unprivileged patients and their families. Her burning faith in the rightness of her cause helped; but it would not have been enough. Her last year in Oxford saw her conversion from the rather vague agnosticism of her family to an earnest but never exclusive evangelical Christianity.[6] She also enrolled in that

4. In retrospect her headmistress at Roedean did discern in her exceptional qualities of leadership.
5. 'Florence' could write to 'Sidney', the Secretary of State for War (Sidney Herbert, kinsman of the Earl of Pembroke), as social equal – and did almost daily.
6. She was slightly disappointed that Pope John Paul II on the first pontifical visit to the United Kingdom, densely crowded with official engagements, could not at short notice accept her invitation to have luncheon at St Christopher's.

most traditional of nursing schools, the Florence Nightingale School of Nursing at St Thomas's Hospital in London.[7] She seemed to be a success; but soon after qualifying she injured her back and was advised to move sideways. Training as a hospital almoner, today called social worker, she started to work in the evenings at St Luke's Hospital; and it was here that the idea and determination to improve the care of the terminally ill and the dying crystallised in her mind.

. . .

The need was glaring. Existing provisions were almost unbelievably meagre. In the acute wards of the famous teaching and voluntary hospitals in London and in other big cities medical and nursing care were excellent; but those requiring terminal or long-term treatment quickly became 'bed-blockers'. Teaching students and training future hospital consultants – no other career in medicine was worth considering – were deemed to require a quick turnover of acutely ill patients. They were expected to get better or to die. With this overriding objective in mind, as soon as possible the 'chrons' were transferred to other institutions 'designated' for that purpose. In truth they were institutions which were unusable for anything else. Some were old workhouses. Others were tuberculosis sanatoriums made redundant by the advent of streptomycin. Many were in remote parts of the country, idyllic for a holiday but inaccessible to family and friends. That was recognised as a drawback; but one could not have everything. Another drawback was that recruitment of medical and nursing staff was difficult, sometimes impossible. In some 'long-stay hospitals' local practitioners took turns to call at irregular intervals. But most disastrously, both in the acute wards of the teaching hospitals and in long-term establishments severe intractable pain was regarded as a sad but unavoidable accompaniment of most forms of dying.

The therapeutic catch-phrase, handed down from generation to generation since late Victorian times, was 'painkillers on demand'. This sounded good. It seemed to mean that patients got painkillers when they needed and therefore demanded them. In practice it meant that they got them too late or not at all. The universally accepted view was that opioids were addictive, dangerous and even sinful. They therefore had to be used sparingly or, preferably, not at all. This was made clear to patients, with the implication that they should 'demand' pain-relieving drugs only when the pain was intolerable. And intolerable meant intolerable. To be spared lengthy homilies, loaded silences and disapproving looks many patients suffered agonies rather than beg for

7. This was the nurses' training school founded by Florence Nightingale after her return from the Crimea. It became a shrine to her high and exacting standards as well as an internationally recognised model of excellence.

another tablet or injection. Nor was the crucial fact appreciated that the *extinction* of pain needs analgesics in far higher doses than the *prevention* of pain. The latter can often be achieved by small doses given at regular intervals. Recognition of this simple fact by the hospice movement would become an advance as momentous as any in the history of pain relief. (In post-operative patients the 'self-regulation' of painkillers is the norm in many surgical wards today and has been shown to require drugs in far lower doses than when these were doled out on so-called demand.) But that was still in the future. Before the introduction of preventive analgesia loving families often discharged relatives in obviously severe mental and physical distress to be nursed at home. But without guidance and support, that too was unsatisfactory. The still prevalent evil reputation of cancer as a desperately miserable end dates from the early decades of the twentieth cetury rather than from the distant past.

...

To teach herself the basics of what was not yet called 'palliative care' Cicely Saunders searched the literature. This did not take long. There was virtually none. Standard medical texts devoted no space to the problem; and no nursing textbooks existed dealing specifically with the terminally ill. A few eminent Victorian doctors had published articles. John Snow of the Cancer Hospital (now the Royal Marsden Hospital) in London, not related to Snow of anaesthetic fame, wrote a paper on pain relief in 'cancerous disease'.[8] He came to the conclusion that available analgesics were 'less than satisfactory'. Moral support was all-important. He praised the hospital chaplaincy service. Most patients were happier at home than in hospital; but nursing at home was often 'fraught with difficulty'. There was no published response. In 1906 the great William Osler reviewed 500 dying patients. 'Ninety suffered great bodily pain or distress of one sort or another, 11 showed mental apprehension, 2 positive terror.' He deplored this state of affairs but had no suggestion as to how it might be improved.[9]

Perhaps the most remarkable forerunner of the hospice movement was a New England family doctor. In 1935, at the age of eighty, Alfred Worcester published the text of three lectures he had delivered to Boston medical students under the title *The Care of the Aged, the Dying and the Dead.*[10] He pointed out that pressures of modern urban life had made the care of the dying within their own families difficult or impossible. Yet institutional

8. J. Snow, 'Opium and Cocaine in the Treatment of Cancerous Disease', *British Medical Journal*, 1896 (21), 718.
9. Sir W. Osler, quoted in N. Sykes, P. Edmonds and J. Wiles (eds), *Management of Advanced Disease*, 3rd edn (London, 2005), p. 15.
10. A. Worcester, *The Care of the Aged, the Dying and the Dead* (Springfield, Illinois, 1935; reprinted, Oxford, 1961).

provisions were inadequate. Ahead of his time he argued that in cancerous patients the relief of pain must take precedence over most other needs. In another pioneering review of 200 patients, published in the prestigious *New England Journal of Medicine*, four medical social workers, Ruth Abrams, Gertrude Jameson, Mary Poehlman and Sylvia Snyder, made an impassioned and well-argued plea for improved terminal care of cancer patients:

> There should be competent medical supervision uninterrupted throughout the course of illness, good nursing service to supplement medical supervision, the services of a medical social worker available for aid and consultation to the doctor, the patient and the family. Hospital facilities sufficient to meet the need of patients in need of such care [should be] within reach of a tumour clinic with laboratory facilities, and the patient's home.[11]

It must have sounded like a pipe-dream. In many parts of the United States, many of those requirements are still not available for those who cannot pay.

. . .

Determined to change all this, Cicely Saunders re-entered St Thomas's Hospital as a medical student.[12] Once again and inevitably perhaps, she did not fit in; but she qualified as a doctor in 1957, at the age of thirty-nine. Her historic paper, 'Dying of Cancer' was published in the *St Thomas's Hospital Gazette* a few months later and 'Care of the Dying' in the *Nursing Times* the next year. Her report on 900 patients treated at St Joseph's Hospice changed the landscape of pain relief and the care of the terminally ill.

> Our study clearly shows that opiates are not addictive for patients with advanced cancer; that the regular giving of opiates does not cause a major problem of tolerance; that giving oral morphine works and that it does so not by causing indifference to pain but by relieving it. Set alongside the prevailing myths of tolerance and addiction, it is not surprising that a patient arriving at St Joseph's should say: 'The pain in the other hospital was so bad that if anyone came into the room, I would scream: "Don't touch me! Don't come near me!" [With] regular treatment with morphine balanced to her need . . . she became alert and cheerful . . . and maintained her composure until her death a few weeks later.[13]

11. R. Abrams, G. Jameson, M. Poehlman and S. Snyder , 'Terminal Care in Cancer. A Study of Two Hundred Patients Attending Boston Clinics', *New England Journal of Medicine*, 1945 (232), 719.
12. It was a friend, Norman Barrett, thoracic surgeon to the hospital, who told her bluntly that if she wanted to persuade doctors, she needed a medical qualification.
13. C. Saunders, 'The Treatment of Intractable Pain in Terminal Cancer', *Proceedings of the Royal Society of Medicine*, 1964 (4), 68.

What Cicely Saunders herself described as the 'founding myth' of palliative medicine was defined for her by a Jewish patient, David Tasma, who had survived the Warsaw ghetto and had settled in England. Friendless and without a family, he was treated for terminal cancer in a well-run but extremely busy surgical ward of a general hospital. She visited him regularly. They discussed better symptom relief, a commitment to openness, respect for different perceptions of the right way of giving appropriate care to dying patients and the need for dedicated 'hospices' as well as for proper home care. He left the sum of £500 which should pay for a window in the hospice that his friend Cicely would eventually establish.[14] Three reports helped. One published by the Marie Curie Foundation in 1952 documented much suffering at home as seen by district nurses.[15] Glyn Hughes' *Peace at the Last* reviewed terminal care nation-wide.[16] J.M. Hinton published a widely quoted paper on the physical and mental distress of the dying in a London teaching hospital.[17] But Cicely Saunders remained the moving and inspiring spirit, mobilising the medical establishment, the Department of Health, the Churches and the general public. The outcome was the opening of St Christopher's Hospice in Sydenham in south London, for the care of 54 in-patients in 1967.[18] A home-care programme was added in 1969. Within a few years the venture was widely recognised as a significant medical advance.[19] Other similar institutions sprang up. Most now maintain teams of nurses and

14. There is now a plain window at St Christopher's inscribed with Tasma's name. It overlooks a car park.
15. Marie Curie Memorial Foundation, *Report on a National Survey Concerning Patients Nursed at Home* (London, 1952).
16. H.L.G. Hughes, *Peace at the Last. A Survey of Terminal Care in the United Kingdom* (London, 1960).
17. J.M. Hinton, 'The Physical and Mental Distress of the Dying', *Quarterly Journal of Medicine*, 1963 (32), 1.
18. St Christopher is no longer on the canonical register of saints in Rome and may have been a mythical person; but as the traditional patron saint of those embarking on a journey, the name seems appropriate.
19. Cicely Saunders' personal achievement was also recognised. Honours bestowed on her over the next half-century included the honorary fellowships of most medical and nursing colleges of Britain, numerous British and foreign honorary doctorates, prestigious international prizes, the DBE and the Order of Merit. (The last she shared with Florence Nightingale, who was among the first to be appointed when the order was instituted in 1901.) Her astonishing correspondence between 1959 and 1999, enlisting the great and the good (and even the not so great and the not so good) in her crusade has been published (see D. Clark, *Cicely Saunders . . . Selected Letters 1959–1999*, Oxford, 2002).

    Dame Cicely died at St Christopher's Hospice while this chapter was being written, on 14 July 2005. She was eighty-seven, a reformer of genius, a caring doctor and a great lady.

doctors which allow patients in potentially severe pain to be nursed in their own home as well as in hospital.[20]

. . .

Hospice care has created the new specialty of palliative medicine.[21] Palliative medicine has taken symptomatic medicine, once treated by doctors as a thinly disguised synonym for second-class medicine, and has made it into a science-based vocation. It now has its specialist journals and comprehensive textbooks.[22] A large literature deals with practical aspects. But it is still one of the most open branches of the healing arts, admitting 'alternative therapies' as often worth exploring. Palliative medicine inevitably embraces the treatment of diseases which are incurable but rejects the negative connotation that has clung to the term for centuries. It recognises the need for special knowledge and skills as well as for a dedicated organisational framework. Recognition is not the same as achievement; but it is the essential first step. And in Britain at least and in many West European countries, the specialty has probably made greater strides in the last quarter-century than any of its more telegenic competitors.

The treatment of pain is inevitably at the heart of palliative care – it is by far the commonest symptom – but what needs treating is, in Cicely Saunders' terminology, 'total pain'. The term implies something almost daunting: pain in all its immense psychological and physical complexity. The outermost shell of treating total pain is the approach to patients, the making and establishing of trust. Communication skills remain basic to effective palliation. Of course many of these skills cannot be learnt, any more than can the sadly maligned 'bedside manner'; but there is now much accumulated knowledge which can help those prepared to distil common sense from sometimes arid academic rules.

Beyond the first approach, there are the psychological and physical factors which are not pains in themselves but can decisively influence the perception of pain and the patient's response to it. In the first category the two great negatives are anxiety and depression, both requiring recognition and expertise to treat. Anxiety can be no more than a niggling uncertainty; but it can grow into panic attacks terrifying to witness and even more terrifying to experience.

20. Hospice Information Service at St Christopher's, *Directory, Hospice and Palliative Care Services in the United Kingdon and Republic of Ireland* (London, 2004).
21. The term comes from the Latin *palliare*, to cloak (as in the pallium of priestly vestment), sometimes to deceive; but it has been used in medicine to describe relieving symptoms rather than curing diseases at least since the fifteenth century.
22. See D. Doyle, G. Hanks and N. MacDonald (eds), *Oxford Textbook of Palliative Medicine*, 2nd edn (Oxford, 1998), and Sykes, Edmonds and Wiles (eds), *Management of Advanced Disease.*

How common it is in advanced disease is still impossible to tell. Often, perhaps usually, it is combined with a degree of depression. The borderline between sadness and depression is uncertain, arguably no more than semantic, but there are now guidelines to its assessment and management. A high proportion of patients admitted to hospices in advanced – or terminal – disease pass through a stage of confusion or even delirium. This too can be a painful experience, requiring expert treatment. But there are also positive psychological modulators of pain, confidence in the caring team being by far the most important. If fear can intensify pain, the opposite is also true. Almost any pain is bearable provided relief is in sight. Good palliative care will exploit those positive modulators to the full.

On the physical side pain in advanced disease is almost always part of a pain syndrome centred on one or several body systems. The pain, in other words, may be a painful cough, a pain associated with nausea and vomiting, the pain of subacute intestinal obstruction, the pain of constipation, the itching and irritation of jaundice, the pain of urinary retention and other kinds of urinary dysfunction, the pain of fatigue and sleeplessness, the pain arising in bones, the pain of neuralgias, the pain which accompanies weakness and haemorrhage, muscle pain, the pain of joint bleeds, the variety of pain associated with heart and lung disease, the pain of difficult swallowing – the list is by no means complete. Palliative care must tackle all those specifically as well as pain as such. It is a difficult task. One inevitably feels that Trousseau's great precept for clinicians about illnesses needs adapting to pain: there are no pains, there are only people in pain.

In advancing disease it may be necessary to establish a hierarchy of needs; and, when there is a conflict of requirements, pain sometimes takes precedence over others. Its associated symptoms, such as agitation, breathlessness and nausea, must also be recognised and alleviated. At its best, palliative care requires both knowledge and dedication; and today it can achieve a high measure of success.[23] A. Tookman summed up its 'secret' as 'attention to detail', not a grandiose sounding but a demanding and yet attainable objective.

. . .

Not all palliative and home-care teams are as effective as they might be. As in all branches of medicine, patients who do not 'fit' into a pre-ordained pattern suffer. But the best of the hospices – and many are humblingly good – deal effectively not only with patients but also with their families and with the needs of their own staff. The last is important. The nursing care of advanced disease can be extraordinarily onerous. It requires the constant support of a

23. Among many patients who have written about their first-hand experience has been the song-writer and pianist Donald Swann (*Swann's Way*, London, 1963).

team. Family and friends too need to be treated with sensitivity. In good hospices there is a minimum of rules. Instead of everything being forbidden unless positively indicated, everything is allowed unless positively contra-indicated. Advanced illness and even death then become what they should be: part of life. To those ancient enough to recall old 'homes for the dying' – many battling valiantly against overwhelming odds – the cheerful atmosphere of an efficient modern hospice is a revelation.

Of course much needs to be done. The faith motivating new ventures is often hard to maintain. Local schoolchildren laid out the gardens of St Christopher's Hospice. The Sydenham fire brigade hung the curtains. This kind of support does not continue indefinitely. As happened to general anaesthesia a hundred and fifty years ago, awareness that an important advance has been made may take decades to trickle down through the profession. Old attitudes to pain linger. Unsupported by their doctor, thousands of distraught families still buy morphine on the black market to ease the suffering of their beloved. In many parts of the country expert advice and trained nursing are still not available. Palliative care is still a long way from being a high political priority. Also as happened a hundred and fifty years ago with antisepsis, countries other than its land of origin are now leading the way. Nevertheless, the hospice movement and the concepts of modern palliation allow this book to end on a note not of despair but of hope.

# EPILOGUE

Hope – but not triumph. A succession of books published in the 1950s and 1960s bore titles like 'Victory over Pain', 'The End of Pain', or 'Pain Defeated'. Such celebrations now seem premature. This may sound ungracious. It is not meant to belittle past achievement. There have been stupendous advances in pain relief. Surgical anaesthesia is among the half-dozen most beneficial and civilising discoveries of the human race. Modern analgesics, some of them of ancient origin, may come a close second. Even since the 1950s the spread of the hospice movement has been a wonderfully welcome development. But pain has not been defeated.

Since the triumphalist years of the mid-twentieth century, two truths have emerged. First, life-span is not the same as health-span. A steel rod (such as a pin in the hip) may function perfectly until one day it breaks. Before that moment it neither bends nor wobbles. Although the cause of the fracture is commonly described as 'metal fatigue', the metal shows no evidence of anything that could be described as tiredness. Even less does it suffer from any of the innumerable mental and physical afflictions of human old age. Human death may one day resemble such a break. But it does not do so yet. Every extension of the human life-span tends to increase the sum total of human suffering. This is especially true of pain. It is not simply a function of ageing, though ageing is one contributor. The two world wars and subsequent conflicts offer an illustration in younger people. In 1915 three-quarters of the British wounded who arrived at advanced field dressing stations in France and needed an amputation subsequently died from shock, sepsis, exposure or a combination of all those. Even in 1918 the figure hovered around 60 per cent. In the last years of the Second World War, largely because of advances in the treatment of shock and sepsis, the post-amputation death rate was down to about 15 per cent. In American military hospitals in Vietnam the mortality had fallen to below 5 per cent. Unfortunately, a significant proportion of the successful amputees, perhaps as many as a third, later developed 'phantom-limb pains'. This proportion had not changed. Nor has there been significant advance in treating the condition. Many of the amputees were still suffering from this often terrible complaint twenty years after their operation. What is true of military surgery is also true of civilian accidents. This is not an argument against saving lives, let alone one in favour of euthanasia. But the divergence between life-span and health-span remains near the top of the agenda.

Second, there is still no form of pain relief that does not entail loss of control over the person's actions and destiny. The loss compared to the pain which the treatment abolishes may be so trivial as to make its very mention laughable. Who would not give up normal sensation over part of their face for an hour or two in order to have a dental extraction without pain? The slight somnolence after some excellent analgesics is barely noticeable. But surrendering sensation over part of one's face for even an hour or a slightly diminished command over one's faculties still represents a loss. When it comes to chronic painful syndromes the balance is often much finer. In the chapter on 'Dolorism' the thesis 'I suffer therefore I am' was held up to ridicule. It deserves to be. But it is not *entirely* ridiculous when not suffering involves a fraction of not being in command. Of course being 'in command' is, even at the best of times, the most ridiculous of human illusions. But, like most self-deceptions, it is an astonishingly powerful one. And it should be respected. A pain may be troublesome, but the victim may decide to bear it uncomplainingly rather than suffer any loss of 'control'. And why indeed should they bear it *uncomplainingly*? Why should patients in pain, even those who have chosen pain as the lesser of two evils, not lament their predicament? A growing impatience with suffering, especially when it is notionally self-inflicted, is one of the least attractive features of modern life. Victorians could often do little to alleviate symptoms; but many were wonderfully sympathetic listeners. It was a great gift and its loss is a matter for regret.

. . .

But most important, there are still many forms of acute as well as of chronic pain which are beyond relief. The agony of migraine, the pain associated with certain kinds of backache, the suffering of osteoporosis (much more common today than in the days before steroids), many neuralgias and the syndromes associated with nerve deafferentiation are often mentioned. Among those which are *hardly ever* mentioned is the pain suffered by addicts during the course of withdrawal from their addiction. It is as near hell as anything suffered by sick people before 'Victory over Pain'. And there is a whole world of pain considered only in passing in the present book.

It was mentioned in the first chapter that the link between mental and physical pain has been taken for granted in all ancient civilisations. Today, once again, advanced researchers emphasise that the two are inseparable and should be treated as one. One day that will happen. But it has not happened for thousands of years and it has not happened yet. Over centuries of Western civilisation grief, fear, anguish, rage, bitterness, jealousy, despair, loneliness and other kinds of mental torment have been portrayed by poets, writers, artists and musicians, at times sublimely. But, in contrast to the investigation of physical pain, the medical exploration of mental suffering is barely a hundred years old; and its practical management has hardly yet begun. To John Milton 'the worst of evils' meant physical pain (though of course he was aware of its spiritual dimension) and the present work has dealt largely with the same subject. But it is impossible to conclude a book on the fight against pain without affirming that no 'Victory over Pain' can be celebrated until the treatment of mental hurt has advanced at least as far as has the treatment of physical suffering.

Once again therefore: hope yes; triumph no.

# BIBLIOGRAPHY

Apart from a few historic papers, this Bibliography is confined to books. References to individual papers can be found in the footnotes.


J. Abernethy, *Introductory Lectures Exhibiting Some of Mr Hunter's Opinion Respecting Life and Diseases* (London, 1819). The best work of the greatest surgical teacher in London of his day. For pain, see especially Chapter 17.

J.J. Abrahams, *Lettsom: His Life, Times, Friends and Descendants* (London, 1933). Fashionable Quaker physician and one of the most colourful characters of the London medical scene during the first half of the nineteenth century. A sympathetic biography.

R. Abrams, G. Jameson, M. Poehlman and S. Snyder, 'Terminal Care in Cancer: A Study of Two Hundred Patients Attending Boston Clinics', *New England Journal of Medicine*, 1945 (232), 719. This paper by four social workers marks a watershed: it was the first objective study published in a prestigious journal dealing with the unmet needs of the dying and their families.

K.B. Absolon, *The Belle Epoque of Surgery: The Life and Times of Theodor Billroth* (Rockville, Madison, 1985). An abridged and highly readable version of the author's massive four-volume biography. No surgeon benefited more than Billroth from the discovery of anaesthesia and the linked adoption of antisepsis.

E.H. Ackerknecht, *Medicine at the Paris Hospitals, 1794–1848* (Baltimore, 1967). A key source book embellished with brilliant drawings by Daumier, some little known.

J.M. Addington-Hall and I.J. Higginson (eds), *Palliative Care for Non-Cancer Patients* (Oxford 2002). An important textbook on a comparatively neglected aspect of palliation.

E.D. Adrian, *The Basis of Sensation. The Action of Sense Organs* (London, 1928). The author, later Lord Adrian, shared the Nobel Prize with Charles Scott Sherrington six years after the publication of this book. The work contains the memorable sentence: 'Whatever our views about the relation of mind and body, we cannot escape the fact that there is an unsatisfactory gap between such events as the sticking of a pin into my finger and the appearance of a sensation of pain in my consciousness. Part of the gap is obviously made up of events the psychological method by itself can tell us nothing at all about.' Nor, at present, can any other method.

Paulus of Aegina, *The Seven Books of Paulus Aeginata*, translated by F. Adams, 3 volumes (London, 1844). A publication of the Sydenham Society.

D.G.C. Allan and R.E. Schofield, *Stephen Hales: Scientist and Philanthropist* (London, 1980).

R. Allendy, *Journal d'un médicin malade* (Paris, 1980). A haunting and terrible modern diary of pain written by a doctor.

N. André, *Observations sur les maladies d l'urèthre et sur plusieurs faits convulsifs* (Paris, 1756). The first description of tic douloureux, quoted by Fothergill.
E. Andrews, 'The Oxygen Mixture. A New Anaesthetic Combination', *Chicago Medical Examiner*, 1868 (9), 656. A historic paper. The addition of pure oxygen to nitrous oxide and ether made prolonged anaesthesia relatively safe.
J. Annas, *Hellenistic Philosophy of the Mind* (Berkeley, 1981).
J. Antall (ed.), *Pictures from the Past of the Healing Arts. The Semmelweis Medical Museum and Library* (Budapest, 1972). Contains several valuable essays in English by Hungarian scholars.
J. Arbuthnot, *An Essay Concerning the Effect of Air on Human Bodies* (London, 1733). On the threshold of pneumatic chemistry.
O. Arnold, *Le Corps et l'âme. La vie religieuse au XIXe siècle* (Paris, 1984).
J. Arnott, *The Question Considered: Is It Justifiable to Administer Chloroform?* (Hunterian Oration) (London, 1843).
G. Arnulf, *L'Histoire tragique et merveilleuse de l'anaesthésie* (Paris, 1989).
E. Aron, *Histoire de l'anaesthésie* (Paris, 1954). Excellent on the early experiments with ether in France.
J.K. Aronson, *An Account of Foxglove and its Medicinal Use, 1785–1985* (London, 1985).
R.S. Atkinson and T.B. Boulton (eds), *The History of Anaesthesia, Proceedings of the Second International Symposium* (London, 1989). Contains many excellent and important contributions, including a paper on the early spread of ether anaesthesia in the United Kingdom by R.H. Ellis.
Augustine, St, *Confessions*, translated by H. Chadwick (Oxford, 1992) 3 volumes. The greatest autobiography ever written. In several passages, the writer discusses the Christian significance of physical pain.
J. Austen, *Sanditon*, ed. M. Drabble (Harmondsworth, 1974).
F. de Backer, *Lourdes et les médecins* (Paris, 1905).
W. Backer, M. Braun and D. Simononsen, *Moses ben Maimon, sein Leben, sein Werk and sein Einfluss*, 2 volumes (Leipzig, 1932).
G. Baglivi, *De praxi medica* (Rome, 1696).
H. Bailey and W.J. Bishop, *Notable Names in Medicine and Surgery* (London, 1946). Charming and idiosyncratic vignettes of some great doctors of the past.
A. Bain, *The Senses and the Intellect* (London, 1855).
A. Bain, *The Emotions and the Will* (London, 1859). A highly influential work which, together with the previous work, firmly established pain as a neurological phenomenon. The author was the founder of the journal *Mind*.
F. Bainbridge, *Remarks on Chloroform in Alleviating Human Suffering* (London, 1848). A neglected pioneering work.
D. Bakan, *Disease, Pain and Sacrifice* (Chicago, 1968).
K. Baker, *Condorcet: From Natural Philosophy to Social Mathematics* (Chicago, 1975). One of the great, tragic and comparatively little known figures of the Enlightenment.
H.K. Barski, *Guillaume Dupuytren, a Surgeon in His Place and Time* (New York, 1984). A sympathetic biography of a none-too-likeable but important figure in preanaesthetic surgery.
D. Bartwell and A.T. Sandison, *Diseases of Antiquity* (Springfield, Illinois, 1937).
F. Batisse, *Montaigne et la médicine* (Paris, 1962). To many French men and French women Montaigne remains an inexhaustible source of wisdom on every topic, even medicine. The present volume suggests that they may be right.
T. Beddoes, *A Letter to Erasmus Darwin, M.D., on a New Method of Treating Pulmonary Consumption and Some Other Diseases Hitherto Found Incurable* (Bristol, 1783). Contains the first mention of ether inhalation, though for the wrong purpose.
T. Beddoes (trans. and ed.), *The Chemical Essays of C.W. Scheele* (London, 1786). Landmarks in the history of chemistry, brilliantly rendered by Beddoes.

T. Beddoes, *Observations on the Nature and Cure of Calculus, Sea Scurvy, Consumption, Catarrh and Fever: Together with Conjectures upon Several Other Subjects of Physiology and Pathology* (London, 1793). A fascinating book, despite the forbidding title, with flashes of insight as well as much inspired nonsense.

T. Beddoes, *A Guide for Self Preservation, and Parental Affection; or Plain Directions for Enabling People to Keep Themselves and their Children Free from Several Common Diseases* (Bristol, 1793). Another of Beddoes's pontifical and unputdownable outpourings.

T. Beddoes, *Alternatives Compared, or What shall the Rich Do to be Safe?* (London, 1797). It was the best of times. It was the worst of times.

T. Beddoes, *Hygeia: or Essays Moral and Medical, on the Causes Affecting the Personal State of our Middling and Affluent Classes*, 3 volumes (Bristol, 1802–3). The essential Beddoes.

T. Beddoes and J. Watt, *Considerations on the Medicinal Use of Factitious Air and on the Manner of Obtaining them in Large Quantities*, 2 volumes (Vol. I by Beddoes; Vol. II by J. Watt) (Bristol, 1974).

T. Beddoes and J. Watt, *Considerations on the Medicinal Use and on the Production of Factitious Airs* (Bristol, 1795). A brilliant work which deserves an annotated reprint.

G.S. Bedford, *Clinical Lectures on the Diseases of Women and Children*, 8th edn (New York, 1867).

G.S. Bedford, *The Principle and Practice of Obstetrics*, 3rd edn (New York, 1863). Contains an 'up-to-date discussion on the vexed question of anaesthesia'.

H.K. Beecher, *The Measurement of Subjective Responses* (Oxford, 1959). Contains the author's description of the responses of the wounded at the Anzio beachhead in Italy in 1944.

L. McCray Beier, *Sufferers and Healers. The Experience of Illness in Seventeenth-Century England* (London, 1987). An excellent book.

B. Bell, *A System of Surgery*, 3 volumes (Edinburgh, 1801). Heavy going but the surgeon's bible in its day.

C. Bell, *The Anatomy and Physiology of Expression* (London, 1806). The first and perhaps the most readable work of the greatest surgical physiologist of his age.

C. Bell, *An Idea of a New Anatomy of the Brain* (Edinburgh, 1811). Privately printed for his friends, it contains Bell's first thoughts on the specificity of nerves.

C. Bell, *Practical Essays* (Edinburgh, 1841). Brilliant.

L. Bending, *The Representation of Bodily Pain in Late Nineteenth-Century English Culture* (Oxford, 2000). Especially good on changing attitudes to pain as the century progressed.

M.R. Bennett, *A History of the Synapse* (Sydney, 2001).

J. Bentham, *Animals and Man in Historical Perspective*, ed. J. and B. Klaitis (London, 1971).

C. Bernard, *Introduction a l'étude de la médecine expérimentale* (Paris, 1865). A classic and the only work of Bernard available in English translation.

C. Bernard, *Leçons sur les anaesthétiques et sur l'asphyxie* (Paris, 1875). Extraordinarily modern and illuminating, like everything the great physiologist wrote.

H. Bernheim, *De la Suggestion dans l'état hypnotique et dans l'état de veille* (Paris, 1884). A classic in its time, discussed by Freud.

V. Berridge and G. Edwards, *Opium and the People* (London, 1981). An excellent work, both scholarly and readable.

A. Bertrand, *Traité du somnambulisme* (Paris, 1923). Interesting and a good read though not particularly original.

J.M. Besson, *La Douleur* (Paris, 1992). A slender little book with good pictures.

W. Beveridge, *Influenza: The Last Great Plague* (Prodist, 1977).

A.K. Bier, 'Versuche über die Cocainisierung des Ruckenmarkes' ('Experiments in the Cocainisation of the Spinal Cord'), *Deutsche Zeitschrift der Chirurgie*, 1899 (51), 361. The beginning of spinal anaesthesia.

H.J. Bigelow, 'Insensibility during Surgical Operations Produced by Inhalation', *Boston Medical and Surgical Journal*, 1846 (35), 309. The first report in print of ether anaesthesia. Several other important papers on early anaesthesia were published over the next few years in the same journal.

J. Bigelow, *Brief Expositions of Rational Medicine* (Boston, 1858).

H.J. Bigelow, *Surgical Anaesthesia: Address and Other Papers* (Boston, 1900).

E.J. Bing (ed.), *The Secret Letters of the Last Tsar: The Confidential Correspondence between Nicholas II and his Mother, Dowager Empress Marie Feodorovna* (New York, 1938). Background to the Tsarevitch's suffering.

J-N. Biraben, *Les Hommes et la peste en France et dans les pays européens et mediterranéens*, 2 volumes (Paris, 1975). Not an easy read but authoritative and evocative.

E. Blackwell, *Pioneer Work in Opening the Medical Profession to Women* (London, 1895).

J. Bland-Sutton, *The Story of a Surgeon* (London, 1930).

E. Blantyre Simpson, *Sir James Y. Simpson* (Edinburgh, 1896).

J.W. Blassingame, *The Slave Community: Plantation Life in the Antebellum South* (Oxford 1972). A brilliant survey and background to some of Chapter 34.

G. Bloch, *Mesmerism. A Translation of the Original Scientific and Medical Writings of F.A. Mesmer* (Los Altos, California, 1980). The writings of Mesmer were never as influential as his personal practice; but this is an excellent translation of his main work. An interesting read.

E. Blom, *Stepchildren of Music (Rousseau, Mme Favart and Mozart)* (London, 1926). A charming book, describing the first performance of *Bastien and Bastienne* in the Mesmers' house on the Landstrasse in Vienna.

T.S.R. Boase, *Death in the Middle Ages: Morality, Judgement and Remembrance* (New York, 1972). A key work for the period.

F. Boissier de Sauvages, *Dissertatio medica de motuum vitalium causa* (Montpellier, 1740).

F.K. Boland, *The First Anaesthetic. The Story of Crawford Long* (Athens, Georgia, 1950).

M. Bonduelle, T. Gelfand and C.G. Goetz, *Charcot, un grand médecin dans son siècle* (Paris, 1995).

J.J. Bonica, *The Management of Pain* (Philadelphia, 1953). The beginning of the Pain Clinic concept by its pioneer, now in its third edition and a doorstop rather than a book.

J. de Bonnefon, 'Faut-il fermer Lourdes?' *Les Paroles françaises et romaines* (1 July 1906), p. 3.

T.N. Bonner, *Becoming a Physician: Medical Education in Britain, France, Germany and the United States, 1750–1945.* An important book on a neglected topic.

M. Booth, *Opium: A History* (New York, 1996). A well-researched and highly readable history.

M. Booth, *Cannabis* (London, 2003). Not perhaps quite as exciting a story as opium; but this is another highly readable history.

D. Bostock, *Plato's 'Phaedo'* (Oxford, 1986).

F. Bottomley, *Attitude to the Body in Western Christendom* (London, 1979).

F. Boureau, *Contrôler votre douleur* (Paris, 1986).

N. Bowditch, *History of the Massachusetts General Hospital*, 2nd edn (Boston, 1872). Pages 215–383 cover the history of the first etherisations.

J. Bowker, *Problems of Suffering in Religions of the World* (Cambridge, 1970). A wide-ranging study covering all major religions.

P. Bowler, *Reconciling Science and Religion. The Debate in Early Twentieth-Century Britain* (Chicago, 2001).

J. Braid, *Neurypnology or the Rationale of Nervous Sleep Considered in Relation to Animal Magnetism.* (1867) Mesmerism put on a practical footing by a serious medical investigator. It came too late: hypnosis, a word Braid invented, as a means of surgical anaesthesia had been killed by ether and chloroform.

F.D. Bratton, *Moses Maimonides. Medieval Modernist* (Boston, 1967).

F.T. Brechka, *Gerard van Swieten and his World* (The Hague, 1970). A fundamental study of an important secondary character in the history of medicine.

J. Breitenstein, *Le Problème de la souffrance* (Strasbourg, 1901). A famous and pioneering book in its day.

P. Brenot, *Les Mots de la douleur* (Bordeaux, 1992).

J. Breuer and S. Freud, *Studien über Hysterie* (Leipzig, 1895).

M. Brion, *Daily Life in the Vienna of Mozart and Schubert*, translated by J. Stewart (London, 1961).

R.C. Brock, *The Life and Work of Astley Cooper* (Edinburgh, 1954). A master surgeon's tribute to another.

W. Brockbank, *Ancient Therapeutic Arts* (London, 1954).

L. Brockliss and C. Jones, *The Medical World of Early Modern France* (Oxford, 1997).

J.H. Brooke, *Science and Religion. Some Historical Perspectives* (Cambridge, 1991).

B. Broomhill, *The Truth about Opium Smoking* (London, 1882). An interesting historical sidelight.

J. Brown, *Horae Subsceviae* (London, 1907). A best-selling Edwardian collection of reminiscences by a doctor which contains 'Rab and his Friends', recalled in Chapter 18.

T. Bryant *The Practice of Surgery* (Philadelphia, 1873). A stout defence of anaesthesia.

W. Buchan, *Domestic Medicine, or the Family Physician* (Edinburgh, 1769). The most successful and widely read book on home doctoring of the late eighteenth century.

W. Buchan, *Observations Concerning the Prevention and Cure of Venereal Diseases* (London, 1796). The heroic age of mercury and opium.

*Bulletin of the History of Medicine*, 'Crawford W. Long, the Pioneer of Ether Anaesthesia', 1942 (12), 191. A centenary commemorative volume.

V. Buranelli, *The Wizard from Vienna – Franz Anton Mesmer* (New York, 1975). An excellent read.

F.C. Burkitt, *The Religion of the Manichaeans* (Cambridge, 1925).

F.C. Burkitt, *Church and Gnosis* (Cambridge, 1932).

F. Burney, *Journals and Letters*: see Hemlow.

A.R. Burr, *Weir Mitchell: His Life and Letters* (New York, 1928).

R. Burton, *The Anatomy of Melancholie*, 6th edn (London, 1652; reprinted New York, 1864).

H. Butterfield, *The Origins of Modern Science, 1300–1800* (London, 1949).

F. Buytendijk, *Uber den Schmerz* (Berne, 1948). A profound if (in, the writer's opinion, often wrong-headed) moral disquisition.

W.F. Bynum, *The Science and the Practice of Medicine in the Nineteenth Century* (Cambridge, 1991).

W.F. Bynum and R. Porter (eds), *William Hunter and the Eighteenth-Century Medical World* (Cambridge, 1985).

P.J.G. Cabanis, *Rapports du physique et du moral de l'homme* (Paris, 1824). A highly influential and brilliantly written book which reviews the then current ideas on the 'value' of pain.

A.W. Campbell, *Historical Studies on the Localisation of Cerebral Function* (Cambridge, 1905).

W.B. Cannon, *Bodily Changes in Pain, Hunger, Fear and Rage* (New York, 1915). A classic.

J. Carnochan, *Amputation of the Entire Lower Jaw* (New York, 1852).

F.F. Cartwright, *The English Pioneers of Anaesthesia* (Bristol, 1952). Excellent biographical essays on Hickman, Beddoes and Davy.

F.F. Cartwright, *The Development of Modern Surgery* (London, 1967). A brief introduction to the subject which conveys the excitement of new surgical developments and the personalities of some of the pioneers.

S. Cartwright, *Natural History of the Prognathous Species of Mankind* (1857), reprinted in *Slavery Defended: The Views of the Old South* (Englewood Cliffs, New Jersey, 1963).

J.H. Cassedy, *Medicine in America: A Short History* (Baltimore, 1991).

E.J. Cassell, *The Nature of Suffering* (New York, 1990).

D. Caton, *What a Blessing She Had Anaesthesia* (London, 1999). The title refers to Queen Victoria's role in popularising chloroform. A good 'insider's' history of obstetric anaesthesia, especially informative on the American scene.

Celsus, *De Medicina*, 3 volumes, with English translation by W.G. Spenser. Unlike most medical texts of Antiquity (like the dauntingly monumental bilingual edition of Galen's works in 20 volumes) which are mostly for specialists and classical scholars, Celsus remains a fascinating read even for those with only a smattering of Latin. The title 'Cicero of Medicine' may not commend him to those who are no fans of the great orator; but the English translation is excellent too. The famous passage about inflammation is in Volume II.

T. Chadwick and W.N. Mann, *The Medical Works of Hippocrates* (Oxford, 1950). The excellent joint work of a classical scholar and an eminent physician.

W. Channing, *A Treatise on Etherisation in Childhood* (Boston, 1848). The introduction of anaesthesia in obstetrics in America.

J.M. Charcot, *Leçons sur les maladies du système nerveux*, 8 volumes (Paris, 1873). A medical classic, the beginning of neurology and neuropsychology and still a wonderful read. The command of detail, both clinical and pathological, is stunning. The different ataxias are described in Volume II, p. 14, one of many highlights.

J.M. Charcot, 'La Foi qui guérit', *Revue hebdomedaire*, 1892 (3 December), 112. Summarises Charcot's sceptical yet oddly ambivalent attitude to the miracles of Lourdes.

G. de Chauliac, *La Grande Chirurgie* (first published in Paris in 1363; reprinted Paris, 1890).

J.W. Chelius, translated by J.F. South, *A System of Surgery*, 2 volumes (London, 1847).

T. Chevalier, *A Treatise of Gunshot Wounds* (London, 1806).

G. Cheyne, *The Natural Method of Cureing* (London, 1742).

C. Cipolla, *Public Health and the Medical Profession in the Renaissance* (Cambridge, 1976).

D. Clark, 'Originating a Movement: Cicely Saunders and the Development of St Christopher's Hospice, 1957–67', *Mortality*, 1998 (3–1), 43. A good account of one of the most important developments of the second half of the twentieth century.

D. Clark (ed.), *Cicely Saunders, Founder of the Hospice Movement. Selected Letters 1959–1999* (Oxford, 2002).

E. Clarke and J. Stannard, *The Human Brain and Spinal Cord. A Historical Study Illustrated with Writings from Antiquity to the Twentieth Century* (San Francisco, 1996).

J. Cocteau, *Opium* (Paris, 1955). The scintillations of a fan and an addict.

J. Collier, *The Ecclesiastical History of Great Britain* (first published 1708–14). The 1840 edition contains a vivid description of the burning of Archbishop Cranmer and other martyrdoms.

J.V. Colquhoun, *Isis Revelata. An Inquiry into the Origin, Progress and Present State of Animal Magnetism*, 2 volumes (Edinburgh, 1836).

A. Conway, *The Conway Letters: The Correspondence of Anne, Viscountess Conway, Henry Moore and the Friends, 1642–1684*, ed. M.H. Nicolson and S. Hutton (Oxford, 1992). A wonderfully revealing correspondence about the soul, the meaning of pain, the possibility of cures and related matters by some of the brightest minds of the Scientific Revolution, including the first description of migraine by an exceptionally articulate sufferer.

B. Cooper, *The Life of Sir Astley Cooper, Bart*, 2 volumes (London, 1843). A reverent portrait by a nephew.

J.F. Cooper, *The Pathfinder, or The Inland Sea* (New York, 1840). Apart from being a wonderful read, it also gives an insight into popular views of the respective sensitivities to pain of different races and the two genders, discussed in Chapter 30.

S. Cooper, *An Introductory Address to the Students of University College* (London, 1844).

R. Cooter (ed.), *Studies in the History of Alternative Medicine* (London, 1988).

H. Corbin, translated by W.R. Trask, *Avicenna. The Visionary Recital* (Dallas, 1980).

J.L. Corning, 'Spinal Anaesthesia and Local Medication of the Cord', *New York Medical Journal*, 1885 (42), 483. The introduction of spinal anaesthesia.

P. Corsi (ed.), *The Enchanted Loom: Chapters in the History of the Neurosciences* (New York, 1991). Contains some splendid contributions.

D. Cotunno, *De ischiade nervosa Commentarius* (Venice, 1764). Contains what may be the first description of a true neuralgia, that is pain arising in a nerve itself, in this case the popliteal.

A. Crabtree, *From Mesmer to Freud: Magnetic Sleep and the Roots of Psychological Healing* (Yale, 1993). A basic source-book.
J.V. Crawford and A.E. Walker (ed.), *A History of Neurosurgery* (Baltimore, 1951). A good multi-author book with an excellent chapter on the history of surgery for pain.
J. Crawford Adams, *Shakespeare's Physic* (London, 1989). There is no better guide to medicine in Shakespeare's time than Shakespeare, and no better guide to Shakespeare's medicine than this book.
C.A. Creighton, *A History of Epidemics in Britain* (London, 1965).
A. Crosby, *Epidemic and Peace, 1918: America's Forgotten Pandemic* (Cambridge, 1976).
A. Crosby, *Ecological Imperialism: The Biological Expansion of Europe, 900–1900* (Cambridge, 1986). A revelatory book.
A. Cunningham, *The Anatomical Renaissance: The Resurrection of the Anatomical Projects of the Ancients* (Aldershot, 1997).
A. Cunningham and A. French (eds), *The Medical Enlightenment of the Eighteenth Century* (Cambridge, 1990).
A.R. Damasio, *The Feeling of What Happens* (London, 2000).
W.C.D. Dampier, *A History of Science and its Relation to Philosophy and Medicine* (New York, 1930). An interesting and, in its day, provocative work.
C. Darwin, *Expression of Emotions in Man and Animals* (London, 1872). A comparatively little-known work, full of interesting insights.
C. Darwin, *The Autobiography of Charles Darwin, 1809–1882* (first published in 1929; London, 1958). Contains fascinating details of Darwin's time as a medical student in Edinburgh.
E. Darwin, *The Botanic Garden. The Poetical works of E. Darwin* (London, 1806).
A. Dastre, *Les Anaesthésiques, physiologie et application chirurgicales* (Paris, 1890). Probably the first clinical textbook of anaesthesia by a professor at the Sorbonne.
A. Daudet, *La Dalou*, in *Oeuvres complètes*, Vol. XI, (Paris, 1967). Probably the most famous, brilliantly written and haunting first-hand account of slowly progressive tabes dorsalis. Proust could not believe that such a work could be composed while the author was under almost continuous medication with morphine; and to the present writer too there is a whiff of uncertainty about the authenticity of the account.
H. Davy, *Researches, Chemical and Philosophical, Chiefly Concerning Nitrous Oxide, or Dephlogisticated Nitrous Air, and its Respiration* (London, 1800). Contains the suggestion of general anaesthesia fifty years before its general introduction.
J. Davy, *Memoirs of the Life of Sir Humphry Davy* (London, 1836). John, Humphry's brother, a quarrelsome man, was deeply offended that he was not consulted by Lady Davy when she commissioned Paris to write the official biography. In fact, apart from a few vivid childhood reminiscences, his memoir does not significantly differ from the official biography.
J. Davy, *Fragmentary Remains of Sir Humphry Davy* (London, 1858). See above. Neither Paris nor Davy makes more than a cursory reference to nitrous oxide.
A.G. Debus, *The English Paracelsious* (New York, 1966). A deeply researched work on a comparatively neglected subject.
A. Deffarge, *Histoire critique des anaesthésies anciens et en particulier des eponges somnifères à base de drogues* (Bordeaux, 1928). An important study of primary sources but not an easy read.
T. Delorme, *La Douleur: un mal à combattre* (Paris, 1999). A prettily illustrated extended magazine article dealing with various aspects of pain.
M-J. Delvecchio-Good, P.E. Brodkin, B. J. Good and A. Kleinman, *Pain as a Human Experience. An Anthropological Study* (Berkeley, 1992).
D. C. Dennett, *Consciousness Explained* (London, 1991). A successful book addressed to the lay reader, infuriating to those who disagree with the explanation.
T. De Quincey, *The Confessions of an English Opium-Eater* (1822; reprinted by Bodley Head, Oxford, 1930). A classic.

T. De Quincey, *Recollections of the Lakes and the Lake Poets* (1834–9 reprinted by Penguin, London, 1970). A more readable and attractive book than the *Confessions.*

R. Descartes, *Oeuvres et lettres* (Paris, 1953). La Pléiade edition remains the best source to Descartes in French.

R. Descartes, *Treatise of Man* (*De l'homme*) translated by T. Steele Hall (Cambridge, Mass, 1972). A landmark in its day and still surprisingly readable in an excellent translation.

W.P. Dewees, *An Essay on the Means of Lessening Pain in Certain Cases of Difficult Parturition* (Philadelphia, 1806). Contains a revealing discussion of racial differences in pain perception.

K. Dewhurst, *Thomas Willis as a Physician* (Berkeley, 1964).

K. Dewhurst (ed.), *Dr Thomas Sydenham (1624–1689): His Life and Original Writings* (Berkeley, 1966). An important work on a great and still neglected doctor.

S.A. Diamond, *Anger, Madness and the Daimonic: The Psychological Genesis of Violence, Evil and Creativity* (Albany, New York, 1996).

J.H. Dible, *Napoleon's Surgeon* (London, 1970).

G. Dick-Read, *Childbirth without Pain* (first published in 1933; New York, 1980). The original gospel of natural childbirth.

S. Doane, *Surgery Illustrated. Compiled from the Works of Cutler, Hind, Velpeau and Blasius* (New York, 1837).

T. Dormandy, *The White Death. A History of Tuberculosis* (London, 2000).

T. Dormandy, *Moments of Truth: Four Creators of Modern Medicine* (London, 2004).

D. Doyle, G. Hanks and N. MacDonald (eds), *Oxford Textbook of Palliative Medicine*, 2nd edn (Oxford, 1998). Probably the most comprehensive and authoritative multi-author textbook on palliative care.

R. Druitt, *The Principles and Practice of Modern Surgery* (Philadelphia, 1842).

E. Du Bois-Reymond, *Untersuchungen über die thierische Electricität* (Berlin, 1848). Contains, among other interesting observations, the description of his electric 'anaesthesiometer', widely used (and misused by Lombroso and H. Ellis) for at least a century to measure sensitivity to pain.

G. Duchenne, 'De l'Ataxie locomotrice progressive', *Archives générales de médecine*, 1858 (5e série, 12), 641. First instalment of the classic description of 'locomotor ataxia' or syphilitic tabes dorsalis.

G. Duhamel, *Vie des martyrs* (Paris, 1917). A deeply moving book inspired by the First World War. Its message of compassion was misused by the 'doloristes' in the 1930s.

P. Dumâitre, *Ambroise Paré, chirurgien de 4 rois de France* (Paris, 1986). A scholarly and stylish modern biography.

J.B. Dumas, 'Recherches de chimie organique', *Annales de chimie et de la physique*, 1934 (46), 113. The first full account of the chemical (but not the anaesthetic) properties of chloroform and the proposal of the name for the compound.

B.M. Duncum, *Development of Inhalation Anaesthesia* (London, 1947). Excellent on Mayow and Beddoes, though in general now out of date.

M.I. Duran-Reynals, *The Fever Bark Tree: The Pageant of Quinine* (New York, 1946).

E. Earnest, *S. Weir Mitchell, Novelist and Physician* (Philadelphia, 1948). A good biography of the founder of American neurology, the coiner of the term 'phantom limb' and an exceptional character.

B. Ebbel (trans. and commentator), *The Papyrus Ebers. The Greatest Egyptian Medical Document* (Copenhagen, 1937).

C.M.N. Eire, *From Madrid to Purgatory: The Art and Craft of Dying in Sixteenth-Century Spain* (Cambridge, 1995). A scholarly work which includes a detailed account of the illness and death of Philip II (but the wrong diagnosis) and of St Teresa of Avila.

J.S. Elliot, *Outline of Greek and Roman Medicine* (London, 1914). A short and elegant introduction.

J. Elliotson, *Numerous Cases of Surgical Operations without Pain* (London, 1843).

J. Elliotson (ed.), *The Zoist: The Journal of Cerebral Physiology and Mesmerism* (London, 1843–56). The highest but belated flowering of hypnotism in surgery.

E.S. Ellis, *Ancient Anodynes: Primitive Anaesthesia and Allied Conditions* (London, 1946). A discursive labour of love about ancient and tribal analgesics and remedies, unfairly dismissed by later workers.

H. Ellis, *A History of Surgery* (London, 2000). An excellent and richly illustrated introduction to the subject, primarily for a medical readership.

R.H. Ellis (ed.), *The Case Books of John Snow* (London, 1994). A valuable source-book for the work of a reticent hero, with an informative introduction by the editor.

R.W. Emerson, *Women* in *The Complete Writings of Ralph Waldo Emerson*, ed. T.H. Johnson (Cambridge, Massachusetts, 1963).

G.L. Engel, '"Psychogenic" Pain and the Pain-Prone Patient', *American Journal of Medicine*, 26 (1959), 899. A landmark paper in the modern approach to chronic pain. The concept of 'pain-proneness' propounded.

I. Epstein (ed.), *Moses Maimonides 1135–1204. Anglo-Jewish Papers in Connection with the Eighth Centenary of his Birth* (London, 1935). Contains some interesting articles.

J. Erichsen, *On the Study of Surgery* (London, 1850).

A. Escohotado, *A Brief History of Drugs. From the Stone Age to the Stoned Age*, translated by K.A. Symington (Rochester, NY, 1999) A racy subtitle to a racy book.

J.N. Estes, *Dictionary of Protopharmacology and Therapeutic Practices, 1700–1850* (Canton, Massachusetts, 1934). An invaluable and authoritative dictionary with a useful list of pharmaceutical symbols.

European Association for Palliative Care, 'Morphine in Cancer Pain: Modes of Administration', *British Medical Journal*, 1996 (312), 823.

P. Fairlie, *The Conquest of Pain* (New York, 1980).

M. Fakhri, *Averrhoes Ibn Rush. His Life, Works and Influence* (Oxford, 2001).

Faria, Abbé, *De la Cause du sommeil lucide* (Paris, 1819).

A.D. Farr, 'Religious Opposition to Obstetric Anaesthesia: A Myth?' *Annals of Science*, 1983 (40), 159. Casts doubt on the generally accepted story of religious opposition to Simpson's ideas (as more or less endorsed in the present book). Of course religious leaders and polemicists did not speak with one voice; and perhaps Simpson did like to cast himself in the role of St George even with not much of a dragon in sight. But who doesn't?

Most Scotsmen thoroughly approved of Simpson. When news came that he had been knighted, Sir Walter Scott wrote to him suggesting that he should choose the image of a newborn babe for his coat of arms with the motto: 'Does your mother know that you're out?'

F.W. Farrar, *The Life of St Augustine*, in *The Life of the Fathers* (Edinburgh, 1889; reprinted London, 1993).

F. Fearing, *Reflex Action: A Study in the History of Physiology and Psychiatry* (Baltimore, 1930).

W. Fergusson, *A System of Practical Surgery* (London, 1846). The last flowering of preanaesthetic and pre-Listerian surgery.

P. Ferris, *Dr Freud, a Life* (London, 1993). Freud played a small supporting role in the introduction of cocaine as a local anaesthetic, as well as being a user. In the mountainous and mostly unreadable Freud literature this book is a pleasure to read as well as being well balanced between total scepticism and unquestioning adoration.

H.L. Fields, *Pain* (New York, 1987). A book for health professionals.

C. G. Finney, *Lectures on Revivals of Religion*, ed. W.G. McLoughlin (reprinted 1835; Cambridge, Massachusetts, 1860).

V. Finney, *Colour. Travels Through the Paintbox* (London, 2002).

E. Fischer, *Aus meinem Leben* (Berlin, 1922). Charming reminiscences of a Nobel laureate and one of the discoverers of the barbiturates.

E. Fischer and J. von Mering, 'Uber eine Klasse von Schlafmitteln', *Therapie der Gegenwart*, 1903 (5), 97. The beginning of the barbiturates.

R.B. Fisher, *Joseph Lister, 1827–1912* (London, 1977). The best Lister biography.

A. Flint, 'Conservative Medicine', *American Medical Monthly*, 1862 (18), 1. A blueprint for a school of medicine.

F.A. Fluchiger and D. Hanbury, *Pharmacography. A History of of the Principal Drugs of Vegetable Origin* (London, 1974).

M. Forschner, *Die stoische Ethik* (Stuttgart, 1981).

W.W. Fortenbaugh (ed.), *Theophrastus of Ephesus: On his Life and Work* (Oxford, 1985).

S. Fothergill, *A Concise and Systematic Account of a Painful Affliction of the Nerves of the Face Commonly Called the Tic Douloureux* (1777; reprinted in London, 1804). A brilliant brief account of an ancient and terrible disease, the first in English. For a century or so the condition in Britain was called 'Fothergill's disease'.

M.J.P. Foucault, *Histoire de la folie à l'âge classique: folie et deraison* (Paris, 1961). Strictly for those who can take Foucault.

A. Fournier, *Syphilis héréditaire tardive* (Paris, 1878). A French medical classic.

G. Fourure, *Malheur et châtiment. Histoire d'une controverse et réflexions théologiques* (Lille, 1955). Dense but thoughtful reflections on the theological problem of pain and suffering.

J. Foxe, *The Acts and Monuments of John Foxe*, ed. J. Prett (London, 1977).

R.G. Frank, *Harvey and the Oxford Physiologists: Scientific Ideas and Social Interaction* (Berkeley, 1980).

M. Frede and A.O. Rorty (eds), *Essays on Aristotle's 'De Anima'* (Oxford, 1992).

R. K. French and A. Wear (eds), *The Medical Revolution of the Seventeenth Century* (Cambridge, 1989).

S. Freud and J. Breuer, *Studien über Hysterie* (Frankfurt an Main, 1891). Psychoanalysis in the making.

M. von Frey: see Handwerken.

Y.D. Frolov, *Ivan Pavlov and his School: The Theory of Conditioned Reflexes*, translated by C.P. Dutt (London, 1937).

R. Fülöp-Miller, *Triumph over Pain* (New York, 1938). At one time regarded as a 'standard' work on the history of anaesthesia and still full of interesting detail; however, much of it has been disproved by modern research.

J.F. Fulton, *Physiology of the Nervous System* (Philadelphia, 1908). The standard textbook of its time.

S. Galbraith, *Dr John Snow (1813–1858) The Early Years* (London, 2002). Excellent on Snow's family background and roots.

S. Garfield, *Mauve* (London, 2000). A well researched and well written short (but long enough) biography of William Perkin.

H.S. Gasser and J. Erlanger, 'The Role Played by the Sizes of the Constituent Fibres of a Nerve Trunk in Determining the Form of its Action Potential Wave', *American Journal of Physiology*, 1927 (80), 522. Sums up the results of many years of fundamental research about nerve transmission.

S. Gaukroger, *Descartes, an Intellectual Biography* (Oxford, 1995).

A. Gauld, *A History of Hypnotism* (Cambridge, 1992).

A. Gauvain-Picard and A. Megnier, *La Douleur de l'enfant* (Paris, 1993). An excellent and sensitive book.

J.P. Gavit, *Opium* (London, 1925).

G. Geison, *Michael Foster and the Cambridge School of Physiology. The Scientific Enterprise in Late Victorian Society* (Princeton, 1978).

P. Ghilioungui, *Magic and Medical Science in Ancient Egypt* (London, 1963).

H.B. Gibson, *Pain and its Conquest* (London, 1982).

O. Gigon, *Sokrates: Sein Bild in Dichtung und Geschichte* (Berne, 1987).

R.C. Gill, *White Water and Black Magic* (New York, 1940). The background to the introduction of curare to modern anaesthetics.

A.M. Goichon, *Introduction à Avicenne* (Paris, 1933).

S.D. Goitein, *A Mediteranean Society* (Berkeley, California 1971). Excellent on Maimonides and his world.

A. Goldscheider, *Über den Schmerz* (Berlin, 1894). A clinician's challenge to the traditional Cartesian concept of pain. The time was not ripe: nobody took much notice.

A. Goldscheider, *Das Schmerz Problem* (Berlin, 1920). An interesting and influential book but now largely superseded.

M. Goldsmith, *Franz Anton Mesmer. The History of an Idea* (London, 1934).

M. Goldsmith, *The Trail of Opium* (London, 1939).

J. Goldstein, 'The Hysteria Diagnosis and the Politics of Anticlericalism in Late Nineteenth-century France', *Journal of Modern History*, 1982 (54), 209. Essential background to the politics of Lourdes.

L.E. Goodman, *Avicenna* (London, 1992). A good short introduction.

T. Gossett, *Race. The History of an Idea in America* (New York, 1963).

W.R. Gowers, *A Manual of Diseases of the Nervous System*, 2 volumes (London, 1886). The most successful British textbook of an emerging specialty.

T.C. Gray and J. Halton, 'A Milestone in Anaesthesia: D-tubocurarine Chloride', *Proceedings of the Royal Society of Medicine*, 1946 (39), 400.

S.H. Greenblatt (ed.), *A History of Neurosurgery, its Scientific and Professional Contexts* (Park Ridge, Illinois, 1992).

S.A. Greenfield, *The Human Brain: A Guided Tour* (London, 1997). Brilliant and provocative.

S.A. Greenfield, *Brain Story* (London, 2000). The book of what must have been a highly viewable television series.

S. A. Greenfield, *The Private Life of the Brain* (London, 2000).

J. Gregory, *Lectures on the Duties and Qualifications of a Physician* (London, 1772). The humane and even uplifting face of eigtheenth-century medicine.

S. Gregory, *Man-Midwifery Exposed and Corrected* (Boston, 1848). A ferocious onslaught on anaesthesia in labour.

J. Gribbin, *The Fellowship: The Story of a Revolution* (London, 2005). Another canter across the scientific revolution of the seventeenth century. Readable; but were old and new really quite so black and white?

M. Griffin, *Seneca: A Philosopher in Politics* (Oxford, 1976). An excellent guide to the Stoic attitude to pain in practice.

R. Griffith, *The Reactionary Revolution: The Catholic Revival in French Literature, 1870–1914* (London, 1966). Important background to Lourdes.

F.M. Freiherr von Grimm, *Correspondance littéraire, philosophique et critique* (Paris, 1829–30). A witty account of the controversy aroused by Mesmer.

M. Grmek, *Raisonnement experimental et recherches toxicologiques chez Claude Bernard* (Paris, 1973). A first-class summary of Claude Bernard's life's work, including his pioneering experiments with curare.

G.B. Gruber, *Einführung in die Geschichte und Geist der Medizin*, 4th edn (Stuttgart, 1952). A perceptive general history.

C.-C. Grüner, *Aphrodisiacus, sive de lue venerea* (Jena, 1789). Contains Marcellus Cumanus' first account of syphilis as observed after the battle of Fornovo on 5 July 1495.

O.C. Gruner, *A Treatise on the Canon of Avicenna* (London, 1930).

G. Guillain, *J.M. Charcot, 1825–93. His Life and Work*, ed. and translated by P. Bailey (London, 1959). A sympathetic and serviceable biography, the only one available in English.

P. Guilleaume, *Médecine, église et foi* (Paris, 1990).

E.J. Gurlt, *Geschichte der Chirurgie*, 3 volumes (Berlin, 1891). Still the best general history of surgery during the centuries before the date of publication.

D. Gutes, *Avicenna and the Aristotelian Tradition* (Leiden, 1988).

D. Guthrie, *A History of Medicine* (London, 1960). Short, civilised, readable.

G. Guthrie, *Commentaries on the Surgery of the War in Portugal, Spain, France and the Netherlands* (London, 1853). An important document but not for the squeamish.

H.W. Haggard, *The Lame, the Halt and the Haggard. The Vital Role of Medicine in the History of Civilisation* (London, 1932).

H.W. Haggard, *Doctors in History* (New Haven, 1934). A lightweight collection of anecdotal chapters by an eminent medical historian.

H. Hale Bellot, *University College London, 1826–1926* (London, 1929). Describes Liston's memorable first operation under anaesthesia.

A.R. Hall, *The Revolution in Science 1500–1750* (London, 1983).

M.B. Hall, *Robert Boyle and Seventeenth Century Chemistry* (Cambridge, 1958).

S. Halliday, *The Great Stink of London* (Stroud, 1999).

H.O. Handwerken (ed.), *Deutschsprächige Klassiker der Schmerzensforschung* (Heidelberg, 1987). Among other interesting articles, it contains Max von Frey's paper on cutaneous pain sensation.

P.W. Harkins, *Galen on the Passions and Errors of the Soul* (Columbus, Ohio, 1963).

B.M. Harris, *The Life Story of J. Marion Sims* (New York, 1959).

R. Harris, *Lourdes. Body and Spirit in a Secular Age* (London, 1999). This deeply researched and beautifully written book succeeds in walking 'the moral and spiritual tightrope' required to assess the events, motivations and achievements of Lourdes.

W.V. Harris, *Restraining Rage. The Ideology of Anger Control in Classical Antiquity* (Cambridge, Massachusetts, 2001). Rage is not pain, but the two are not light-years removed from each other and this pioneering book deals comprehensively with the former in the classical world. Much of it, especially Stoic attitudes, is highly relevant to pain relief.

M-R. Hayoun, *Maimonides. Sa Vie* (Paris, 1999). An excellent modern biography.

A. Hayter, *Opium and the Romantic Imagination* (London, 1968).

H. Head, 'The Afferent Nervous System from a New Aspect', *Brain*, 1905 (68), 99. A landmark paper.

D. Healy, *The Creation of Psychopharmacology* (Cambridge, Massachusetts, 2002).

J. Hemlow (ed.), with G.G. Falle, A. Douglas and J.A. Bourdais de Charbonnière, *The Journals and Letters of Fanny Burney* (Mme d'Arblay) (Oxford, 1975). A superb scholarly edition. Volume VI contains the famous description of the author's mastectomy.

W.E. Henley, *A Book of Verses* (London, 1888). Contains the poet's reminiscences of his days as a tuberculous patient.

A.I. Herschel, *Maimonides* (Berlin, 1932; translated by J. Neugroschl, New York, 1988). A labour of love by a young scholar, it is still arguably the best biography of the great doctor.

E.R. and J.R. Hilgard, *Hypnosis in the Relief of Pain* (Los Altos, California, 1975). The most comprehensive modern treatment of the subject.

J. Hilton, *Rest and Pain* (London, 1863).

J.M. Hinton, 'The Physical and Mental Distress of the Dying', *Quarterly Journal of Medicine*, 1963 (32), 1. A paper which influenced the hospice movement.

R.M. Hodges, *A Narrative of Events Connected with the Introduction of Sulphuric Ether into Surgical Use* (Boston, 1891). Probably the most balanced and reliable account of the murky early history of ether anaesthesia.

D. Hoizey, *A History of Chinese Medicine* (Edinburgh, 1993).

A. Holger-Maehle, 'The Pharmacology of 19th Century Patent Medicine', *Pharmacy in History*, 1988 (30), 3.

A. Holger-Maehle, *Drugs on Trial. Experimental Pharmacology and Therapeutic Innovations in the 18th Century* (Amsterdam, 1999). An interesting and important contribution on early work on opium, quinine and substances tried to dissolve renal stones, an antidote to the prevailing idea that drug trials are a modern invention.

F.L. Holmes, *Lavoisier and the Chemistry of Life. An Exploration of Scientific Creativity* (Madison, Wisconsin, 1987). This is a detailed and penetrating study of Lavoisier's work and publications but says little about his life, times or personality.

O.W. Holmes, 'Anaesthesia'. Letter in E. Warren, *Some Account of Letheon*, 2nd edn (Boston, 1847). The famous letter from Oliver Wendell Holmes, addressed to Morton, in which he suggests the term 'anaesthesia' and 'anaesthetic'.

O.W. Holmes, *Currents and Countercurrents in Medical Science* (Boston, 1861).

V. Horsley, 'Intracranial Neurectomy of the Second and Third Division of the 5th Nerve', *Annals of Surgery*, 1892 (42), 524.

H.L.G. Hughes, *Peace at the Last. A Survey of Terminal Care in the United Kingdom* (London, 1960). A pioneering study.

J.H. Hutchinson, *Practical Paediatric Problems* (London, 1980). Discusses use of barbitruates in neonatal jaundice.

A. Huxley, *Brave New World* (first published in 1932; Vintage Future Classics Edition, London, 2005).

J.-K. Huysmans, *A Rebours* (Paris, 1882).

J.-K. Huysmans, *Les Foules de Lourdes* (1905; reprinted Grenoble, 1993).

B. Inglis, *The Opium War* (London, 1976). A gruesome and splendid read.

B. Inglis, *The Diseases of Civilisation* (London, 1981).

B. Issekutz, *Die Geschichte der Arzneimittelforschung* (Budapest, 1971). Excellent on the precursors and beginnings of aspirin.

L. Iversen, 'Chemical Identification of a Natural Opiate Receptor Agonist in the Brain', *Nature*, 1975 (258), 567.

C.T. Jackson, *A Manual of Etherisation Comprising also a Brief History of the Discovery of Anaesthesia* (Boston, 1861). A historic but wholly mendacious account of the events.

M. Jackson, *Pain: The Science and Culture of Why We Hurt* (London, 2002). A highly readable if somewhat fragmented and arbitrary canter around current topics related to pain.

R. Jackson, *Doctors and Diseases in the Roman Empire* (Norman, Oklahoma, 1988).

J. Jacobi (ed.), *Paracelsus, Selected Writings*, translated by N. Guterman, with an introduction by C.G. Jung (London, 1951). A sympathetic selection of Paracelsus' work, useful to those who might want to judge him for themselves rather than take the word of Prince Charles or the present author.

J.H. James, *On the Causes of Mortality after Amputation of the Limbs* (Worcester, 1850). The state of the art a few years before Lister and the general acceptance of anaesthesia.

J.H. James, *Chloroform versus Pain and Paracentesis of the Bladder* (London, 1870).

L. Jardine, *Ingenious Pursuits: Building the Scientific Revolution* (London, 1999). A highly readable survey of the scientific revolution in England in the seventeenth century.

M. Jay, *Emperors of Dreams. Drugs in the Nineteenth Century* (Sawry, Cambs, 2000). An interesting and well-informed book.

W.A. Jayne, *The Healing Gods of Ancient Civilisations* (New Haven, 1925). A profound and comprehensive work.

D. Jeffreys, *Aspirin. The Remarkable Story of a Wonder Drug* (London, 2004). Despite mildly irritating lapses into tabloidese ('this was all cutting-edge stuff'), this is a great story told with zest. The cut-throat rivalries within the gangsterdom known as the international pharmaceutical industry are riveting.

L. Jerphagnon, *Pascal et la souffrance* (Paris, 1956).

W.D. Jordan, *White over Black: American Attitudes Toward the Negro* (New York, 1969).

S.B. Karch, *A History of Cocaine* (London, 2003). Six historic essays with provocative Introduction on the beginnings of the cocaine trade.

A. Karlen, *Plague's Progress* (London, 1995). This very accessible book for the general reader provides good background to Snow's work on cholera.

K.D. Keele, *Anatomies of Pain* (Oxford, 1957).

K.D. Keele, *William Harvey* (London, 1965). A good biography that focuses on the work rather than the life.

K.D. Keele and R. Smith (eds), *The Assessment of Pain in Man and Animals* (Edinburgh, 1962).
G. Keith, *Fads of an Old Physician* (London, 1897). Delightful reminiscences of one of Sir J. Y. Simpson's close friends and a guest at the dinner party which launched chloroform.
J. Kerner, *Franz Anton Mesmer aus Schwaben, Entdecker des tierischen Magnetismus* (Frankfurt am Main, 1856). A mine of otherwise unavailable biographical information.
G. Keynes *The Apologie and Treatise of Ambroise Paré* (London, 1951). One of the great surgeon-historian's less well-known works.
G. Keynes, *The Life of William Harvey* (Oxford, 1966). A classic which complements Keele's biography.
T.E. Keys, *The History of Surgical Anaesthesia* (New York, 1945). An excellent scholarly book.
M.S. Khan, *Islamic Medicine* (London, 1986).
L.S. King, *The Medical World of the Eighteenth Century* (Chicago, 1958).
L.S. King, *The Road to Medical Enlightenment, 1650–1695* (London, 1970).
L.S. King, *Transformations in American Medicine: from Benjamin Rush to William Osler* (Baltimore, 1991).
D. King-Hale, *Erasmus Darwin* (London, 1963).
A. Kleinman, *Suffering, Healing and the Human Condition* (London, 1988).
R. Knoett, *Herman Boerhaave: Calvinist Chemist and Physician* (Amsterdam, 2002). A useful update on a great teacher set in his time.
C. Koller, 'Über die Verwendung des Cocain zur Anaesthetisierung des Auges', *Wiener Medizinische Wochenschrift*, 1894 (34), 1309. The discovery of local anaesthesia: a historic paper.
Z. Kövecses, *Metaphor and Emotions. Language, Culture and Body in Human Feeling* (Cambridge, 2000).
E. Kratzman, *Die neuere Medizin in Frankreich* (Leipzig, 1846). The French medical scene on the eve of surgical anaesthesia, seen by a francophile German doctor.
F. Krause, 'Resection des Trigeminus innerhalb der Schadenkoile', *Verhalten der Deutschen Gesellschaft der Chirurgie*, 1892 (21), 199. The first successful resection of the trigeminal (Gasserian) ganglion for neuralgia.
F. Kudlien, *Der Beginn des medizinischen Denken bei den Griechen* (Zurich, 1967).
T. Kuhn, *The Structure of Scientific Revolutions*, 2nd edn (Chicago, 1970).
J. Lacarrière, *Les Hommes ivres de Dieu* (Paris, 1975). A well argued yet unconvincing critique of martyrdom.
J. de Lacassagne, *Guy de Maupassant et son mal* (Paris, 1951). The painful attacks of tabes.
R. Lane Fox, *The Classical World. An Epic History from Homer to Hadrian* (London, 2005). A grand yet intimate introduction to our ancestral world.
J.N. Langley, *The Autonomic Nervous System* (Cambridge, 1921). A landmark book.
D.J. Larrey, *Mémoires de chirurgie militaire et campagnes*, 4 volumes (Paris, 1812–14). A superb source-book.
J. Latta, *A Practical System of Surgery*, 3 volumes (Edinburgh, 1795). A standard textbook of its day.
P. Laurie, *Drugs* (London, 1967).
C.H. La Wall, *Four Thousand Years of Pharmacy. The Curious Lore of Drugs and Medicines* (London, 1927).
C. Lawrence, *Medicine in the Making of Modern Britain, 1700–1920* (London, 1994). A penetrating study of a neglected aspect of social history.
O. Leaman, *Moses Maimonides* (London, 1990). An interesting interpretation of the significance of Maimonides outside Judaism.
D. Le Breton, *Anthropologie de la douleur* (Paris, 2000). A pioneering and interesting work which somehow fails to illuminate.
J. LeFanu, *The Rise and Fall of Modern Medicine* (London, 1999). A brilliant book, especially good on the fall.
E. Le Garrec, *Mosaique de la douleur* (Paris, 1991). An idiosyncratic but stimulating book.

I. Leighton, *The Aspirin Age* (London, 1950).

A. Le Pelletier de la Sarthe, *Histoire de la révolution médicale du XIXe siècle* (Paris, 1854). An almost contemporary account of the French medical scene during the ether controversy.

R. Leriche, *The Surgery of Pain* (*La Chirurgie de la douleur*, 1937 translated by A. Young, Paris, London, 1939). A ground-breaking book by a great surgeon and great humanist.

R. Leriche, *La Chirurgie à l'ordre de la vie* (Zeluck, 1945).

R. Leriche, *Souvenirs de ma vie morte* (Paris, 1956). A beautiful book of reminiscences published posthumously but unrevealing of a very private person.

E. Lesky, *Die Wiener Medizinische Schule im 19. Jahrhundert* (Graz, 1965). An indispensable source-book about medical events in Vienna during the time of Vienna's medical pre-eminence in Europe.

I.B. Levitan and L.K. Kaczmarek, *The Neuron: Cell and Molecular Biology*, 2nd edn (New York, 1996).

A.G. Levy, *Chloroform Anaesthesia* (London, 1922). The author was one of the first anaesthetists appointed to Guy's Hospital and a student of anaesthetic history. A good book with an important chapter on the causes of death under chloroform.

C.S. Lewis, *The Problem of Pain* (New York, 1944). A personal Christian exegesis, short, concise and beautifully written, but for the converted.

T. Lewis, *Pain*, facsimile of the first edition, edited by Lady L. Lewis (London, 1981). An elegant long essay mainly about the excessive sensitivity induced around an area of inflammation.

O.K. Linde (ed.), *Pharmakopsychiatrie in der Wandel der Zeit* (Landau in der Pfalz, 1988).

G.A. Lindeboom, *Hermann Boerhaave, The Man and His Work* (London, 1968). A sympathetic and well-researched biography.

R. Liston, *Lectures on the Operations in Surgery* (Philadelphia, 1846). Surgery on the eve of anaesthesia and antisepsis.

W.K. Livingston, *Pain Mechanisms. A Physiological Interpretation of Causalgia and its Related States* (New York, 1943). An important study of phantom limbs and other forms of neurogenic pain.

J.D. Loeser (ed.), *Bonica's Management of Pain* (Philadelphia, 2001). The first edition of this multi-author textbook was ground-breaking. In its third edition, lifting the volume can cause a back injury. What practical purpose such tomes serve is a mystery.

N. Longmate, *King Cholera: The Biography of a Disease* (London, 1966). Background to Snow's work as an epidemiologist.

J.L. and H.C. Lord, *Defence of Dr Charles T. Jackson's Claims to the Discovery of Etherisation, Disproving the Claims Set Up in Favour of W.T.G. Morton* (Boston, 1848).

I.S.L. Loudon, *Medical Care and the General Practitioner 1750–1850* (Oxford, 1986). An important, deeply researched book.

I.S.L. Loudon, *The Tragedy of Childbed Fever* (Oxford, 2001). A beautifully written pioneering monograph.

P.C.A. Louis, *Researches on the Effects of Blood Letting* (Boston, 1836). This ground-breaking book which put medical statistics on the map should have put an end to indiscriminate blood-letting but did not. The book is still more revered in the Anglo-Saxon world than in France.

K. Lucas, *The Conduction of the Nervous Impulse*, revised by E.D. Adrian (London, 1917). The first edition (1906) was one of the publications of the Seminal Years. The revision contains fundamental studies about the effect of morphine and of local anaesthetics on the transmission of nervous impulses.

J.S. Lundy, 'Intravenous Anaesthesia: Preliminary Report of the Use of Two New Thiobarbiturates', *Proceedings of the Mayo Clinic*, 1935 (10), 536. The beginning of Pentothal.

H.M. Lyman, *Artificial Anaesthesia and Anaesthetics* (New York, 1881).

W.G. MacCallum, *William Stewart Halsted, Surgeon* (Baltimore, 1930). An exercise in hagiography.

W. MacDougall, *Introduction to Social Psychology* (London, 1908). One of the important publications of the Seminal Years. It contains the first reference to the 'fight or flight' reaction mediated by the sympathetic nerves.

P.A. McGrath, *Pain in Children. Nature, Assessment and Treatment* (New York, 1990).

G. Macilwain (ed.), *Memoirs of John Abernethy*, 2 volumes (London, 1854). Abernethy was the most popular teacher of surgery in his day.

H. McQuay and A. Moore, *An Evidence-based Resource for Pain Relief* (Oxford, 1998).

B. McQuitty, *The Battle for Oblivion: The Discovery of Anaesthesia* (London, 1969).

F. Magendie, *Leçons sur les fonctions et les maladies du système nerveux* (Paris, 1839). A still readable classic.

J.P. Malgaigne, *Manuel de médecine operatoire fondée sur l'anatomie normale et l'anatomie pathologique* (Paris, 1949). An important textbook in its day, containing a description of cutting the infraorbital nerve for trigeminal neuralgia.

B. Mandeville, *A Treatise of the Hypochondriack and Hysterick Diseases* (London, 1730; reprinted Hildesheim, 1981). Great.

C.C. Mann and M.L. Plummer, *The Aspirin Wars. Money, Medicine and 100 Years of Rampant Corruption* (New York, 1991). This is still the best book about the extraordinary story of aspirin.

H. Mann, *Lectures on Education* (Boston, 1855).

R.D. Mann (ed.), *The History of the Management of Pain: From Early Principles to Present Practice* (Park Ridge, New Jersey, 1988).

J. Mann, *Murder, Magic and Medicine* (Oxford, 1992).

Marcus Aurelius, *Meditations*, 2 volumes, translated by A.S.L. Farquaharson (Oxford, 1944).

T. Martin: see St Thérèse de Lisieux.

H. Martineau (as 'An Invalid'), *Life in the Sick Room* (London, 1844).

H. Martineau, *Letters on Mesmerism* (London, 1845). The author espousing John Elliotson's teaching.

Marx, Alexander, *Studies in Jewish History and Booklore* (New York, 1969). Contains much valuable material on Moses Maimonides and the historical background to his life.

R.K. Massie, *Nicholas and Alexandra* (London, 1968). The moving account of the illness of Alexis, the last Tsarevitch, and of Rasputin's hold on his parents.

K. May, *Winnetou*, 3 volumes (Leipzig, 1907). This marvellous boys' classic, which was translated into many languages and sold by the million, is set on an imaginary American 'Frontier' and is a hymn to heroic endurance. It was written by a man who reputedly never crossed the Atlantic.

P. Mazliak, *Avicenne et Averrhoës. Médicine et biologie dans la civilisation islamique* (Vuibert/Paris, 2002). Excellent short introduction to the two great Islamic doctors.

C. Mazzoni, *Saint Hysteria: Neurosis, Mysticism and Gender in European Culture* (Ithaca, New York, 1996). A sparkling book on the psychology of religious women.

C. Medwick, *Teresa of Avila* (London, 2000). A fine biography of the saint who, despite a debilitating illness and mystical raptures, transformed the Carmelite Order and created some of the greatest works of Western mysticism: the nearest perhaps to pain as inspiration. The book usefully points to other key readings on the subject.

C.D. Meigs, *Females and Their Diseases* (Philadelphia, 1848). Contains on p. 40 the memorable passage: 'If we scan [woman's] position amidst the ornate circles of a Christian civilisation, it is easy to perceive that her intellectual force is different from that of her master and lord . . . The great administrative faculties are not hers. She plans no sublime campaigns, leads no army in battles, nor fleets to victory. The forum of no theatre is for her silver voice . . . She discerns not the courses of the planets . . . Home is her place, except when like the star of the day, she deigns to issue forth . . . to exhibit her beauty and her grace and to scatter her smiles upon all that are worthy to receive such a boon . . . She has a head almost too small for intellect and just big enough for love.'

C.D. Meigs, *Obstetrics, The Science and the Art* (Philadelphia, 1849). A much-quoted

textbook by a highly respected obstetrician and teacher who strongly opposed anaesthesia.
R. Melczak and P.D. Wall, 'Pain Mechanisms: A New Theory', *Science*, 1965 (150), 971. The beginning of the gate control theory.
R. Melczak and P.D. Wall, *The Challenge of Pain* (London, 1996).
W.R. Merrington, *University College Hospital and its Medical School: A History* (London, 1976). The story updated.
A. Mesmer, *Mémoire sur la découverte du Magnétisme Animal* (Geneva, 1779). Mesmer's most important work.
A. Miles, *The Edinburgh School of Surgery before Lister* (London, 1918).
A.H. Miller, 'The Pneumatic Institution of Thomas Beddoes at Clifton', *Annals of Medical History*, 1931 (55), 253.
Mitchell, S. Weir, *Injuries of Nerves and Their Consequences* (Philadelphia, 1872). An important historical book in which the term 'phantom limb' makes its appearance.
Mitchell, S. Weir, 'Remarks on the Effects of the Anhalonium Lewini (The Mescal Button)', *British Medical Journal*, 1896 (2), 569. A brilliant first description of the effects of mescaline.
C. Mondale, *Women. Their Nature and Their Diseases* (Chicago, 1878).
H. Mondor, *Dupuytren*, 2nd edn (Paris, 1945). A somewhat reverential account of a great but far from likeable surgeon.
M. de Montaigne, *Essais*, ed. B. Villey (Paris, 1926). Human – or at least Gallic – destiny in a nutshell.
J. Moore, *A Method of Preventing or Diminishing Pain in Several Operations in Surgery* (London, 1784). A pioneering book describing anaesthesia by nerve compression, almost entirely ignored in its day.
P. Moore, *Blood and Justice* (London, 2003). The story of the first blood transfusions in London and Paris in the seventeenth century.
L.T. More, *The Life and Works of the Honourable Robert Boyle* (Oxford, 1944). Contains Boyle's experience with Valentine Greatraks, the 'Stroker'.
D. Morgan, 'What Mesmer Believed', *Journal of the National Council for Psychotherapy and Hypnotherapy*, 1994 to follow. A balanced and sympathetic summary of Mesmer's doctrines.
D.B. Morris, *The Culture of Pain* (Los Angeles, 1991). An influential and humane book outlining the cultural significance of pain.
R.J. Morris, *Cholera 1832: The Social Response to an Epidemic* (London, 1876).
W.G. Mortimer, *History of Coca, the 'Divine Plant of the Incas'* (New York, 1901).
V. Mott, *Pain and Anesthetics* (Washington, 1863). Written by the celebrated New York surgeon at the request of the US Sanitary Commission for the guidance of medical officers in the US Army. In a sense, the first textbook of anaesthesia.
B. Moynahan, *The Faith. A History of Christianity* (London, 2003).
The Mozarts, *The Letters of Mozart and his Family*, translated by Emily Anderson (London, 1938). Includes numerous references to Mesmer.
J. Müller, *Handbuch der Physiologie des Menschen*, 3rd edn (Berlin, 1844). An immensely influential text, translated into most European languages. It formulated clearly the different functions of different nerves.
S. Natoli, *L'esperienza del dolore. Le forme del patire nella cultura occidentale* (Milan, 1988).
H. Neal, *The Politics of Pain* (New York, 1978). An important book. The author's revelations did much to kindle an interest in the management of pain.
P. Nemo, *Job et l'excès du mal* (Paris, 1986). A very cerebral but interesting modern analysis of the Book of Job.
L.W. Nevius, *The Discovery of Modern Anaesthesia. By Whom Was it Made?* (New York, 1894). A book dedicated to Gardner Quincy Colton.
G.A. and G. Newbold, *Handbook of Medical Hypnosis*, 3rd edn (London, 1968).
W. Noordenbos, *Pain* (Amsterdam, 1959).

V. Nutton, *Ancient Medicine* (London, 2005).

A. Oakley, *The Captured Womb. The History of the Medical Care of Pregnant Women* (Oxford, 1984).

D. O'Leary, *How Greek Science Passed to the Arabs* (London, 1984). One of the most intriguing chapters in Europe's cultural history.

J.M.D. and O.E.H. Olmstead, *Claude Bernard and the Experimental Method in Medicine* (London, 1964). The best work on Bernard in English.

C.D. O'Malley, *Andreas Vesalius of Brussels* (Berkeley, California, 1964). The best biography of the great anatomist.

W. Osler, *The Principles and Practice of Medicine* (Baltimore, 1892). The doctor's bible for decades, translated into at least twenty languages.

A.R.G. Owen, *Hysteria, Hypnosis and Healing. The Work of Jean Martin Charcot* (London, 1971). Good on Charcot's publications.

F.R. Packard, *The Life and Times of Ambroise Paré* (New York, 1926). A sound and useful but pedestrian work.

D. Padfield, *Perceptions of Pain* (Stockport, 2003). Patients attending a pain clinic were asked to describe their pain in terms of a visual image. Fascinating.

W. Pagel, *Paracelsus: An Introduction to Philosophical Medicine in the Era of the Renaissance* (London, 1982). A useful corrective to the low opinion of Paracelsus expressed in the present book. It was written by a charming, learned and much respected but slightly dotty medical historian, pathologist and admirer of Paracelsus.

W. Pagel, *The Smiling Spleen: Paracelsianism in Storm and Stress* (Basle, 1984).

J. Paget, Sir, 'The Discovery of Anaesthesia', *Eclectic Magazine*, 1880 (94), 219. An English view, no axe to grind.

J. Paget, Sir, *Studies of Old Case Books* (London, 1891). One of the most engaging of surgical memoirs.

S. Paget, *Sir Victor Horsley, a Study of his Life and Work* (New York, 1920).

*Pain*, The most important specialist journal dealing with current research on pain, started publication in 1975.

A. Paré, *Oeuvres complètes*, Malgaigne edn, 6 volumes (Geneva, 1970). An impressive scholarly edition.

J.A. Paris, *The Life of Sir Humphry Davy* (London, 1831). The official life written at the behest of Lady Davy.

K. Park, *Doctors and Medicine in Early Renaissance Florence* (Princeton, 1985).

F.A. Pattie, *Mesmer and Animal Magnetism. A Chapter in the History of Medicine* (New York, 1994).

T. Percival, *Medical Ethics* (Manchester, 1804). A ground-breaking book, the first of its kind in modern times. The management of pain is considered in some depth.

M.S. Pernick, *A Calculus of Suffering. Pain, Professionalism and Anaesthesia in Nineteenth-Century America* (New York, 1985). A pioneering and indispensable book as well as an eye-opener to evolving attitudes to pain.

C. Pert, *Molecules of Emotion: Why You Feel the Way You Feel* (New York, 1997). A racy, chatty, infuriating, exhilarating autobiographical canter around the beginning of endorphins, one of the most important developments in the neurosciences and in pain perception, with an Appendix on 'Bodymind Medicine Resources and Practitioners'. Alternative medicine *in excelsis.*

M.J. Peterson, *The Medical Profession in Mid Victorian London* (Berkeley, California, 1978). Well describes the social stratification of the medical profession in Victorian England.

C. Petonnet, *On est tous dans le brouillard* (Paris, 1979).

N. Pirogoff, *Grundzüge der allgemeinen Kriegschirurgie* (Lepzig, 1864). An important textbook of military surgery, which contains a description of rectal anaesthesia and also advocates Listerian antisepsis.

J.-B. Pontalis, *Zwischen Traum und Schmerz* (Frankfurt, 1998).

K.R. Popper and J.C. Eccles, *The Self and its Brain* (New York, 1977). An eloquent exposition of the 'dualist' view of mind and body, now widely out of favour.

M. Porkert and C. Ullman, *Chinese Medicine: Its History, Philosophy and Practice* (New York, 1988).

R. Porter, *Health for Sale: Quackery in England, 1650–1850* (Manchester, 1989). The best book on an important topic.

R. Porter, *Doctor of Society: Thomas Beddoes and the Sick Trade in Late Eighteenth Century England* (London, 1991). A brilliant panoramic view rather than a biography.

R. Porter (ed.), *Medicine in the Enlightenment* (Amsterdam, 1995).

R. Porter, *The Greatest Benefit to Mankind: A Medical History of Humanity from Antiquity to the Present* (London, 1997). An impossible task almost accomplished: a massive and authoritative work, the best single-volume history of medicine.

R. Porter, *Enlightenment* (London, 2001). A sparkling book about the period in which the author probably felt most at home.

R. Porter, *Flesh in the Age of Reason* (London, 2004). How the idea of man made in the image of God was transformed into the notion of the self-made man. Porter's last published book.

P. Pottier, G. Maloney and J. Desautels, *La Maladie et les malades dans la Collection Hippocratique* (Quebec, 1989). Contains interesting comments on pain in Greek tragedies.

D'A. Power, 'The Medical Institutions of London. The Rise and Fall of the Private Medical Schools', *British Medical Journal*, 1895 (1), 388. The only account of the Hunterian Medical School in Windmill Street which John Snow attended.

J. Priestley, *Experiments*, 2 volumes (London, 1790). A classic, describing the isolation of oxygen and other gases.

J. Pringle, *A Discourse on Different Kinds of Air* (London, 1773). One of Beddoes's inspirations, which turned him towards pneumatic medicine.

C. de Puységur, comte, *Rapport des cures operées à Bayonne par le Magnétisme Animal*, translated by M.M. Titerow (Illinois, 1970).

C. Quétel, *History of Syphilis*, translated by J. Braddock and B. Pike (Baltimore, 1990). Deeply researched work, focused especially on the French scene and on the disease rather than on famous victims. No mention of Schubert, Wolf, Schumann or Ibsen's heroes; but excellent passages on Stendhal, Daudet and other nineteenth-century French authors.

V.S. Ramachandran and S. Blakeslee, *Phantoms in the Brain* (London, 1998).

P. Rang and M.M. Dale, *Pharmacology*, 2nd edn (Edinburgh, 1992).

H.R. Raper, *Man against Pain: The Epic of Anaesthesia* (New York, 1945). Excellent on the American scene.

*Rapport des commissaires chargés par le Roi de l'examen du magnétisme animal*, ed. J.S. Bailly (Paris, 1784). Includes A-L. de Jussieu's intelligent and fair dissenting opinion.

J. Renoir, *Renoir My Father*, translated by R. and D. Weaver (London, 1962). Background to Chapter 44, one of a handful of biographies of great artists worthy of their subject: a joy.

R. Rey, *Histoire de la douleur* (Paris, 2000). There is no other 'history of pain' and this is a very learned tome. It is also very heavy going, mainly perhaps because of an almost total lack of paragraphs. It is inevitably most illuminating on French thought.

N.P. Rice, *Trials of a Public Benefactor* (New York, 1859; reprinted in facsimile by the Australian Society of Anaesthetists, 1995). The biography of Morton commissioned by Morton himself, entirely self-serving and unreliable but revealing of the man. The author was never able to collect his fee in full.

B.W. Richardson, 'The Life of John Snow, M.D.', in J. Snow *On Chloroform*, etc. (q.v.). For long this was the only biography of Snow. Written by a friend and admirer, it is perhaps more an extended obituary than a biography; and it is not free from error. But it remains a valuable and attractive first-hand memoir of a great man.

B.W. Richardson, 'John Snow, M.D., A Representative of Medical Science and Art in the

Victorian Era', *The Asclepiad*, 1887 (15), 247. An abridged and slightly revised version of B.W. Richardson's *Life*, above.
J.M. Riddle, *Dioscorides on Pharmacy* (Austin, Texas, 1985).
J. Ridley, *Bloody Mary's Martyrs* (London, 2001). A lucid and vivid account of an episode from the Second Age of Martyrs, both horrifying and uplifting.
G.B. Risse, *Mending Bodies, Saving Souls: A History of Hospitals* (Oxford, 1999). A rich survey of a vast field using the technique of historical snapshots which focuses on individual hospitals and individual patients to illustrate general trends and developments. The book takes the story from classical times to the modern 'Biotechnical Showplaces', institutions of which the author rightly disapproves. Hospices and palliative care will presumably be mentioned in future editions.
J.M. Rist, *Stoic Philosophy* (Cambridge, 1969).
J.M. Rist, *Augustine* (Cambridge, 1994).
J. Robinson, *A Treatise on the Inhalation of the Vapour of Ether, for the Prevention of Pain in Surgical Operations* (London, 1847). The beginning of ether anaesthesia in Britain.
V. Robinson, *Victory over Pain* (London, 1947). The title should specify 'physical' pain and have a tiny question mark tucked into it somewhere; but otherwise this is a good scholarly short history of anaesthesia up to the date of publication, especially strong on the nineteenth and early twentieth centuries.
J. Rochart, *Histoire de la chirurgie française au XIX siècle* (Paris, 1875).
F.C. Rose, *Neurology of the Arts* (London, 2004).
F. Rosner, *Medicine in the Bible and the Talmud: Selection from Classical Jewish Sources* (New York, 1977).
K. Rudolph, *Gnosis* (Edinburgh, 1983).
B. Rush, *Selected Writings*, ed. D.D. Runes (New York, 1947).
J. Russier, *La Souffrance* (Paris, 1963). A thoughtful essay on life, art and suffering.
E. Sachs, *The History and Development of Neurological Surgery* (New York, 1952).
V. Sackville-West, *The Eagle and the Dove: A Study in Contrasts. St Teresa of Avila, St Thérèse of Lisieux* (New York, 1944). A sparkling double biography and study of female sainthood.
C. Saunders, 'The Treatment of Intractable Pain in Terminal Cancer', *Proceedings of the Royal Society of Medicine*, 1964 (4), 68. An early paper on the aims and achievements of the hospice movement.
C. Saunders and R. Kastenbaum, *Hospice Care on the International Scene* (New York, 1997).
S.H. Schneider, *Drugs and the Brain* (New York, 1996). An excellent book.
J.M. Scott, *The White Poppy* (London, 1969).
D. Seale, *Vision and Stagecraft in Sophocles* (Chicago, 1982). A detailed study of the pain in *Philoctetes.*
D.A.E. Shephard, *John Snow, Anaesthetist to a Queen and Epidemiologist to a Nation* (Cornwall, Ontario, 1995). This is a perceptive and scholarly biography of a great but elusive Victorian.
C. Sherrington, *The Integrative Action of the Nervous System* (New Haven, 1961). A classic since its first appearance in 1906.
R.H. Shryrock, *The Development of Modern Medicine* (London, 1948). Good on the changing social status of doctors.
B. Silliman, *Elements of Chemistry* (New York, 1830). Volume II contains the famous passage which announces the discovery of chloroform by Samuel Guthrie. Neither Guthrie nor Silliman was aware of the anaesthetic potential of the substance.
J.Y. Simpson, *Anaesthesia in Surgery. Does it Increase or Decrease Mortality?* (Edinburgh, 1848). The first of Simpson's salvoes (lightly sprinkled with statistics) in favour of chloroform. Still a breathtaking read.
J. Y. Simpson, *Works*, 3 volumes (Edinburgh, 1871–72). Volume II contains the famous passage about the discovery of chloroform as an anaesthetic; but the combination of surgical and Scottish chutzpah in many of the other essays too is awe-inspiring.

J.M. Sims, 'Discovery of Anaesthesia', *Virginia Medical Monthly*, 1877, p. 81. The distinguished obstetrician endorsing Long's claim as the true discoverer of ether anaesthesia.

F. Skey, *Operative Surgery* (London, 1850). Describes the first recorded Caesarean section under ether anaesthesia.

H.A. Skinner, *The Origin of Medical Terms* (Baltimore, 1949). In the writer's opinion the best reference book of its kind.

P. Slack, *The Impact of Plague on Tudor and Stuart England* (London, 1985).

S. Smiles, *Lives of Boulton and Watt* (London, 1865). Contains much useful information on Beddoes and his circle in Bristol.

M.C. Smith, *A History of Minor Tranquilisers* (New York, 1981). An excellent account of the Valium epidemic which quotes Norman Mailer's 'The natural role of 20th century man is to be anxious' (from *The Naked and the Dead*).

P. Smith, *Arrows of Mercy* (Toronto, 1969). The history of curare in anaesthesia.

T. Smith, *An Inquiry into the Origin of Modern Anaesthesia* (Hartford, Connecticut, 1867). The account of the ether controversy by Lizzie Wells's champion.

W.D. Smith, *The Hippocratic Tradition* (Ithaca, New York, 1979).

J. Snow, 'On the Inhalation of the Vapour of Ether in Surgical Operations', a lecture. *Lancet*, 1847 (1), 551. Snow's first publication on surgical anaesthesia.

J. Snow, 'On the Inhalation of Chloroform and Ether, with a Description of an Apparatus', *Lancet*, 1848 (1), 177.

J. Snow, 'On Narcotism by the Inhalation of Vapours', *London Medical Gazette*, a sequence of 15 papers published during 1848 and 1849 in volumes 6, 7, 8 and 9.

J. Snow, 'On the Prevention of Cholera', *London Medical Gazette* 1853 (3), 559.

J. Snow, *On Chloroform and Other Anaesthetics: Their Action and Administration* (London, 1858; reprinted by the Wood Library Museum of Anaesthesiology, Park Ridge, Illinois, 1988). Posthumous collection of papers by a great and modest man, sympathetically edited and annotated by his friend, B. Ward Richardson (q.v.).

J. Snow, *Case Books*, 3 volumes. Manuscripts held by the Royal College of Physicians of London. Transcribed to print by Dr R.H. Ellis (London, 2001). A superb contribution to the history of anaesthesia.

S. Snyder, *Drugs and the Brain* (New York, 1996). Concise and readable.

Sophocles, *Philoctetes*, edited by F. Storr (London, 1924).

E. Soubeiran, 'Recherches sur quelques combinaisons du chlore', *Annales de chimie et de physique*, 1831 (48), 113. The discovery of chloroform independently from Guthrie and Liebig.

A. Soubiran, *Le Baron Larrey, chirurgien de Napoléon* (Paris, 1966). A doctor's sympathetic biography of a great colleague and something of a legend.

H. Spencer, *Principles of Psychology* (London, 1855). This neglected classic is still an interesting read. Great survival value is ascribed to reflex frowning since it enables the threatened animal to fight with the sun shining in its face. In the Darwinian tradition.

S. Stanhope Smith, *An Essay on the Causes and Varieties of Complexion and Figure of the Human Species* (1787; ed. W.D. Jordan, reprinted Cambridge, Massachusetts, 1965).

P. Stanley, *For Fear of Pain. British Surgery, 1790–1850* (Amsterdam, 2003). The history of preanaesthetic and pre-Listerian surgery could be grim reading; but this book is written with perception, sympathy and restraint: it is also wonderfully well researched. Essential reading for anybody interested in pain or the surgical past. The present writer is greatly indebted to it.

D. Stansfield, *Thomas Beddoes M.D.: Chemist, Physician, Democrat* (Dordrecht, 1984). An excellent biography of an elusive man.

E.C. Stanton, *The American Woman and Womanhood, 1820–1920* (New York, 1972).

R.A. Sternbach, 'Survey of Pain in the United States. The Nuprin Pain Report', *Clinical Journal of Pain*, 1986 (2), 49. An important paper which revealed the astonishing frequency of pain syndromes in the general population.

E.C. Stewart, *The Hospitals and Surgeons of Paris* (Philadelphia, 1843).

W.K. Stewart and L.W.Fleming, 'Perthshire Pioneer of Anti-inflammatory Agents', *Scottish Medical Journal*, 1987 (32), 146. The life of Thomas J. Maclagan.

J.E. Stock, *Memoirs of the Life of Thomas Beddoes with an Analytical Account of his Writings* (London, 1811). A far too discreet source-book with a list of Beddoes's publications.

E. Stone, 'An Account of the Success of the Bark of the Willow in the Cure of Agues', *Philosophical Transactions* of the Royal Society (London, 1763), p. 67. The beginning of aspirin.

P.G. Strange, *Brain Biochemistry and Brain Disorders* (New York, 1992).

D. Streatfield, *Cocaine* (London, 2002). Contains much interesting information, including one of the few printed accounts of Halsted's addiction.

K. Sudhoff, *The Earliest Printed Literature on Syphilis, being Ten Tracts from the Years 1495–1498* (Florence, 1925).

J. Sutherland, *Mrs Humphry Ward, Eminent Victorian, Pre-eminent Edwardian* (Oxford, 1990). A first-class biography.

J.R. Swann, *Academic Scientists and the Pharmaceutical Industry* (Baltimore, 1988).

T. Sydenham, *The Works of Thomas Sydenham*, translated by J. Greenhill, with life by R.G. Latham, 2 volumes (London, 1848 and 1850). These scholarly volumes were published under the auspices of the Sydenham Society. The first edition of *Opera omnia* was published in London in 1683. Sydenham was an excellent writer as well as a great clinician and his descriptions of several diseases, especially of gout and fevers, could not be bettered today.

N. Sykes, P. Edmonds and J. Wiles (eds), *Management of Advanced Disease*, 3rd edn (London, 2005). This is still probably the best-edited multi-author text available.

J. Syme, *Observations in Clinical Surgery* (Edinburgh, 1861). A revealing book by a great but elusive surgeon, Lister's teacher and father-in-law.

T. Szasz, *Pain and Pleasure. A Study of Bodily Feelings* (Syracuse, New York, 1988). A much-quoted clever book, introducing the concept of 'painmanship'.

R. Tallis, *Hippocratic Oaths: Medicine and its Discontents* (London, 2004). A disturbing but not a despairing book about the changing face of medicine: beautifully written.

F.L. Taylor, *Crawford W. Long* (New York, 1889). Affectionate biography by Long's daughter.

O. Temkin, *Galenism. The Rise and Fall of Medical Doctrines* (Ithaca, New York, 1973).

J. Teppe, *Apologie pour l'anormale, ou Manifeste du dolorisme* (Paris, 1937). The dolorist manifesto. Awful.

St Teresa of Avila, *The Life of Saint Theresa of Avila by Herself*, translated by J.M. Cohen (London, 1957). Not the most recent but the best translation, now available in paperback.

A.S. Thelwall, *The Iniquities of the Opium Trade with China* (London, 1839).

St Thérèse de Lisieux (Therèse Martin), *Story of a Soul: The Autobiography of St T. de Lisieux*, translated by J. Clarke, O.C.D., 3rd edn (Washington, 1996). A devotional autobiography, published after the author's death and still a bestseller in 40 languages: devotion apart, a moving testimony to human courage.

C.J.S. Thompson, *The Mystery and Art of the Apothecary* (London, 1929).

M.A. Thouret, *Recherches et doutes sur le magnétisme animal* (Paris, 1784). The most telling contemporary criticism of Mesmer.

J. Thuillier, *M. Charcot à la Salpêtrière* (Paris, 1993). The best and most up-to-date biography of a great doctor.

M.M. Titerow, *Foundations of Hypnosis from Mesmer to Freud* (Springfield, Illinois, 1970).

A. de Tocquville, *Democracy in America* (first published in Paris, 1836), 2 volumes (New York, 1945), still a wonderful read.

W. Topham and W. Squire Ward, *Account of a Case of Successful Amputation of the Thigh under Mesmerism* (London, 1842). The trigger of a fierce controversy.

J.P. Trevelyan, *The Life of Mrs Humphrey Ward* (London, 1923). A reverential but elegant biography.

T. Trotter, *A View of the Nervous Temperament. A Practical Enquiry into the Increasing Prevalence, Prevention and Treatment of Those Diseases Commonly Called Nervous, Bilious and Liver Complaints, Indigestion, Low Spirits, Gout, etc.* (London, 1807). A perceptive contemporary view by an exceptionally intelligent and literate doctor of the first great age of addiction.

W. Trotter and H.M. Davies, 'Experimental Studies in the Innervation of the Skin', *Journal of Physiology*, 1909 (38), 134. A ground-breaking paper.

R.G. Twycross and S.A. Lack, *Therapeutics in Terminal Care*, 2nd edn (London, 1990).

M. Ullman, *Islamic Medicine* (Edinburgh, 1978).

P.U. Unschuld, *Medicine in China: A History of Ideas* (Berkeley, 1985).

E.S. Valenstein, *Unkind Cuts: The Rise and Decline of Psychosurgery and Other Radical Treatments for Mental Illness* (New York, 1986). A still deeply frightening overview.

W.H. Van Buren, 'Amputation at the Hip Joint', *Transactions of the New York Academy of Medicine*, 1847 (1), 123. The impact of ether on the most demanding amputation.

J.R. Vane, 'Inhibition of Prostaglandin Synthesis as a Mechanism of Action for Aspirin-like Drugs', *Nature*, 1971 (231), 232. The beginning of aspirin's second innings.

C. de Vaumorel, *Aphorismes de M. Mesmer* (Paris, 1785).

E. Verg, G. Plumpe and H. Schultheis, *Milestones* (Bayer AG, 1988). A self-congratulatory Bayer publication on its own origins.

Vesalius, *On the Fabric of the Human Body*, translated by W.I. Richardson and B. Carman (San Francisco, 1998). A splendidly produced three-volume (originally seven) re-translation of one of the milestones in the history of medicine and science.

E.L. Vigée-Lebrun, Mme, *Memoirs*, translated by H. Shelley (London, n.d.). The wonderfully catty memoirs of Marie Antoinette's favourite painter and an eyewitness to the furore aroused by Mesmer.

J. Vinchon, *Mesmer et son secret* (Toulouse, 1971). Mesmer cleverly distorted by Freudian spectacles. Infuriating.

P. Vinten-Johansen, H. Brody, N. Paneth, S. Rahman and M. Rip, with the assistance of D. Zuck, *Cholera, Chloroform and the Science of Medicine. The Life of John Snow* (Oxford, 2003). In terms of depth of research and documentation this impressive work easily surpasses (without eclipsing) earlier books on Snow. It is the remarkable, perhaps unique, product of collaboration between five professors in different disciplines on the staff of a university (Michigan State). Few if any scars of multi-authorship show, while the benefits are obvious. But Snow was a very private individual and as a person he remains somewhat elusive. Nevertheless, this is an essential addition to the literature of anaesthesia and public heath in the nineteenth century.

A.Violon, *La Douleur rebelle* (Paris, 1992).

J. de Voragine, *Legenda Aurea* (Heidelberg, 1984). Modern edition of the most detailed, harrowing and at times uplifting account of the sufferings of the early Christian martyrs.

A. Vyrubova, *Memoirs of the Russian Court* (New York, 1923). The best primary source for the illness of the Tsarevitch and Rasputin's ascendancy, especially concerning the illness at Spala described in Chapter 38.

A. Waddington, *The Medical Profession in the Industrial Revolution* (Dublin, 1984).

P.D. Wall, *Pain, the Science of Suffering* (London, 1999). A short, interesting and humane book by a leading figure of twentieth-century pain research, now available in paperback.

P.D. Wall and M. Jones, *Defeating Pain: The War against a Silent Epidemic* (New York, 1991). A readable book on the modern approach to pain pioneered by the first author.

P.D. Wall and R. Melzack (eds), *Textbook of Pain*, 4th edn (Edinburgh, 2001). This once ground-breaking textbook now has 100+ authors and needs a weightlifter to lift it.

J. Waller, *Fabulous Science. Fact and Fiction in the History of Scientific Discovery* (Oxford, 2002). Casts doubt on some myths of scientific discovery, sometimes with justice. In the case of Snow and cholera the myth seems to be built up in order to be demolished. In the case of Fleming and penicillin there is more than a grain of truth in the debunking.

But for an idea to be quickly superseded (as was true of antisepsis) is a sign of importance in science, not of the reverse.

D.M. Walmsley, *Anton Mesmer* (London, 1967). The best biography in English.

J. Wardrop, *Aneurysm and its Cure by a New Operation* (London, 1828). A sensation in its day. The author discusses exsanguinations, that is bleeding of almost the whole blood volume, as a possible means of anaesthesia. Fortunately the method did not catch on.

E. Warren, *Letheon*, 2nd edn (Boston, 1847). Contains Oliver Wendell Holmes's famous letter addressed to Morton in which Holmes suggests the term 'anaesthesia'.

E. Warren, *The Life of John Collins Warren, Compiled Chiefly from his Autobiography and Journals*, 2 volumes (Boston, 1860). E. Warren was J.C. Warren's son and present at Morton's historic demonstration of ether anaesthesia.

J.C. Warren, *Etherisation, with Surgical Remarks* (Boston, 1848). An account by one of the central figures of the introduction of ether. It is remarkably sympathetic and humane from a man reputed to be a superb craftsman but cold and unfeeling.

W. Waugh, *John Charnley. The Man and the Hip* (Berlin, 1990).

C. Webster, *The Great Instauration: Science, Medicine and Reform, 1626–1660* (London, 1975). A sparkling book.

W.H. Welch, *The Introduction of Surgical Anaesthesia* (Boston, 1908). A brief but Olympian survey by the eminent microbiologist and first Professor of Medical History in the United States.

F. Werfel, *The Song of Bernadette* (London, 1942). Escaping from Nazi Europe the German Jewish writer was sheltered for a few days in the Shrine in Lourdes. His fine novel, fictionalised but not a historical travesty, was a repayment. Werfel was not directly responsible for the unspeakable Hollywood film concocted from the novel.

W. Whitman, *Leaves of Grass*, New American Library edition (New York, 1958).

G. Whitteridge, *William Harvey and the Circulation of Blood* (London, 1971). Contains much new information.

G.M. Wickens (ed.), *Avicenna: Scientist and Philosopher, A Millenary Symposium* (London, 1952). Six authoritative essays on various aspects of Avicenna's life and work. The one by Professor A.J. Arberry on Avicenna's life and times is particularly valuable, even giving a sample of the great doctor's rather feeble poetry.

J. Wilde, *The Hospital, a Poem in Three Books, Written in the Devon and Exeter Hospital* (Norwich, 1809). Not in the Milton class perhaps but evocative of a chronic patient's experience in a large English provincial hospital in the preanaesthetic, pre-Listerian and pre-Florence Nightingale era.

T. Willis, *The Anatomy of the Brain and Nerves, Five Treatises* (originally printed in London in 1681. Facsimile edited by W. Feindel, Montreal, 1965). An exemplary re-edition of a ground-breaking work on the function of the brain and nerves.

C. Wilson, *The Invisible World: Early Modern Philosophy and the Invention of the Microscope* (Princeton, 1995).

G. Wilson, *One Grand Chain: The History of Anaesthesia in Australia, 1846–1962*, Vol. I (Melbourne, 1995). An important and deeply researched source.

G. Winger, F.G. Hofmann and J.H. Woods, *A Handbook on Drug Abuse and Alcohol: Biomedical Aspects* (New York, 1992).

A. Winter, *Mesmerise: Power of the Mind in Victorian Britain* (Chicago, 1998).

L.R. Wolberg, *Medical Hypnosis*, 2 volumes (New York, 1948).

R. J. Wolfe, *W.T.G. Morton and the Introduction of Surgical Anaesthesia* (San Anselmo, California, 1990). A massive, scholarly and much-needed exercise in debunking; surely a 'definitive' verdict. It cleverly brings the period as well as the characters alive. Some deserve commemoration more than others.

A. Worcester, *The Care of the Aged, the Dying and the Dead* (Springfield, Illinois, 1935; repr. Blackwells, Oxford, 1961). A profound and passionate work published decades ahead of its time based on three lectures given to medical students in Boston. The author, a general practitioner with a lifetime's interest in advanced disease, was eighty at the time.

World Health Organisation, *Cancer Pain Relief and Palliative Care* (Technical Report Series 804) (Geneva, 1990).

F. Yossoupov, Prince, *Rasputin* (New York, 1927). Background to Chapter 38.

F. Yossoupov, Prince, *Lost Splendour* (London, 1953).

M. Zborovski, *People in Pain* (San Francisco, 1969). A pioneering and much-quoted anthropological study.

P. Ziegler, *The Black Death* (New York, 1969). A readable account of the most devastating epidemic in European history (expressed in percentage terms of population). The 'Spanish flu' of 1918–19 killed many more people in absolute numbers.

C. Zimmer, *Soul Made Flesh: The Discovery of the Brain – and How It Changed the World* (London, 2005). Despite the pot-boiling title – the brain was not 'discovered' and, sadly perhaps, the non-discovery did not change the world – this is a sparkling survey of the seventeenth-century scientific revolution, its antecedents and consequences, by a distinguished American science writer. Focusing on the life and work of Thomas Willis, a great doctor as well as arguably the founder of modern neuroscience, it brings to life the leading figures of that revolution – Harvey, Hooke, Hobbes, Boyle, Wren – with sideways glances at Descartes, Paracelsus, Sydenham and many others. A great read and a mine of interesting information.

E. Zola, *Le Docteur Pascal* (Paris, 1887). A classic description of general paralysis of the insane (GPI), a late complication of syphilis, in fiction. The most arresting appearance of the congenital form of the disease is in Ibsen's play, *Ghosts.*

E. Zola, *Mes Voyages, Lourdes, Rome, Journeaux inédits,* edited by René Ternois (Paris, 1958). The best modern edition of Zola's impassioned polemic against Lourdes. In contrast to his advocacy of Dreyfus, Zola's diatribes against Lourdes often relied on a wilful distortion of the truth. In terms of sales, *Lourdes* was nevertheless his most successful novel.

S. Zweig, *Mental Healers: Franz Anton Mesmer, Mary Baker Eddy, Sigmund Freud,* translated by E. and C. Paul (New York, 1932). The three characters had something in common but all three must have turned in their graves when Zweig yoked them together in a single volume. The writing is majestic.

# INDEX

With few exceptions individual works of writers, painters and other artists are not listed; users are advised to consult the entry for the author. Names of classical Romans are given under the anglicised form or the English one by which they are best known. Arabic names are indexed under the first letter of the whole name, e.g. Harun al-Rashid under 'H' and Abu'l Kasim Mansur under 'A'.


Abbasid dynasty 61, 63
Abbot, Gilbert 218
Abd al Rahman III 67
Aberdeen, Lord 229
Abernethy, John 174
Abraham 49
Abrams, Ruth 495
Abu Ali al-Hussayn ibn 'Abdallah *see* Avicenna
Abu Ali al-Hussayn ibn 'Abdallah ibn Sina *see* Avicenna
Abu'l Kasim Mansur *see* Firdausi
Abu'l-Qasim Khalaf ibn Abbas al-Zahravi *see* Albucasis
Abu-l-Walid Mohammedibn Ahmad ibn *see* Averroës
Académie royale 124, 134, 195
Academy of Science, Augsburg 142
acetaminophen 457
acetaniline 361
acetone 246
acetylcholine 406–7
acetylcholinesterase 407
acetylsalicycilic acid *see* aspirin
Achilles, Greek hero 41, 90n
acupuncture 5, 14, 478, 485
Adam, biblical character 29
Adonis 26
adrenaline (epinephrine) 406
Adrian, E.D. (later Lord) 402
Adventism 287
Aeneas 11, 27, 64n, 90n
aerometer 119n
Aesculapius 11 and n, 13, 15, 32, 279n,
Agesandros of Rhodes 90
Agrippina the Younger 19
Alacoque, Marguerite Marie 123, 440
Albert, Prince 274, 338n
Alberti, Leon Battista 94
Albucasis 70 and n
Allbutt, Sir Clifford 368, 376
alchemy 62, 92, 101, 156, 159
Alcmene 2
Alder Wright, C.D. *see* Wright
Aldine Press 89
Alexis, Tsarevitch 391
algesics 305
algometer 304 and n
D'Alembert, Jean le Rond 145
Alexander the Great 19, 20n, 41
Alexandra, Russian empress 391
Alexandria 25, 37, 55n, 58n
Alexandrovna, Grand Duchess Xenia 3
Al-Fadr 68
alkali 62, 256 and n
alkaloids 20n
Allcock, Rutherford 196
Allen, John 215
Allen and Hanbury's 250
Allenby, Field Marshal Lord 428 and n
Alma-Tadema, Sir Laurence 380
al-Mansur Qualawun 72n
Al-Mu-tadidi Hospital, Baghdad 63
Aloysius, St (in Rome) 101

Alston, Dr Charles 250
alternative medicine 93, 138n
Alzheimer, Alois 329–30 and n
Amarra 427n
ambrosia 27
ambulance, flying 172n
Amélie, queen of Louis-Philippe of the French 445
American Medical Association 283, 289
  Committee of Surgery 282
American Society of Dental Surgeons 285
amphetamines 407, 415
amputation 14, 31, 71, 80, 103, 105–7, 170 and n, 172n, 175, 189–90, 194, 196, 199, 223, 227, 229, 232, 240–1, 271, 325n
  breast 197
  finger 237
  limbs: at the ankle 176n; at the hip 168, 175; above the knee 241; below the knee 190, 237; at the shoulder 234; mid-thigh 228; of the thigh 385; traumatic 483
  penis 197
amyl nitrite 346
amylobarbitone 420
amytal 414
anaemia 435, 447, 456
anaesthesia 12, 13n, 14, 15, 26n, 28 and n, 31, 32–3, 34, 39, 78, 81, 102, 106–8, 111, 151, 168, 176 and n, 189, 192–3, 193–5, 199, 205 and n, Ch. 22 *passim*, 238, 241, 245–7 and n, 255n, Ch. 27 *passim*, Ch. 28 *passim*, 282–6, 288, 293–5, 312n, 365
  apparatus 272–3
  and chloroform 192, 273n; *see also* chloroform
  cyclopropane and curare 461
  and ether 202, 204–6, 222–4, 235–7, 239–41, 270, 283; *see also* ether
  Epidural 435
  first child delivered under chloroform by Simpson 246
  general 26, 101, 170, 185, 192, 202, 208, 210, 214, 220, 226, 228n, 238, 241, 275n, 307, 313
  inhalation 113, 192, 204–5, 217–19, 224, 277, 286
  intravenous 117–19
  local 192, 307, 365, 370–2, 374–5, 469, 471, 481
  spinal 373
  *see also* Dudley, Thomas Hickman; Simpson, Sir James Young; Snow, John
anaesthetic machine 436
analgesia, congenital 438, 485
analgesics 62, 63, 138, 305–6, 370, 415, 419–20, 42
Anderson, Thomas 335
André, Nicolas 424
Andrews, Edmund 205n, 460
aneurysms 168–9 and n, 257, 231n
angina 83, 305n, 308, 343–6, 467–9, 675–7
Anicia, Juliana, daughter of Anicius Olibrius 25
Anicius Olibrius, Roman Emperor of the West 25n
animal magnetism 137f
anodynes 16, 26, 31, 69, 102, 134, 223, 229
Anrep, Vassaily Konstantinovich 371
Anthony, St 54 and n, 184
antiepileptic drugs 428
anticoagulants 180n
antipyrine 361
antisepsis 13n, 170 and n, 176n, 185, 192, 335–8, 372n, 499
aperient 15
aphasia 328–9
Aphrodite 11, 91
Apollo, classical god 37
Apollo of Belvedere, statue 90
Apollodorus 25
Apollonia, early Christian martyr 38, 55 and n
apomorphine 11n
apothecaries 57, 75, 96, 100–1, 132, 134, 154, 342, 352
Apothecary's Company, the 250, 331 and n,
appendicitis 48, 317, 332, 333, 338, 434, 438
Apuleius 28
Aquapedante, Hyeronimus Fabricius ab 107
*Arabian Nights, The* 3, 72 and n
arachidonic acid 458
Arany, János 236
d'Arblay, Alexandre 171–2 and n
d'Arcel 147
Argyll-Robertson, Douglas 316n
Aristotle 20, 65n, 76, 79, 114n, 122, 309, 469
Arianism 58 and n
Arius 46n
Armstrong-Jones, Sir Ronald 341
Arnald of Villanova 69n
Arnold, Dr Thomas 200 and n, 382
Arnold, Matthew 379

Arnold, Thomas 379
arsenic 155n
Artemis, St 54 and n
Asa, King 53
Aschenbrandt, Theodor 369
asepsis 170 and n, 185, 337
Ashkham, Mary 361n
Aspern, battle of 483
asphyxia and asphyxiation 193 and n, 235, 269, 278, 283
aspirin 362n, 364, 383n, 453, 455–8
Aspro 455
asthma 154, 158
astrology 76, 79, 140
astronomy 77, 147
Athanasius 46n
Athanasianism 57–8 and n
atheism 326
Athene, classical goddess 293
Athenodoros of Rhodes 90
Atkinson's Baby Preservative 253
*Atropa belladonna see* deadly nightshade
*Atropa mandragor see* mandragora atropine
Atropine 26, 30, 33n, 255
Augustine of Hippo, St 50, 59
Augustus II, king of Poland and elector of Saxony 30
Aurelia, mother of Julius Caesar 13n
Aurelius, Marcus, Roman emperor *see* Marcus Aurelius
Aurelius Sabinus, official under Diocletian 37n
*auto-da-fès* 109
autoimmune disease 108
autonomic nervous system 317, 467, 477
Autosuggestion 153
Ave Marias 82
Avenzoar 71 and n
Averoës 64n, 71
Avicenna 9, 64–7 and n, 92, 191

Babinski, Josef 319 and n
Baden 149
Baden-Baden 304
Bach, Johann Sebastian 544
Baeyer, Johann Adolph 412
Bailly, Jean Sylvan 147–8
Baker, General Sir George 240
Balard, Antoine Jerome 345–8 and n
Balfour, Andrew 243n
Balfour, Arthur 380
Ball, John 250
Balzac, Honoré de 137n, 257
Banting, Frederick 340, 414n
Barbara, Saint 413
Barber Surgeons' Company 103
'barbital' 414
barbiturates, intravenous 415n
barbituric acid 413–15
Barker, Dr Fordyce 258
Barrett, Howard 492
Barrett, Norman 495n
Barth, Professor Joseph 144
Bartholomaeus Anglicus 29
Bartolinus, Thomas 191
Basil, Saint 56
Bassano, Jacopo 54n
Bastian, Hector 329 and n
Bayer's 360–4, 453–6
Bayer, Friedrich 360
Bazille, Jean 445
Beardsley, Aubrey 347
Beatrice, Princess, daughter of Queen Victoria 275n
Beaudelaire, Charles 238
Beauharnais, General 149n
Beaumarchais, Pierre Augustin Caron de 148
Beaumont, Elie de 211, 224
Beaumont, William 211–12 and n
Becket, St Thomas à 55
Beddoes, Thomas 151–2, 155–61, 162, 164–7
Beddoes Jr, Thomas 166 and n
Beddoes, Anna, *née* Edgeworth 157, 161, 166 and n
Bedford, Dr Gunning S. 292
Beer, E. 470
Beeton (Isabella), Mrs 254
Behring, Emil von 461
Bell, Benjamin 190
Bell, Charles 32, 169 and n, 175, 201, 243n, 257n, 301, 426
Bell, George 175
Bell's palsy 169n, 426n
Bellini, Giovanni 17n
Bellini, Vincenzo 371
Benda, Julien 438
Benedek, I. 29, 31
Benedict of Nursia 57
Bennett, A.E. 464
Bennett, Alexander Hughes 367, 370
Bennett, John Hughes 243n, 306
Bentham, Jeremy 291
Bergson, Josef 237
Bernadette Soubirous *see* Soubirous
Bernard, Christian 475

Bernard, Claude 306–9, 317, 328, 354, 365, 463–4 and n
Bernard, Mme 307
Bernard of Clairvaux, St 53, 57
Bernays, Martha 368
Bernhard, Sara 374
Bernheim, Hippolyte 326
Bernheims, Alexander, Josse and Gaston 446, 448
Bernini, Gianlorenzo 121n
Berry, Duc de 30
Berteaux, Henri 326
Bibesco, Prince George Valentin 446
Bible, The 59n, 85, 89, 98n, 113, 200, 248
Bichat, Marie François Xavier 131, 469
Biedermeyer style 233n
Bier, August 373
Bigelow, Henry Jacob 210, 218, 220–1, 223, 226, 227, 290 and n
Bigelow, Jacob 227
Bigelow, Mary 227
Billroth, Theodor 336
Bismarck, Otto von 306, 320
Black, Joseph 151, 155, 243n
Black Death 83–5 and n, 454n
Blackwell, Elizabeth 276n
Blackwell, Emily 276n
Blaise, St 54–5 and n
blindness, 12 and n, 37n, 55, 64, 141–4, 426, 443
Blizzard, Sir William (Sir Billy Fretful) 176
Boccaccio, Giovanni 106 and n
Boë, Sylvius de la 128 and n
Boerhaave, Hermann 126–7 and n, 131–2, 243n
Bonaparte dynasty 445
Bonaparte, Jerome 240
Bonaparte, Joséphine 149 and n, 172n
Bonaparte, Marie-Louise 172n
Bonaparte, Mathilde 240
Bonaparte, Napoleon *see* Napoleon
Bonica, John 488
Bonnard, Pierre 449
Bonnefoix, Mme 254
Bonnefon, Jean de 325
Boot, Jesse 457n
Booth, Luce, Claire 155n
Boots, the chemist 457
Boott, Dr Francis 227–8 and n, 269
Bordeu, Théophile de 126–7 and n
Boris Godunov, operatic hero 344n
Borodino, battle of 172n, 191 and n, 392
Botha, Louis 454n
Botkin, S.P. 393, 404
Botticelli, Sandro 95
Boulton, Matthew 157n
Bourbon kings of France 320n, 354
*see also under* specific kings
Boyd, Mary 429
Boyle, Robert 82n, 102, 115n, 117–20 and n, 151
Brahms, Johannes 337n
Braid, James 198
Brande, John 252
Braque, Georges 399
Brasenose College, Oxford 117, 380
Brasevolus 461n
Braun, Heinrich 374
Bremner und Söhne, surgical instrument makers 191n
Brettauer, Josef 371
Brewster, C. Starr 220–2
Briand, Aristide 20
British Medical Association 93n, 390–1
*British Medical Journal, The* 289n, 376, 477
Britten, Benjamin 44n
Broca, Peter Paul 328 and n
Brodie, Sir Benjamin 197, 280–1
Broglie, Princesse Joséphine de 378
Brooke, Arthur 106
Brougham, Henry, Lord 196
Brouillet, Pierre-André 319
Brown, John 182–3, 243n
Browne, J. Collis 252
Browne, Lancelot 114
Browning, Elizabeth Barrett 137n
Browning, Robert 137n
Brown-Séquard, Charles Edouard 311 and n
Brücke, Ernst 304 and n
Brufen 457
Brunschwig, Hyeronymus 104
Brunton, Thomas Lauder 305, 344–6
Brutus the Elder 44n, 64n
Buchan William 351
Budd, Dr George 262n
Buffon, Georges-Louis Leclerc, comte de Buffon 131, 297
Bullein, William 107
Bulwer-Lytton, Edward 137n
Burke and Hare 243n
Burke, Edmund 160
Burne-Jones, Edward 380
Burnett's Pharmacy 216
Burney, Dr Charles 171n
Burney, Francis 'Fanny', later Mme d'Arblay 171–4 and n, 238
Burton, Robert 96, 138n

Butler, Alfred 241
Byron, George Gordon, Lord 134
Byzantine Empire 61, 67

Cabanis, Pierre 131
Cabral, Pedro Alvarez 100
Cadge, William 228–30
Caelius Aurelianus 30
Caesarean section 13 and n, 241
Caesar, Julius 13n, 35n, 64n
Cagliostro, Alessandro, 'Count' (Giuseppe Balsamo) 145
Cahn, Arnold 361
Caillebotte, Gustave 446
Caius, John 103n
Cajal, Santiago Ramon y 399–402
Calaucha, Father Antonio de la 352
Calckar, Ian Stephan van 95
Caligula, Roman emperor 27
callisthenics 14n
Calvin, John 112–13 and n
cancer 48n, 163, 172n, 174n, 283, 284n, 315, 337 and n, 375, 419, 439, 466, 475, 494–6
  of the bone 48
  childhood 475
  *see also* leukaemia
Candler, Griggs 375
cannabis (*indica*) 10n, 27, 63, 66, 72, 75, 341 and n
Cannon, Walter 308n, 410
Capuchin monks 133
carbon dioxide 159, 162, 163, 193, 264, 271–2, 307
carbon monoxide 159, 162 and n
carboxyhaemoglobin 162n
Carl of Habsburg, Archduke 483
Carlyle, Thomas 359 and n
Carmody, Francis X. 435
Cartwright, F.F. 192–3
Cartwright, Samuel A. 295
Caruso, Enrico 455n
Catherine I, Tsarina 30
Cato, the Elder, Marcus Porcius 28, 294
Caton, Robert 312–13 and n
Catullus, Gaius Valerius 21
cauterisation 28, 71 and n, 77, 104, 107, 124, 191n, 283, 427
Cavaradossi, Mario, operatic hero 344
Cavendish, Henry 154, 346n
Caventou, Joseph 257, 354
'Celebrated Pain Killer', Perry Davis' 252
Celsus, Aurelius Cornelius 21–2 and n, 26, 64, 89, 350, 458
Cerberus 12
cerebrospinal fluid 373, 472
Ceres 21n
Cézanne, Paul 447
Chaliapine, Feodor 484n
Chambord, Henri Charles Ferdinand, Comte de 320n
Chambord, Paul 308
Chamisso, Adalbert von 135
Champier, Symphorien 96,112
Channing, Walter 209
Chaplin, Charlie 455
Charcot, Jean 320
Charcot, Jeanne 320
Charcot, Jean-Martin 313, 317–21 and n, 325–30, 443
Charigot, Aline (Mme Renoir) 446
Charles I, king of England 114 and n
Charles II, king of England 115n, 124–5, 352
Charles V, holy Roman emperor 95n
Charles VIII, king of France 98
Charles IX, king of France 104n
Charles X, king of France 168n, 176, 194
Charles, prince of Wales 93n
Charnley, John 475
Chateaubriand, François Auguste René 134
Châtelet, Hyppolite 382
Chaucer, Geoffrey 65n, 77, 105, 138
Chauliac, Guy de 69n, 74n, 76, 79–82
Chekhov, Anton 399
childbed fever 127, 183, 238n, 266, 432
chloral hydrate 308, 348
chlorodyne 252
chloroform 198n, Ch. 25, 269–71, 273 and n, 274–5 and n, 283, 285, 290, 298, 307, 308, 312n, 334, 338, 483
chlorpromazine 474
cholera 251–2 and n, Ch. 27 *passim*, 326, 342
Christina, queen of Sweden 121 and n
Christison, John 244–5, 254
Christison, Sir Robert 368
Christopher, St 54 and n
Churchill, Frederick 229–30
Churchill, Winston 383
Clark, Sir James 338n
Cleopatra, queen of Egypt 29
*Clostridium tetani* 387
Cicero, Marcus Tullius 22, 42 and n
cinchona 127, 131, 352–4, 365–6
Civiale, Jean 210n
Clarendon, Edward Hyde, 1st earl of 344
Clark, James 274, 280–1

claudication, intermittent 468–9
Claudius, Matthias 136n
Cleanthes 43
Clement VI, Pope 85 and n
Cline Jr, Henry 178–9
Clive of India, Robert 133 and n
Cloquet, Hyppolite 198
Clouston, Thomas 341
Clowes, William 103
Club des hashishiens 418n
cocaine 304, Ch. 36 *passim*, 384, 407, 455
Cocteau, Jean 10n
Codeine 20n, 257
*Codex Anicia* 25n
Coga, Arthur 118n
Coleridge, Samuel Taylor 128, 134, 157, 161, 164, 167, 259
colic 332 and n, 403, 417, 441
Collier, Jeremy 39 and n
Coloman the Bookish, king of Hungary 75
Colt, Sam 203–4
Colt Firearms Manufactory, Hartford 204n
Colton, Gardner Quincy 208, 214
comas 239, 278, 419, 430, 475
Conan Doyle, Arthur 376
conditioned reflexes 404–5
Condorçet, Marie Jean Antoine Nicoles de Caritat, marquis de 125 and n
Confucianism 14
*Conium maculatum see* hemlock
Conolly, John 340
Constantine I, the Great 46
Constantius 58
Cook, Captain James 99n
Cooper, Sir Astley 176, 196, 210
Cooper, Samuel 175
Cope, Jonathan 353
Copeland, Dr Samuel 197
Copernicus, Nicolaus 99n
Corbus, Valerius 101–2
Corday, Charlotte 160n
cordotomy 470–1
Corfield, Major 197
Corning, James Leonard 373
Corot, Camille 320
Cortés, Hernan 97n, 100
cortisone 458, 474
Coryton, Mr, schoolmaster to Humphry Davy 161n
Cosmas, St 37–40
Coward, Noel 455n
Cowley, Abraham 367n
Cowley, Dorothy 488
Cranach, Lucas, the Elder 17n, 101
Cranmer, Thomas, archbishop of Canterbury 39 and n
Crawford, Jane 203n
Crispi, Don Pedro Nolasco 367n
Crito, Platonic character 32
Cromwell, Oliver 127, 129, 350n
Crooke, John 13n
Cros, Madeleine, Mme de 320
Crusades 53, 68 and n, 75, 128
Cruveilhier, Jean 328 and n
Cullen, William 126–7 and n, 155, 243n
Culpeper, Nicholas 350 and n
cupping 73, 76, 80, 124, 184, 286
'curare' 307–8, 462–5
curarine 463–4
CURB 420
Curzon, George Nathaniel, 1st marquis 380
Cushing, Harvey 402, 428
Cutler, Thomas 270
cyclopropane 462
cycloxigenase 2 (COX-2) 458–9

Dale, Henry 39, 406
Damian, St 37–9
Dante Alighieri 64 and n
Davies, George 455
Darwin, Charles 293
Darwin, Erasmus 157 and n, 161, 163
Dawson of Penn, Bertrand, 1st viscount 338n
Daumier, Honoré Victorin 320
Davey, John 339
David, Old Testament king of Israel 42, 82n
Davies, H.M. 410
Davis, Perry 252
Davy, Humphry 157, 161 and n, 162–5, 171n, 200, 256n, 373
Day, Thomas 157n
deadly nightshade 33 and n
Debussy, Claude 348n
Decius, Roman emperor 35–6, 54n
Defoe, Daniel 132
Degas, Edgar 449n
Deïanira 11
Déjérine, Jules 317 and n, 408
delirium 183–4
delirium tremens 259
Delius, Frederick 315n
Delorme, Julien 238
Delson, Charles 148
dementia praecox *see* schizophrenia
Demosthenes 28
Dempsey, Jack 414

demyelinating dieases 305
Denis, Jean Baptiste 118
Derosne, Jean-Pierre 254
Desault, Pierre-Joseph 145
Descartes, René 105, 120–2 and n, 27, 126, 309, 481, 485–6
Dewar, James 154n
Dewees, Dr 293
dextroamphetamine 416
diabetes 32, 109, 317, 414n, 484
diaphoretic 109
diaphoretic chambers 99
Dick Read, Grantley 478
Dickens, Catherine 281
Dickens, Charles 196 and n, 221, 281
Dickinson, Emily 284
Dickenson, Samuel 155
*dictamnus see* dittany
Dicte, Crete 27
Diderot, Denis 145
Diebold, Arthur 454
Dieffenbach, Johann Friedrich 235
digitalis 127 and n
Diocles of Carystos 24
Diocletian, Roman emperor 36–8 and nn, 46, 54n, 55n
Diogenes 64
Dionysus 17
Dioscorides, Pedatius 3, 25–8, 29, 64 and n
diphtheria 317, 461 and n
Disprin 456–7
disseminated sclerosis 305
dittany 27
Doctour of Physic (Chaucerian character) 64–5n, 77 and n
dolorism 437, 466, 501
Domitian, Roman emperor 35, 42n
*Don Quixote*, Miguel de Cervantes 21
Donne, John 96
Donnet , Abbé, later archbishop of Bordeaux 306 and n
dopamine 408
Dostoyevsky, Fyodor 137n
'Dot, Admiral' 454n
Douglas-Wilson, Ian 443
Douthwaite, Arthur 456
Dover's powder 252
Dreser, Heinrich 363
Dreyfus, Alfred 327
Dubois, Antoine 172–5, 210 and n
Dubois, Paul 238
Dubois obstetric forceps 172
du Bois-Reymond, Emil 303–4n
Duchenne, Guillaume (Duchenne de Boulogne) 316 and n
Duchesne, Joseph *see* Quercetanus
du Deffand, Marie-Anne de Vichy-Chamrond, marquise de 145
Dudley, Benjamin Winslow 202
Dudley, Thomas 196
Duhamel, Georges 437
Duisberg, Carl 455
Dumas, Jean Baptiste 247 and n, 360–1, 345n
du Maurier, George 137 n
Duncan, Matthews 245–6
Dupuytren, Guillaume 168–70 and n, 176, 210
Dupuytren's contracture 168n
Duran-Ruel, Paul 445
Dürer, Albrecht 94

Ebers, Georg Moritz 12 and n
    Ebers papyrus 12–13 and n
Eclectic Sect 283
Edgeworth family 157
    Lovell 157n, 164
    Maria 157
Edward II, king of England 82n
Edward VI, king of England 13n
Edward VII, king of England 338, 460
Eglin, Raphael 153
Eichendorf, Paul 42
Eichengrün, Arthur 362
Eichrodt, Ludwig 233n
Einhorn, Alfred 374
Einstein, Albert 170n
Eisenmenger, Victor 462
electroconvulsive therapy 464
Elgar, Edward 374
Elizabeth, queen mother 338n
Elliott, T.R. 406
Elliotson, John 195–8 and n, 200
Ellis, Havelock 304 and n
Elsholtz, Johannes Sigmund 119n
Emerson, Ralph Waldo 284
Empedocles 65n
Empson, Charles 261n
endorphins 486
enkephalins 486
Epictetus (philosopher) 41–2 and n
epilepsy 54. 82, 272, 319 and n, 419
Erb, Wilhelm Heinrich 410
Erichsen, John 280
Erlenmeyer, Albrecht 369
Erskine, John Thomas, 8th earl of Mar 254
erysipelas 54 and n, 184

Esdaile, James 197
Esmarch, Friedrich von 190n
ether 3, 102, 151, 159, 195, 203–7, 216f, 227, 234f, 245–7, 269n, 270f, 276f, 282f, 285
ethyl oxide 246
Eugénie, empress of the French 358
Eusebius (church historian) 37, 47
Eustachian tube 95–6 and n
Eustachius (Eustachio), Bartolomeo 95
Eusthathius of Sebasteia, Bishop 56
Evelyn, John 119
Everett, Edward, 202
Eylau, battle of 1 and n

Fabian, Pope 36
Fabiola, St 56
Fabricius ab Aquapendente, Hyeronimus 107
Failleret, Félicien 239
Falkland, Lucius Bentinck Cary, 10th viscount 240
Fallersweisen, Paschal von 236
Fallopius (Fallopio), Gabriele 95n, 97–8 and n, 111 and n
Faraday, Michael 356
Fatimid dynasty 68
Fergusson, Sir William 27, 168 and n, 244n, 274
Ferguson's of Southwark 258
Fernel, Jean 89n
Ferriar, John 340
Ferrier, David 312–13 and n
Festetich, Count György 251
Feuchtersleben, Ernst 139n
Fildes, Luke 476
Firdausi (Abu'l Kasim Mansur) 13
Fischer, Emil 413–14
Fitzgerald, Scott 440
Fleischl-Marxow, Ernst von 369
Flint, Austin 288
Flourence, Jeanne de 211
Flourens, Marie Jean Pierre 247n, 311–12
forceps 168n, 178, 238
Forest, John W. De 291
Fornovo, battle of 98
Foster, Sir Michael 403
Fothergill, Samuel 424–5 and n, 434
Fourneaux, Ernest 374
Foxe, John 40
foxglove 127 and n
fractures 47, 170
  compound 325
France, Anatole 42
Francis I, king of France 104
Francis of Assisi, St 1
Francis of Lorraine, holy Roman emperor 102
Francis, William 372n
Franck, François 468
Franklin, Benjamin 133, 147 and n, 223
Freeman, Walter 464
freezing spray 191 and n
Freiburg babies 429
'Fretful, Sir Billy' *see* Blizzard
Freud, Sigmund 20 and n, 42, 137n, 140n, 148, 319, 330, 368f
Frey, Max von 410
Friedensheim Home of Rest 492
Fritsch, Gustav Theodor 312 and n
Frobenius, W.G. 102
Frontinus, Sextus Julius 29
Frost, Eben H. 217–19
Fuller, John 270

Gachet, Doctor 449
Gabrielle (Renoir's model) *see* Renard 448 and n
Gage, Philip 312
Galen 24, 26 and n, 28, 62, 64–6 and n, 79, 89 and n, 92, 95, 103
Gall, Franz Joseph 311 and n
Galton, Francis 304n
Galvani, Luigi 140 and n
Garcia, Manuel 462n
Gardner, Augustus 283
Gardner, Eliza Hannah 193
Garnier, Charles 320
Garnier, Jeanne 49
Garraway's Coffee House 250
Gartner, Gustav 370
Gaskell, Walter 305, 409
Gasser, Johann Laurentius 423n
Gasserian ganglion 423, 426
Gate Control theory 482f
Gauss, Carl 433
Gay-Lussac, Joseph Louis 256 and n
Geoffrin, Marie Thérèse, Mme 145
George III, king of England 26
George IV, king of England 153, 159
George V, king of England 338n, 395
George VI, king of England 468
George's Hospital, London, St 169, 190, 240, 241, 258 270
Gérardin, Jean-Philippe 195
Gerhard, Charles 357
Gersdorff, Hans von 104
Gessner, Father Johann 139
Ghiberti, Lorenzo 94

Gibbon, Edward 35n
Giddy, Davis 157 and n, 161n, 166 and n
Gide, André 409, 438
Gilbert, Davis *see* Giddy
Gill, Richard 464
Gillies, Sir Harold 462
Girardin, Marc 306
Girtin, Thomas 135
Gladstone, William Ewart 346, 356
Gleyre, Académie 445
Gluck, Christoph Willibald von 139, 145
Göbbels, Josef, Dr 456
Godfrey's Cordial 253
Goethe, Johann Wolfgang von 2, 44, 134, 156, 255, 311n
    *Heidenröslein* 2
Golden Mortimer, W. 375
Golgi, Camillo 399f, 400n
Goldscheider, A. 410
Gonville and Caius College, Cambridge 103n
Gonville Hall, Cambridge 12, 103n
Goodenough, Dr, character from Thackeray 196
Gordon, Alexander 183n
Gould, Marshal 218
Gowers, Sir William 427
Goya, Francisco de 400
Gracchus, Gaius Sempronius 17
Grant, George R. 202
granulation tissue 47
Gray, Thomas Cecil 465
Greatrakes, Valentine 142n
Greene, Graham 455n
Greener, Hannah 277
Greenhill, Gulielmus Alexander 130, 160
Greenhill, Susan 402n
Gregory, James 244
Grey, Thomas de, 5th baron Walsingham 240
Griffith, Harold 465
Grimm, Friederich Melchior von 145
Grotto del Cane 3, 194
Grünewald, Matthias (Mathis) 91
Grunpeck, Joseph 98n
Guiliano da Sangallo 90
Guillotine, Dr Joseph Ignace 147 and n, 258
Guinther von Andernach, Johannes 89 and n
Guitry, Lucien 374
Gull, Dr William 280
Gustavus Adolphus, king of Sweden 121n
Guthrie, George 240
Guthrie, James 247n
Guthrie, Samuel 247n

Hadrian, Roman emperor 35n
haemoglobin 162n
haemophilia 275n, 391n
haemorrhages 77, 175, 190, 194
    cerebral 262n
haemorrhoids 70, 77, 142, 240
    strangulated 113
Haggard, Rider 388
Hahnemann, Samuel 286
Hales, Stephen 126, 309 and n
Hall, Marshall 309–10
Hall, Sir John 283
Haller, Albrecht von 126–7 and n
Hallett, Ellen 406
Hals, Frans 121 and n
Halsted, William Stewart 371f, 372n
Hamilton, Frank H. 288
Hamilton, James 244
Hamilton, Lady Emma 135, 199
Hamilton, Sir William 199
Hammurabi, king of Babylon 17
Hand, Judge 454
Hannibal 19
Hardcastle, Dr William 262
Harris, Ruth 324n
Harrison, John P. 283
Harun al-Rashid 61, 72n
Harvey, William 96, 111–12 and n, 113–17, 123, 126
Hasskarl, Justus Charles 365
Hawkins, Caesar 281
Hawkins, John 251
Haydon, John 217
Haydn, Franz Joseph 132, 139
Haydn, Michael 139
Head, Henry 409f
Heberden, William 344 and n
Hecquet, Philippe 134
Hector 64n
Heider, Moritz 236
Heine, Heinrich 484
helenium 10 and n
Helen of Troy 10 and n
Hell, Maximilian 140–1
hellebore, black 25
*Helleborus niger see* hellebore
Hellingshausen, Count Julius von 231
Helmholtz, Hermann von 302–3 and n, 481
Helmont, Jean-Baptiste van 151n
Hemingway, Ernest 440, 455n

hemlock 31–3, 43n, 81
water hemlock 78
henbane 26, 63, 81
Hench, Philip 474
Henri II, king of France 104n
Henri III, king of France 104n
Henri IV, king of France 20n, 120n
Henry VII, king of England 89n
Henry VIII, king of England 30, 39
Hepp, Paul 361
*Herbarum vivae eicones* (Otto Brunfels) 96
Herbert, Sidney 492n
Hercules 2, 11, 13
Herodotus 12 and n, 27, 129
Hero of Alexandria 258
heroin (diacetylmorphine) 363, 435, 486
Hewitt, Frederick William 460
Heyden, Friedrich von 357
Heyfelder, Johann Ferdinand 234
Heywood, C.F. 218
Hickman, Henry Hill 4, 192–5, 198, 200
Hilary of Poitiers, St 58–60
Hildegard, St 29–30
Hill, Benjamin 283
Hilton, John 48 and n, 105n
Hinton, J.M. 496
Hirsch, Aloysius 423n
Hisdai ibn Shaprut 67
Hitzig, Eduard 312 and n
HIV virus in haemophilia 391n
Hocheisen, Franz 433
Hodgkin, Thomas 262n
Hoffmann, Albert 417–18
Hoffmann, E.T.A. 137n
Hoffmann, Felix 362
Hoffmann, Friedrich 102, 127 and n
Hofmann, August Wilhelm 356
Hogarth, William 42n
Hohenheim, Theophrastus Philippus Bombastus *see* Paracelsus
Holbein, Hans 94
Hölderlin, Johann Christian Friedrich 134
Hollyer, Thomas 108n
Holmes, Gordon 409
Holmes, Oliver Wendell 183n, 223, 290
Holmes, Sherlock 376
Holton, John 465
Hölty, Ludwig 135
homeopathy 286
Homer 10, 11n, 17, 54, 64n
Hooke, Robert 117 and n
Hoo Loo 173–81
Hooper, John, bishop of Gloucester 40
Horace, Quintus Horatius Flaccus 17, 64n
Horsley, Sir Victor 2, 426–8
hospices 491f
Howell, W.H. 406
Hua T'o 9, 14 and n
Hugh de Lucca 78
Hughes, Glyn 496
Hugo, Victor 137n, 184
Humboldt, Baron Alexander von 256, 463
humours 66, 92, 98
Hunayn ibn Ishaq 61
Hunter, Charles 258
Hunter, John 169 and n, 190
Huss, John 99n
Hutchinson, Sir Jonathan 133
Hutten, Ulrich von 98n
Huxley, Aldous 284, 418
Huxley, Dr Harry 383, 384
Huysmans, Joril-Karl 327
hydrotherapy 287, 292
hyoscamine 26 and n
hyoscine 30
*Hyoscyamus* 26, 66, 107
*Hyoscyamus niger see* henbane
hypnosis 3, 34, 137, 150, 197, 478
hypnotherapy 478
hysteria 141, 196, 276, 319 and n, 326–7, 388, 442

Iapyx 27, 90n
Ibn Sinna *see* Avicenna
Ibn Zuhr *see* Avenzoar
Ibsen, Henrik 315–16n
Ignatius of Antioch 38
*Iliad, The* 61
immunosuppressants 447
inflammatory response 458
Ingrassia, Giovanni 96n
Ingres, Dominique 378, 445
Innocent X, Pope 352
Intocostrin 464
iodine 346
iodoform 246
ipecacuanha 132
Iri, Keeper of the Royal Rectum 13
Isiodorus, bishop of Seville 29

Jackson, Charles Thomas 209–14, 215, 221, 224
Jackson, Chevalier 462
Jackson, John Hughlings 328–9 and n
Jackson, James 209–10
Jackson, Susan, *née* Bridge 211
Jacob (biblical character) 28
Jacomet, Police Superintendent 322–3

James I, king of England 101, 114
James II, king of England 118
James the Apostle 53
James, Henry 349
James, William 383
Jameson, Gertrude 495
Jefferson, Thomas 133, 289, 298
Jena, battle of 149n
Jenner, Edward 128 and n, 190
Jerome, St 56
Jerome of Prague 99
Joan of Arc 29
Job 53, 98n
Johannes Guinther von Andernach *see under* Guinther von Andernach
John XXI, Pope 74 and n
John of Arderne 80 and n, 128
John of Gaddesdon 82n
John of Gaunt 80
John of Jeruslem, St, Knights of 75
John Paul II, Pope 123, 492
Johnson, Enid 465
Johnson, Henry 241
Johnson, Dr Samuel 126, 133
Johnson and Johnson 457
Jokai, Mor 347
Jones, Dr John 133–4
Jones, Thomas 250
Joseph II, holy Roman emperor 145n
Julian, Don Antonio 367n
Julius II, Pope 90
Jungken, Johann Christian 234
Junot, Monsieur 445
Jussieu, Jean-Pierre de 147, 148n
Jussieu, Joseph de 367n
Justin, St 35, 38

Kafka, Franz 455n
Kaiser, The *see* Wilhelm II
Kandinski, Vassily 399
Kant, Immanuel 42 and n
Keats, John 135
Keith, George 245–6
Kekule von Stradonitz, Friedrich 412
Keller, Ferdinand 19
Kellogg, Henry 287
Kennedy, Emory 244
Kennedy, John 419
Kerr, Norman 339
Kershaw, Dr Edward 385f
Key, Aston 179–80
King, John 161
Kinglake, Dr Robert 162, 164
Kingsley, Charles 251
Kipling, Rudyard 374
Knight, T.A. 194
Knowles, James 13n
Knox, John 243n
Koch, Robert 114n, 235, 266, 473
Kolbe, Hermann 357
Koller, Carl 370f
Köllicker, pupil of Johannes Müller 302
Kraepelin, Emil 329 and n
Kreuser, Albert Heinrich 237
Krönig, Bernhardt 429
Kuhlmauss, Johann 211
Kuhn, Franz 462
Kusmaul, Adolf 233n

*Ladies' Home Journal* 429
Laennec, René-Théophile-Hyacinthe 114n, 263, 280
Lafarge, Dr Hyacinth 258
Landor, Walter Savage 157
Lane, Sir Arbuthnot 357 and n
Langenbeck, Konrad Johann 235
Langley, John Newport 409
Langridge, Mr, Liverpool veterinary surgeon 241
Laocoön, The (Hellenistic Greek sculpture, first century) 91
Lao-Tse 14
Larrey, Baron Dominique 1 and n, 172–4 and n, 191 and n, 195
laryngitis, tuberculous 256
Lateran, the, papal palace 90
Latham, R.G. 129n
Latta, Thomas 265n
laudanum 92–3 and n, 128–34 and n, 169, 183, 249, 251, 252–7, 348
*see also* opium
Lauder, Thomas 305
Lavoisier, Antoine Laurent 147, 154 and n, 155, 160, 161n, 346
Laycock, Tom 347
Ledger, Charles 366
Leeuwenhoek, Antoni van 115–17
Legallois, Jean 312 and n
Leigh, John 254
Lemmon, William Thomas 374
Lenau, Nikolaus 315n
Lenin, Vladimir 405
Leonardo da Vinci 94
Leontius, bishop of Antioch 56
Leopardi, Count Giacomo 135
Leopold, Prince, later duke of Albany 275n
Leopold II, holy Roman emperor 145n
Leopold Salvator, Archduke 332

Leo the Great, Pope 50
leprosy 53
Leriche, René 425, 466–8, 485
Lermontov, Mikhail Yurevich 135
Leroy, Charles 145, 147
Leroux, Henri 354
Lettsom, John Coakley 262n
leukaemia 243n, 475
Levant Company 249
levodopa 475
Lévy, Monsieur 445
Lewis, Cecil Day 440
Liebig, Justus von 154n, 247n, 256n, 302, 348
Linacre, Thomas 89n, 103n
Lincoln, Abraham 296, 455
Linné (Linnaeus), Carl von 341
*Genera plantarum* 18
Lister, Joseph 14n 169–70 and n, 177, 185, 191, 200, 228 and n, 336, 460
Liston, Robert 170, 175n, 197, 208, 209f, 232 and n, 235, 242, 244n, 245, 270, 276
Livingstone, Dr David 483
Livy, Titus Livius 44 and n
Lloyd, Mr, Newcastle surgeon 277
Lloyd George, David 395n, 455
lobotomy, frontal 352n, 471
Locke, John 129, 130
Loewi, Otto 406f
Lockwood, Harold 454n
Locock, Dr 274
Lombroso, Cesare 304 and n
Long, Crawford Williamson 202f, 217
Long, James 202
Long, Mary Caroline, *née* Swain 203
Lonsdale, Miss 328
Lord's Prayer, The 34
Louis XIV, king of France 425
Louis XV, king of France 20
Louis XVI, king of France 134
Louis, dauphin of France 352
Louis XVIII, king of France 168n
Louis Philippe, 'King of the French' 238, 445 and n
Lower, Robert 117 and n, 118n
Löwig, Karl Jacob 355
Lowry, David 203
LSD (lysergic acid) 418
Lucian 21
Lucretia 44
Lucullus, Lucius Licinius 18
Ludwig, Friedrich Karl 305n
Ludwig, Karl 302, 304–5 and n
Lugo, Juan de 352
Lugol, Jean Guillaume, 347
Lunar Society 157
Lundy, John 414
Luther, Martin 89, 112
lysergic acid *see* LSD
Lycias, prefect under Diocletian 38
Lyman, Dr Henry M. 293
Lyon, Amy *see* Hamilton, Lady Emma

MacEwen, William 401
Maclagan, Thomas 357
McBurney, Charles 337 and n
McDowell, Ephraim 202 and n
*McClure's Magazine* 429
McNeil Laboratories 457
Madame Butterfly 344n
Magellan, Ferdinand 75
Magendie, François 239, 257 and n, 278, 307, 426
Magill, Ivan 462 and n
Mahler, Gustav 461n
Maillol, Aristide 449
Maimonides 68f and n
Mallebranche, Nicolas 142n
Malgaigne, Joseph-François 237
Malpighi, Marcello 116
mandragora 21f, 27–30, 63, 78, 107
mandrake *see* mandragora
Manet, Edouard 383, 415n
Mani 49–50
Manichaeism 47, 80
Mansfield, Hector 197
Mansur ibn Ishaq 63
Mantegna, Andrea 94
Mar, Earl of *see* Erskine 254
Marat, Jean-Paul 160 and n
Marcellus, Roman Christian soldier 37
Marcet, Alex 177
Marchais, Jean 256
Marchaux, Julien 238
Marcus Aurelius, Roman emperor 42 and n, 69
Maréschal, Jean-Baptiste 425
Margaret of Antioch, St 55 and n
Maria Josepha, archduchess of Austria 139n
Maria Theresa, empress 102, 132, 140, 141, 143, 145 and n
Mariani, Angelo 374
Marie Antoinette, queen to Louis XVI of France 145, 146 and n, 149, 445
Marie Curie Foundation 496
Marius, Gaius 33
Markham, Clemens 465

Martindale, William 375
Marx, Karl 238, 303
Matas, Rudolph 372, 462
Matisse, Henri 449
Mattioli, Pier Andrea 97 and n
Maudsley, Sir Henry 340
Maupassant, Guy de 315
Maurier, George du 390
Mauroy, Antoine 118n
Maxentius, Marcus Aurelius Flavius, 46
Maximilian Joseph III, elector of Bavaria 142
Maximilianus, Roman Christian soldier 36
May, Karl 45 and n
Maydl, Karel 461
Mayo brothers 337 and n,
Charles 414
Mayow, John 151f, 162
Mayrhofer, Johann 130n, 136
meconium 19
Medici, Marie de', queen to Henri IV of France 20n
Meggison, Dr Tom 277
Melampus 25
Melczak, Ronald 482
Melville, James 156
Menelaus 10
meperidine (pethidine) 417
Mepropbamate 417
Meer, Fritz ter 456
Merck & Co, Darmstadt 369, 414
Mercy, Florimund count d'Argenteau 145
Meredith, George 388
Mering, Josef von 413
Merskey, Harold 480
Merton, William, Lord 240
Mesmer, Franz Anton and mesmerism Ch 15 *passim*, 195f, 217
Mestivier, Louis 331
Meyer, Herr Bieder 233 and n
Meynert, Theodor 408
Michael, St 54
Michelangelo Buonarroti 17n, 90
microscopes and microscopists 115 and n, 265
Mildmay, Sir Henry 240
Milena, Grand Duchess 391
Miller, James 246, 278
Millerites 287
Millinger, J.S. 29
Milton, John 502
Mistinguette (Jeanne Bougeois) 375
Mitchell, Silas Weir 105 and n
Mitchil, Samuel Latham 162
Modigliani, Amadeo 348n
Mohammed, prophet of Islam 49, 63
Mohammed ibn Zakariya-al-Razi *see* Rhazes
Molière, Jean-Baptiste Poquelin 125, 148
Moline, Charles 148
Monardes, Nicolas 100
Mondale, Christabel 200
Mondeville, Henri de 69n, 77f and n
Monet, Claude 18, 445
Monro family, the 243n
Montagna, Mazzeo della 106
Montaigne, Michel de 108, 109 and n
Montegazza, Paolo 367
Montezuma 97n
Moore, James 190 and n, 192
Moore Jr, John 190 and n
Moore Sr, John 190 and n
Moreno, Thomas M. y Maiz 370
Morisot, Berthe 445
Morpheus, classical god of dreams 255n
morphine 20n, 23n, 249, 250, 255f and n, 308, 315, 429, 486
Morse, Samuel 212 and n
Mort, Ben 292
Morton, William Thomas Green 210f, 213f, 216f, 227
Moshe ben Maimon 68
Mott, Valentine 168 and n, 175
Moynihan, Lord 337 and n, 383, 462
Mozart family 138n
Mozart, Wolfgang 138 and n, 141, 143
Mucius Scaevola, Gaius 44
mulberry tree, 26, 78
Müller, Johannes 302–5
Munch, Edvard 318, 378
Munch, Sophie 378
Munro, Marilyn 420
Munro, Thomas 340n
Murchison, Dr 262n
Murdoch, William 355
Mussorgski, Modest 344n
Mythras, Roman and Persian god 35n

naloxone 486
Napier, Admiral Sir Charles 240
Napoleon I 3n, 49, 72n, 135, 138 and n, 148 and n, 158 and n, 171n, 172n, 306, 317 and n, 445
Napoleon III 321n, 324n, 438
Nasser, President of Egypt 422n
National Health Service 477
Natural Childbirth Movement 478
Neal, Helen 475, 487

Neal, John 475
Nelson, Horatio, Lord 135, 153n, 199
'Nembutal' 420
Neocaesaria, bishop of, under Diocletian 37
neonatal narcosis 435
neonatal asphyxia 435
nepenthe 11 and n, 16
Nero, Roman emperor 19, 25 and n, 34 and n, 47
nerves
    central nervous system 2, 435
    fibres 304
    motor commands 304
    motor functions 32
    tracts 302, 303
Nessus 11
neuromuscular transmission 304, 462
neurones 399, 400
neuropathy 109, 316
    diabetic 30, 109
neuropraxia 3
neurosyphilis 316, 469
neurotransmitters 406f, 474
Newton, Isaac 101, 118, 119n
Nicholas I, Russian tzar 239
Nicholas II, Russian tzar 391f
Nicholls, Thomas 287, 292
Nicot, Jean 100f
Niefer, Jacob 13n
Niemann, Albrecht 367
Nietzsche, Friedrich 383
Nightingale, Florence 185, 239
nitric oxide 152, 346n
nitrous gas 152
nitrous oxide, 3, 151, 162, 346n, 203–4 and n, 206 and n, 209, 213–15, 222, 269n
Noah, biblical character 17 and n, 49
nociceptors 483
non-steroidial anti-inflammatories (NSAIDs) 458
Norris, James 340n
'novocaine' 374
Noyes, Henry 371
Numa Pompilius, second king of Rome 13n
*Nymphaea ample* 11n
*Nymphaea caeruleum* 11n

O'Dwyer, Joseph 461
Odyssey 10, 11 and n
Offenbach, Jacques 320
Offray de la Mettrie, Julien 127
Oldenburg, Henry 115
Old Man of the Mountains, the 75
ophthalmoscope 303
Opinius, Lucius 17
opium 17n, 19, 23–5, 31, 66, 75–6, 93 and n, 105, 118, 125n, 127–31 and n, 191, 249, 259–64 and n, 365
    intravenous 117–18 and n, 119n
    *see also* laudanum;
opioid receptors 472
Oporinius, Joannes 93
Orange, William of 120
Orinoco Indians 463
Orpheus 64n
Ortega y Gasset 455
O'Shaughnessy, William 265n, 342
Osiris, Egyptian god 17
Osler, Sir William 342, 371
Osterlin, Franzl 141
Osterwald, Peter van 142
Ovid, Publius Ovidius Naso 21 and n, 26, 64n
Oxygen 114, 153–5, 159, 162–3 and n, 205 and n, 215, 269, 272

Paget, Sir James 174, 176
*Pain*, journal 487
pain mechanisms 479, 480f
pain prevention 494
pain syndromes 498
Palmer, John 229
Pandora 203
*Papaver rhoeas* 18 and n
*Papaver somnierum* 18 and n
Papper, Emmanuel 464
Paracelsus (Theophrastus Hohenheim) 22, 92–5 and n, 122,
Paradies, Maria Theresa von 143–4
Paré, Ambroise 104–5 and n, 106–7 and n, 189, 422n
Parham, F.W. 462
Paris Commune 320
Parker, George, Earl of Macclesfield 353
Parker, William 208
Parkinson, James 314 and n, 328
Parry, Sir Hubert 454n
Pascal, Blaise 172
Pasteur, Louis 114n, 170n, 266, 307, 320, 335
Patin, Guy 98n
Patroclus 11, 90n
Paul the Apostle 35, 50
Paul, Frank Thomas 337
Pavlov, Ivan Petrovich 211 and n, 305, 404f, 482

Pavy, F.G. 339
Payne, Captain 199
Peachey, Sir Harry 268n
Pearson, Richard 159
Peel, Sir Robert 280–1
Pegenstecher, Johann 355
Pelletier, Pierre-Joseph 257
Pemberton, J.S. 375
penicillin 317n, 473
Penn, William 303
Pepys, Samuel 108n, 118
Perityphlitis 332
Perkin, William Henry 356, 357n
Pert, Candace 486
Pétain, Philippe 327
Peter, St, Synoptic Gospel of 2n, 39
Peter the Apostle 34
Peter the Great 30
Petit of Paris 30
Petit, Louis 446
Petrie, Miss 246
Petrus Hispanus *see* John XXI, Pope
Péyramale, Abbé Dominique 322
phantom limbs 105 and n, 422n, 484, 501
breast 422n, 484
phenacetin 349, 361, 457
phenobarbital 419
Philip II, king of Spain 109–110 and n
Philip III, king of France 77
Philip III, king of Spain 110
Philip IV, king of France 77
Philip V, king of France 86
Philip II of Macedon 28
phlogiston 153–4
Picasso, Pablo 399, 448
Pickford, James 278
Pierson, Abel 290
Pindar 11
pineal gland 122 and n
Pinel, Philippe 149, 318
Piper, Priscilla 460
Piria, Raffaelo 355
Pirogoff, Nikolai Ivanovich 239
Pisarev, D.I. 404
Pissarro, Camille 445
Pitcairn, Archibald 243n
Pitha, Franz von 233
Pitt the Younger, William 156, 160
Pius IX, Pope 323
Pizarro, Francisco 97n
placebo effect 138–9 and n
Plato 28n, 32, 64n
Pliny the elder 4, 10 and n, 13n, 17, 21, 28, 32, 90, 93n
Pliny the younger 35n
Plutarch 33, 35n
Poe, Edgar Allan 137n, 286
Poehlman, Mary 495
Polycarp, St, bishop of Smyrma 38
Polydamma, queen of Egypt 10
Polydoros of Rhodes 90
Pomet, Jean-Baptiste 134
Pompey, Gnaeus Pompeius Magnus 35n
'Pond's Extract, the Universal Pain Extractor' 253
Popper, Sir Karl 412n
Porsena, Lars 44
Porter, John B., 282
Porter, Roy 350n
postherpetic neuralgia 442
positive pressure respiration 462
Potain, Jules 320
Pravaz, Dr Charles-Gabriel 258
Priestley, Joseph 153f and n, 157n, 160 and n, 346n
Priestley Jr, Joseph 161, 164
Primrose, Sir Archibald John, 4th earl of Rosebery 346
Prisca, wife of Diocletian 36
Prochaska, Georg 309n
Proteus, king of Argos 25
'protopathic' sensation 410
Proust, Marcel 399
Puccini, Giacomo 344n
Puerto Rico 97n
Pushkin, Aleksandr Sergeyevich 134
Puységur brothers 148
Count Maximilien 148
Marquis de 148

Quercetanus 93
Quincey, Thomas de 129, 134
Quincke, Heinrich 373

Ra, Egyptian god 12
Rachel, biblical character 28
Raleigh, Sir Walter 463
Rameau, Jean-Philippe 44
Ransome, John 229
Rasputin, Gregory Efinovich 392f
Reckitt and Colman 456
reflex action 309 and n, 403f
reflexology 378
Régnault, Felix 326
Reichstadt, Napoléon François-Joseph Charles, Duke of (King of Rome) 172n
Reiss, Ludwig 360
Rembrandt (Harmensz) van Rijn 96, 448

Remus 35n
Renard, Gabrielle 448 and n
Renoir, Auguste 444f
Renoir, Claude (Coco) 447
Renoir, Jean 446f
Renoir, Pierre 450f
Repin, I. 405
Reuben, biblical character 28
Reynaud, Theophile 13n
Rhazes 89–91 and n
rheumatoid arthritis 447, 458, 474
Richard I, the Lionheart, king of England 68, 191 and n
Riggs, John Mankey 209, 213, 214
Rimbaud, Jean Arthur 437
Rippl-Ronai, Josef 449
Rivière, Renée 449
Robbins, Richard 18
Robertson, Charles 307
Robespierre, Maximilien 149, 160
Robinson, James 228, 269
Roche, St 54 and n
Roget, Peter Mark 157, 164
Rokitansky, Carl 139n, 177, 232
Romberg, Moritz 316–7 and n
Rome, King of, son of Napoleon I *see* Reichstadt
Rommel, E., Field Marshal 416
Romulus, king of Rome 35n
Roon, Professor Helmuth von 255
Roosevelt, Theodore 380
Rosebery, Lord *see* Primrose
rose 81, 109
  bush 323
  oil 104, 107
Rosetti, Dante Gabriel 348
Rostand, Edmond 454
Rothschild family 448
Roulant, Jean-Jacques 463
Rousseau, Jean-Jacques 139n
Rousseau, Père purveyor of (Lancaster Black Drops) 133
Rousset, François 13n
Roussignault, Pierre 463
Roussy, Gustave 408
Rouvroy, Louis de, duc de Saint-Simon
Roux, Emile 307
Rovenstine, Emery 464
Rowbotham, E.S. 462
Rudder, Pierre de 325n
Rufus of Ephesus 23n
Rugby School 200 and n
Rumpff, Carl 360f
Runge, Friedlieb Ferdinand 355
Rush, Dr Benjamin 288 and n, 289
Russel Reynolds, Sir John 342
Ryle, J. C. 262n
Rynd, Dr Francis 258

Sacher-Masoch, Leopold 85
Sachsen-Altenburg, Eduard von 233
Sadler, James 157, 159
Sahagun, Bernadino de 100
Sandoz 418
St Christopher 496n
St Christopher's Hospice, Sydenham, London 496–7
St Columba's Home, London 492
St Luke's Hospital for the Dying Poor, London 492
St Joseph's Hospice, London 492, 495
Saint-Simon, duc de *see* Rouvroy
Saladin 68 and n, 191 and n
Salah-ed-din Yussuf ibn Ayub *see* Saladin
Samaritan, the Good 54
'Samuel's Herculean Embrocation' 253
sassafras 100
saturday-night palsy *see* neuropraxia
Saunders, Cicely 492f
Sauvage, Boissier de 127 and n
Sayre, Dr Lewis 225
Scaevola *see* Mucius Scaevola
Scarborough, Dr Charles 125
Scarlatti, Alessandro 44, 121n
Schade, Rudolf 448
Scheele, Karl Wilhelm 154f and n
Schiele, Egon and wife, Edith 454n
Schiller, Johann Christoph Friedrich 134–5
schizophrenia 329, 475
Schlegel, August Wilhelm von 134
Schober, Franz 136n
Schomburgh, Robert and Richard 463
Schönlein, C. F. 212
Schopenhauer, Arthur 137n
Schuh, Franz 139n, 232
Schubert, Franz Peter 2, 100n, 136 and n
  *Heidenröslein* 2 and n
Schuller, M. 470
Schumann, Robert 315n, 448
Schwann, Theodor 31 and n
Schwarz, Adalbert 231
Scott, Michael 93n
Scott, Walter 134
Scott, William 208n
Scribonius Largus 19
Scruton, Harold 456
Sebastian, St 54 and n
Sechenov, I.M. 305, 404
Selkirk, Alexander 132

Selwyn, John 118n
Semmelweis, Ignác 114n, 170n, 177, 183, 238, 266
Seneca, Lucius Annaeus 41, 43 and n, 64n, 69
Sepsworth, Maud 379n
Seraphim Hospital, Stockholm 279
Serturner, Friedrich Wilhelm Anton 255f
Servetus, Michael 112f
Severino, Marco Aurelio 191
Sextus, son of Tarquin Superbus 44n
Seymour, Jane 13n
Shakespeare, William 3, 26, 29, 96, 106, 128 and n, 148, 297, 351n
Shaw, Bernard 338n, 405n
Sheldon, Dr, Archbishop of Canterbury 302n
Shelley, Percy Bysshe 134–5
Shephard, David A.E. 262n
Sherrington, Charles Scott 402f
Sibbald, Robert 243n
Siddall, Elizabeth 348
Sienkiewitz, Henryk 35n
Sigismond, Holy Roman Emperor 76
Signorelli, Luca 94
Silliman, Benjamin 247
Sims, James Marion 206n, 295
Simpson, Alexander 279n
Simpson, David 243
Simpson, Jessie, *née* Grindley 244f, 279n
Simpson, Mary, *née* Jarvie 243, 245
Simpson, Sir James Young 192, 243f, 269, 279 and n, 280 and n, 298
Sisley, Alfred 445
Sisters of Charity, Ireland 492
Sixtus II, Pope 35
Sixtus IV, Pope 73
Skey, Frederic 241
Skoda, Josef 139n, 232
Slade, Conrad 448n
sleeping sponge 78, 79, 81
smallpox 128, 139n
Smetana, Bedrich 315
Smiles, Samuel 263
Smith, Edwin 12
Smith, J.H. 215
Smith, Samuel Stanhope 294
Smith, Truman 224
Snow, Frances, *née* Ashkham 261n
Snow, John 261f and n, 279, 460
Snow, John, of the Cancer Hospital 494
Snow, William 261n
Snyder, Sylvia 50, 495
Socrates 19, 31–2, 64n
*Solananceae* 26
Solomon, Old Testament king of Israel 51, 98n
somnambulism 148, 150, 198
somnolence 501
Sophocles 11
Sorrel, Julia, Mrs T. Arnold 379
Soubeiran, French chemist 247n
Soubirous, Bernadette 322f and n, 327
Soubirous, François 321
Soubirous, Louise, *née* Casterot 322
Soubirous, Toinette 321
Soult, Nicolas Jean de Dieu 190n
Southey, Robert 157, 161, 164
Spallanzani, Abbé Lazzaro 155
Spanish 'flu 453, 473
sphincters 403
Sproat, William 264
Spruce, Richard 365
Squibb, E.R. 464
Squire, Peter 228, 342
Stahl, Georg Ernst 126–7 and n, 153 and n
Stalin, Joseph 211n
Stanislavski, Konstantin 39, 399
Stanton, Elizabeth Cady 290
Starling, Ernest 406
Steevens, G.W. 388
Steinbüchel, Robert von 430
Stephen I, St, king of Hungary 73n
Stephen, Palatine, Archduke 233
Stephens, Alexander 202
Stewart, James 224
St Martin, Alexis 211–12
Stone, Rev. Edward 350f
'Stovaine' 374
Störck, Professor Anton Freiherr von 143–4
Stowe, Harriet Beecher 297
Strachey, Lytton 200n
Street's Infant Quieter 253
Stricker, Solomon 370
Strindberg, August 383
strychnine 308
*Strychnos toxifera* 463
Suetonius 27
Sulzbach, Eck von 154n
Sumerians 48
Sushruta 15
Sutherland, John 379n
Swann, Donald 498n
Sweetnam, Captain Bruce 241
Swieten, Gerhard van 102, 132, 139n, 140
Swengali (in fiction) 390
Sybil, Cumaean 11

Sydenham, Thomas 22n, 93n, 129–34 and n
Sydenham, William 129
Sylvester II, Pope 73n
Sylvius, Jacobus 89
Syme, E.S. 198n
Syme, James 168, 170n, 176 and n, 182–3, 243n,
sympathectomy 468
sympathetic ganglia 468
sympathetic nerves 468
Symphorosa, early Christian martyr 38
synapse 403, 482

tabes 316–18
tabetic crisis 470
Tacitus, Gaius Cornelius 18, 38
Tait, Robert Lawson 318, 335
Talbor, Robert 352
Talleyrand-Périgord, Charles Maurice de 145 and n
Taoism 14 and n
Tarquinius Superbus 44 and n
tartar, vitriolated 132
Tasma, David 496
Tasso, Tarquato 464
Taylor, Isaac 258
Telemachus 10
Telemann, Georg Philipp 344
Temple Emmet, Mrs C. 434
Tenney, Glen 217
Tenochtitlan 97n
Teppe, Julien 437f
Thackeray, William Makepeace 196 and n
'thalamic syndrome' 409
thalidomide 459
Theresa of Avila, St 110n
Thérèse of Lisieux, St 53, 123, 324n
Thiersch, Carl 336
Thislethwayte, Warden Robert 351n
Thomson, Sir St Clair 192n
Thoth, Egyptian god 13
thrombosis 3n, 460
tic douloureux 422f
Tieck, Ludwig 137n
*Timaeus*, Plato's 28n
Tintoretto, Jacopo Robusti 54n
Titian, Tiziano Vecellio 33n, 44n, 54n, 109
Titus Flavius Vespasianus, Roman emperor 23n
Tocqueville, Alexis de 291, 296
Toledo, Francisco Hernandes de 100
Tombs Prison, New York 221
Tonga, King of *see* Tupou II
Tookman, A.J. 498
Topham, William 196–7
Torres, Antonio de 97n
Tosca, Floria 344
tourniquet 105, 189–90 and n, 484
Townsend, S.D. 218
Townshend. Chauncey Hare 137n
tracheostomy 461
Trajan, Roman emperor 23, 35n
transcutaneous peripheral nerve stimulation (TEPS) 485
Travers, Benjamin 278
Trendelenburg, Friedrich 461
Trevelyan, Janet, *née* Ward 381
trigeminal neuralgia 428
Trilby, fictitious character 390
Triptolemus, Roman mythological character 21n
Trollope, Anthony 379
Trotter, Thomas 135
Trotter, Wllfred 410
Trousseau, Armand 57, 387, 461n, 498
Troy and the Trojan War 10, 22, 35n, 91
Tuffier, Theodore 374
Tupou II of Tonga, King George
Turner, William, dean of Wells 30, 96
Tutankhamon, pharaoh 28
twilight sleep 429f
Tylenol 457
typhoid 284n
typhus 97n

Uffizi, museum 91n
ultrasound 138n
Ulysses 10
Ummayid dynasty 67, 70

vaccination 191
Vaccinov, first child in Russia to be vaccinated 246
Valeria, daughter-in-law of Diocletian 36
Valerian, Roman emperor 36
Valéry, Paul 438
Valerius Tarsus 22
Van Gogh, Vincent 449n
Vane John 460
Vasco da Gama 75
Vaughan, Victor 454
Vauquelin, Louis-Nicolas 354
Velpeau, Alfred 237
Venables, James M. 204
Verdi, Guiseppe 344n
Verlaine, Paul Marie 318
Verocchio, Andrea del 95
Vesalius, Andreas 95 and n, 107

Vicary, Thomas 103
Victoria, Queen of England 198n, 192 and n, 200 and n, 248, 274–5 and n, 279n
Victoria Cross 376
Victoria, Tomás Louis de 344
Vigée-Lebrun, Mme 445
Vioxx 459
Virgil, Publius Vergilius Maro 11, 27
Virchow, Dr Rudolf 234, 305–6 and n, 463
vitamin B deficiency 317
vitamin D deficiency 387
*Vitis sylvestris* 17
*Vitis vinifera* 17
Vitus, St 79 and n
*Vogue* magazine 471
Volkman, Richard von 336
Voltaire, François-Marie Arouet 99n, 145
Vuillard, Edouard 449
Vulpian, Alfred Edmé 318
Vyrubova, Anna 393

Wadham College, Oxford 351 and n
Wagner, Richard 220
Wagram, battle of 149n
Wakley, Thomas 196
Waldie, David 246
Wall, Patrick D. 282, 482
Wallace, Edgar 455
Waller, Augustus 310 and n
Walpole, Cutler, ficticious 338n
Walsingham, Lord *see* Grey
Ward, Dorothy 384, 385
Ward, William Squire 197
Ward Richardson, Sir Benjamin 30, 262n, 269, 273n
Ward, Mrs Humphrey 379f
Ward, Tom 381
Ward, Theodore 380
Warren, John Collins 210–11, 215, 218, 222
Wassermann reaction 316n
Watteau, (Jean-) Antoine 256 and n
Watt family 157, 161
  Gregory 159, 161
  James 157 and n, 159, 161, 162
Waugh, Evelyn 380
Wayne, F.G. 454
Wedgwood, family 157, 161
  Josiah 157n
  Thomas 133, 161
Weiss, Albertus 30
Weiss, William 454n
Wellbegone, John 241
Welch, William Henry 305, 472
Wellcome, Sir Henry 192
Wellesley, Arthur, 1st duke of Wellington 133, 135
Wells, Elizabeth, *née* Dainton 209, 214–15, 222
Wells, Horace 195, 208–12, 213–15, 224, 226
Wepfer, Johann Jakob 314
Wernicke, Carl 329 and n
Wesselenyi, Miklos 233
White, Charles 295
White, Lowell 488
Whitehead, Henry 268
Whitman, Walt 288
Whitridge Williams, J. 435
Whytt, Robert 243n
Wilberforce, William 165
Wilde, Oscar 380, 389
Wilder, Thornton 442
Wilhelm II, German emperor 395n, 435
Willis, Thomas 95n, 117 and n, 302 and n, 339n, 467n
Wilson family, 177
  George 177–9
  Jessie 177
Wilson, President Woodrow 454
Winckelman, Johann Joachim 255
Winslow, Jacob Benignus 467
Winslow's Soothing Syrup, Mrs 253
Wiseman, Richard 103 and n
Withering, William 127 and n, 157n
Wodehouse, P.G. 455n
Wöhler, Friedrich 412n
Wolf, H.G. 471
Wolf, Hugo 315n
Wolfart, Dr Ernst 150
Wolff, Adalbert 237
Wolff, Albert 445
Womball, James 196
Wombwell, Sir George 240
*Woman's Home Companion* 429
Wood, Dr Alexander 258
Woodall, John 103
Wooster, Bertie, fictitious character 455
Worcester College, Oxford 134
Worcester, Alfred 494
Wordsworth, William 134, 157
Wren, Sir Christopher 115n, 117 and n, 302n
Wright, Lewis O. 465
Wright, C.R. Alder 363

Xenophon 19
Yale, University 457
Yossoupov, Prince Felix 392f
Young, John 250

Zeno of Citium 41
Zeno of Elea 41
Zeus 10 and n
Zola, Emile 327
Zoroaster 35n
Zozes, pharaoh 13
Zweifel, Paul 435
Zyklon B gas 456